Biorefinery of Alternative Resources: Targeting Green Fuels and Platform Chemicals

Sonil Nanda • Dai-Viet N. Vo •
Prakash Kumar Sarangi
Editors

Biorefinery of Alternative Resources: Targeting Green Fuels and Platform Chemicals

Springer

Editors
Sonil Nanda
Department of Chemical and Biological Engineering
University of Saskatchewan
Saskatoon, Saskatchewan, Canada

Dai-Viet N. Vo
Center of Excellence for Green Energy and Environmental Nanomaterials
Nguyễn Tất Thành University
Hồ Chí Minh City, Vietnam

Prakash Kumar Sarangi
Directorate of Research
Central Agricultural University
Imphal, Manipur, India

ISBN 978-981-15-1803-4 ISBN 978-981-15-1804-1 (eBook)
https://doi.org/10.1007/978-981-15-1804-1

This Springer imprint is published by the registered company Springer Nature Singapore Pte Ltd.
The registered company address is: 152 Beach Road, #21-01/04 Gateway East, Singapore 189721, Singapore

Preface

The production and overwhelming usage of fossil fuels and petrochemicals have resulted in several environmental concerns such as increased greenhouse gas emissions, global warming and pollution. Moreover, the production and consumption of fossil fuels and their derivatives are prodigiously increasing due to rapid urbanization, industrialization, population growth and improvements in day-to-day lifestyle. Due to the many adverse effects of fuels and chemicals derived from fossil resources on the environment and ecosystems, it has become highly imperative to find alternatives that are not only environmentally friendly but also fully or partially biodegradable, cost-effective, feasible. Moreover, these green alternatives, usually referred to biofuels, biochemical and biomaterials should be promising to make a paradigm shift in the consumer market to replace the fossil fuels and petrochemicals.

Renewability, carbon neutrality, abundancy, reasonably priced as well as non-competency to food, fodder and arable lands are other attributes of a potential bioresource to generate biofuels and biochemicals. Some of such potential bioresources include lignocellulosic biomass (e.g., agricultural crop residues, forestry residues and energy crops), microalgae, municipal solid wastes, industrial effluents, cattle manure and other organic refuse. This book is a compilation of twenty chapters, which discusses the potential of such bioresources to produce biofuels, platform chemicals and other bio-based products. Several thermochemical and biological conversion technologies are described in terms of their conversion pathways to biofuels and biochemical through biomass-to-liquid (e.g., pyrolysis, liquefaction, fermentation and mechanical extraction), biomass-to-gas (gasification and anaerobic digestion) and gas-to-liquid (Fischer-Tropsch catalysis). This book provides the up-to-date information on the production and utilization of biofuels and biochemical, biomass conversion routes (thermochemical, hydrothermal, biological, mechanical and physicochemical), reforming technologies as well as techno-economic and life-cycle assessment studies.

Chapter 1 by Arun and Dalai gives an overview of the opportunities, prospects and challenges in the biofuel sector in the current market scenario. A detailed analysis on the risk factors association with technological innovation in the biofuels sector is the main objective of this chapter.

Chapter 2 by Pasin et al. describes the bioconversion potential of lignocellulosic biomass to bioethanol. The composition of cellulose, hemicellulose and lignin as well as their biosynthesis in different agricultural crop residues and hydrolysis are provided.

The pretreatment procedures, enzymatic hydrolysis and microbial fermentation of biomass to second-generation bioethanol are described.

Chapter 3 by Rana and Parikh highlights the catalytic conversion of ethanol to several value-added chemicals such as acetaldehyde, acetic acid, acetone, ethylene, butanol, 1,3-butadiene and ethyl acetate. The role of catalysts, catalyst supports, metal-support interactions, oxygen storage capacity, acidity and basicity of catalyst in the conversion of ethanol to platform chemicals are described.

Chapter 4 by Phung and Busca summarizes the reaction pathways in bioethanol conversion such as dehydration, oxidation, steam reforming, dehydrogenation and Guerbet reaction to base chemicals and fuel derivatives. A highlight on different catalysts, catalytic reactions and ethanol-derived products are made.

Chapter 5 by Nanda et al. throws light on butanol and propanol as the next-generation biofuels. The fuel chemistry, production technologies from petrochemicals and biomass as well as biotechnological developments in the fermentation of lignocellulosic biomass to produce butanol and propanol are provided.

Chapter 6 by Sarangi et al. makes a review of the industrial applications and production pathways of biomethanol as a biofuel and biochemical. The industrial applications, technical challenges, future perspectives and several production pathways of biomethanol are summarized.

Chapter 7 by Naira et al. analyses various thermochemical and biological conversion routes of lignocellulosic and algal feedstocks to produce biofuels and biochemicals for use in energy, food, pharmaceutical, cosmetic and textile industries. The co-production technologies of biofuels and biochemicals from sugars (pentose and hexose) and lignin as well as the generation of bioactive components such as lipids, carbohydrates and proteins from algae have been described.

Chapter 8 by Koshin et al. discusses the main biorefining approaches and concepts in the thermochemical and biological conversion of rice husk and nutshells into valuable gaseous, liquid and solid biofuel products. The physicochemical properties and geographical distribution of rice husk and nutshells as well as their industrial relevance of their fuel and chemical products are highlighted.

Chapter 9 by Singh et al. reviews the thermochemical (e.g., pyrolysis, liquefaction, torrefaction and gasification) and biological (enzymatic saccharification and fermentation) conversion technologies of *Miscanthus* as an energy crop to biofuels. The value-added applications of *Miscanthus*, especially in pulp and papermaking, biocomposites and biochemical production are also discussed. The physicochemical properties of bio-oil and biochar generated from *Miscanthus* are also described for fuel and material applications.

Chapter 10 by Suryawanshi et al. reviews the challenges, opportunities, recent developments and techno-economic feasibility of hydrothermal liquefaction and pyrolysis of lignocellulosic biomass for bio-crude oil production. The scope of the chapter also extends to bio-oil upgrading technologies, value-added chemical production and application of novel catalysts in hydrothermal liquefaction and pyrolysis of waste biomass.

Chapter 11 by Masoumi et al. described the production of biocrude oil via hydrothermal liquefaction of microalgae. Special attention is also given to the effects of process parameters on hydrothermal liquefaction to improve the bio-oil yield. Several bio-oil upgrading techniques are explored to remove the heteroatoms using various heterogeneous acid catalysts.

Chapter 12 by Mahari et al. makes a review of the recent advancements in co-pyrolysis as a next-generation conversion technology for energy and material recovery from biomass. The chapter also includes a discussion on the impact of process parameters on co-pyrolysis, technical challenges, opportunities, application and combustion performance of the resulting synthetic liquid fuel products.

Chapter 13 by Volli et al. deals with the conversion of biomass and organic residues to bio-oil through thermochemical technologies. The technical developments towards improving the bio-oil yields, fuel properties, influence of process parameters and reactor configurations are discussed in details.

Chapter 14 by Trinh et al. reviews the pros and cons of biomass pyrolysis technology along with opportunities for environmental benefits and circular economy. The integration of molecular modeling with actual experiments has been highlighted as a new paradigm for mechanistic studies to design hybrid catalysts and enhance the bio-oil yield and upgrading processes. The chapter also introduces the novel sonochemical technique in biomass treatment and conversion.

Chapter 15 by Samart et al. describes novel approaches in the conversion of hydroxymethylfurfural using heterogeneous catalysts to advanced fuels and chemicals. The chapter also discusses the synthesis of hydroxymethylfurfural from sugars along with its conversion through oxidation, photocatalytic and electrochemical, hydrogenation, Oxidative amidation and reductive amination and polycondensation reactions.

Chapter 16 by Devi and Dalai provides the information about different pathways available for the conversion of glycerol into specialty chemicals. Various approaches and strategies have been discussed to investigate the effects of reaction parameters, i.e., temperature, pressure, catalyst type, reaction time, type of solvent on glycerol conversion and product selectivity.

Chapter 17 by Siang et al. summarizes the recent advances in catalytic steam reforming of glycerol to produce syngas. The yield in terms of catalytic design using various catalysts, supports and promoters, operating conditions are described. The mechanistic pathways and kinetic models are provided to describe glycerol reaction rates.

Chapter 18 by Minh et al. describes the thermodynamic aspect of biogas reforming under different conditions. Some significant works related to catalyst design, kinetic and mechanistic studies of biogas reforming processes are described.

Chapter 19 by Nayak et al. displays an extensive report on the opportunities and challenges in biodiesel production and its compatibility in present diesel engines without any engine modification. The experimental data quantifying the performance, emissions and combustion analysis of biodiesel are also summarizes.

Chapter 20 by Bhatia et al. describes the progress in microbial fuel cells in terms of structural modification, substrates utilization, modes of operation, supplementation of different microbial communities, challenges and opportunities.

This chapter provides insights on the technological advancements and utilities of the microbial fuel cells.

The editors are grateful to all the authors for contributing their scholarly materials to develop this book. Our sincere thanks to Springer Nature for the editorial assistance in preparing this book.

Saskatoon, Saskatchewan, Canada Sonil Nanda
Hồ Chí Minh City, Vietnam Dai-Viet N. Vo
Imphal, Manipur, India Prakash K. Sarangi

Contents

Editors and Contributors

About the Editors

Sonil Nanda is a Research Associate at the University of Saskatchewan in Canada. He received his Ph.D. degree in Biology from York University in Toronto and has worked as a postdoctoral researcher at York University and the University of Western Ontario in Canada. His research areas are related to advanced biofuels and biochemicals, thermochemical and biological conversion technologies, generation of hydrothermal flames for hazardous waste treatment, biochar-based agronomy, bioremediation, as well as carbon capture and sequestration. He has published more than 80 peer-reviewed journal articles, 25 book chapters and 80 conference proceedings. He is the editor of several books published by Springer, Elsevier and CRC Press. He is also the guest editor of several Special Issues of esteemed journals, namely the *International Journal of Hydrogen Energy*, *Chemical Engineering & Technology*, *Waste and Biomass Valorization*, *SN Applied Sciences*, *Biomass Conversion and Biorefinery*, *Chemical Engineering Science* and *Topics in Catalysis*.

Dai-Viet N. Vo is the Director of the Center of Excellence for Green Energy and Environmental Nanomaterials at Nguyen Tat Thanh University in Vietnam. He received his Ph.D. degree in Chemical Engineering from the University of New South Wales in Australia and has worked as a postdoctoral fellow at the University of New South Wales in Sydney and Texas A&M University in Qatar. His research interests include Fischer-Tropsch synthesis, novel catalyst characterization, catalytic reforming and other clean fuel technologies. He has published more than 100 peer-reviewed journal articles, 8 book chapters and 80 conference proceedings. He is the editor of several books for Elsevier, Springer and CRC Press as well as guest editor of Special Issues of *the International Journal of Hydrogen Energy, Comptes Rendus Chimie, Arabian Journal of Chemistry, Waste and Biomass Valorization, SN Applied Sciences, Topics in Catalysis, Chemical Engineering & Technology, Journal of Chemical Technology & Biotechnology, Biomass Conversion and Biorefinery, Chemical Engineering Science* and *Frontiers in Energy Research.*

Prakash K. Sarangi is a Scientist with specialization in Food Microbiology at the Central Agricultural University in India. He received his Ph.D. degree in Microbial Biotechnology from Ravenshaw University in India. His research focuses on bioprocess engineering, renewable energy, biochemicals, biomaterials and rural development. He has published more than 50 peer-reviewed journal articles and 15 book chapters. He is the editor of several books for Springer, CRC Press and IK International Publishing House. He is a fellow of the Society for Applied Biotechnology, and a life member of the Biotech Research Society of India; Society for Biotechnologists of India; Association of Microbiologists of India; Orissa Botanical Society; Medicinal and Aromatic Plants Association of India; Indian Science Congress Association; Forum of Scientists, Engineers & Technologists; and International Association of Academicians and Researchers.

Contributors

Sumaiya Zainal Abidin Faculty of Chemical and Natural Resources Engineering, Universiti Malaysia Pahang, Kuantan, Pahang, Malaysia

Prince Nana Amaniampong Institut de Chimie des Milieux et Matériaux de Poitiers (IC2MP), Université de Poitiers, Poitiers, France

Khursheed B. Ansari Department of Chemical Engineering, Aligarh Muslim University, Aligarh, Uttar Pradesh, India

Naveenji Arun Department of Chemical and Biological Engineering, University of Saskatchewan, Saskatoon, Saskatchewan, Canada

Arghya Banerjee Department of Chemical Engineering, Birla Institute of Technology and Science (BITS) Pilani, Pilani, Rajasthan, India

Franco Berruti Department of Chemical and Biochemical Engineering, University of Western Ontario, London, Ontario, Canada

Latika Bhatia Department of Microbiology and Bioinformatics, Atal Bihari Vajpayee University, Bilaspur, Chhattisgarh, India

Nguyen Thanh Binh Faculty of Chemistry, Vietnam National University, Cầu Giấy, Hà Nội, Vietnam

Phan Minh Quoc Binh Vietnam Petroleum Institute, Vietnam National Oil & Gas Group, Hà Nội, Vietnam

Venu Babu Borugadda Department of Chemical and Biological Engineering, University of Saskatchewan, Saskatoon, Saskatchewan, Canada

Guido Busca Department of Civil, Chemical and Environmental Engineering, University of Genova, Genova, Italy

Chin Kui Cheng Faculty of Chemical and Natural Resources Engineering, Universiti Malaysia Pahang, Kuantan, Pahang, Malaysia

Juliana da Conceição Infante Departamento de Bioquímica e Imunologia, Universidade de São Paulo, Ribeirão Preto, São Paulo, Brazil

Ajay K. Dalai Department of Chemical and Biological Engineering, University of Saskatchewan, Saskatoon, Saskatchewan, Canada

Duy Quang Dao Institute of Research and Development, Duy Tân University, Thanh Khê, Đà Nẵng, Vietnam

Sutapa Das Department of Chemical Engineering, Indian Institute of Technology Guwahati, Guwahati, Assam, India

Ana Sílvia de Almeida Scarcella Departamento de Bioquímica e Imunologia, Universidade de São Paulo, Ribeirão Preto, São Paulo, Brazil

Paula Zaghetto de Almeida Departamento de Bioquímica e Imunologia, Faculdade de Medicina de Ribeirão Preto, Universidade de São Paulo, Ribeirão Preto, São Paulo, Brazil

Maria de Lourdes Teixeira de Moraes Polizeli Departamento de Bioquímica e Imunologia, Faculdade de Medicina de Ribeirão Preto, Universidade de São Paulo, Ribeirão Preto, São Paulo, Brazil

Bruna Rego de Vasconcelos Department of Chemical and Biotechnological Engineering, Université de Sherbrooke, Sherbrooke, Québec, Canada

Parmila Devi Department of Chemical and Biological Engineering, University of Saskatchewan, Saskatoon, Saskatchewan, Canada

Asmaa Drif INCREASE, Université de Poitiers, Poitiers, France

Shin Ying Foong Pyrolysis Technology Research Group, School of Ocean Engineering, Universiti Malaysia Terengganu, Kuala Terengganu, Terengganu, Malaysia

Anjani Ravi Kiran Gollakota Department of Safety, Health and Environmental Engineering, National Yunlin University of Science and Technology, Douliou, Taiwan, Republic of China

Vaibhav V. Goud Department of Chemical Engineering, Indian Institute of Technology Guwahati, Guwahati, Assam, India

Pham Thanh Huyen School of Chemical Engineering, Hanoi University of Science and Technology, Hai Bà Trưng, Hà Nội, Vietnam

Aishah Abdul Jalil School of Chemical and Energy Engineering, Universiti Teknologi Malaysia, Johor, Malaysia

Surachai Karnjanakom Department of Chemistry, Rangsit University, Pathumthani, Thailand

Suwadee Kongparakul Department of Chemistry, Thammasat University, Pathumthani, Thailand

Anton P. Koskin Boreskov Institute of Catalysis SB RAS, Novosibirsk, Russia

Janusz A. Kozinski Department of Chemical Engineering, University of Waterloo, Waterloo, Ontario, Canada

Su Shiung Lam School of Ocean Engineering, Universiti Malaysia Terengganu, Kuala Terengganu, Terengganu, Malaysia

Minh Thang Le School of Chemical Engineering, Hanoi University of Science and Technology, Hai Bà Trưng, Hà Nội, Vietnam

Wan Adibah Wan Mahari School of Ocean Engineering, Universiti Malaysia Terengganu, Kuala Terengganu, Terengganu, Malaysia

R. Mahesh Department of Biosciences and Bioengineering, Indian Institute of Technology Guwahati, Guwahati, Assam, India

Soumen K. Maiti Department of Biosciences and Bioengineering, Indian Institute of Technology Guwahati, Guwahati, Assam, India

Shima Masoumi Department of Chemical and Biological Engineering, University of Saskatchewan, Saskatoon, Saskatchewan, Canada

Doan Pham Minh Université de Toulouse, IMT Mines Albi, Centre RAPSODEE, Albi, France

Purna Chandra Mishra School of Mechanical Engineering, Kalinga Institute of Industrial Technology (KIIT University), Bhubaneswar, Odisha, India

Venkateswara R. Naira Department of Biosciences and Bioengineering, Indian Institute of Technology Guwahati, Guwahati, Assam, India

Sonil Nanda Department of Chemical and Biological Engineering, University of Saskatchewan, Saskatoon, Saskatchewan, Canada

Biswajeet Nayak Department of Mechanical Engineering, Einstein Academy of Technology and Management, Bhubaneswar, Odisha, India

Swarup Kumar Nayak School of Mechanical Engineering, Kalinga Institute of Industrial Technology (KIIT University), Bhubaneswar, Odisha, India

Trinh Duy Nguyen Center of Excellence for Green Energy and Environmental Nanomaterials, Nguyễn Tất Thành University, Hồ Chí Minh City, Vietnam

Muhamad Mat Noor Faculty of Mechanical Engineering, Universiti Malaysia Pahang, Pekan, Pahang, Malaysia

Suraj K. Panda Department of Biosciences and Bioengineering, Indian Institute of Technology Guwahati, Guwahati, Assam, India

Parimal A. Parikh Department of Chemical Engineering, Sardar Vallabhbhai National Institute of Technology, Surat, Gujarat, India

Thiago Machado Pasin Departamento de Bioquímica e Imunologia, Universidade de São Paulo, Ribeirão Preto, São Paulo, Brazil

Thanh Khoa Phung School of Biotechnology, International University, Vietnam National University, Hồ Chí Minh City, Vietnam

Mihir Kumar Purkait Department of Chemical Engineering, Indian Institute of Technology Guwahati, Guwahati, Assam, India

Paresh H. Rana Department of Chemical Engineering, Government Engineering College, Bhuj, Gujarat, India

Rachita Rana Department of Chemical and Biological Engineering, University of Saskatchewan, Saskatoon, Saskatchewan, Canada

Prasert Reubroycharoen Department of Chemical Technology, Chulalongkorn University, Bangkok, Thailand

Nurul Asmawati Roslan Faculty of Chemical and Natural Resources Engineering, Universiti Malaysia Pahang, Kuantan, Pahang, Malaysia

Chanatip Samart Department of Chemistry, Faculty of Science and Technology, Thammasat University, Pathumthani, Thailand

Prakash K. Sarangi Directorate of Research, Central Agricultural University, Imphal, Manipur, India

Herma Dina Setiabudi Faculty of Chemical and Natural Resources Engineering, University Malaysia Pahang, Kuantan, Pahang, Malaysia

Chi-Min Shu Department of Safety, Health and Environmental Engineering, National Yunlin University of Science and Technology, Douliou, Taiwan, Republic of China

Tan Ji Siang School of Chemical and Energy Engineering, Universiti Teknologi Malaysia, Johor, Malaysia

Arshdeep Singh Department of Chemical and Biochemical Engineering, University of Western Ontario, London, Ontario, Canada

Pravin G. Suryawanshi Department of Chemical Engineering, Indian Institute of Technology Guwahati, Guwahati, Assam, India

Ahimee Hernandez Torres Université de Toulouse, IMT Mines Albi, Centre RAPSODEE, Albi, France

Thi Tuong Vi Tran Department of Chemistry, Faculty of Science and Technology, Thammasat University, Pathumthani, Thailand

Quang Thang Trinh Cambridge Centre for Advanced Research and Education in Singapore (CARES), Campus for Research Excellence and Technological Enterprise (CREATE), Singapore, Singapore

Dang Thanh Tung Vietnam Petroleum Institute, Vietnam National Oil & Gas Group, Hà Nội, Vietnam

Aleksey A. Vedyagin Boreskov Institute of Catalysis SB RAS, Novosibirsk, Russia

Dai-Viet N. Vo Center of Excellence for Green Energy and Environmental Nanomaterials, Nguyễn Tất Thành University, Hồ Chí Minh City, Vietnam

Vikranth Volli Department of Safety, Health and Environmental Engineering, National Yunlin University of Science and Technology, Douliou, Taiwan, Republic of China

Inna V. Zibareva Boreskov Institute of Catalysis SB RAS, Novosibirsk, Russia

Growth of Biofuels Sector: Opportunities, Challenges, and Outlook

1

Naveenji Arun and Ajay K. Dalai

Abstract

The demand for biofuels is increasing due to the uncertainties in the supply of fossil fuels, increased pollution hazards, rural economic growth, and the necessity to control carbon emissions. It is important to seek an alternate fuel resource to curb the carbon emissions. The transition from fossil fuel refineries to sustainable biorefinery can be clearly noticed in the present era. The dependency on food-based crops, which leads to the food versus fuel issue, has been addressed with the search for alternative and sustainable nonfood feedstocks such as lignocellulosic biomass. Until date, the methodology to unlock the full potential of lignocellulosic biomass is still in infancy. The migration from food-based biofuel feedstocks to lignocellulosic feedstocks and other organic wastes needs technological innovation and policies that could provide financial support. The detailed analysis on the risk factors associated with technological innovation in the biofuels sector is relatively less, which is the main objective of this chapter.

Keywords

Biofuels · Lignocellulosic biomass · Biorefining

N. Arun · A. K. Dalai (✉)
Department of Chemical and Biological Engineering, University of Saskatchewan, Saskatoon, Saskatchewan, Canada
e-mail: ajay.dalai@usask.ca

S. Nanda et al. (eds.), *Biorefinery of Alternative Resources: Targeting Green Fuels and Platform Chemicals*, https://doi.org/10.1007/978-981-15-1804-1_1

1.1 Introduction

Globally, the consumption of renewable and low-carbon biofuels is on rise owing to the international policies that focus on decreasing the greenhouse gas (GHG) emissions. Biofuels derived from lignocellulosic biomass, oil seeds, microalgae, and other organic wastes such as municipal solid wastes, cattle manure, industrial effluents, and sewage sludge have tremendous potentials to meet the future clean energy demands of the world (Nanda et al. 2015). Biofuels can be derived through the thermochemical conversion (e.g., pyrolysis, liquefaction, and gasification) and biochemical conversion (e.g., anaerobic digestion and fermentation) of organic wastes (Nanda et al. 2014, 2017b). The usable forms of biofuels are mostly in the form of liquid (e.g., bio-oil, biodiesel, microalgal oil, bioethanol, and biobutanol) and gaseous (e.g., syngas, producer gas, biohydrogen, and biomethane) forms (Nanda et al. 2016a). However, the bio-oils derived from the pyrolysis of biomass and organic wastes require catalytic upgrading (hydrotreating) to be transformed into synthetic transportation fuels (Arun et al. 2015, 2017).

The most commonly used classification of biofuels is presented in Fig. 1.1 (Raud et al. 2019). Based on the production process, biofuels can be classified as primary biofuels (from biomass in natural form) and secondary biofuels (from biomass in processed form). The first-generation biofuels involve the usage of edible feedstocks, which create the "food versus fuel" argument worldwide (Nanda et al. 2018). The second-generation biofuels addressed this issue as they are mostly produced from nonedible biomass through thermochemical and biochemical processes. Algae and other aquatic biomass are classified as third-generation biofuels and offer advantages such as low land usage, high lipid content, and high atmospheric CO_2 uptake capabilities (Correa et al. 2019). Genetic modification of third-generation biofuels can lead to the forecasted fourth-generation biofuels, which are mostly under research and developmental phase. Table 1.1 gives the examples of different generations and categories of biofuels. The advantages of different generations of biofuels are summarized in Table 1.2.

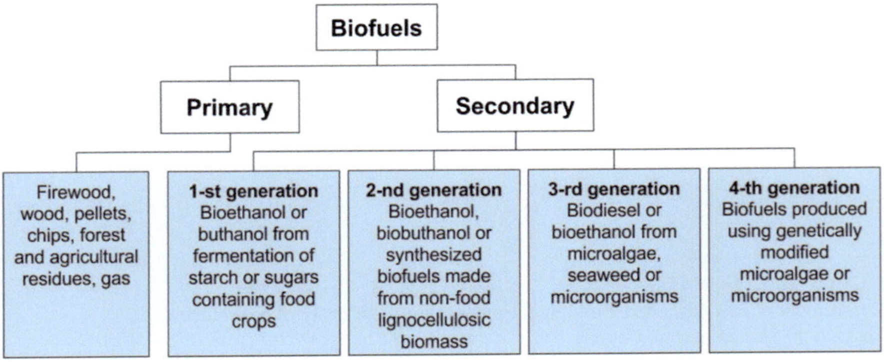

Fig. 1.1 Classification of biofuels according to processing technology and biomass type (Reproduced with permission from Raud et al. 2019)

The sustainable biorefining of biomass can also result in the production of a broad spectrum of marketable products such as biofuels, biochemicals, biomaterials, bioadditives for food and feed, as well as heat and power (Hassan et al. 2019; Arun and Dalai 2019). During the Paris climate conference (COP21) in 2015, nearly 200 countries decided to limit global warming below 2 °C (Chen et al. 2016). Many studies have reported about the "food versus fuel" debate, and a transition toward cellulosic alcohols has gained equal attention as cellulosic ethanol has the potential to lower the GHG emission by 90% in comparison to gasoline.

At the current population growth rate, it is estimated that the earth's population in 2030 will be around 8.5 billion people (Hassan et al. 2019). In 2016, the annual GHG emission was measured at 51.9 gigatons of CO_2 equivalent ($GtCO_2e$), while the target is to reduce the emissions by 11–13.5 $GtCO_2e$ per year by 2030. To control the catastrophic damages caused by anthropogenic activities, the United Nations agreed on 17 sustainable development goals for 2030 (Hassan et al. 2019). Several criteria considered for evaluating the socioeconomic and environmental benefits of biofuel production systems are shown in Fig. 1.2.

According to the Food and Agriculture Organization (FAO), the total world consumption of agricultural products will be 60% higher in 2050 in comparison to the consumption in 2005. Another estimation by FAO indicates that by 2050, additional 70 million hectares of cultivated land will be required to meet the food demands of the future generation. Hence, there is a clear indication of competition between urbanization and agriculture. The urban expansion can potentially result in loss of agricultural lands and it is estimated that the global loss of agricultural lands will be 1.8–2.4% by 2030.

The European Bioeconomy Strategy was launched by the European Commission in 2012, which was themed on "Innovating for sustainable growth: A bioeconomy for Europe" (European Environment Agency 2019). After assessment by the European Union Commission in 2017, the scope of the current action plan was found to be inadequate for the sustainable development of the biorefinery sector in

Table 1.1 Different generation and categories of biofuels (Reproduced from Kamani et al. 2019 with permission from The Royal Society of Chemistry)

Generation	Source/substrate	Product
Primary	Firewood, wood chips, pellets, animal waste, forest, and crop residues	Used in unprocessed form, mainly for heating, cooking, and electricity purposes
Secondary	Seeds, grain, and sugars	Bioethanol/biobutanol (by the fermentation of starchy or sugar-rich crops), biodiesel (by the transesterification of plant oils)
Secondary	Lignocellulosic biomass	Bioethanol/biobutanol (using enzymatic hydrolysis), methanol, mixed alcohol, green diesel (by thermochemical processes) and biomethane (by anaerobic digestion)
Tertiary	Algae and seaweed	Biodiesel and bioethanol from algae and seaweeds, hydrogen from microorganisms and green algae

Table 1.2 Advantages and disadvantages of different generations of biofuels (Reproduced with permission from Abdullah et al. 2019)

Topic	First-generation biofuel	Second-generation biofuel	Third-generation biofuel	Fourth-generation biofuel
Competition with food crops	Made from edible oil and starch feedstock	No food-energy conflict	No food-energy conflict	No food-energy conflict
Land footprint	Requires arable land	Require arable land or forests	Non-arable land can be used for cultivation	Non-arable land can be used for cultivation
Conversion to biofuels	Easy conversion	Need sophisticated downstream processing technologies due to high contents of hemicelluloses and lignin	Easy conversion due to increased hydrolysis and/or fermentation efficiency	Easy conversion due to increased hydrolysis and/or fermentation efficiency
Water footprint	Potable water is required for cultivation	Potable water is required for cultivation	Waste, saline, and non-potable water also can be used	Waste, saline, and non-potable water also can be used
Environment friendliness	Using pesticides and fertilizers are of the main concerns	No expenditure on fertilizer or pesticides. Deforestation is a concern	CO_2 fixation, wastewater treatment, and no expenditure on fertilizer are benefits. Ecological concerns such as marine eutrophication are a disadvantage	CO_2 fixation and wastewater treatment are the benefits, but release of genetically modified organisms is a main concern
Commercialization	Commercially produced	Commercially produced	Insufficient biomass production for commercialization	Insufficient biomass production for commercialization
Sustainability	Use of natural resources such as water and land is not conservative	Does not preserve ecology due to deforestation concerns	Does not have favorable economics	There are concerns about the release of genetically modified organisms to the environment and ecological risks
Nutrient requirements	Using pesticides and fertilizers is of the main concern	Does not require any fertilizer treatment	Large carbon and nitrogen sources are required. Solar energy is only available during daytime. Nutrients can be recycled in the process	Large carbon and nitrogen sources are required. Solar energy is only available during daytime. Nutrients can be recycled in the process

Harvesting	Harvesting is done by hand or machine picking	Harvesting is done by hand or machine picking	Harvesting of microalgae is expensive and complicated	Harvesting of microalgae is expensive and complicated
Regulation	The regulations are fairly clear	The regulations are fairly clear	No regulation is available for marine cultivation	No regulation is available for marine cultivation. Furthermore, strict regulation is for the intended release of GM algae
Financial input	The capital cost is fairly low	The capital cost is fairly low	The initial cost for large-scale cultivation is too high	The initial cost for large-scale cultivation is too high
Environmental condition	Parameters such as temperature and humidity must be within a suitable range	Parameters such as temperature and humidity must be within a suitable range	Can be cultivated in harsh environmental conditions such high pH, salinity, and high light intensities	Can be cultivated in harsh environmental conditions such high pH, salinity, and high light intensities

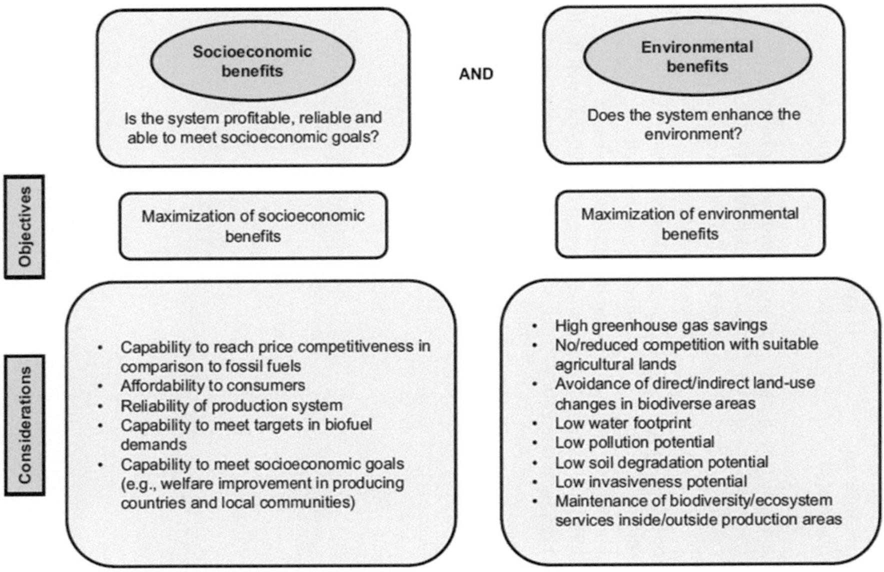

Fig. 1.2 Criteria to be considered when evaluating the socioeconomic and environmental benefits of biofuel production systems (Reproduced with permission from Correa et al. 2019)

the Europe. More recently, Brazil, the USA, Canada, the European Union, and many Asian countries have started posing strict legal mandates for the usage of biofuels on commercial scale.

According to International Energy Outlook (2016), the world's energy consumption is expected to increase by approximately 48% between 2012 and 2040. The production of conventional biodiesel involves the usage of fertilizers, mechanical equipments (which uses fossil fuels), and arable land. According to the Innovation Outlook, Advanced Biofuels-IRENA, the basic energy input to produce conventional biodiesel comes from fossil fuels. Therefore, the production cost for conventional biodiesel is US $1.6 per liter in comparison to US $1.2 per liter for diesel. In European Union, the annual turnover of current biofuel economy is €2.3 trillion and 18.5 million people are employed in the biofuel sector (Hassan et al. 2019). The S2Biom project supported by the European Commission indicates that the bio-based products will have a market worth of €40 million by 2020 and are projected to have a 4% annual growth rate. Many biofuel policies struggle to address the integrations between company's perspective and policy development. The success of biofuel policies has been primarily based on climate change and community support. It is estimated that nearly 67 biorefineries based on lignocellulosic biomass are in current operation globally. Tables 1.3 and 1.4 illustrate the main products that can be obtained by the hydrothermal conversion of lignocellulosic and non-lignocellulosic biomasses, respectively.

The commercialization of biofuels sector involves high capital expenditure (CAPEX), and it is important to integrate the biochemical processes with the existing

Table 1.3 Summary of main structures and major compounds of lignocellulosic biomass (Reproduced from Usman et al. 2019 with permission from The Royal Society of Chemistry)

Lignocellulosic biomass	Percentages of main structures			Major compounds into hydrothermal conversion aqueous products
	Cellulose (wt%)	Hemicellulose (wt%)	Lignin (wt%)	
Crop straw and husk	29.2–46.0	18.2–36.4	15.0–28.2	Volatile fatty acids, capronic acid, lactic acid, furfurals, sugars, alcohols, and cyclopentenone
Newspaper	44.2	17.8	26.8	Volatile fatty acids, furfurals, sugars, alcohols, and phenols
Orange pomace	14.3	6.3	3.3	Acetic acid, 5-hydroxymethylfurfural, furfurals, ethanol, acetone, butanone, and alkyl derivatives
Recycled paper	60.8	14.2	8.4	Volatile fatty acids, furfurals, sugars, alcohols, and cyclopentenone
Spent grain	18.5	26.5	19.1	Cyclopentenones, carboxylic acids, pyrazines, and ketones
Sugarcane bagasse	56.0	4.6	36.4	Volatile fatty acids, phenols, furfurals, sugars, alcohols, and cyclopentenone
Switchgrass	32.8	23.7	18.2	Volatile fatty acids, phenols, furfurals, sugars, alcohols, and cyclopentenone
Woody biomass	38.6–63.6	7.7–20.2	17.6–32.7	Phenols, furfurals, glycolic acid, acetic acid, alcohols, and cyclopentenone

Table 1.4 Summary of main structures and major compounds of non-cellulosic biomass (Reproduced from Usman et al. 2019 with permission from The Royal Society of Chemistry)

Non-lignocellulosic biomass	Percentages of main structures			Major compounds into hydrothermal conversion aqueous products
	Protein (wt%)	Carbohydrates (wt%)	Lipids (wt%)	
Dried distillers' grains	42.2	35.0	22.4	Cyclopentenones, carboxylic acids, pyrazines, ketones, and oxygenated aromatics
Food waste	15.0–25.2	41.3–62.1	13.4–30.2	Volatile fatty acids, 5-hydroxymetgylfurfural, furfural, ethanol, ketones, alkyl derivatives of 2-cyclopenten-1-one
Macroalgae	12.2–30.9	54.3–83.6	0.9–6.2	Nitrogenous compounds, long-chain fatty acids, glycerol, alcohols, and acetone
Microalgae	8.1–71.4	4.2–57.1	2.4–40.1	Volatile fatty acids, phenols, pyrazines, benzenes, alkanes, and fatty acids
Mixed cultural algae	27.2	17.9	5.7	Short-/long-chain organic acids, amino acids, phenols, urea, N-heterocyclic compounds, acetamide, and ketones
Sewage sludge	27.6–33.4	3.3–4.0	6.6–13.8	Volatile fatty acids, benzene, acetic acid, carbonic acid, alkenes, phenolic, and aromatic compounds

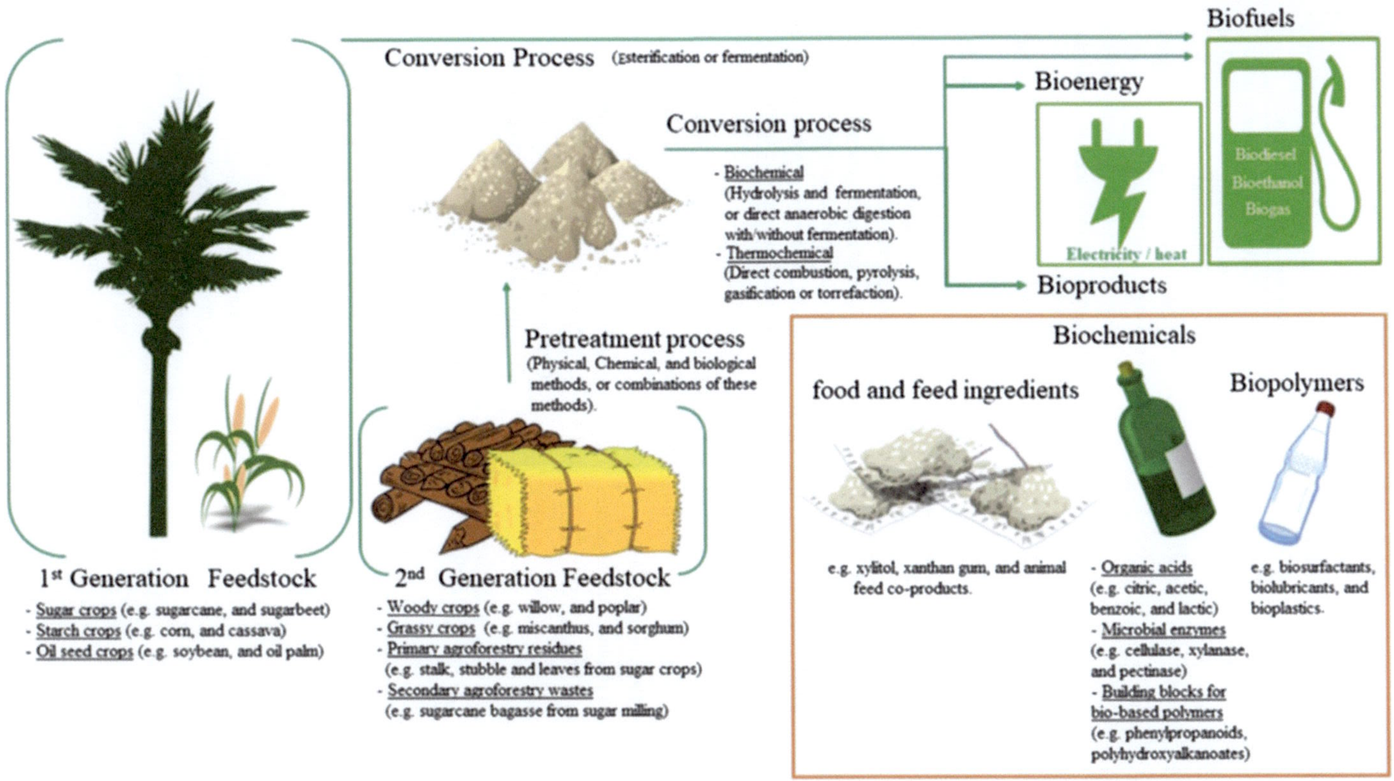

Fig. 1.3 Schematic diagram showing differences between first- and second-generation lignocellulosic feedstocks, valorization processes, and end products (Reproduced with permission from Hassan et al. 2019)

refinery setup to produce biofuel products with minimum operational cost (Fig. 1.3). In the future, it is imperative to consider the energy efficiency of biofuel blends and the influence of carbon taxation policy, and they needed to be included in the analysis. The policy makers should consider and analyze the risks involved in the commercialization of technologies that produce advanced biofuels. The identified potential risks are related to the management processes, market conditions, and profitability. The developed policies should find a balance between the translation of technology and business perspective. Most of the green energy innovation policies have attempted to address the optimal balance between technological push and supply-demand pull in green technologies. This chapter provides insights into the present state of energy demand, search for novel feedstocks, and potential risks associated in the commercialization of novel biofuels production technologies.

1.2 Biofuel Scenario in Canada and the World

After agreeing to the 2015 Paris Climate Conference (COP21), the Government of Canada has taken initiatives to curb the greenhouse gas (GHG) emissions (Nanda et al. 2016b). To achieve the Clean Fuel Standard and achieve the climate change objectives, discussions with provincial and federal governments and indigenous peoples and rural communities are in swift progress. According to the International Energy Agency (IEA), to meet the Paris commitment of keeping the global warming below 2 °C, the consumption level of biofuels should triple by 2030. On an average, about 140–150 patents are filed annually and this clearly indicates a worldwide hunger for advanced biofuel systems.

Owing to Canada's extensive forestry and agricultural resources, the prospects for advanced biofuels seem to be promising. In Canada, the Renewables Fuels Regulations Act was enacted in August 2010. According to this act, the gasoline pool must contain 5% by volume of renewable fuel, and in the diesel pool, the renewable fuel content should be 2% by volume. With the depression of oil prices in 2015 in Canada, the energy sector continued to account for 20% and 18% of the gross domestic product of Alberta and Saskatchewan, respectively (Mondou et al. 2018). According to Dragojlovic and Einsiedel (2015), biofuels generated from nonfood crops are recently gaining societal acceptance in the USA compared to corn-based ethanol. Moreover, the communal perception on the climate change and its associated risks and threat are the key predictor of the defiance toward biofuels in the USA.

In Canada, the annual consumption of biodiesel has increased by 100% in comparison to 2010 (123 million liters). The Federal Renewable Fuels Regulations also mandates the minimal blending of renewable fuels to decline the average lifecycle carbon intensity (CI) of fuels over time. The energy density of ethanol is about 33% lower than gasoline and it means that consumers must purchase more of ethanol to meet the energy requirements. This also indicates greater distribution cost and higher tax rates. It must be noted that blending ethanol (5–10 vol%) can increase the energy efficiency of the vehicle by 1%. In addition, blending with ethanol also permits refineries to produce gasoline blend stock with lower octane number, which can potentially reduce the greenhouse gas intensity. In future, it is imperative to consider the energy efficiency of biofuel blends and the influence of carbon taxation policy and they needed to be included in the analysis.

Biomass from lignocellulosic materials (e.g., agricultural crop residues and woody biomass), algae, and oilseed crops can be converted to biofuels through applicable thermochemical and biochemical conversion routes. The oil extraction from oilseed crops and algae undergoes transesterification to produce biodiesel. The de-oiled algal biomass can undergo supercritical water gasification due to its high moisture content to produce syngas (Reddy et al. 2014; Okolie et al. 2019; Yadav et al. 2019). The syngas is further converted to refined fuels such as green diesel, green kerosene, and other hydrocarbons through Fischer-Tropsch catalysis. Fermentation of syngas using acidogenic bacteria can produce bioethanol. Lignocellulosic biomass and de-oiled algae can also undergo pyrolysis to produce bio-oil, biochar, and producer gas. The sugars in lignocellulosic biomass can be recovered using enzymatic hydrolysis and fermentation to produce alcohols, e.g., bioethanol and biobutanol (Fig. 1.4), or through gasification and Fischer-Tropsch catalysis to produce alcohols and hydrocarbon (Fig. 1.5) (Nanda et al. 2014, 2017a).

Algae-based biofuels seem to have the potential to offer a sustainable pathway for bioenergy and bioproducts (Yadav et al. 2019). Alga is an energy density biological source and can accumulate lipids up to 50% of dry cell weight (U.S. DOE 2010). As a part of advanced biofuel systems, algal biofuels have gained attention. The catalytic cracking process for the biofuels production from algal oil is energy intensive and it is important to analyze the energy return on investment (EROI). In chemical and enzymatic processes, reusability of catalyst is an important factor and various studies have been done to develop a cost-effective catalyst for chemical and enzymatic conversion. Algal biofuel production process possesses high productivity (per acre) because it is not based on food resources and can be cultivated on open ponds and wastewater. It is imperative to cultivate the specific type and strain of algae (e.g., microalgae, macroalgae, and cyanobacteria) for the successful algae-based biofuels production process. Although algal biofuels industry seems promising, it is limited by several factors such as biomass cultivation, availability of advanced processing facilities, technical challenges, and logistic issues (Chen et al. 2016). There are many uncertainties in the availability and commercial usage of fossil fuels after 2040. In the last two decades, the oil price had an average volatility of 30% per year. According to the UK Production Capacity Outlook to 2030, the design and development of cost-effective catalysts for fuel upgrading technologies such as Fischer-Tropsch process, hydrotreating, or transesterification seem to be a primary challenge in process commercialization.

As discussed earlier, biofuel feedstocks can be categorized into three categories such as (a) biomass produced on marginal lands, (b) agricultural crop residues and forestry biomass, and (c) cattle manure and organic wastes (Junginger et al. 2006). It is globally estimated that these waste and biomass categories can supply bioenergy accounting up to 200 EJ, 100 EJ, and 100 EJ, respectively. However, an access to complete integrated biofuels process assessment is unavailable. The estimation

Fig. 1.4 The agricultural residues used for the production of bioethanol (Reproduced from Kamani et al. 2019 with permission from The Royal Society of Chemistry)

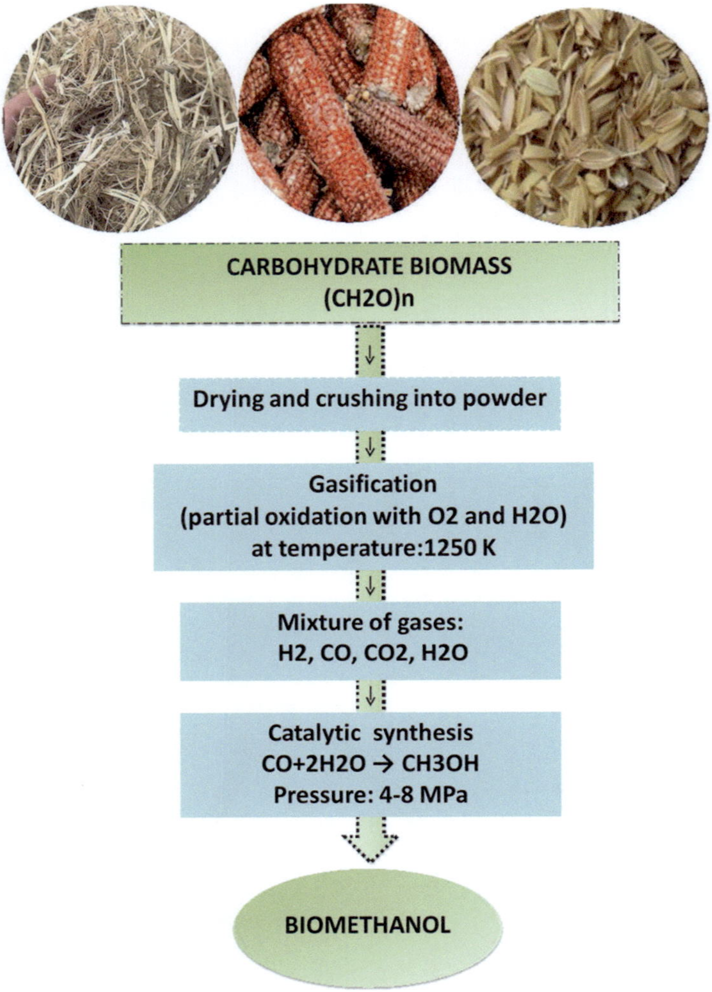

Fig. 1.5 Production of biomethanol from carbohydrate biomass by gasification (Reproduced from Usman et al. 2019 with permission from The Royal Society of Chemistry)

studies done by several international organizations such as USEPA, Stockholm Environmental Institute, and the Intergovernmental Panel on Climate Change (IPCC) provide different conclusions (100–500 EJ per year in 2050) on the reliability of biomass to meet the future energy demands (Junginger et al. 2006). The major difference in the analysis can be attributed to the uncertainty in land availability and yield limits.

1.3 Policies and Their Impacts on the Success of Advanced Biofuels: Scenario in Canada and the World

The governmental commitment to policies and the durability of the policies are of concern for the commercialization of advanced biofuels market. The impact of policy changes is drastic on the venture capital companies in comparison to refineries based on fossil fuels. For the deployment of futuristic biofuels production processes, the uncertainties associated with the government policies need to be reduced. In Canada, many private and federal organizations provide financial support for the commercialization of biofuels sector. The Canadian Foundation for Innovation (CFI) provides financial support for the development of research infrastructure.

The Sustainable Development Technology Canada (SDTC) is a foundation funded by the government of Canada and it provides nonrepayable funds for the development of novel processes and technologies (in the pre-commercial phase) that focus to curb GHG emissions. Through the Next-generation Biofuels Fund (NGBF), about $500 million was sanctioned to private research centers. Recently, Canada's Networks of Centers of Excellence funded a research initiative called BioFuelNet Canada, which specifically focused on the Agriculture and Agri-Food Canada, Canadian Forest Services, and Transport Canada. Like BioFuelNet Canada, the Cellulosic Biofuel Network was initiated in 2010. The Renewable Industries Canada (RIC) and the Advanced Biofuels Canada (ABC) are the two organizations focusing on the advanced biofuels sector in Canada. The industries that produce conventional biofuels are represented by Renewable Industries Canada, and industries in western Canada (which primarily produce biodiesel) are represented by Advanced Biofuels Canada. Both these organizations have their own unique expertise and policy capabilities. For practical implementation of low-carbon fuel standard (LCFS), skill set development of lifecycle analysis of GHG emissions of fuels is important.

Decision-making step in the commercialization of novel biofuels technologies is strongly dependent on the ability of the government and corporates to take risk without developing aversion considering the challenges in getting higher returns after commercial runs. Tables 1.5 and 1.6 summarize some notable biofuel projects around the world and European Union, respectively. Scientifically, the term risk can refer to challenges, barriers, or constraints and the source of risk can be external (e.g., political, economic, social, and technological) and/or internal (management and operations in an organization). The nontechnical challenges and biases usually hinder the commercial development of bioenergy rather than technical issues (McCormick and Kaberger 2007). The nontechnical risk factors are the factors related to (a) policies; (b) supply, price, and demand of feedstocks; (c) fluctuations in the prices of fossil fuels; (d) food versus fuel debate; and (e) lack of refinery to process advanced feedstocks.

Pries et al. (2016) performed a detailed analysis on the different risks and their influence on the commercialization of advanced biofuels technology. The risks could be classified as political, economic, social, technological, primary, and support activities risks. The primary political risk arises from the support from the

Table 1.5 Main biofuel projects around the world (Reproduced with permission from Su et al. 2017)

Company or region	Cultivation method	Technology	Products and scale
LiveFuels (USA)	Open ponds, open water systems	Grows and harvests algae in estuarine environments using natural systems to keep cost low. The company has established pilot operations across the USA	Algal fuel
OriginOil Inc. (USA)	Open ponds	Developing a technology to transform algae into true competitor to petroleum	Diesel, gasoline, jet fuels, plastics, and solvents
PetroSun	Open ponds	Extracting algae oil in the farm and transporting to biodiesel refinery	Algae oil (4.4 million gallons annually)
Neste Oil (Europe)	Open ponds	It is the only company that commercially produces high-value renewable fuels with more than 10 raw materials in which NEXBTL yields renewable diesel at about two million tons annually	Biofuels, plant oil, and algae oil
Seambiotic (Mediterranean)	Open ponds	Cooperating with power plant and cultivating algae with flue gas in Israel	Food, chemicals, and biofuels
Aquaflow Bionomics (New Zealand)	Open ponds	Gaining feedstocks from algae derived from wastewater purification	Biocrude oil
Biofuel System (UK)		Producing biofuels with seaweed	Biogasoline or biodiesel
A2BE Carbon Capture (USA)	Closed systems	Establishing carbon capture cycle system. CO_2 emitted from industries is captured and used in algae cultivation	Biofuels
GreenFuel Technologies (USA)	Closed systems	Cultivating microalgae with industrial waste gas	Drop-in biofuels
Solazyme, Inc. (USA)	Closed systems	Cultivating microalgae in dark in large storage tank and providing energy in growth medium with sugar	Biofuels

(continued)

Table 1.5 (continued)

Company or region	Cultivation method	Technology	Products and scale
Algenol Biofuels (USA)	Closed systems	Located in desert. Cultivating microalgae on large scale with seawater and CO_2 to produce ethanol. Cultivating blue-green algae and blue algae simultaneously based on patent, fixing nitrogen, and reducing cost	Ethanol (yields amount to 9000 gallon/acres in 2013)
Sapphire Energy (USA)	Closed systems	The company's molecular platform can convert sunlight and CO_2 into renewable carbon neutral fuels	Green crude oil. Suitable for infrastructures such as automobile, refinery, and pipeline
Inventure Chemical Technology (USA)	Closed systems	The patented mixed supercritical fluid processing technology allows for the direct conversion of a host of waste biomass materials to fuels, chemicals, and non-gasification intermediates	Ethanol, butanol, triglycerides, fatty acids, dimethyl ether, and isoprene
Solena (USA)	Closed systems	Plasma techniques are used in gasification of algae and other organic matters. It can generate electricity from microalgae treated with gasification	Establishing a 40 MW power plant in cooperation with Sunflower Company of Kanas
Solix Biosystems (USA)	Closed systems	The core of Solix's intellectual property consists of its proprietary AGS technology. AGS is an extended surface area culture system that incorporates Solix's proprietary Lumian™ photobioreactor panels	Renewable chemicals and fuels, personal care products, and nutritionals
XL Renewables (USA)	Closed systems	Patented algae production technology—Simgae for simple algae. Phyco is currently developing a commercial production facility near Phoenix, Arizona. XL Renewable has formed Phyco BioSciences, Inc. to commercialize algae biomass as a food and industrial crop	Biofuels and bioproducts

(continued)

Table 1.5 (continued)

Company or region	Cultivation method	Technology	Products and scale
Aurora Algae	Open ponds	Using genetically modified algae to produce biodiesel based on patented techniques from UCB. The algae owned by the company show light color with light penetration for the purpose of high yields. Aurora Algae has developed the world's first commercial-scale photosynthetic platform for sustainable algae-based product development	Eicosapentaenoic acid, proteins, biodiesel, etc.
Cellena (USA)	Photobioreactors and open ponds	Cellana's patented ALDUO™ system consists of a series of photobioreactors coupled with open ponds. Cellena possesses the techniques for oil extraction from seaweed without chemicals or oil extraction process. To date, over 20 metric tons of whole algae (dry weight) have been produced. Since 2009, Cellana has operated its Kona Demonstration Facility, a 6-acres state-of-the-art production and research facility in Hawaii	Biofuels, Omega-3 eicosapentaenoic acid, and docosahexaenoic acid and animal feed
LanzaTech	Closed systems	LanzaTech's microbial gas fermentation technology in which microorganisms and microalgae are cultivated in captured CO and CO_2 yields fuels and chemicals	Bioethanol, aviation biofuels, bioproducts
BioProcess Algae	Closed systems	Applying biofilm techniques to cultivate microalgae possessing the Grower Harvester™ harvesting methods based on patented technology	Animal feed, nutritionals, and transportation fuels
Solix Biosystems, Inc.	Closed photobioreactor panels	The company's production in algal growth system is seven times as much as that in open ponds	Renewable chemical products and fuels

(continued)

Table 1.5 (continued)

Company or region	Cultivation method	Technology	Products and scale
Heliae	Closed systems and pond system	Heliae possesses Volaris™ commercial production platform, AMP™ strains rapid screening system, and Helix™ algal strains cultivation system. In 2010, Heliae built up a 20-acre microalgae cultivation commercial demonstration plant and invested 13 million dollars in plant expansion	Biofuels and bioproducts

governing political parties and their proposed biofuel regulations. The economic risk primarily originates from the feedstock market price. The technological risks pose less threat in comparison to the other listed threats. The risk associated with profitable operations is the most noted primary activities risk as analyzed.

1.4 Outlook for a Sustainable Future

Energy can be considered as the most influential commodity in supply-demand concept worldwide. Over the last decade, the annual increase in global CO_2 emission was 2.5% and this has resulted in the increase in world's temperature by 2 °C (Friedlingstein et al. 2014). The consumption of renewable and low-carbon biofuels is on the rise owing to the international policies that focus on decreasing the greenhouse gas emissions. The land use patterns can vary among nations and different lands such as marginal, degraded, agricultural, and pastureland should be integrated to the current land use patterns. The introduction of biomass into the available lands must happen gradually without affecting the agricultural productivity in any land. The introduction of bioenergy production technologies must be able to provide additional income in the rural regions.

For the growth of biofuel crops, the sustainability and carbon footprint involved during the growth need attention. Canada is well positioned with resources and availability of marginal lands to produce lignocellulosic biomass-based biofuels. The first-generation and second-generation feedstocks are not substantial enough to meet the present energy demands, and development of leading technology using third-generation feedstocks is imperative. Due to the suboptimal climate in Canada, photosynthetic microalgae cannot provide a complete solution for the futuristic energy demand. Based on the current microalgae-based biofuels production technology, the net energy yield is 0.93. In other words, an input energy of 0.93 MJ is required for every 1 MJ of energy produced using photosynthetic microalgae. The usage of heterotrophic microalgae-based technology for biofuels production offers several advantages such as higher productivity, reduction in capital investments, and minimal operating cost. For the development of commercially viable process,

Table 1.6 The Bio-based Industries Joint Undertaking (BBI-JU) funded projects to support lignocellulose biorefining industry in the European Union (Reproduced with permission from Hassan et al. 2019)

Project and website	Start date	End date	BBI-JU contribution (€)	Focus
BIOFOREVER (https://www.bioforever.org)	2016	2019	9,937,998	Demonstrate the commercial viability of lignocellulosic biorefining (from woody biomass) for the chemical industry
BIOSKOH (http://bioskoh.eu)	2016	2021	21,568,195	Demonstrate the first of a series of new second-generation biorefineries in Europe
EUCALIVA (http://eucaliva.eu)	2017	2021	1,795,009	Create a whole value chain from lignin using *Eucalyptus* waste
GRACE (http://www.grace-bbi.eu)	2017	2022	12,324,632	Explore the potential of the nonfood industrial crops as a source of biomass for the bioeconomy
GreenSolRes (http://www.greensolres.eu)	2016	2021	7,451,945	Demonstrate the commercial viability of converting lignocellulosic biomass to levulinic acid
HyperBioCoat (http://www.hyperbiocoat.eu)	2016	2019	4,617,423	Develop biodegradable polymers derived from food processing by-products
iFermenter (https://ifermenter.eu)	2018	2022	3,997,825	Conversion of forestry sugar residual streams to antimicrobial proteins by intelligent fermentation
LIBRE (http://www.libre2020.eu)	2016	2020	4,566,560	Lignin-based carbon fibers for composites
LigniOx (http://www.ligniox.eu)	2017	2021	4,338,374	Lignin oxidation technology for versatile lignin dispersants
LIGNOFLAG (http://www.lignoflag-project.eu)	2017	2022	24,738,840	Produce bioethanol involving a bio-based value chain built on lignocellulosic feedstocks
PEFerence (http://peference.eu)	2017	2022	24,999,610	Produce furan dicarboxylic acid, a bio-based building block to produce high-value products
SSUCHY (https://www.ssuchy.eu)	2017	2021	4,457,194	Produce sustainable structural and multifunctional biocomposites from hybrid natural fibers and bio-based polymers
SWEETWOODS (https://sweetwoods.eu)	2018	2022	20,959,745	Produce and deploy high purity lignin and affordable platform chemicals through wood-based sugars

(continued)

Table 1.6 (continued)

Project and website	Start date	End date	BBI-JU contribution (€)	Focus
UNRAVEL (https://www.bbi-europe.eu/projects/unravel)	2018	2022	3,603,545	Develop advanced pretreatment, separation, and conversion technologies for complex lignocellulosic biomass
US4Greenchem (https://www.us4greenchem.com)	2015	2019	3,457,602	Combined ultrasonic and enzyme treatment of lignocellulosic feedstock as substrate for sugar-based biotechnological applications
VALCHEM (http://www.valchem.eu)	2015	2018	13,125,941	Value-added chemical building blocks and lignin from wood
WoodZymes (https://www.woodzymes.eu)	2018	2021	3,253,874	Extremozymes for wood-based building blocks: from pulp mill to board and insulation products
ZELCOR (http://www.zelcor.eu)	2016	2020	5,256,993	Zero-waste lignocellulosic biorefineries by integrated lignin valorization

modification of cell metabolism may be essential. However, a risk-based assessment is essential to understand the potential negative impact on local ecosystems.

The interaction of bioenergy sector with parameters such as CO_2 sequestration, food production for consumption, nature preservation, and biodiversity needs detailed analysis. A detailed model illustrating this interaction will provide improved understanding on the growth of biofuels sector in future. The technological development has direct influence on the capability of biofuels sector. The two main components of technological development are production of final products (energy carriers) from biomass and transport of these energy carriers. The usage of perennial crops such as hybrid poplar, *Miscanthus*, and sugarcane should be promoted in comparison to annual agricultural products due to the environmental and economic upper edge.

As mentioned by Scaife et al. (2015), some technological factors regarding biofuels need to be addressed such as whether the optimal technology is regionally specific or universal. Moreover, whether the technology be used to produce a broad spectrum of biofuels and biochemicals is still questionable. Additionally, the use of biological systems should address their utility based on photosynthetic or heterotrophic growth of microorganisms. The biomass harvesting processes, initial products extraction, and biomass conversion into biofuels at a cost-effective and energy-efficient manner are desirable.

1.5 Conclusions

The microalgae-based biofuels processes still need techno-economic analysis before commercialization. The third-generation biofuels production technology is still in research and developmental phase, and especially, microalgal biofuels are generally overlooked owing to the limitations in availability of literature supporting the success of microalgal in other countries. The present economy is highly fossil based and under the condition of fluctuations in oil prices. The dominance of fossil fuels is based on various aspects primarily due to policies established in the past, industrial infrastructure, and mind-set of customers to find fossil fuels more reliable than advanced biofuels. Biofuels have promising potentials for a sustainable bioeconomy that can not only reduce greenhouse gas emissions but also reduce the pollution problems and global warming. It is highly imperative to explore more on the research and development of biofuel feedstocks and the conversion technologies.

References

Abdullah B, Muhammad SAFS, Shokravi Z, Ismail S, Kassim KA, Mahmood AN, Aziz Md. MA (2019) Fourth generation biofuel: a review on risks and mitigation strategies. Renew Sust Energ Rev 107:37–50

Arun N, Dalai AK (2019) Life-cycle assessment of biofuels produced from lignocellulosic biomass and algae. In: Nanda A, Sarangi PK, Vo DVN (eds) Fuel processing and energy utilization. CRC Press, Florida, USA, pp 177–185

Arun N, Maley J, Chen N, Sammynaiken R, Hu Y, Dalai AK (2017) NiMo nitride supported on γ-Al_2O_3 for hydrodeoxygenation of oleic acid: Novel characterization and activity study. Catal Today 291:153–159

Arun N, Sharma RV, Dalai AK (2015) Green diesel synthesis by hydrodeoxygenation of bio-based feedstocks: Strategies for catalyst design and development. Renew Sust Energ Rev 48:240–255

Chen M, Smith PM, Wolcott MP (2016) U.S. biofuels industry: a critical review of opportunities and challenges. BioProd Business 1:42–59

Correa DF, Beyer HL, Fargione JE, Hill JD, Possingham HP, Thomas-Hall SR, Schenk PM (2019) Towards the implementation of sustainable biofuel production systems. Renew Sust Energ Rev 107:250–263

Dragojlovic N, Einsiedel E (2015) What drives public acceptance of second-generation biofuels? Evidence from Canada. Biomass Bioenergy 75:201–212

European Environment Agency (2019) Innovating for sustainable growth: a bioeconomy for Europe, EC, 2012. Communication from the Commission to the European Parliament, the Council, the European Economic and Social Committee and the Committee of the Regions. https://www.eea.europa.eu/policy-documents/innovating-for-sustainable-growth-a. Accessed 25 Apr 2019

Friedlingstein P, Andrew RM, Rogelj J, Peters GP, Canadell JG, Knutti R, Luderer G, Raupach MR, Schaeffer M, van Vuuren DP, Le Quere C (2014) Persistent growth of CO_2 emissions and implications for reaching climate targets. Nat Geosci 7:709–715

Hassan SS, Williams GA, Jaiswal AK (2019) Moving towards the second generation of lignocellulosic biorefineries in the EU: Drivers, challenges, and opportunities. Renew Sust Energ Rev 101:590–599

International Energy Outlook 2016. U.S. Energy Information Administration 2016. https://www.eia.gov/outlooks/ieo/pdf/0484(2016).pdf. Accessed 25 Apr 2019

Junginger M, Faaij A, Rosillo-Calle F, Wood J (2006) The growing role of biofuels - opportunities, challenges and pitfalls. Int Sugar J 108:618–629

Kamani MH, Eş I, Lorenzo JM, Remize F, Roselló-Soto E, Barba FJ, Clark JH, Khaneghah AM (2019) Advances in plant materials, food by-products, and algae conversion into biofuels: use of environmentally friendly technologies. Green Chem 21:3213–3231

McCormick K, Kaberger T (2007) Key barriers for bioenergy in Europe: economic conditions, know-how and institutional capacity, and supply chain co-ordination. Biomass Bioenergy 31:443–452

Mondou M, Skogstad G, Bognar J (2018) What are the prospects for deploying advanced biofuels in Canada? Biomass Bioenergy 116:171–179

Nanda S, Azargohar R, Dalai AK, Kozinski JA (2015) An assessment on the sustainability of lignocellulosic biomass for biorefining. Renew Sust Energ Rev 50:925–941

Nanda S, Golemi-Kotra D, McDermott JC, Dalai AK, Gökalp I, Kozinski JA (2017a) Fermentative production of butanol: perspectives on synthetic biology. New Biotechnol 37:210–221

Nanda S, Kozinski JA, Dalai AK (2016a) Lignocellulosic biomass: a review of conversion technologies and fuel products. Curr Biochem Eng 3:24–36

Nanda S, Mohammad J, Reddy SN, Kozinski JA, Dalai AK (2014) Pathways of lignocellulosic biomass conversion to renewable fuels. Biomass Conv Bioref 4:157–191

Nanda S, Rana R, Sarangi PK, Dalai AK, Kozinski JA (2018) A broad introduction to first, second and third generation biofuels. In: Sarangi PK, Nanda S, Mohanty P (eds) Recent advancements in biofuels and bioenergy utilization. Springer Nature, Singapore, pp 1–25

Nanda S, Rana R, Zheng Y, Kozinski JA, Dalai AK (2017b) Insights on pathways for hydrogen generation from ethanol. Sustain Energ Fuel 1:1232–1245

Nanda S, Reddy SN, Mitra SK, Kozinski JA (2016b) The progressive routes for carbon capture and sequestration. Energ Sci Eng 4:99–122

Okolie JA, Rana R, Nanda S, Dalai AK, Kozinski JA (2019) Supercritical water gasification of biomass: a state-of-the-art review of process parameters, reaction mechanisms and catalysis. Sustain Energ Fuel 3:578–598

Pries F, Talebi A, Schillo RS, Lemay MA (2016) Risks affecting the biofuels industry: a U.S. and Canadian company perspective. Energy Policy 97:93–101

Raud M, Kikas T, Sippula O, Shurpali NJ (2019) Potentials and challenges in lignocellulosic biofuel production technology. Renew Sust Energ Rev 111:44–56

Reddy SN, Nanda S, Dalai AK, Kozinski JA (2014) Supercritical water gasification of biomass for hydrogen production. Int J Hydrogen Energ 39:6912–6926

Scaife MA, Merkx-Jacques A, Woodhall DL, Armenta RE (2015) Algal biofuels in Canada: status and potential. Renew Sust Energ Rev 44:620–642

Su Y, Song K, Zhang P, Su Y, Cheng J, Chen X (2017) Progress of microalgae biofuel's commercialization. Renew Sustain Energy Rev 74:402–411

U.S. DOE (2010) National Algal Biofuels Technology Roadmap. U.S. Department of Energy, Office of Energy Efficiency and Renewable Energy, Biomass Program

Usman M, Chen H, Chen K, Ren S, Clark JH, Fan J, Luo G, Zhang S (2019) Characterization and utilization of aqueous products from hydrothermal conversion of biomass for bio-oil and hydro-char production: A review. Green Chem 21:1553–1572

Yadav P, Reddy SN, Nanda S (2019) Cultivation and conversion of algae for wastewater treatment and biofuel production. In: Nanda S, Sarangi PK, Vo DVN (eds) Fuel processing and energy utilization. CRC Press, Florida, USA, pp 159–175

Bioconversion of Agro-industrial Residues to Second-Generation Bioethanol

2

Thiago Machado Pasin, Paula Zaghetto de Almeida, Ana Sílvia de Almeida Scarcella, Juliana da Conceição Infante, and Maria de Lourdes de Teixeira de Moraes Polizeli

Abstract

Bioenergy is the term used for energy produced from lignocellulosic biomass. Most of the residues produced from agricultural and industrial activities present high levels of lignocellulose. They are mostly formed by rigid structures mainly containing hemicellulose and cellulose intermixed by lignin. These macromolecules are linked by covalent and hydrogen bonds, thus forming a complex architecture, which give a great resistance to their hydrolysis. This hampers the subsequent production of fermentable sugars and their fermentation to produce second-generation bioethanol. The technologies to obtain second-generation bioethanol, independent of the plant source, involve the hydrolysis of polysaccharides from the biomass in order to generate sugars that can be fermented by yeasts. This chapter addresses the importance of biomass for the production of green fuels. In this chapter, the potential of different lignocellulosic biomasses, especially the agricultural crop residues, is described. The composition of the main molecules forming the cell wall of different plants is provided. The enzymes that are involved in the deconstruction of plant cell walls as well as the release of fermentable sugars are discussed. The pretreatment and fermentation of biomass for the second-generation ethanol production by yeasts are described. Some

T. M. Pasin · P. Z. de Almeida · A. S. de Almeida Scarcella · J. da Conceição Infante · M. L. T. M. Polizeli (✉)
Departamento de Bioquímica e Imunologia, Faculdade de Medicina de Ribeirão Preto, Universidade de São Paulo, Ribeirão Preto, São Paulo, Brazil
e-mail: polizeli@ffclrp.usp.br

S. Nanda et al. (eds.), *Biorefinery of Alternative Resources: Targeting Green Fuels and Platform Chemicals*, https://doi.org/10.1007/978-981-15-1804-1_2

challenges concerning the technologies are considered, but, on the other hand, some alternatives are also pointed out.

Keywords
Agro-industrial residues · Second-generation bioethanol · Biorefinery · Green fuels · Enzymes · Saccharification

2.1 Introduction

Biomass refers to any matter of plant origin, which can be processed to provide more elaborate bioenergetic and chemical forms suitable for end use. Around the world, more than 146 billion tons of residues are annually available, but only a portion (up to 100 million) of it is being used for energy and biofuel production (Ayres 2014). The rest of these residues is usually burnt or left behind, which may lead to the worsening of the greenhouse effect. A few of the most biomass-producing countries in the world are Brazil, China, Indonesia, Russia, and the USA (Table 2.1).

Food, fiber, or wood production generates biomass, which could be useful for energy production using mature technologies. Currently, the final biomass energy produced is 50 EJ totaling 14% of the final energy used in the world, but it has a potential to be increased up to 150 EJ by 2035. A total supply of 38–45% is estimated to be originated from waste and residues from agriculture and the remaining supply would be shared by the energy crops, forestry, and residual products (IRENA 2014). According to the World Bioenergy Association (WBA 2014), the estimated potential of residues from agriculture in order to produce energy varies between 17 and 128 EJ. There is a high potential of using residues from agriculture, especially in Asia and the Americas because of the different crops highly produced in these regions. There is an enormous potential for exploiting the use of biomass in order to produce larger quantities of products with high value.

Table 2.1 Comparative production of crops in different countries

Crop	Country	Production (million tons/year)	References
Barley	Russia	20.4	Worldatlas & Graphic Maps (2017)
Coffee	Brazil	2.8	FAS (2018a)
Cotton	China	5.9	FAS (2018b)
Maize	USA	370.5	FAO (2017)
Oil palm	Indonesia	40.5	Iskandar et al. (2018)
Potato	China	99.0	FAO (2019)
Rice	China	210.1	Muthayya et al. (2014)
Soybean	USA	119.5	FAO (2017)
Sugarcane	Brazil	758.5	FAO (2017)
Wheat	China	134.4	Worldatlas & Graphic Maps (2019)

Due to the various possible sources of lignocellulosic biomass, this chapter is focused on describing the structure and characteristics of its major components, such as cellulose, hemicellulose, pectin, and lignin present in the primary and secondary cell walls. The resins, fatty acids, phenols, tannins, nitrogenous compounds, and mineral salts, e.g., calcium, potassium, and magnesium on a smaller scale, can also be found, which depends on the plant species (Neureiter et al. 2002). Subsequently, the chapter presents a description of different biomass sources that can be used for the production of bioenergy, followed by the major enzymatic systems that can degrade plant cell walls and their mechanism of action on biomass, especially in sugarcane, for the production of second-generation bioethanol.

2.2 Composition of Lignocellulosic Biomass

Cellulose (23–50 wt% of the dry matter of lignocellulosic biomass) is a linear homopolymer containing up to 15,000 units of β-D-glucose bound by β-1,4-glycosidic linkages (Nanda et al. 2015). It has reducing and nonreducing ends, and it is extremely resistant to degradation (Michelin et al. 2013). The strength of cellulose is due to many hydroxyl groups on the glucose structure that contribute to the formation of massive intramolecular bounds (linkages among glucose units of the same molecule) and intermolecular hydrogen bounds (among glucose units of adjacent molecules), which are responsible for stiffness. The intramolecular bonds are responsible for the formation of fibrils, which are highly ordered structures. According to the degree of organization of the bonds between the cellulose chains, the structure can be crystalline (highly ordered) or amorphous (less ordered). The amorphous regions can absorb water more easily and are more susceptible to enzymatic action.

Hemicellulose is a heteropolysaccharide (15–45 wt% of dry lignocellulosic material) with branched chains of monosaccharides, mainly including aldopentoses (xylose and arabinose) and aldohexoses (glucose, mannose, and galactose). This macromolecule also contains deoxyhexoses and acids, such as β-D-galacturonic acid, D-4-*O*-methylglucuronic acid, and β-D-glucuronic acid (Polizeli et al. 2005). The variety of bonds and branching, as well as the presence of different monomeric units, contributes to the complexity of the hemicellulosic structure and its different conformations. Unlike cellulose, hemicellulose has low molecular mass (100–200 glycosidic units) and does not contain crystalline regions, which makes it easier to hydrolyze in non-drastic conditions (Polizeli et al. 2005).

There are different types of hemicellulose, such as arabinoxylan, acetylglucuronoxylan, arabinan, arabinogalactan, xylan, galactomannan, xyloglucan, galactoglucomannan, and glucomannan. Hemicellulose is classified according to its sugar chain composition. Thus, the term hemicellulose does not denote a defined chemical compound, rather, a set of polymeric components present in fibrous plants where each component has different properties. The major hemicellulose constituent is the xylan, which is a xylose β-1,4-polymer with several branched residues (Polizeli et al. 2005). The fermentation of pentoses is not yet as developed as the processes involving glucose. Mannan is the second most abundant component of the hemicellulosic fraction being widely found in woods (gymnosperms), tubers, seeds, and grains in different compositions structure and complexity. The main chain of mannan consists of mannose residues bound by β-1,4-linkages or a combination of mannose and glucose residues associated by a same type linkage. In addition, the major chain of mannan may have side chains attached to α-1,6-galactose residues (De Marco et al. 2015).

Pectins are heterogeneous polysaccharides, which form the middle lamella being the largest component inside it. These polysaccharides are adhesive materials in the extracellular portion found in the higher plant cells located in the primary walls. The structure consists of axial bonds with acid units of α-1,4-D-galacturonic and contains molecules of galactose, arabinose, xylose, and L-rhamnose as the side chains (Polizeli et al. 2013).

The biochemical structure of lignin (10–30 wt% of dry lignocellulosic material) cannot be related to sugar molecules, and it is not suitable for bioconversion or fermentative routes. However, this fraction plays a key role in the success of hydrolysis technology, since it blocks the access to cellulose. The structure of lignin has a three-dimensional polymeric structure formed by *p*-propylphenol units with methoxy substituent on the aromatic ring joined by ether bonds that crosslink each other. This macromolecule is originated through sinapyl, coumaric, and coniferyl alcohols.

2.3 Commonly Available Agricultural Crop Residues in the World

2.3.1 Barley and Coffee Residues

Barley (*Hordeum vulgare*) is one of the main cereal sources produced in the world for humans and livestock consumption. Every year, an area of around 50 million hectares is harvested, being the fourth major cereal produced in the world. The most abundant staple crops cultivated in the world are wheat (~200 million hectares), rice (~170 million hectares), and maize (~145 million hectares). Russia, Australia, Ukraine, and Canada are the major barley-producing countries (WAP 2019). Barley and beer residues are similar to other lignocellulosic materials being composed of cellulose, hemicellulose, and lignin. It is reported that the percentage of cellulose and hemicellulose in such materials is high (Table 2.2), which can be hydrolyzed and fermented aiming the production of biofuels and biochemicals.

The coffee tree is a plant of permanent culture that belongs to the Rubiaceae family and to the genus *Coffea* such as *Coffea arabica* and *Coffea canephora*. Around 8.9 million tons of coffee are produced over an area of approximately 10.6 million hectares (FAO 2018). The major coffee producers in the world are Brazil, Vietnam, and Colombia (Al-Abdulkader et al. 2018). One ton of bark is generated per each ton of beans of coffee that are processed (Saenger et al. 2001). The coffee pulp is the main residue originated in the wet processing of the mature coffee,

Table 2.2 Cellulose, hemicellulose, and lignin composition of some of the most widely found lignocellulosic feedstocks in the world

Residues	Cellulose (wt%)	Hemicellulose (wt%)	Lignin (wt%)	References
Barley bagasse	23.0	32.7	24.4	Tamanini and Hauly (2004)
Coffee husks	22.7	14.7	12.4	Souza et al. (2001)
Cotton straw	40.7	10.5	15.4	Silanikove et al. (1988)
Corn cob	31.7	34.7	20.3	Tamanini and Hauly (2004)
Oil palm fiber	41.9	29.5	30.0	Sudiyani et al. (2013)
Potato peels	55.2	11.7	14.2	Lenihan et al. (2010)
Rice straw	38.0	27.0	8.0	Santos et al. (2012)
Soybean fibrous	16.0	23.0	28.0	Heck et al. (2002)
Sugarcane bagasse	43.0	26.0	22.0	Rocha et al. (2011)
Wheat straw	35.0	24.0	25.0	Zhu et al. (2005)

representing approximately 40 wt% of the dry weight of the coffee bean, which is rich in cellulose and hemicellulose (Table 2.2), proteins, minerals, and appreciable amounts of tannins, caffeine, and potassium. Currently all these raw materials could be used in developed processes to generate value-added and bulk chemicals products such as active secondary metabolites, unicellular proteins, enzymes, and organic acids (Pandey et al. 2000). Another alternative is the use of these wastes for bioenergy production.

2.3.2 Sugarcane, Maize, Rice, and Wheat Residues

Sugarcane, maize, rice, and wheat residues generate most of the lignocellulosic residues in the world (Saini et al. 2015). One of the most famous plant being used nowadays is the sugarcane (*Saccharum* ssp.). It is a C4 crop that has been used for a long time in the sugar production, but currently bioethanol and electricity have become the most famous value-added energy products generated through bioconversion and burning, respectively (Cotrim et al. 2018). The total harvested area for sugarcane in the world is estimated to be around 26.7 million hectares and the major producers are Brazil, India, and China. After the processing of the sugarcane, large quantities of bagasse are generated what presents good amounts of cellulose and hemicellulose. These factors make it an interesting, renewable, and cheap resource to produce biochemicals and biofuels (Bhatia and Paliwal 2011).

There are several varieties of commercial sugarcane in the world. In Brazil, the SP80-3280 variety is one of the most studied. The composition of the cell wall of SP80-3280 variety sugarcane is reported having fundamental importance for understanding the enzymes necessary for its saccharification (De Souza et al. 2013). In recent years, the search for sugarcane with high levels of sugars for first-generation bioethanol and the accumulation of biomass for second-generation bioethanol production have increased. This sugarcane is known as *energy cane*, which presents the particularity of having a higher fiber content, being distributed in the fractions of lignin, cellulose, and hemicellulose. The *energy cane* comes from genetic improvements of the sugarcane (*Saccharum spontaneum*) to increase the energy availability and adaptation to different cultivation environments (Tew and Cobill 2008).

Maize (*Zea mays*) is a well-known cereal being considered the main input in animal protein production, in human nutrition, and also in the biofuel production. The total area harvested each year in the world is estimated to be around 145 million hectares and the major maize producers in the world are the USA, China, and Brazil (FAO 2018). From Table 2.2 it can be inferred that the maize residues have small amounts of lignin, which makes it interesting for the use in the bioenergy and in the production of chemicals.

Rice consists of two main species such as *Oryza glaberrima* and *Oryza sativa*. It is a plant of the grass family and is among the products intended for human consumption, being second in importance, only behind wheat. In some parts of the world, especially in Asia, rice is staple food crop (CONAB 2018). It is harvested from over 163 million hectares in more than 100 countries (Laborte et al. 2017). The major rice producers are China, India, and Indonesia (FAO 2018). During the rice cultivation, huge amounts of rice straw are generated along with leaf blades, stems, leaf sheaths, and husks as the leftovers. These residues are one of the largest lignocellulosic wastes in the world. Among the most rice straw producers is Asia which alone generates 667.6 million tons of rice straw that consists of 91.3% of the global rice production (Saini et al. 2015). This residue is also an important source with similar contents of hemicellulose, lignin, and cellulose when compared with other organic residues, which could be better used for the production of biofuels and biochemicals.

Wheat is considered a staple food and cereal in all over the world (Shewry 2009; Balkovič et al. 2014). The common wheat, *Triticum aestivum*, is the most widely grown in the world, over an arable land about 220.4 million hectares (FAO 2018). During the wheat harvesting process, straw is generated at a rate of 2.5–7.5 tons/hectare annually under rigorous farming conditions. Cellulose, hemicelluloses, and lignin make up the most part of the agricultural residues in a similar rate. However, wheat straw presents pectin and proteins, whereas rice straw has silica within its tissues (Sarkar et al. 2012).

2.3.3 Cotton

Cotton is a staple fiber found across the world. It belongs to the genus *Gossypium* within the Malvaceae family. Cellulose is the main component of the cotton bolls. It is estimated that 25 million tons of cotton are annually produced in the world over an area that estimates about 2.5% of the total arable land at a global scale. The cotton cultivation area in the world is accounted around 33.6 million hectares per year and the major cotton producers are China, India, and the USA (Daisy et al.

2018). After harvesting the cotton, huge amounts of residues are left in the field, such as side branches, stalks, bolls, seeds with adhering cotton lint, leaves, etc. (Huang et al. 2012). The total amount of residues (e.g., cotton straw, cotton sticks, cotton wood) generated from the cotton harvesting can range from 5 to 7 tons/hectare. The quality of these materials that are generated from the cotton harvesting presents a high variability as commonly observed in crop wastes. However, these residues are also important sources of lignin, cellulose, and hemicellulose as shown in Table 2.2.

2.3.4 Oil Palm Residues

Oil palm is a vegetable oil extracted from the mesocarp of *Elaeis guineensis*. Palm oil is a common cooking ingredient in Brazil, Africa, and Asia. Oil palm trees present a highly efficient land use with a large yield when compared with other crops used for oil extraction. It is estimated that a total harvested area of about 17 million hectares of mature oil palm plantation produces a total of 62.6 million tons of oil palm per year. The major oil palm producers are Indonesia, Malaysia, and Nigeria (EPOA 2018). Europe is the major importer of oil palm for using it as a precursor for biodiesel production (Soeriaatmadja and Leong 2018). In addition to its massive potentials for biodiesel production, all the oil palm residues generated after the oil extraction could also be used to produce the bioethanol since it is composed of cellulose, hemicellulose, and lignin (Table 2.2).

2.3.5 Potato Residues

Potato (*Solanum tuberosum*) is a staple food crop cultivated in most countries across the world. Potatoes are considered the fifth largest crop for human consumption after wheat, rice, maize, and barley. In 2016, the total world potato production and harvested area were estimated at 376 million tons and 19 million hectares, respectively (FAO 2018). The three largest potato producers in the world are China, India, and Russia. Potato processing industry generates massive amounts of organic wastes. It is estimated that one quarter of the total weight of potato is obtained as wastes from the food processing plants. These wastes can be used, for example, as a carbon source for yeast during alcohol fermentation to produce bioethanol. The potato peel waste contains adequate amount of starch, hemicellulose, cellulose, lignin (Table 2.2), and residual carbohydrates, which make potato peel a suitable feedstock for ethanol production (Ojewumi et al. 2018).

2.3.6 Soybean Residues

The soybean (*Glycine max*) is known to be widely cultivated because of its comestible bean that presents several different uses. The worldwide soybean production was estimated around 337 million tons in 2017–2018 (WASDE 2018). The major producers of soybean in the world are the USA, Brazil, and Argentina. Based on the global production, it is possible to affirm that these three countries are responsible by approximately 85% of the total soybean production (FAO 2015). During the recovery of the protein from the soybean seeds, large amounts of residues are generated. The residues generated are composed of 16 wt% cellulose and 23 wt% hemicellulose (Table 2.2). Previous studies have estimated that approximately 10,000 tons per year of these by-products are generated and only a small portion is used for animal feeding, while the rest is mainly disposed in the environment (Heck et al. 2002).

2.4 Enzymatic Saccharification of Lignocellulosic Residues

2.4.1 Cellulases

Cellulases are glycosyl hydrolases that act in the cleavage of cellulose and can be classified into endoglucanases (EG), exoglucanases (CBH), and β-glucosidases (BG) (Payne et al. 2015; Wei and Mcdonald 2016). Some cellulases are organized in complexes called cellulosomes. These structures are produced by anaerobic bacteria (Artzi et al. 2017) and fungi (Haitjema et al. 2017) as a strategy to improve the hydrolysis efficiency. They are composed of a complex scaffolding as structural subunit and various enzymatic subunits. Many cellulases and hemicellulases are attached to structures called carbohydrate binding module (CBM). These structures have the property to attach the enzyme to the substrate. CBMs are not necessary to the hydrolytic function and there is no evidence that they improve the action of cellulases (Várnai et al. 2014).

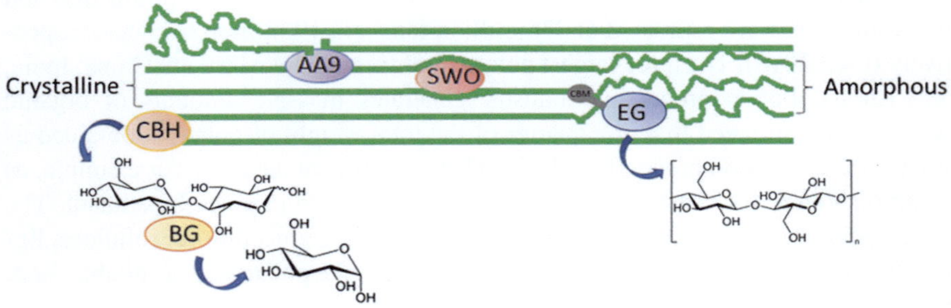

Fig. 2.1 Cellulase action on cellulose crystalline and crystal structure. Endoglucanase (EG) in blue at the amorphous region with a CBM attached to the crystalline region releasing smaller cellulose chains. Cellobiohydrolase (CBH) in orange is attached to the crystalline terminal releasing cellobiose. β-Glucosidase (BG) in yellow cleaves the cellobiose to glucose. Auxiliary activity 9 enzyme (AA9) in purple acts on the crystalline region leaving a rupture in the cellulose chain. Swollenin (SWO) in pink acts between the cellulose chains resulting in a less crystalline structure

Endo-β-1,4-glucanases (EC 3.2.1.4) hydrolyze the glycosidic bonds mainly at the amorphous regions of cellulose (Fig. 2.1). They are distributed along 16 out of 134 glycoside hydrolase (GH) families and are characterized by shorter loops. Thus, this feature enables the active site fissures to reach accessible sites of cellulose chain. The endoglucanases act in the interior of cellulose, resulting in long-chain oligomers. When compared with other cellulases, they have a faster dissociation rate and show the best cellulose liquefaction results, thus decreasing the chain length and consequently the viscosity (Badieyan et al. 2012; Boyce and Walsh 2015; Juturu and Wu 2014).

Exoglucanases or cellobiohydrolases act in the end of crystalline cellulose chains to release cellobiose (Fig. 2.1). They act efficiently in cellulose hydrolysis and are divided into EC 3.2.1.91 (acting in the nonreducing ends) and EC 32.1.176 (acting in the reducing ends). These enzymes have long loops usually forming tunnels around the catalytic residues; thus the substrate is directed along these tunnels to encounter the active site (Obeng et al. 2017; Wilson and Kostylev 2012).

β-Glucosidases (BGs, EC 3.2.1.21) break down cellobiose to glucose (Fig. 2.1). Several β-glucosidases are also able to catalyze the reverse reaction by transglycosylation in which glucose molecules are transferred to another glucose molecule or cellobiose to yield different oligosaccharides. β-Glucosidases are susceptible to product inhibition, especially at high-biomass conditions. Thus, it is one of the biggest bottlenecks for the total cellulose conversion to glucose. The addition of large amounts of β-glucosidases in enzymatic cocktails has the extra goal to reduce the product inhibition exhibited by cellobiose on cellobiohydrolases and endoglucanases (Andríc et al. 2010; Singhania et al. 2013).

The most reported cause to product inhibition on fungal β-glucosidases is via competitive inhibition, but it is also described as one of the noncompetitive and mixed mechanisms. The majority of β-glucosidases belong to the 1 and 3 GH families, but they are also found in families 5, 9, and 30. The enzymes belonging to family GH1 are 10 to 1000-fold more tolerant to glucose when compared to GH3 β-glucosidases. The structural analysis shows that glucose tolerance is strongly correlated with active site accessibility (Andríc et al. 2010; De Giuseppe et al. 2014; Singhania et al. 2013).

2.4.2 Auxiliary Activity 9 Enzyme and Swollenin Protein

Auxiliary activity 9 (AA9) enzyme is a fungal lytic polysaccharide monooxygenase (LPMO), a cooper-dependent enzyme that is added to the most commercial cocktails. AA9 boost the action of cellulases in lignocellulosic material hydrolysis. It promotes the oxidative cleavage of glyosidic bounds of C1 or C4, thus disturbing cellulose crystallinity and giving access to canonical cellulases, consequently improving the overall hydrolysis yield (Fig. 2.1). The oxidation promoted by LPMO is dependent on an electron donor, e.g., ascorbic acid, cellobiose dehydrogenase, and lignin or photocatalytic systems. The oxidation occurs only in the presence of

O_2 or H_2O_2 as the co-substrate (Bertini et al. 2018; Möllers et al. 2017; Müller et al. 2018; Walton and Davies 2016).

Swollenins (SWOs) are nonenzymatic proteins with high homology to plant expansins. This class of proteins has the property to cause deagglomeration of crystalline cellulose. SWOs act in the breakage of hydrogen bonds between the cellulose chains, thus loosening the structure, increasing the accessibility of canonical cellulases and enzymatic hydrolysis (Fig. 2.1). SWOs can also disrupt hemicellulose, promoting a better solubilization and acting synergistically with hemicellulases (Cosgrove 2000; Gourlay et al. 2013; Kang et al. 2013; Kim et al. 2014; Santos et al. 2017).

2.4.3 Hemicellulases

Hemicellulases are classified as the class of enzymes responsible for depolymerizing the hemicellulose. In order to achieve the complete degradation of hemicelluloses, the action of several enzymes belonging to GH families is important. The most significant enzymes of these complex are discussed as follows (Polizeli et al. 2005; Saha 2003).

Xylan, a xylose β-1,4-polymer, is a major constituent of hemicellulose. For the degradation of xylan, two main enzymes are necessary, namely, endo-β-1,4-xylanase and β-1,4-xylosidase (Fig. 2.2). Endo-β-1,4-xylanases (EC 3.2.18) or endoxylanases catalyze the hydrolysis of xylan to xylooligosaccharides. This enzyme acts in the internal backbone of xylan promoting the

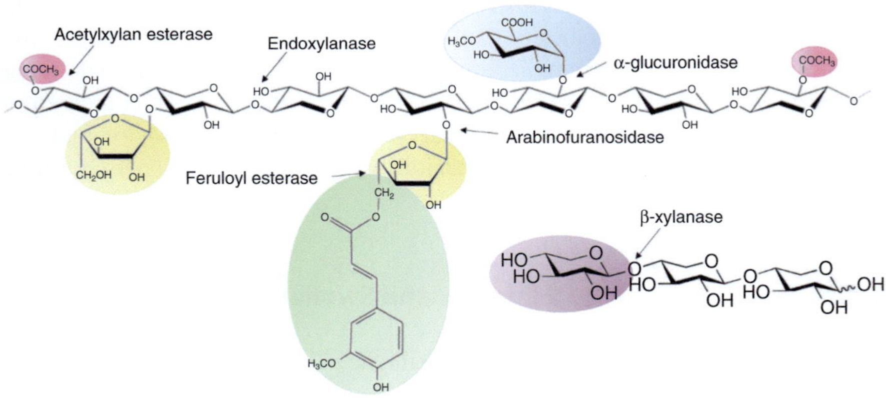

Fig. 2.2 Simplified structures of hemicelluloses with ramifications and the enzymes that act in their hydrolysis. Acetyl substitutions cleaved by acetyl xylan esterase are shown in pink. Arabinan removed from xylan chain by arabinofuranosidase is shown in yellow. Ferulic acid detached to arabinan by feruloyl esterase is shown in green. 4-*O*-Methyl-D-glucuronic acid removed by α-glucuronidase is shown in blue. The nonreductive end of an oligosaccharide cleaved by β-xylanase to obtain xylose monomers is shown in purple. The xylan backbone cleaved in the interior regions by endoxylanase releasing xylooligosaccharides is shown by the non-highlighted region

hydrolysis of β-1,4-glycosidic linkages. Exo-β-1,4-xylosidase (EC 3.2.1.37) or β-1,4-xylosidase acts in the nonreducing ends of xylan structure or xylooligosaccharides including xylobiose, thus releasing D-xylose monomers (Burlacu et al. 2016; Heinen et al. 2018; Polizeli et al. 2005).

The removal of the ramifications attached to xylan chain is promoted by a complex of enzymes. The most important enzymes are α-L-arabinofuranosidase, α-glucuronidase, acetyl xylan esterase, and feruloyl esterase (Fig. 2.2). α-L-Arabinofuranosidase (EC 3.2.1.55) or arabinofuranosidase cleaves the nonreductive end of the arabinan in the positions *O*-2 or *O*-3 releasing oligomers (Juturu and Wu 2013). α-Glucuronidase (EC 3.2.1.139) hydrolyses the 1-2-glycosidic bound between xylan and the side chain of 4-*O*-methyl-D-glucuronic acid, thus releasing glucuronic acid (Shallom and Shoham 2003). Acetyl xylan esterase (EC3.1.1.72) removes the acetyl substitutions attached on xylose releasing acetic acid (Tenkanen et al. 1992). Feruloyl esterase (EC3.1.1.73) or feruloyl xylan esterase hydrolyses the ester bond between ferulic acid and arabinose (Dilokpimol et al. 2016).

The mannan-degrading enzymes are β-mannosidase (EC 3.2.1.25), β-mannanase (EC 3.2.1.78), and β-glucosidase (3.2.1.21). There are also additional enzymes such as α-galactosidase (EC 3.2.1.22) and acetyl mannan esterase (EC 3.1.1.6) required to remove side-chain substituents, thus creating more sites for enzymatic hydrolysis (Fig. 2.3a–f) (Moreira and Filho 2008). According the Carbohydrate-Active Enzymes Database (CAZy), the mannanases are classified into GH families 5, 26, and 113 (Cantarel et al. 2009; Cruz 2013). These GH families have double-displacement mechanism holding anomeric configuration (De Marco et al. 2015).

The complete hydrolysis of the mannan chain demands the synergy of exo-acting and endo-acting hydrolases together with accessory enzymes to reach complete hydrolysis. The heterosynergy is the synergistic action between main-chain and side-chain enzymes (β-mannanase and α-galactosidase) and homosynergy between two main-chain enzymes (β-mannosidase and β-mannanase) or between two side-chain enzymes (acetyl mannan esterase and α-galactosidase) reported in the degradation of mannan (Moreira and Filho 2008).

It is possible to visualize the presence and the size of the β-mannanase following the molecular markers using the technique of sodium dodecyl sulfate-polyacrylamide gel electrophoresis (SDS-PAGE) and zymography. Multiple bands stained for β-mannanase activity (Fig. 2.3g, h) can be detected using these techniques, which migrate at molecular masses ranging from 20 to 50 kDa. The variety of forms is usually described for hemicellulases from bacteria and fungi as the result of posttranslational modifications and differential messenger RNA (mRNA) processing. These mannanases can be isozymes, allozymes, and products of different alleles in the same gene or same molecules with different posttranscriptional modifications (De Marco et al. 2015).

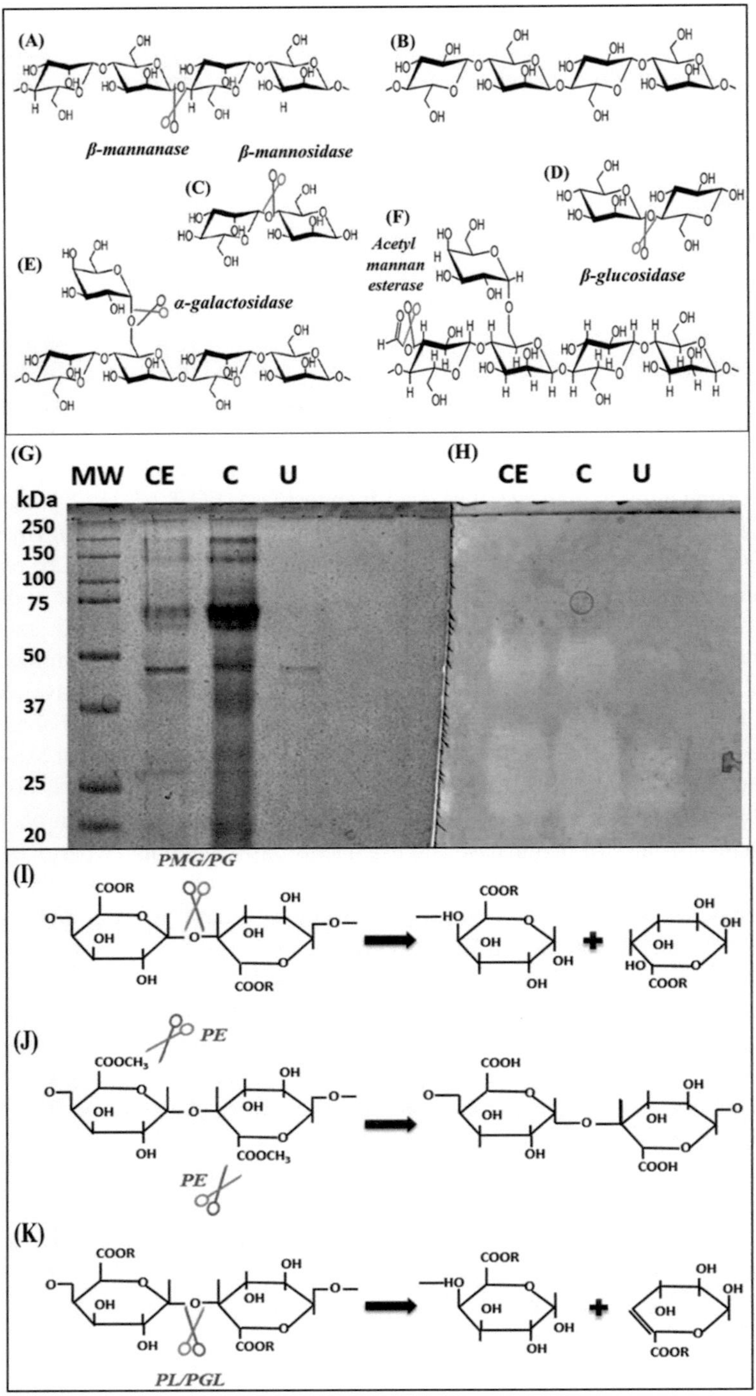

Fig. 2.3 Structures of mannans and pectins. Zymography of mannanase activity. (**a**) Mannan, (**b**) glucomannan, (**e**) galactomannan, and (**f**) galactoglucomannan. The mannan is hydrolyzed by

2.4.4 Pectin-Degrading Enzyme System

Pectinases (EC 3.2.1.15) are a set of complex enzymes that degrade pectin in plant cell wall (Fig. 2.3i–k). Pectin-degrading enzymes are capable of degrading pectic substances per de-esterification (esterases) or depolymerization (lyases and hydrolases) reactions (Tariq and Latif 2012) and include polygalacturonases (PGs), pectinesterases (PEs), pectate lyases (PGLs), and pectin lyases (PLs) (Ahlawat et al. 2009; Polizeli et al. 2013). Polygalacturonases (PGs) are hydrolases, which cleave α-1,4-glycosidic linkage in homopolygalacturon backbone. They are classified into GH family 28 according to the Carbohydrate-Active Enzymes Database (CAZy) classification.

Endopolygalacturonases (EC 3.2.1.15) randomly attack the α-1,4-glycosidic bonds of the polysaccharide chain, thus releasing galacturonic acid oligomers. On the other hand, exopolygalacturonase type I (EC 3.2.1.67) acts in the nonreducing end and the D-galacturonic acid is hydrolyzed. Di-galacturonate is released as a result of the action of exopolygalacturonase type II (EC 3.2.1.82) in the nonreducing end of polygalacturonic acid (Khan et al. 2013). Pectinesterases (EC 3.1.1.11) are esterases that catalyze the methoxyl group of pectin forming pectic acid (Amin et al. 2017). Endopectin lyases (EC 4.2.2.10) accomplish trans-elimination reaction of pectin. Endopectate lyases (EC 4.2.2.2) and exopectate lyases (EC 4.2.2.9) cleave internal and nonreducing end of α-1,4-polygalacturonic acid, respectively, via β-elimination reaction (Yang et al. 2018).

2.4.5 Delignifying Enzymes

Delignifying enzymes are used for biomass pretreatment to remove lignin prior to bioconversion of biomass. As the lignin is removed or ruptured, the hemicellulose and cellulose fibrils become exposed. Thus, the glycoside hydrolases can have a better access to the substrate. In addition, the delignifying enzymes can remove some inhibitors (mainly phenolics), which interfere in the fermentation process (Kudanga and Roes-Hill 2014).

The most important delignifying enzymes are the laccases (EC1.10.3.2). They are multi-copper enzymes crucial to lignin degradation and the removal of phenolics. They can be divided into different enzyme families according to Sirim et al.

Fig. 2.3 (continued) β-mannanase, while α-galactosidase releases galactose and acetyl mannan esterase releases acetyl groups. The mannobiose and glucomannobiose are hydrolyzed by (**c**) β-mannosidase and (**d**) β-glucosidase, to generate the monosaccharides mannose and glucose. SDS-PAGE (12%) and zymography of mannanase activity in crude extract, stained with Coomassie Brilliant Blue G-250 (**g**) and 0.1% Congo red (**h**). Molecular weight (MW), crude extract (CE) of soybean husk, crude extract concentrated (C) tenfold 30.000 NMWC in the concentration system QuixStand Benchtop (GE Healthcare), crude extract ultrafiltered (U) obtained by the same system as above. (**i**) R = H (PG) and CH3 (PMG); (**j**) PE and (**k**) R = H (PGL) and CH3 (PL). The scissors indicate the pectinases acting on the pectic substrates. Polygalacturonase (PG), polymethylgalacturonase (PMG), pectinesterase (PE), pectate lyase (PGL), and pectin lyase (PL)

Table 2.3 Classification of laccase families, sequences, and structures

Superfamily	Homologous families	Proteins	Structures
A (Basidiomycete)	4	201	13
B (Ascomycete)	6	421	6
C (Insect)	8	168	0
G (Plant)	5	333	0

Reference: Sirim et al. (2011)

(2011) as shown in Table 2.3. Laccases act by promoting the oxidation of one electron in a wide variety of substrates, notably the lignin phenylpropanoids (Mate and Alcalde 2017). Other important enzymes are the heme-peroxidases. They catalyze hydrogen peroxide-dependent oxidative degradation of lignin of phenolic (as manganese peroxidase-EC1.11.1.13) and in the non-phenolics (as lignin peroxidase-EC1.11.1.14) substrates (Plácido and Capareda 2015).

2.5 Bioconversion of Biomass to Second-Generation Bioethanol

Among several alternative biomasses described in this chapter, the production of bioethanol from sugarcane (*Saccharum* spp.) is very advantageous. Sugarcane is rich in sucrose and allows the fermentation of this sugar by microorganisms resulting in first-generation (1G) bioethanol. On the other hand, the resulting lignocellulosic biomass (bagasse) can be used to produce second-generation (2G) bioethanol. To improve each step of the 2G ethanol production, it is necessary to know the composition of lignocellulosic residues.

Although the main targets for 2G ethanol production are sugarcane in Brazil (Polizeli et al. 2011, 2016, 2017) and corn in the USA (Li et al. 2018), there are studies reporting other biomasses, which were previously described in this chapter such as barley (Yang et al. 2015) and coffee (Nguyen et al. 2019). The key elements in the production of 2G bioethanol from sugarcane bagasse are shown in Fig. 2.4 and discussed below.

2.6 Pretreatment of Lignocellulosic Biomass

Pretreatment of biomass may contribute to enhance the cell wall hydrolysis. It alters the chemical composition and physical structure of the substrates, since it aims to separate the carbohydrates from the lignin matrix. This process of separation benefits the enzymatic access to the cellulose and hemicellulose. The presence of lignin allows the unproductive binding and inactivation of the cellulases, thus making the high 2G ethanol yield practically impossible (Srinorakutara et al. 2013).

Some effects of the pretreatments are (a) breakage of the lignin structure and its connection with cellulose and hemicellulose, (b) improvement in the access of cellulases to

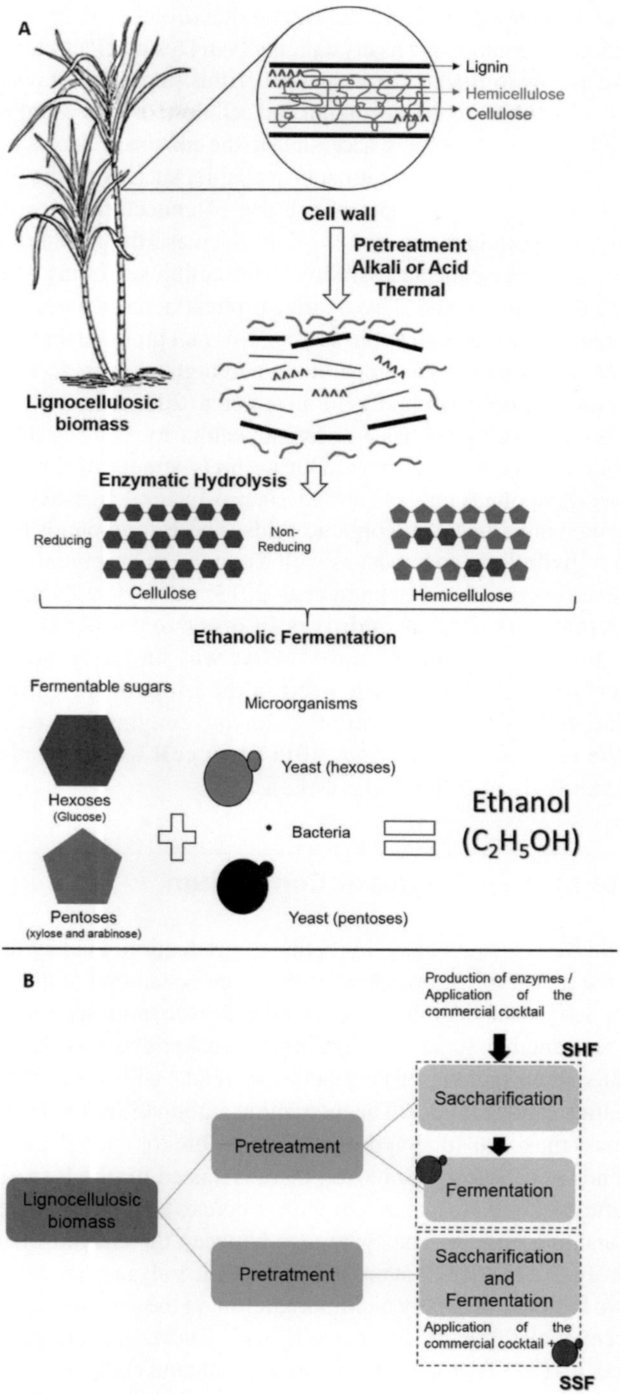

Fig. 2.4 Scheme of the 2G bioethanol production process including pretreatment and bioconversion of lignocellulosic biomass

cellulose due to the removal of hemicellulose and lignin, and (c) decrease in the degree of polymerization of cellulose and its crystallinity (van Dyk and Pletschke 2012; Nanda et al. 2014). According to Buckeridge et al. (2019), this step makes it possible to break down several covalent bonds between lignin and cellulose or hemicellulose, thus making the lignocellulosic biomass more accessible to the enzymatic attack.

There are broad ranges of pretreatments available, such as acidic, basic, or thermal. The acid pretreatment consists of the use of concentrated or diluted acids, associated with temperatures above 120 °C to decrease the rigidity of lignocellulosic biomass and increase the accessibility to the cellulose. In this type of pretreatment, hemicellulose is removed. The alkaline pretreatment removes lignin from the biomass and causes less damage to the sugars. Thermal pretreatment, also known as autohydrolysis, uses water, high temperature, and high pressure to break the lignocellulosic matrix (Michelin et al. 2014; Polizeli et al. 2015).

Each of these pretreatments has a different peculiarity, causing different chemical modifications as mentioned above. During pretreatment of the cellulosic residues, inhibitory compounds may be formed, such as furfural, hydroxymethylfurfural (HMF), phenolic compounds, and organic acids. These compounds have an impact on the action of hydrolytic enzymes, as well as on the microorganisms involved in the fermentative process (Srinorakutara et al. 2013; van Dyk and Pletschke 2012).

The surfactants are used as additives in order to avoid the adsorption of enzymes to their substrate in an unproductive way and to enhance the hydrolysis. The surface active agents are used in the bioconversion in the following ways: (a) during pretreatment, (b) during enzymatic saccharification, and (c) for the recycling of enzymes after plant cell wall degradation (Giese et al. 2012; van Dyk and Pletschke 2012).

2.7 Obtaining an Enzymatic Consortium

In order to achieve a sustainable 2G bioethanol production, the key to increase the efficiency of the saccharification is the thorough understanding of the plant cell wall structure since enzymatic degradation of lignocellulosic biomass is required to yield most of the fermentable sugars. According to Buckeridge and de Souza (2014), plant cell walls show regions of polymers that relate with other polymer chains, such as xyloglucan and cellulose. The interaction among the macromolecules of cell walls of biomass makes its total hydrolysis impossible for endoglucanases to attack glycosidic bonds of cellulose. Moreover, there is a need for the formulation of efficient enzymatic cocktails with enzymes with endo/exo characters as well as branching and debranching feature. The synergism between the enzymes responsible for cell wall hydrolysis considers that the structure of the polysaccharides present in the lignocellulosic biomass may play an important role in the kinetic hydrolysis properties (Mohanram et al. 2013; De Souza et al. 2013; Buckeridge and de Souza 2014).

De Souza et al. (2013) proposed a theoretical scheme composed of the enzymes necessary for the saccharification of the cell wall of sugarcane, which included hemicellulases, cellulases, β-galactosidases, lichenases, and other accessory enzymes.

Commercial enzymatic cocktails available for the depolymerization of lignocellulosic materials are loosely defined as a complex mixture containing about 80–200 proteins, including hemicellulases, cellulases, and accessory enzymes (Mohanram et al. 2013). To improve the performance of enzyme cocktails, it took years of bioprospecting studies, optimization of strains through genetic engineering for the production of enzymes, and development of pretreatment strategies since these are the major technical challenges in an efficient 2G ethanol production process.

The enzymes act in synergy for substrate degradation for ethanol production. van Dyk and Pletschke (2012) as well as Li et al. (2014) have demonstrated that the understanding of synergism is of fundamental importance to clarify the mechanism of the action of the enzymes alone and the interaction between them. The enzymatic saccharification is beneficial when compared to other methods because it is a more specific and cleaner process besides the reduction of the formation of inhibitory compounds. During the enzymatic hydrolysis of lignocellulosic biomass, oligosaccharides, disaccharides, and monomers are formed. They cause the inhibition of β-glucosidase and β-xylosidase associated with the saccharification as previously mentioned in this chapter. High concentrations of these sugars may have an impact on the hydrolytic efficiency of the lignocellulosic residue (van Dyk and Pletschke 2012).

For the bioconversion of lignocellulosic residue to fermentable sugars, it is necessary to decrease the cost of enzymes, in addition to studies aiming to enhance the efficiency of the enzymes used in the cell wall hydrolysis. In this way, an effective tool to reduce the cost of the various hydrolytic enzymes would be their immobilization. This process leads to an economy in the process of producing biofuels besides allowing the recovery and reuse of the enzymes (Borges et al. 2014; Maitan-Alfenas et al. 2015).

2.8 Fermentation for the Production of Second-Generation Bioethanol

Alcoholic fermentation consists of the transformation of the sugars from the hydrolysate into ethanol, CO_2, and energy due to the catalytic action of the yeast or bacteria. There are two modes of operations for bioethanol production of lignocellulosic biomass, namely, separate hydrolysis and fermentation (SHF) and simultaneous saccharification and fermentation (SSF) (Fig. 2.4b). Efforts at process integration are still constrained by obstacles. There is a need of industrial microorganisms capable to assimilate pentoses and hexoses as well as the control in the formation and elimination of inhibitory compounds resulting from the whole process. Another factor that must be studied is the standardization of the ideal conditions for the integration of two or more processes to be simultaneously performed in a single bioreactor (Ramos et al. 2015).

The pentoses are fermented by yeasts of the genus *Candida*, *Komagataella*, and *Meyerozyma*, but studies with these yeasts are still limited. The hexoses are assimilated by yeasts *Saccharomyces cerevisiae*, which have high fermentative efficiency.

This efficiency is due to the rapid growth of yeasts as well as the ability to metabolize sugars. In addition, they have the ability to produce and consume ethanol and are tolerant to high concentrations of this compound and resistant to low levels of oxygen. Another bacterial producer of bioethanol, *Zymomonas mobilis*, has shown promising results for the large-scale production of bioethanol.

2.9 Conclusions

The dependence on nonrenewable and highly polluting energy sources has led to seek alternative fuel sources. Clean energy has the advantage to lower CO_2 emissions, providing energy capable of reducing the production of greenhouse gases, maintaining environmental sustainability, mitigating climate change, and stimulating the global economy. Thus, the growing demand for biofuels has motivated the use of biological materials, specifically photosynthetic organisms, as the raw materials.

It is known that billions of tons of plant-based residues are obtained across the world. However, only a small fraction of this organic waste is used for bioenergy and bioproducts. Lignocellulosic biomass is mainly constituted by a polysaccharide-lignin complex with varied composition depending on the plant variety. The deconstruction of this intricate complex is one of the main bottlenecks in the conversion of biomass to biofuels. The development of novel technologies is necessary to make the hydrolysis and saccharification economically viable. Genetic modifications of plants resulting in a vegetal with increased content of biodegradable and fermentable saccharides are some of the perspectives. Another focus in plants is to increase the fiber content, as the *energy cane*, considering 2G ethanol production from sugarcane.

Another problem confronted with 2G ethanol production is the biochemical structure of the lignin because it blocks the access to cellulose. Various pretreatments of biomass alter the chemical/physical structure of the plant cell walls, thus separating the polysaccharides from lignin and allowing better enzymatic access to cellulose, hemicellulose, and pectin. One of the problems of such pretreatments is the formation of inhibitory compounds as furfural, hydroxymethylfurfural, phenolic compounds, and organic acids, which inhibit enzymatic action and microbial activity during bioconversion. Hydrothermal pretreatment and steam explosion are alternative strategies to acid, alkaline, and other pretreatments. Delignifying enzymes may contribute as a biomass pretreatment for bioethanol production since the removal of lignin exposes hemicellulose and cellulose fibrils, and consequently, glycoside hydrolase can have better access to the substrates.

In all cases, it is important to have an in-depth and detailed knowledge of the structure of the biomass cell constituents and how they interact aiming at the production of fermentable sugars. Immunofluorescence techniques using antibodies against the macromolecules of the biomass cell wall have given excellent know-how, and a glycolic code of sugarcane walls has also been reported in the literature.

Enzymatic saccharification of raw or pretreated biomass for the liberation of fermentable sugars by yeast is advantageous when compared to the chemical methods because of being a selective, cleaner, and less energy-intensive process, which produces less amounts of inhibitory biological compounds. For this purpose, commercial enzyme cocktails obtained from recombinant microorganisms, such as bacteria and filamentous fungi, are used. Enzymatic hydrolysis process is largely successful in many cases, but problems, such as the viability of the strain, mycotoxins, contamination, genetic instability, reproducibility, inhibition of the enzyme by end products, and other molecular complications, have been reported. Some proposed alternatives include bioprospecting studies of mesophilic and thermophilic strains and optimization of strains through genetic engineering for the production and thermostabilization of enzymes.

Efficient enzymatic cocktails include the cellulases, hemicellulases, pectinases, ligninases, and, accessory enzymes and proteins, such as auxiliary activity 9 and swollenin. Synergism is expected between the hydrolitic enzymes, auxiliary activity enzymes, and swollenins because they can act together favoring the oxidative cleavage of glycosidic bonds and deagglomeration of crystalline cellulose, thus giving access to cellulases and facilitating enzymatic hydrolysis. Besides, the debranching of hemicellulose is favored by the use of enzymatic cocktail. The addition of ligninolytic enzymes in the cocktail helps in removing some phenolic compounds that may interfere in the final fermentation process. Usually, the combinatory action of enzymes reverts to higher levels of fermentable sugars. Thus, the understanding of the synergic mechanisms between the hydrolytic enzymes is fundamental to improve the saccharification process.

A perturbing factor for large-scale bioconversion studies, which still needs attention, is the high cost of commercial hydrolytic enzymes. An alternative could be the immobilization or co-immobilization of the enzymes. This process sometimes results in enzymatic hyperactivation and allows various reuses of the biocatalysts (enzyme and support). Furthermore, as may be concluded, various studies must be developed aiming to achieve the total success of the lignocellulosic biomass conversion in fermentable sugars. Besides, it is a clean process with many advantages that need recognition.

Acknowledgments The Laboratory of Microbiology and Cell Biology of Faculdade de Filosofia, Ciências e Letras de Ribeirão Preto, Universidade de São Paulo, Brazil, has been supported by Fundação de Amparo à Pesquisa do Estado de São Paulo (FAPESP 2014/50884-5 and 2018/07522-6). Maria de Lourdes Teixeira de Moraes Polizeli is a research fellow of Conselho de Desenvolvimento Científico e Tecnológico (CNPq, 301963/2017-7). Thiago Machado Pasin, Paula Zaghetto de Almeida, Ana Sílvia de Almeida Scarcella, and Juliana da Conceição Infante are recipients of the Coordenação de Aperfeiçoamento de Pessoal de Nível Superior (CAPES, Finance Code 001) scholarship. We thank Mariana Cereia for the language review of the chapter.

References

Ahlawat S, Dhiman SS, Battan B, Mandhan RP, Sharma J (2009) Pectinase production by *Bacillus subtilis* and its potential application in biopreparation of cotton and micropoly fabric. Process Biochem 44:521–526

Al-Abdulkader AM, Al-Namazi AA, AlTurki TA, Al-Khuraish MM, Al-Dakhil AI (2018) Optimizing coffee cultivation and its impact on economic growth and export earnings of the producing countries: The case of Saudi Arabia. Saudi J Biol Sci 25:776–782

Amin F, Bhatti HN, Bilal M, Asgher M (2017) Improvement of activity, thermo-stability and fruit juice clarification characteristics of fungal exo-polygalacturonase. Int J Biol Macromol 95:974–984

Andríc P, Meyer AS, Jensen PA, Dam-Johansen K (2010) Reactor design for minimizing product inhibition during enzymatic lignocellulose hydrolysis: I. Significance and mechanism of cellobiose and glucose inhibition on cellulolytic enzymes. Biotechnol Adv 28:308–324

Artzi L, Bayer EA, Moraïs S (2017) Cellulosomes: Bacterial nanomachines for dismantling plant polysaccharides. Nat Rev Microbiol 15:83–95

Ayres RU (2014) Economic growth. In: Bubble Economy: Is Sustainable Growth Possible? MIT Press, Cambridge, Massachusetts; London, England, pp 277–306

Badieyan S, Bevan DR, Zhang C (2012) Study and design of stability in GH5 cellulases. Biotechnol Bioeng 109:31–44

Balkovič J, Velde van der M, Skalský R, Xiong W, Folberth C, Khabarov N, Smirnov A, Mueller ND, Obersteiner M (2014) Global wheat production potentials and management flexibility under the representative concentration pathways. Glob Planet Chang 122:107–121

Bertini L, Lambrughi M, Fantucci P, De Gioia L, Borsari M, Sola M, Bortolotti CA, Bruschi M (2018) Catalytic mechanism of fungal lytic polysaccharide monooxygenases investigated by first-principles calculations. Inorg Chem 57:86–97

Bhatia L, Paliwal S (2011) Ethanol production potential of *Pachysolen tannophilus* from sugarcane bagasse. Int J Biotechnol Bioeng Res 2:271–276

Borges DG, Baraldo Junior A, Farinas CS, de Lima Camargo Giordano R, Tardioli PW (2014) Enhanced saccharification of sugarcane bagasse using soluble cellulase supplemented with immobilized β-glucosidase. Bioresour Technol 167:206–213

Boyce A, Walsh G (2015) Characterisation of a novel thermostable endoglucanase from *Alicyclobacillus vulcanalis* of potential application in bioethanol production. Appl Microbiol Biotechnol 99:7515–7525

Buckeridge MS, De Souza AP (2014) Breaking the "glycomic code" of cell wall polysaccharides may improve second-generation bioenergy production from biomass. BioEnergy Res 7:1065–1073

Buckeridge MS, Grandis A, Tavares EQP (2019) Disassembling the glycomic code of sugarcane cell walls to improve second-generation bioethanol production. In: Ray R, Ramachandran R (eds) Bioethanol production from food crops. Elsevier, Amsterdam, pp 31–43

Burlacu A, Cornea CP, Israel-roming F (2016) Microbial xylanase: a review. Sci Bull XX:335–342

Cantarel BL, Coutinho PM, Rancurel C, Bernard T, Lombard V, Henrissat B (2009) The carbohydrate-active enzymes database (CAZy): an expert resource for glycogenomics. Nucleic Acids Res 37:233–238

CONAB (2018) Companhia Nacional de Abastecimento—National Supply Company, 2018. http://www.conab.gov.br.

Cosgrove DJ (2000) Loosening of plant cell walls by expansins. Nature 407:321–326

Cotrim CA, Soares JSM, Kobe B, Menossi M (2018) Crystal structure and insights into the oligomeric state of UDP-glucose pyrophosphorylase from sugarcane. PLoS One 13:1–13

Cruz AF (2013) Mannan-degrading enzyme system. In: Fungal enzymes. CRC Press, Boca Raton, pp 233–257

Daisy M, Rajendran K, Amanullah MM (2018) Effect on microbial population, quality parameters and green fodder yield of leguminous crops under Bt cotton intercropping system. Int J Curr Microbiol App Sci 7:332–337

De Giuseppe PO, Souza TDACB, Souza FHM, Zanphorlin LM, Machado CB, Ward RJ, Jorge JA, Furriel RDPM, Murakami MT (2014) Structural basis for glucose tolerance in GH1 β-glucosidases. Acta Crystallogr Sect D Biol Crystallogr 70:1631–1639

De Marco JCI, De Souza Neto GP, Castro CFS, Michelin M, Polizeli MLTM, Ferreira Filho EX (2015) Partial purification and characterization of a thermostable β-mannanase from *Aspergillus foetidus*. Appl Sci 5:881–893

De Souza AP, Leite DCC, Pattathil S, Hahn MG, Buckeridge MS (2013) Composition and structure of sugarcane cell wall polysaccharides: implications for second-generation bioethanol production. Bioenergy Res 6:564–579

Dilokpimol A, Mäkelä MR, Aguilar-Pontes MV, Benoit-Gelber I, Hildén KS, De Vries RP (2016) Diversity of fungal feruloyl esterases: Updated phylogenetic classification, properties, and industrial applications. Biotechnol Biofuel 9:231

EPOA (2018) European Palm Oil Alliance. https://www.palmoilandfood.eu/en/palm-oil-production

FAO (2015) Soybeans, Production/Crops/World. Food and Agriculture Organization of the United Nations. http://www.fao.org/faostat/en/#data/QC

FAO (2017) Food and Agriculture Organization of the United Nations. http://www.fao.org/faostat/en/#rankings/countries_by_commodity

FAO (2018) Food and Agriculture Organization of the United Nations. http://www.fao.org/faostat/en/#data/QC

FAO (2019) Food and Agriculture Organization of the United Nations. http://www.fao.org/faostat/en/#search/potato

FAS (2018a) Foreign Agricultural Service—Department of Agriculture of the United States. https://apps.fas.usda.gov/psdonline/circulars/coffee.pdf.

FAS (2018b) Foreign Agricultural Service—Department of Agriculture of the United States. https://www.fas.usda.gov/data/china-cotton-and-products-annual-2

Giese EC, Pierozzi M, Dussán KJ, Chandel AK, Da Silva SS (2012) Enzymatic saccharification of acid-alkali pretreated sugarcane bagasse using commercial enzyme preparations. J Chem Technol Biotechnol 88:1266–1272

Gourlay K, Hu J, Arantes V, Andberg M, Saloheimo M, Penttilä M, Saddler J (2013) Swollenin aids in the amorphogenesis step during the enzymatic hydrolysis of pretreated biomass. Bioresour Technol 142:498–503

Haitjema CH, Gilmore SP, Henske JK, Solomon KV, De Groot R, Kuo A, Mondo SJ, Salamov AA, Labutti K, Zhao Z, Chiniquy J, Barry K, Brewer HM, Purvine SO, Wright AT, Hainaut M, Boxma B, van Alen T, Hackstein JHP, Henrissat B, Baker SE, Grigoriev IV, O'Malley MA (2017) A parts list for fungal cellulosomes revealed by comparative genomics. Nat Microbiol 2:1–8

Heck JX, Hertz PF, Ayub MAZ (2002) Cellulase and xylanase production by isolated amazon Bacillus strains using soybean industrial residue based solid-state cultivation. Braz J Microbiol 33:213–218

Heinen PR, Bauermeister A, Ribeiro LF, Messias JM, Almeida PZ, Moraes LAB, Vargas-rechia CG, De Oliveira AHC, Ward RJ, Filho EXF, Kadowaki MK, Jorge JA, Polizeli MLTM (2018) GH11 xylanase from *Aspergillus tamarii Kita*: Purification by one-step chromatography and xylooligosaccharides hydrolysis monitored in real-time by mass spectrometry. Int J Biol Macromol 108:291–299

Huang W, Bai Z, Hoefel D, Hu Q, Lv X, Zhuang G, Xu S, Qi H, Zhang H (2012) Effects of cotton straw amendment on soil fertility and microbial communities. Front Environ Sci Eng 6:336–349

IRENA (2014) Remap 2030 Global Bioenergy Supply and Demand Projections

Iskandar MJ, Baharum A, Anuar FH, Othaman R (2018) Palm oil industry in South East Asia and the effluent treatment technology—a review. Environ Technol Innov 9:169–185

Juturu V, Wu JC (2013) Insight into microbial hemicellulases other than xylanases: a review. J Chem Technol Biotechnol 88:353–363
Juturu V, Wu JC (2014) Microbial cellulases: engineering, production and applications. Renew Sustain Energy Rev 33:188–203
Kang K, Wang S, Lai G, Liu G, Xing M (2013) Characterization of a novel swollenin from *Penicillium oxalicum* in facilitating enzymatic saccharification of cellulose. BMC Biotechnol 13:42
Khan M, Nakkeeran E, Umesh-Kumar S (2013) Potential application of pectinase in developing functional foods. Annu Rev Food Sci Technol 4:21–34
Kim IJ, Lee HJ, Choi I-G, Kim KH (2014) Synergistic proteins for the enhanced enzymatic hydrolysis of cellulose by cellulase. Appl Microbiol Biotechnol 98:8469–8480
Kudanga T, Roes-Hill M (2014) Laccase applications in biofuels production: current status and future prospects. Appl Microbiol Biotechnol 98:6525–6542
Laborte AG, Gutierrez MA, Balanza JG, Saito K, Zwart SJ, Boschetti M, Murty MVR, Villano L, Aunario JK, Reinke R, Koo J, Hijmans RJ, Nelson A (2017) RiceAtlas, a spatial database of global rice calendars and production. Nat Sci Data 4:170074
Lenihan P, Orozco A, O'Neill E, Ahmad MNM, Rooney DW, Walker GM (2010) Dilute acid hydrolysis of lignocellulosic biomass. Chem Eng J 156:395–403
Li J, Zhou P, Liu H, Xiong C, Lin J, Xiao W, Gong Y, Liu Z (2014) Synergism of cellulase, xylanase, and pectinase on hydrolyzing sugarcane bagasse resulting from different pretreatment technologies. Bioresour Technol 155:258–265
Li YH, Zhang XY, Zhang F, Peng LC, Zhang DB, Kondo A, Bai FW, Zhao XQ (2018) Optimization of cellulolytic enzyme components through engineering *Trichoderma reesei* and on-site fermentation using the soluble inducer for cellulosic ethanol production from corn stover. Biotechnol Biofuels 11:49
Maitan-Alfenas GP, Visser EM, Guimarães VM (2015) Enzymatic hydrolysis of lignocellulosic biomass: converting food waste in valuable products. Curr Opin Food Sci 1:44–49
Mate DM, Alcalde M (2017) Laccase: a multi-purpose biocatalyst at the forefront of biotechnology. Microb Biotechnol 10:1457–1467
Michelin M, Polizeli MLTM, Ruzene DS, Silva DP, Teixeira JA (2013) Application of lignocellulosic residues in the production of cellulase and hemicellulases from fungi. In: Polizeli MLTM, Rai M (eds) Fungal enzymes. CRC Press, Boca Raton, pp 31–64
Michelin M, Ruiz HA, Silva DP, Ruzene DS, Teixeira JA, Polizeli MLTM (2014) Cellulose from lignocellulosic waste. In: Gopal K, Mérillon RJ-M (eds) Polysaccharides. Springer International Publishing, Basel, Switzerland, pp 1–33
Mohanram S, Amat D, Choudhary J, Arora A, Nain L (2013) Novel perspectives for evolving enzyme cocktails for lignocellulose hydrolysis in biorefineries. Sust Chem Process 1:15
Möllers KB, Mikkelsen H, Simonsen TI, Cannella D, Johansen KS, Bjerrum MJ, Felby C (2017) On the formation and role of reactive oxygen species in light-driven LPMO oxidation of phosphoric acid swollen cellulose. Carbohydr Res 448:182–186
Moreira LRS, Filho EXF (2008) An overview of mannan structure and mannan-degrading enzyme systems. Appl Microbiol Biotechnol 79:165–178
Müller G, Chylenski P, Bissaro B, Eijsink VGH, Horn SJ (2018) Biotechnology for biofuels the impact of hydrogen peroxide supply on LPMO activity and overall saccharification efficiency of a commercial cellulase cocktail. Biotechnol Biofuels 11:209
Muthayya S, Sugimoto JD, Montgomery S, Maberly GF (2014) An overview of global rice production, supply, trade, and consumption. Ann N Y Acad Sci 1324:7–14
Nanda S, Mohammad J, Reddy SN, Kozinski JA, Dalai AK (2014) Pathways of lignocellulosic biomass conversion to renewable fuels. Biomass Convers Bioref 4:157–191
Nanda S, Maley J, Kozinski JA, Dalai AK (2015) Physico-chemical evolution in lignocellulosic feedstocks during hydrothermal pretreatment and delignification. J Biobased Mater Bioenerg 9:295–308
Neureiter M, Danner H, Thomasser C, Saidi B, Braun R (2002) Dilute-acid hydrolysis of sugarcane bagasse at varying conditions. Appl Biochem Biotechnol 98:49–58

Nguyen QA, Cho EJ, Lee D-S, Bae H-J (2019) Development of an advanced integrative process to create valuable biosugars including manno-oligosaccharides and mannose from spent coffee grounds. Bioresour Technol 272:209–216

Obeng EM, Adam SNN, Budiman C, Ongkudon CM, Maas R, Jose J (2017) Lignocellulases: a review of emerging and developing enzymes, systems, and practices. Bioresour Bioprocess 4:16

Ojewumi ME, Job AI, Taiwo OS, Obanla OM, Ayoola AA, Ojewumi EO, Oyeniyi EA (2018) Bio-conversion of sweet potato peel waste to bio-ethanol using *Saccharomyces cerevisiae*. Int J Pharm Phytopharm Res 8:46–54

Pandey A, Soccol CR, Nigam P, Soccol VT (2000) Biotechnological potential of agro-industrial residues. I: Sugarcane bagasse. Bioresour Technol 74:69–80

Payne CM, Knott BC, Mayes HB, Hansson H, Himmel ME, Sandgren M, Stahlberg J, Beckham GT (2015) Fungal Cellulases. Chem Rev 115:1308–1448

Plácido J, Capareda S (2015) Ligninolytic enzymes: a biotechnological alternative for bioethanol production. Bioresour Bioprocess 2:23

Polizeli MLTM, Rizzatti ACS, Monti R, Terenzi HF, Jorge JA, Amorim DS (2005) Xylanases from fungi: properties and industrial applications. Appl Microbiol Biotechnol 67:577–591

Polizeli MLTM, Corrêa ECP, Polizeli AM, Jorge JA (2011) Hydrolases from microorganisms used for degradation of plant cell wall and bioenergy. In: Buckeridge MS, Goldman GH (eds) Routes to Cellulosic Ethanol. Springer, New York, NY, pp 115–134

Polizeli MLTM, Damásio ARL, Maller A, Cabral H, Polizeli AM, Rai M (2013) Pectinases produced by microorganisms: properties and applications. In: Polizeli MLTM, Rai M (eds) Fungal enzymes. CRC, Boca Raton, pp 327–351

Polizeli MLTM, Peralta RM, Bracht A, Michelin M, Somera AF (2015) Enzymes prospection from fungi and biomass pretreatment for biorefinery application. In: Silva RN (ed) Mycology: current and future developments. Bentham Science Publishers, Sharjah, pp 57–81

Polizeli MLTM, Vici AC, Scarcella ASA, Cereia M, Pereira MG (2016) Enzyme system from *Aspergillus* in current industrial uses and future applications in the production of second-generation ethanol. In: Gupta VK (ed) New and future developments in microbial biotechnology and bioengineering: *Aspergillus* system properties and applications. Elsevier, Amsterdam, pp 127–140

Polizeli MLTM, Somera AF, de Lucas RC, Nozawa MSF, Michelin M (2017) Enzymes involved in the biodegradation of sugarcane biomass: challenges and perspectives. In: Buckeridge MS, De Souza AP (eds) Advances of basic science for second generation bioethanol from sugarcane. Springer International Publishing, New York, NY, pp 55–79

Ramos LP, da Silva L, Ballem AC, Pitarelo AP, Chiarello LM, Silveira MHL (2015) Enzymatic hydrolysis of steam-exploded sugarcane bagasse using high total solids and low enzyme loadings. Bioresour Technol 175:195–202

Rocha GJM, Gonçalves AR, Oliveira BR, Olivares EG, Rossell CEV (2011) Steam explosion pretreatment reproduction and alkaline delignification reactions performed on a pilot scale with sugarcane bagasse for bioethanol production. Ind Crop Prod 35:274–279

Saenger M, Hartge EU, Werther J, Ogada T, Siagi Z (2001) Combustion of coffee husks. Renew Energy 23:103–121

Saha BC (2003) Hemicellulose bioconversion. J Ind Microbiol Biotechnol 30:279–291

Saini JK, Saini R, Tewari L (2015) Lignocellulosic agriculture wastes as biomass feedstocks for second-generation bioethanol production: concepts and recent developments. Biotech 5:337–353

Santos FA, De Queiróz JH, Colodette JL, Fernandes SA, Guimarães VM, Rezende ST (2012) Potencial da palha de cana-de-aucar para produção de etanol. Quim Nova 35:1004–1010

Santos CA, Filho JAF, Donovan AO, Gupta VK, Tuohy MG, Souza AP (2017) Production of a recombinant swollenin from *Trichoderma harzianum* in *Escherichia coli* and its potential synergistic role in biomass degradation. Microb Cell Factories 16:83–93

Sarkar N, Ghosh SK, Bannerjee S, Aikat K (2012) Bioethanol production from agricultural wastes: an overview. Renew Energy 37:19–27

Shallom D, Shoham Y (2003) Microbial hemicellulases. Curr Opin Microbiol 6:219–228
Shewry PR (2009) Wheat. J Exp Bot 60:1537–1553
Silanikove N, Danai O, Levanon D (1988) Composted cotton straw silage as a substrate for *Pleurotus* sp. cultivation. Biol Wastes 25:219–226
Singhania RR, Patel AK, Sukumaran RK, Larroche C, Pandey A (2013) Role and significance of beta-glucosidases in the hydrolysis of cellulose for bioethanol production. Bioresour Technol 127:500–507
Sirim D, Wagner F, Wang L, Schmid RD, Pleiss J (2011) The laccase engineering database: a classification and analysis system for laccases and related multicopper oxidases. Database 2011:1–7
Soeriaatmadja W, Leong T (2018) European ban on palm oil in biofuels upsets Jakarta, KL. Straitstimes. 2018 Available at https://www.straitstimes.com/asia/se-asia/european-ban-on-palm-oil-in-biofuels-upsets-jakarta-kl. Accessed 20 Aug 2010
Souza AL, Garcia R, Pereira OG, Cecon PR, de C Valadares-Filho S, Paulino MF (2001) Composição químico - bromatológica da casca de café tratada com amônia anidra e sulfeto de sódio. Rev Bras Zootec 30:983–991
Srinorakutara T, Suttikul S, Boonvitthya N (2013) Effect of different pretreatment methods on enzymatic saccharification and ethanol production from sugarcane shoots and leaves. J Food Sci Eng 3:309–316
Sudiyani Y, Styarini D, Triwahyuni E, Sudiyarmanto, Sembiring KC, Aristiawan Y, Abimanyu H, Han MH (2013) Utilization of biomass waste empty fruit bunch fiber of palm oil for bioethanol production using pilot—Scale unit. Energy Procedia 32:31–38
Tamanini C, Hauly MC (2004) Resíduos agroindustriais para produção biotecnológica de xilitol Agro-industrial residues in biotechnological production of xylitol. Semin Ciências Agrárias 25:315–330
Tariq A, Latif Z (2012) Isolation and biochemical characterization of bacterial isolates producing different levels of polygalacturonases from various sources. Afr J Microbiol Res 6:7259–7264
Tenkanen M, Puls J, Poutanen K (1992) Two major xylanases of *Trichoderma reesei*. Enzym Microb Technol 14:566–574
Tew T, Cobill RM (2008) Genetic improvement of sugarcane (*Saccharum* spp.) as an energy crop. In: Vermerris W (ed) Genetic improvement of bioenergy crops. Springer, Gainesville, FL, pp 249–272
van Dyk JS, Pletschke BI (2012) A review of lignocellulose bioconversion using enzymatic hydrolysis and synergistic cooperation between enzymes-Factors affecting enzymes, conversion and synergy. Biotechnol Adv 30:1458–1480
Várnai A, Mäkelä MR, Djajadi DT, Rahikainen J, Hatakka A, Viikari L (2014) Carbohydrate-binding modules of fungal cellulases: occurrence in nature, function, and relevance in industrial biomass conversion. Adv Appl Microbiol 88:103–165
Walton PH, Davies GJ (2016) On the catalytic mechanisms of lytic polysaccharide monooxygenases. Curr Opin Chem Biol 31:195–207
WAP, World Agricultural Production (2019). http://apps.fas.usda.gov/psdonline/circulars/production.pdf
WASDE, World Agricultural Supply and Demand Estimates (2018). https://www.usda.gov/oce/commodity/wasde/Secretary_Briefing.pdf
WBA, World Bioenergy Association (2014) Global Bioenergy Statistics. http://toe.worldbioenergy.org/content/wba-gbs
Wei L, Mcdonald AG (2016) A review on grafting of biofibers for biocomposites. Materials (Basel) 9:303–325
Wilson DB, Kostylev M (2012) Cellulase processivity. Methods Mol Biol 908:93–99
Worldatlas & Graphic Maps (2017) The leading barley producing countries in the World. https://www.worldatlas.com/articles/the-leading-barley-producing-countries-in-the-world.html
Worldatlas & Graphic Maps (2019) Top wheat producing countries. https://www.worldatlas.com/articles/top-wheat-producing-countries.html

Yang M, Kuittinen S, Zhang J, Vepsäläinen J, Keinänen M, Pappinen A (2015) Co-fermentation of hemicellulose and starch from barley straw and grain for efficient pentoses utilization in acetone–butanol–ethanol production. Bioresour Technol 179:128–135

Yang Y, Zhang Y, Li B, Yang X, Dong Y, Qiu D (2018) A *Verticillium dahliae* pectate lyase induces plant immune responses and contributes to virulence. Front Plant Sci 9:1–15

Zhu SD, Wu Y, Yu Z, Liao J, Zhang Y (2005) Pretreatment by microwave/alkali of rice straw and its enzymatic hydrolysis. Process Biochem 40:3082–3086

Catalytic Transformation of Ethanol to Industrially Relevant Fine Chemicals

3

Paresh H. Rana and Parimal A. Parikh

Abstract

The industrial developments in the nineteenth century was due to coal energy, while oil contributed majorly in the twentieth century. However, a question about future energy source remains unanswered. With increased population, rapid industrialization, and rising concerns for environment preventative measures, total dependence on fossil fuel resources is not viable. Hence, immediate attention is required to develop green and sustainable resource, thus stimulating research into biomass as an alternative to conventional sources. After coal, oil, and gas, biomass is world's fourth largest energy source and is estimated to be equivalent about 14% of global primary energy. There is a need to recognize a suitable species derived from biomass, which can produce valuable chemicals in order to mitigate dependency on fossil resources. Bioethanol is one of such biomass-derived platform solvents with a potential to be used as a fuel and a sustainable feedstock of a variety of chemicals. This chapter deals with the conversion of ethanol to many value-added chemicals such as acetaldehyde, acetic acid, acetone, ethylene, butanol, 1,3-butadiene, and ethyl acetate through catalytic routes. It also discusses the role of supports, metal–support interactions, oxygen storage capacity, acidity and basicity of catalyst in the product distribution, and catalytic activity. The challenges and prospects of ethanol conversion to different chemicals are also described.

Keywords

Biomass · Ethanol · Catalysts · Commodity chemicals

P. H. Rana (✉)
Department of Chemical Engineering, Government Engineering College, Bhuj, Gujarat, India
e-mail: ranaph78@gmail.com

P. A. Parikh
Department of Chemical Engineering, Sardar Vallabhbhai National Institute of Technology, Surat, Gujarat, India

S. Nanda et al. (eds.), *Biorefinery of Alternative Resources: Targeting Green Fuels and Platform Chemicals*, https://doi.org/10.1007/978-981-15-1804-1_3

3.1 Introduction

Fossil fuels are used to meet the energy requirements and produce different chemicals that make our lives comfortable. However, rampant utilization of fossil feedstocks led to their scarcity and global warming. To conquer these drawbacks with fossil raw materials, in past decades constant efforts have been made to investigate the sustainable route for synthesis of variety of chemicals starting with renewable resources. In this regard, biomass can be considered a reliable alternative source to replace fossil-based feedstocks for producing fuels and chemicals (Nanda et al. 2015). It is abundant, renewable, and readily available and causing less emission of CO_2 upon utilization. Thus, it has lower impact on the environment and synthesizes chemicals via environmentally benign pathways (Huber et al. 2006).

Several potential platform chemicals such as levulinic acid, ethanol, sorbitol, furfural, hydroxymethylfurfural, furans, and γ-valerolactone are derived from biomass through various processes. Among them, ethanol is considered as valuable molecules due to its numerous usages. It is well-known that bioethanol is obtained from different edible biomass feedstocks via fermentation process. However, rising concern towards the future food availability and ecological system leads to produce ethanol from biomass-derived nonfood feedstocks. Presently, bioethanol contributes almost 90% of biofuel production and its annual production reached over 100 billion liters (Hu et al. 2018). Ethanol finds its major application in blending with gasoline owing to high research octane number, low toxicity to environment, and comparable energy content. Research shows that 5–27% petroleum can be saved via blending of ethanol (Hu et al. 2018; Mika et al. 2018).

Apart from blending with gasoline, ethanol is used as a solvent for substances including flavoring and coloring agents, perfumes, and medicines. Apart from these, it can also be used in dyes and intermediates, as additives in paint industry, in agricultural chemicals, and as odor agents as shown in Fig. 3.1 (Mika et al. 2018). The blending of ethanol with gasoline is limited to 5–10%. Therefore, with higher

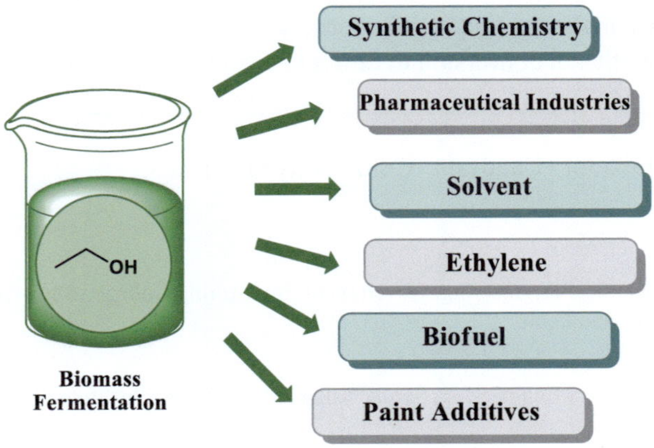

Fig. 3.1 Many industrial utilities of ethanol

throughput and reduced price, it is expected that ethanol will be used for the synthesis of industrially important chemicals such as acetaldehyde, acetic acid, ethylene, 1,3-butadiene, acetone, 1-butanol, diethyl ether, and propylene, which are currently derived from nonrenewable resources (Angelici et al. 2013; Sun and Wang 2014). For instance, acetaldehyde is commercially produced via catalytic oxidation of ethylene in energy-intensive Wacker process, generating chlorinated wastes (Čičmanec et al. 2018). Ethylene, an important compound for petrochemical industries, can be obtained via dehydration of ethanol.

Catalytic transformation of ethanol to different value-added chemicals is discussed in this chapter. Different ethanol conversion technologies specifically to acetic acid, acetaldehyde, acetone, 1,3-butadiene, and 1-butanol are discussed in this chapter. The conversion of ethanol to hydrocarbons emphasizes on the advancement of catalyst and conceptions about reaction mechanism were discussed. The chief aim of this chapter is to provide basic idea about ethanol conversion and identification of different reactions aiming future research.

3.2 Conversion of Ethanol to Small Oxygenates

Chemical building blocks can be produced using chemical or fermentation process. Acetic acid is the only product which is commercially produced from fermentative route and other chemicals produced via one- or two-step fermentation process (Weusthuis et al. 2011). However, our focus is on conversion of ethanol to value-added chemicals through chemical process. Rass-Hansen et al. (2007) have used ethanol as a feedstock for intermediate chemicals such as ethylene, acetic acid, and acetaldehyde. Furthermore, they argued that value-added chemicals from ethanol would be more economical as compared to its use as a fuel additive.

This section discusses the direct conversion of ethanol to various intermediates or by-products (Fig. 3.2) such as acetaldehyde (production projected to 1.4 million tons by 2022) (Garbarino et al. 2019), acetic acid (12×10^6 ton per year) (Mostrou et al. 2018), 1,3-butadiene (global demand of 10.5 million tons in 2010 and predicated 3% increase in rate per year) (Baylon et al. 2016), ethyl acetate (global production 2.6 million tons in 2010) (Men'shchikov et al. 2014), butanol (produced at 1.2 billion gallons per year) (Bharathiraja et al. 2017), and ethylene (annual production of 140 million tons) (Hulea 2018).

3.2.1 Acetaldehyde

Acetaldehyde is a versatile chemical that can be used in the production of industrial chemicals, such as acetic acid, ethyl acetate, acetic anhydride, n-butanol, crotonaldehyde, ethyl acetate, and many others (Ob-eye et al. 2019). Until the 1960s, acetaldehyde was produced by hydration of acetylene using sulfuric acid and mercuric sulfate as catalyst. However, mercuric sulfate used in this process is toxic. Furthermore, ethylene was used for the synthesis of acetaldehyde. In this process,

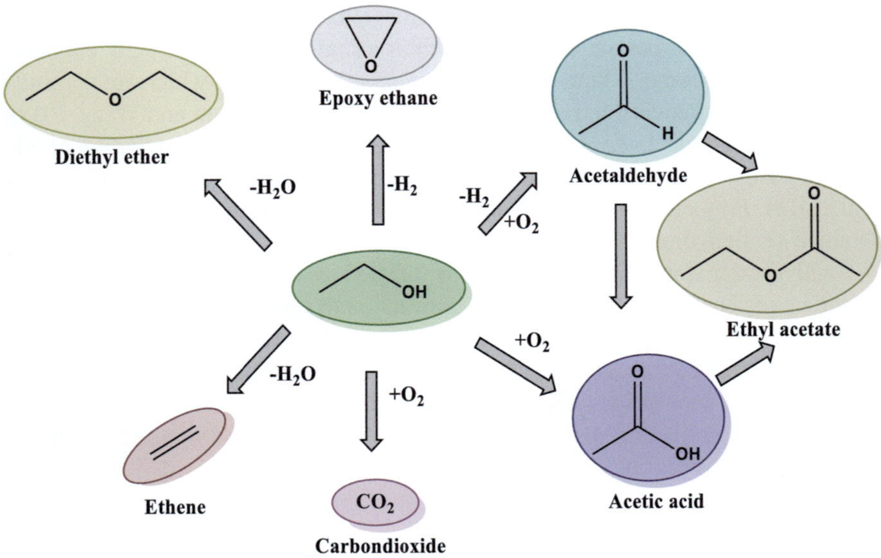

Fig. 3.2 Reaction network involved in ethanol transformations

ethylene is first hydrated to ethanol, which subsequently undergoes dehydrogenation or partial oxidation to produce acetaldehyde in the presence of air using silver gauze catalyst at 450 °C under 0.3 MPa pressure. Presently, acetaldehyde is commercially produced from catalytic oxidation of ethylene using $PdCl_2/CuCl_2$ in water in the presence of oxidants (Angelici et al. 2013; Ob-eye et al. 2019). This process generates chlorinated wastes and consumes more energy to purify air and treat wastewater, and also it requires expensive titanium in the reactor tubing (Čičmanec et al. 2018). Dehydrogenation of ethanol with or without air is environmentally and economically favorable for acetaldehyde production. In the past few years, intensive studies have been carried out for this reaction over various types of catalysts.

Ethanol reaction was performed over Au/CeO_2 surface using temperature programmed desorption (TPD), Fourier transform infrared (FTIR) spectroscopy, and in steady-state catalytic conditions and observed that product selectivity changes significantly with temperature. At 300 °C and 500 °C, acetaldehyde and acetone were main product, respectively (Sheng et al. 2004). Takei et al. (2011) have studied Au nanoparticles (Au NPs) supported on 23 metal oxides for gas phase oxidation, which shows product distribution to be depending on nature of supports. Acid or base metal oxides showed more than 95% selectivity towards acetaldehyde when temperature was higher than 200 °C, while for less than 200 °C, complete oxidation occurred to CO_2 and H_2O with p-type semi-conductive metal oxides such as MnO_2, Co_3O_4, and CeO_2. Both acetaldehyde and acetic acid were obtained with n-type semi-conductive metal oxides, e.g., ZnO and V_2O_5. In p-type semi-conductive metal oxide supports, high availability of oxygen species at surface enables complete oxidation of ethanol.

Zheng and Stucky (2006) employed Au/SiO_2 for aerobic oxidation of ethanol and studied effects of particle size on ethanol conversion and product selectivity at 200 °C. They observed that small Au nanoparticles of 3.5 nm did not revealed high ethanol conversion as shown 6.3 nm Au nanoparticles. Lower ethanol conversions of 24% and 22% were observed with 3.5 nm and 8.2 nm Au nanoparticles, respectively. However, they exhibited high selectivity to acetaldehyde (90%) than that of 6.3 nm particles (75%). They also demonstrated that increased loading of Au on SiO_2 led to increased selectivity towards ethyl acetate (90%).

Guan and Hensen (2009) employed Au nanoparticles supported on mesoporous and conventional silica for gas phase oxidation of ethanol. They checked effect of both the size of Au nanoparticles and the structure of support with and without oxygen. Irrespective of Au nanoparticle size, in oxidative conditions dehydrogenation rate of ethanol was substantially higher as compared to non-oxidative conditions. In the absence of oxygen, selectivity to acetaldehyde was near to 100% and above 90% at 350 °C and 400 °C, respectively. Au nanoparticles with 4.9 and 5.8 nm supported on SBA-15 and SBA-16, respectively, exhibited higher ethanol conversion (95%), indicating optimal particle size for reaction. In oxidative conditions, with Au/MCM-41 highest 20% ethanol conversion and 90% selectivity towards acetaldehyde were obtained at 200 °C. However, no conversion was observed at same temperature in non-oxidative conditions. The turnover frequency (TOF) study showed completely different relation between TOF and Au nanoparticle size, where Au nanoparticles with <7 nm size displayed almost similar activity and particle size >10 nm have a higher activity per surface atom in the presence of oxygen. The results of this study clearly demonstrated that catalytic performance relies on the nature of support and particle size of Au.

Mielby et al. (2014) used encapsulated Au nanoparticles in zeolites silicate-1 (Au/Recryst-S1) for catalytic gas phase oxidation of bioethanol (10 wt%) and compared its performance with other 3 Au/silicate-1, Au/meso-silicate-1 and Au/APS-silicate-1 catalysts. They found that Au/Recryst-S1 was found better as compared to other used catalysts and obtained ethanol conversion (50%) with 98% selectivity towards acetaldehyde at 200 °C and has high site time yield of acetaldehyde. This was ascribed to three-dimensional distribution and metal–support interfacial sites, which played a vital role in the catalytic activity.

Zhang et al. (2015) demonstrated solvent-free synthesis of silicate-1 (S1) encapsulating Au–Pd bimetallic nanoparticles (Au–Pd@S1). They found that Au–Pd@S1catalyst achieved higher ethanol conversion (92.8%) and outperformed the reported Au/Recryst-S1 catalyst (Mielby et al. 2014). This can be due to higher activity of bimetallic Au–Pd compared to Au nanoparticles in oxidation reactions. They have also studied influence of water in bioethanol oxidation using Au–Pd@S1 and Au–Pd/S1 (prepared by conventional hydrothermal route) and reported that Au–Pd@S1 catalyst exhibited higher catalytic activity (82% ethanol conversion) as compared to conventionally prepared Au–Pd/S1 catalyst (2% ethanol conversion) in the presence of 90% water. The activity of Au–Pd@S1 was due to the hydrophobicity of the S1 zeolite, which enabled to efficiently separate the ethanol and water, hindering access of water to the Au–Pd particles in the zeolites micropores.

Rana and Parikh (2017a) have used SiO_2, CeO_2, and ZrO_2 as supports for Au and Ag nanoparticles for ethanol (10 wt%) oxidation in the presence of air and evaluated catalytic behavior between 200 °C and 350 °C and 5 atm with the gas hourly space velocity (GHSV) of 18,000 mL/g_{cat}/h. Among the used catalyst, Au supported on ZrO_2 and CeO_2 exhibited higher conversion of ethanol to 38% and 34%, respectively, with 72% and 83% selectivity towards acetaldehyde, respectively, than other catalysts. This can be attributed to strong interaction of Au nanoparticles with supports. Ag/ZrO_2 displayed lower ethanol conversion (27%) as compared to Au/ZrO_2 and Au/CeO_2. This was due to oxidation of Ag^0 under oxidative condition, which may lead to deactivate Ag for reaction. Weaker interaction between Ag and ZrO_2 and non-transfer of electron charge were also responsible. In Au/ZrO_2, the presence of Au leads to reduce ZrO_2 surface (Zr^{4+} to Zr^{3+}) partially, thus creating oxygen deficiencies on surface of ZrO_2. This occurrence resulted into the increased density of anionic vacancies on surface of ZrO_2 and spillover of oxygen to Au from ZrO_2 surface. Higher availability of active oxygen atom resulted into enhanced catalytic activity.

Rana and Parikh (2017b) further investigated ethanol oxidation with Ag/CeO_2 as catalyst describing importance of metal–support interactions. Ag/CeO_2 catalyst exhibited excellent results in terms of acetaldehyde selectivity (>90%) and time on stream (>36 h) in comparison to reported catalytic systems at reaction conditions (200–350 °C, 5 atm with GHSV of 18,000 mL/g_{cat}/h). The presence of Ag reduces CeO_2 surface, thus creating crystal defects and activation of lattice oxygen of CeO_2 support. Here Ag/CeO_2 interface allows transfer of atomic oxygen from surface of ceria to Ag particles. This intense interaction at interface resulted into reduced bond energy of Ce–O in CeO_2 and increased positive charge on Ag.

Rana and Parikh (2018) have extended ethanol oxidation with series of mixed oxides $Ce_xZr_{1-x}O_2$ (x = 0.25, 0.5 and 0.75) as supports for Au nanoparticles and investigated role of oxygen supply/storage capacity for this reaction. Results demonstrated that increased ethanol conversion was attributed to the insertion of Zr to Ce framework, which led to enhance oxygen storage capacity (OSC) and thus increases supply of oxygen to the reaction. The changes in lattice parameter and defects in structure of Ce were observed with increased Zr content in ceria. This changed in lattice parameter also shrinks cell volume; later stress induced by decreased unit cell volume lowers activation energy for oxygen ion diffusion within lattice and favors the reduction on Ce surface. As compared to Au/ZrO_2 (Rana and Parikh 2017a) catalyst which exhibited 72% selectivity towards acetaldehyde, 85–90% acetaldehyde selectivity was achieved with Au/$Ce_xZr_{1-x}O_2$. This can be due to high strength of basic sites of mixed oxide than bare ZrO_2 and was also due to the presence of water which blocked the Lewis acid site and resulting dehydrogenation of ethanol (Rahman et al. 2016; Rana and Parikh 2017b, 2018).

Simakova et al. (2010) examined TiO_2, SiO_2, and Al_2O_3 on Au nanoparticle catalysts for ethanol oxidation in gas phase and reported unusual behavior of Au/TiO_2, which showed double peak in temperature range of 80–280 °C. At 125 °C, active oxygen species was formed at the catalytic sites, which was responsible for catalytic activity at low temperature. In the presence of H_2, active oxygen species was generated under mild reaction conditions (Simakova et al. 2010; Sobolev et al. 2012). Active species recommended at lower temperature were surface O^-, surface O_2–, anions, or peroxides (Sobolev et al. 2012; Sobolev and Koltunov 2015). At higher temperatures, a sharp decrease in the activity was observed. This can be ascribed to decay, desorption, or suppressed concentration of oxygen species (Simakova et al. 2010; Sobolev et al. 2012). Oxygen is promoting dehydrogenation reaction at high temperature for oxidative route. In contrast, Holz et al. (2014) proposed that acetaldehyde formation followed a Mars–van Krevelen mechanism at high temperature using Au/TiO_2. Herein, the presence of highly dispersed Au nanoparticles on TiO_2 led to reduced TiO_2, resulting in more oxygen vacancies, and the created oxygen vacancies over TiO_2 act as active sites for the reaction.

In the ethanol gas phase oxidation, vanadium oxide has been widely studied as major catalyst. Metal oxides such as TiO_2 (Jørgensen et al. 2009; Sobolev et al. 2013; Kaichev et al. 2016); hydrotalcite, Al_2O_3, TiO_2, and SBA-15 (Hidalgo et al. 2016); and SiO_2, Al_2O_3, and ZrO_2 (Beck et al. 2012) have been used as supports for V_2O_5. Among all, TiO_2 was found to be better in demonstrating catalytic performance. This can be attributed to several factors such as:

1. Ethanol oxidation mainly depends on the monolayer of the polymeric form of vanadium and does not depend on the presence of the bulk form of V_2O_5 (Sobolev et al. 2013).
2. Redox mechanism involved ethanol oxidation where the reactant oxidizes by the oxidized surface of catalyst and this surface re-oxidized via oxygen of gas phase.

During the reaction on the V_2O_5/TiO_2, Ti^{4+} state was maintained, whereas a reversible reduction of V^{5+} cations resulted in V^{4+} and V^{3+} under reaction conditions (Beck et al. 2012; Kaichev et al. 2016). Nearly 15 wt% V_2O_5/TiO_2 catalyst exhibited excellent catalytic performance, which produced higher than 90% selectivity towards acetaldehyde and 95% ethanol conversion in oxidation of aqueous ethanol (50 wt%) at 180–185 °C (Jørgensen et al. 2009). Furthermore, in this reaction at 165 °C and with low gas velocity, above 80% selectivity towards acetic acid was obtained.

Ob-eye et al. (2019) investigated a single-step synthesis of acetaldehyde through dehydrogenation of ethanol with Cu, CE, Co, and Ni metal doped on commercially activated carbon. Under specified reaction conditions, Cu/ACC (activated carbon catalyst) exhibited 65% ethanol conversion and 96% acetaldehyde selectivity at 350 °C owing to its higher Lewis acidity than Ce, Co, and Ni doped on activated carbon and confirmed that dehydrogenation of ethanol is favored in Lewis acid sites. On the other hand, Ni/ACC catalyst was found potentially suitable to produce ethylene via ethanol dehydration at 400 °C.

Garbarino et al. (2019) studied ethanol dehydrogenation over $Cu/ZnAl_2O_4$ catalyst and observed that the presence of Cu over support surface was essential to obtain higher selectivity towards acetaldehyde via suppressing the acidic/basic site of support that favors formation of diethyl ether by ethanol dehydration. At higher temperature (>400 °C), selectivity towards acetaldehyde decreased as more thermodynamically stable products formed from acetaldehyde.

Synergetic effect of metal–support was investigated for oxidation of ethanol in gas phase using $M/MgCuCr_2O_4$ (M = Au, Pd, Pt, Ag, and Cu). In aerobic conditions, facile oxidation of Cu and Ag occurred, which resulted into their lower catalytic activity. $Pt/MgCuCr_2O_4$ and $Pd/MgCuCr_2O_4$ were found to be more active for ethanol oxidation and less selective towards acetaldehyde. These can be ascertained to cleavage of C–C bond of ethanol at moderately low temperature. The synergy between Au^0 and Cu^+ in case of $Au/MgCuCr_2O_4$ catalyst plays a vital role. It increased acetaldehyde selectivity up to 96% with 100% conversion of ethanol at 250 °C. The role of Cu^+ is to activate O_2, which facilitated cleavage of O–H bond in the reaction. This indicates that the activation of O_2 worked as the basic sites (Liu et al. 2015). The performance of pertinent catalyst systems reported in literature is shown in Table 3.1. Due to variations in reaction conditions, direct comparison of catalytic activity cannot be made.

3.2.2 Acetic Acid and Ethyl Acetate

Carbonylation of methanol, which relies on fossil sources, is the only established route to meet the current demands of acetic acid. Numerous heterogeneous catalysts have been explored for its synthesis via aerobic ethanol oxidation in gas and liquid phase. In majority of catalytic systems, pure oxygen or air is used as the oxidant. Acetic acid generally produced in two steps from oxidation of ethanol. Ethanol is first oxidized to acetaldehyde and subsequently to acetic acid.

Au, Pt and Pd supported $MgAl_2O_4$ catalyst for liquid phase oxidation of ethanol (5 wt% in water) and found that all three catalyst exhibited similar activity but Au nanoparticles were more selective towards acetic acid than Pt and Pd nanoparticles at 150 °C and 3 MPa (Christensen et al. 2006). Further studies with supported Au catalysts confirmed that the catalytic performance depends on the size of Au nanoparticle (Sun et al. 2008) and nature of supports (Christensen et al. 2006; Jørgensen et al. 2007; Tembe et al. 2009). For instance, Au nanoparticles supported on TiO_2 and ZnO displayed similar catalytic activity, while Au/Al_2O_3 showed poor performance (Tembe et al. 2009).

Takei et al. (2015) used Au/NiO as the catalyst for ethanol oxidation and obtained high conversion with NiO support in comparison to other oxides. They checked the effects of different metal cations as dopant into NiO and found Cu to enhance semiconductivity, which resulted into higher selectivity to acetic acid and maintained ethanol conversion. Moreover, Fe- and Co-doped catalyst exhibited lower ethanol conversion, while acetic acid selectivity remained almost unchanged.

Gorbanev et al. (2012) studied the effects of different supports for $Ru(OH)_x$ for aerobic oxidation of aqueous ethanol (2.5–50 wt%) at high temperatures and oxygen pressure. There was little impact of ethanol concentration on the product distribution at applied reaction conditions. On the other hand, to achieve a high yield of acetic acid, temperature greater than 125 °C was essential, which depends on the

Table 3.1 Various catalysts reported for synthesis of acetaldehyde from oxidation of ethanol in gas phase

Catalyst	Metal loading (wt%)	Ethanol in feed (%)	Temperature (°C)	Pressure (atm)	Ethanol/O_2/carrier gas (vol%)	Ethanol conversion (%)	Acetaldehyde selectivity (%)	Time on stream (h)	GHSV ($mLg_{cat}^{-1} h^{-1}$)	References
Ag/CeO_2	1	10	350	5	0.02:20.99:78.99	36	87	36	18,000	Rana and Parikh (2017b)
Ag/ZrO_2	2	99.9	250	–	Inert atmosphere	24	31	6	W/F = 38 min	Freitas et al. (2016)
$Au/Ce_{0.25}Zr_{0.75}O_2$	1	10	350	5	0.02:20.99:78.99	46	90	36	18,000	Rana and Parikh (2018)
Au/CeO_2	1	100	200	1	1:2 (ethanol:O_2)	31.8	56	–	257,120	Sheng et al. (2004)
$Au/MgCuCr_2O_4$	0.5	100	250	–	1.5:3:95.5	92	90	10	100,000	Liu et al. (2015)
Au/SiO_2	1	10	350	–	9-10/0/90	8	90	–	12,000	Gazsi et al. (2011)
Au/ZrO_2	1	10	350	5	0.02:20.99:78.99	38	72	4	18,000	Rana and Parikh (2017a)
Cu/AlMgO-P	10	100	300	–	N_2 used as carrier gas	60	100	6	31.1 h^{-1} (WHSV)	Petrolini et al. (2019)
$Cu/ZnAl_2O_4$	30	96	350	–	N_2 used as carrier gas	96	90	–	10,000	Garbarino et al. (2019)
$PdO/m\text{-}ZrO_2$	0.6	100	175	4	3.0:19.4:77.6	80	10	15	10,200	Letichevsky et al. (2015)
V_2O_5/TiO_2	15	50	215	2.7	3.3 (O_2/ethanol molar ratio)	100	43	–	32,960	Jørgensen et al. (2009)

type of supports. Over-oxidation of ethanol occurred at about 200 °C led to a decreased selectivity towards acetic acid. Under the reaction conditions of 5 wt% aqueous ethanol, 175 °C, 30 bar pressure, and 3 h of reaction time, an increased order of ethanol conversion and selectivity to acetic acid followed the order: $Ru(OH)_x/TiO_2 < Ru(OH)_x/Al_2O_3 < Ru(OH)_x/MgAl_2O_4 < Ru(OH)_x/CeO_2$. In the case of $Ru(OH)_x/CeO_2$, increased loading of Ru on ceria surface resulted into larger particle sizes, which subsequently decreased the catalytic activity. PdO/Al_2O_3 and $PdO/m\text{-}ZrO_2$ catalysts employed for ethanol oxidation selectively produced acetic acid. $PdO/m\text{-}ZrO_2$ catalyst was found to be more active than PdO/Al_2O_3 owing to the oxygenate species which are adsorbed and spillover from surface of $m\text{-}ZrO_2$ to PdO and vice versa (Letichevsky et al. 2015). Thus, atomic oxygen plays a vital role in the ethanol conversion with reducible oxide (ZrO_2). Similar observation was made with Au-supported CeO_2, ZrO_2, and SiO_2 catalysts for ethanol oxidation. Moreover, Au/ZrO_2 and Au/CeO_2 exhibited higher activity owing to their reducibility (Rana and Parikh 2017a).

Li and Iglesia (2007) employed $Mo_{0.61}V_{0.31}Nb_{0.08}O_x/TiO_2$ catalyst, which yielded more than 90% of acetic acid at complete ethanol conversion at 237 °C and 1.6 MPa with aqueous ethanol and oxygen. They observed that water promotes the rate of acetic acid formation via the sequential oxidation of acetaldehyde. The decreased rate of acetaldehyde led to the production of stable acetic acid consequently preventing its oxidation. Furthermore, $MoV_{0.3}Nb_{0.12}Te_{0.23}O_x$ mixed oxide was found to be superior to $Mo_{0.61}V_{0.31}Nb_{0.08}O_x/TiO_2$ catalyst because it demonstrated higher synthesis rate for acetic acid at 290 °C and atmospheric pressure. In contrast, Sobolev and Koltunov (2011) argued that the presence of water is not essential for higher selectivity of acetic acid.

Xiang et al. (2017) investigated the effects of different Cu content in CuCr catalyst for acetic acid production from ethanol in the absence of oxygen and concluded that reaction proceeds in two steps, such as (a) dehydrogenation of ethanol to acetaldehyde and (b) reaction of adsorbed aldehydes with surface hydroxyl groups to produce acetic acid. Both the reactions are performed on the surface of Cu^0 and/or Cu^+ species. The presence of Cr content enhanced the selectivity to acetic acid by providing more surface oxygen.

Acetic acid was produced in liquid phase aerobic oxidation of bioethanol with Au/CuM_2O_4 (M = Fe, Cr, Al) at 140 °C and 3 MPa, and the role of metal–support interactions was investigated (Hu et al. 2018). $Au/CuFe_2O_4$ catalyst exhibited higher ethanol conversion (63%) and selectivity towards acetic acid (81%) as compared to other catalysts. This can be attributed to enhanced O_2 activation and ethanol owing to the existence of negatively charged $Au^{\delta-}$ and redox active Fe^{+2} species in $Au/CuFe_2O_4$ catalyst.

Ethyl acetate is mainly used as a solvent in a variety of industries. Industrially, it is synthesized by several routes such as (a) esterification between ethanol and acetic acid, (b) reaction of acetic acid with ethylene, and (c) disproportion of acetaldehydes through Tishchenko reaction. However, corrosiveness and toxicity of used reagents are the major drawbacks for these methods. The production of ethyl acetate from ethanol occurs in one-pot procedure. It is an alternative process and considered as an environmentally and economically viable method. It is produced via an oxidative route and a dehydrogenation route. In oxidative route, 2 mol of water molecules are generated while hydrogen is produced in dehydrogenative route. Dehydrogenation pathway has more commercial value as it produces hydrogen as the coproduct without the use of oxygen, thus avoiding flammability of mixtures (Sánchez et al. 2011). However, the separation of products from reaction mixture becomes a major issue for both processes.

Inui et al. (2004) used Cu·Zn·Zr·AlO catalyst for the dehydrogenation of ethanol to ethyl acetate. Catalyst ($Cu:ZnO:ZrO_2:Al_2O_3$ = 12:1:2:2 molar ratio) showed about 84% ethyl acetate selectivity and 66% conversion of ethanol at 220 °C, thus favoring the reaction mechanism by suppressing selectivity of methyl ethyl ketone to less than 5%. It is considered as an undesirable by-product that forms azeotropic mixture with ethyl acetate. In this case, coupling between ethanol and acetaldehyde occurred at mixed metal oxide surface instead of Cu metal site.

Physically mixed Cu–ZnO–Al_2O_3 or PdO/m-ZrO_2 anchored over m-ZrO_2, t-ZrO_2, and a-ZrO_2 catalyst was employed for the synthesis of ethyl acetate from ethanol with and without oxygen (Gaspar et al. 2010). Three different phases of Zr oxide (m-ZrO_2, t-ZrO_2, and a-ZrO_2) with Cu–ZnO–Al_2O_3 were tested in the dehydrogenative route, while PdO/m-ZrO_2 was used in oxidative route. Among all the tested catalysts, m-ZrO_2 and t-ZrO_2 exhibited similar catalytic activity in terms of ethanol conversion and selectivity to acetaldehyde and ethyl acetate, while a-ZrO_2 was found to be less selective towards ethyl acetate due to its lower basicity. In both oxidative and dehydrogenative routes, the condensation reaction between acetaldehyde and ethanol or ethoxide species occurred on the surface of Zr.

Sánchez et al. (2005) studied the conversion of ethanol to ethyl acetate using Pd (1 wt%) supported on various oxides such as SiO_2, Al_2O_3, ZnO, SnO_2, and WO_3–ZrO_2 at 250 °C and 1 MPa. Results showed that the catalytic activity and product distribution change with the nature of support. Pd@ZnO and Pd@SnO_2 were found to be most efficient catalysts for the production of ethyl acetate at moderate ethanol conversion. The reaction mechanism of dehydrogenative route studied with commercial Cu/ZnO/Al_2O_3 (CZA) was physically mixed with ZrO_2, CeO_2, Al_2O_3, and SiO_2. The bare CZA catalyst exhibited lower ethanol conversion and produced acetaldehyde as the main product. The physical mixture of CZA and oxides showed higher rate of ethanol consumption and increased selectivity to ethyl acetate when compared to the bare CZA. Among the used catalysts, CZA with ZrO_2 was found to be superior in catalytic performance exhibiting higher rate of ethanol consumption and ethyl acetate selectivity. This was attributed to the higher basic sites of ZrO_2. The proposed reaction mechanism involved (i) the formation of acetaldehyde on bare CZA, (ii) the attack of ethoxide or ethanol to produce hemiacetal, and (iii) the dehydrogenation of ethyl acetate. It was observed that the basicity of oxides plays a vital role in the formation of ethoxide, which can be considered as the rate determining step (Zonetti et al. 2011).

Carotenuto et al. (2013) applied three commercial Cu-based catalysts for dehydrogenative/oxidative route and evaluated the catalytic performance by varying the operating conditions. In the dehydrogenative route, at lower residence time (about 0.15 gh mol^{-1}), 300 °C, and atmospheric pressures, acetaldehyde was the main product with 65% selectivity and around 85% ethanol conversion with CuCrAl. However, addition of small amounts of O_2 ($O_2/CH_3CH_2OH = 0.6$) favored higher selectivity towards acetaldehyde (90%) at temperatures less than 200 °C. About 90% selectivity towards ethyl acetate and ethanol conversion around 45% were observed when the reaction was carried out at higher pressure (20 bar), longer residence time, and 220 °C by co-feeding 3.66×10^{-3} mol/h of H_2 (25 cm^3/min of a mixture of 6% H_2 in N_2). Herein, the role of H_2 was not only to inhibit the generation of acetaldehyde by lowering its concentration but also to retain the catalyst in its reduced form.

The literature relating to the oxidative route of ethyl acetate is very limited. Gaspar et al. (2009) employed PdO/SiO_2 catalyst for the synthesis of ethyl acetate in oxidative route and observed that the reaction mechanism followed similar steps as demonstrated by Inui et al. (2004). The oxidative dehydrogenation of ethanol formed acetaldehyde, which reacted with ethanol or ethoxide species for producing ethyl acetate. The results indicate almost similar reaction mechanism of oxidative and dehydrogenative routes. However, the data generated by them does not clearly indicate where the condensation reaction occurred. They also observed that the change in ethanol consumption rate had influenced the formation of acetaldehyde and ethyl acetate. Low ethanol consumption rate was the rate limiting step for acetaldehyde, while high consumption rate was for condensation reaction (Gaspar et al. 2009).

3.2.3 Acetone

Acetone is primarily used in the production of methyl methacrylate and meta acrylic acid monomers, which are highly in demand nowadays in the manufacturing of bisphenol-A (BPA polycarbonate) as the commercial solvent in the synthesis of drugs and polymers and as thinner in paint industries. Currently, acetone is produced either from cumene process or dehydrogenation of 2-propanol. These processes derive their feedstock from fossil resources, which are associated with many environmental issues. For example, benzene and propylene used in cumene process are considered as fossil raw materials. Acetone can be synthesized from ethanol as mentioned in Eq. (3.1):

$$CH_3CH_2OH + H_2O \rightarrow CH_3COCH_3 + CO_2 + 4H_2 \tag{3.1}$$

Many heterogeneous catalytic systems are reported for synthesis of acetone from ethanol. In this context, Murthy (1988) examined Fe_2O_3–ZnO, Fe_2O_3–CaO, and Fe_2O_3–Mn and found that ZnO-promoted iron oxide catalyst exhibited better catalytic activity and stability. Nakajima et al. (1994) employed 24 oxides for dehydration or dehydrogenation of ethanol in the presence of water. At 440 °C, complete ethanol conversion and 94% acetone selectivity were achieved using Fe–Zn catalyst. Furthermore, they observed that oxides having both acidic and basic properties yielded higher acetone. Similarly, $Cu/La_2Zr_2O_7$ catalyst having both acidic and basic sites exhibited higher selectivity to acetone (70%) and complete ethanol conversion at 400 °C. This indicates that the presence of both acidic and basic sites is essential for the formation of acetone (Bussi et al. 1998).

Zinc–calcium oxide with different ratio is prepared by co-precipitation method and investigated in vapor phase reaction of ethanol to acetone. With Zn:Ca = 9:1 ratio, highest ethanol conversion (100%) and selectivity towards acetone (86%) were obtained at 425 °C. The increased amount of water (from 20% to 90%) in reaction mixture resulted in higher ethanol conversion from 45% to 90% and increased acetone selectivity from 15% to 74% at 450 °C. These results revealed that the presence of water had positive impact on ethanol conversion and selectivity for acetone (Baghiyev et al. 2011). Liu et al. (2013) employed mixed oxide of $Zn_xZr_yO_z$ for isobutene production from ethanol and observed that acetone was a major product at temperature range of 400–475 °C. A further increase in temperature (>475 °C) resulted in a decreased selectivity of acetone, but the selectivity to isobutene increased. It was suggested that acetone synthesis occurred via ethanol dehydrogenation and self-aldol condensation of acetaldehyde.

Iwamoto (2015) employed two catalysts (Sc/In_2O_3 and Y_2O_3–CeO_2) and demonstrated different reaction mechanism followed by the formation of acetone from ethanol. In case of Sc/In_2O_3 catalyst, acetaldehyde was first oxidized by either water or hydroxyl groups present on the surface of catalyst to produce acetic acid; subsequently ketonization takes place to generate acetone and CO_2. The condensation of acetaldehyde with Y_2O_3–CeO_2 produced ethyl acetate, which decayed to give acetic acid and ethene. Acetone formation was followed by same ketonization step as reported with Sc/In_2O_3.

Rodrigues et al. (2013) proposed the reaction mechanism by employing $Cu/ZnO/Al_2O_3$ (CZA) and ZrO_2. ZrO_2 and CZA surfaces generate ethoxide by the adsorption of ethanol, which later generated ethoxide when the oxides surface migrated to CZA and dehydrogenated to give acetaldehyde. Consecutively, the spillover or sorption of acetaldehyde from CZA to ZrO_2 surface and interaction with lattice oxygen of ZrO_2 induced acetate, which further converted to acetone, H_2, and CO_2. Similar reaction mechanism was proposed by de Lima et al. (2017) with Ag–CeO_2. Insertion of Ag increased reducibility of CeO_2; similar phenomenon was reported by Rana and Parikh (2017b).

Further investigation of reaction mechanism and effect of acid–base properties of catalyst were carried out with $Zn_xZr_{1-x}O_{2-y}$ catalysts by del Silva-Calpa et al. (2017). They concluded that the inserted Zn led to an increased basicity of m-ZrO_2. The increased basicity of $Zn_xZr_{1-x}O_{2-y}$ mixed oxide resulted in higher selectivity towards small oxygenate molecules compared to m-ZrO_2. They also observed that in the absence of water, acetone was not formed. Other works also supplemented that water is essential for acetone formation (Nakajima et al. 1994; Baghiyev et al. 2011).

The reaction mechanism for acetone formation was similar to those proposed by Rodrigues et al. (2013) and de Lima et al. (2017). Recently, Rodrigues et al. (2017) employed different catalysts such as $Ce_{0.75}Zr_{0.25}O_2$ (CZ), $Cu/ZnO/Al_2O_3$ (CZA),

ZrO_2, CZA + CZ, and CZA + ZrO_2 for this reaction and presented detailed mechanism using analytical techniques such as thermal desorption spectroscopy (TPD), temperature-programmed surface reaction (TPSR), and infrared mass spectrometry (IR-MS). The obtained results showed that oxygen vacancies on oxide surface increased reducibility and water dissociation activity of catalyst. This ultimately resulted in higher acetone selectivity. The acidity of catalyst not only promoted the adsorption of oxygenated species but also prevented the dehydration of ethoxide.

3.3 Conversion of Ethanol to Hydrocarbons

3.3.1 Ethylene

Ethylene is an important platform chemical used to produce polymers (e.g., polyethylene, polyvinylchloride, and polystyrene), ethylene glycol, ethylene oxide, ethyl benzene, and many other chemicals. It is mainly obtained from petroleum-derived feedstocks through thermal cracking. The continuous depletion of fossil fuel resources has led to an increased price of ethylene. With increased throughput and decreased cost, bioethanol conversion to ethylene looks more attractive and viable (Sun and Wang 2014). The typical reaction for the production of ethylene through dehydration of ethanol is shown in Eq. (3.2):

$$CH_3CH_2OH \rightarrow C_2H_4 + H_2O \tag{3.2}$$

Thermodynamic conditions of the reaction favor the production of diethyl ether as a by-product in lower temperature range. The generation of ethylene is endothermic and occurs in the temperature range of 200–450 °C. Hence, a major challenge is to develop catalysts which are capable to selectively produce ethylene at lowest possible temperature (Angelici et al. 2013). HZSM-5 (Phillips and Datta 1997) and γ-Al_2O_3 (Phillips and Datta 1997; Zhang et al. 2008) were found to be prominent catalysts in terms of activity and selectivity towards ethylene. HZSM-5 yielded higher ethylene selectivity (95%) and ethanol conversion (>90%) at 300 °C in comparison to γ-Al_2O_3 that produced 80% selectivity towards ethylene and 85% conversion of ethanol at 450 °C. It indicated that HZSM-5 worked efficiently at lower temperature, while γ-Al_2O_3 required higher temperatures (Zhang et al. 2008). High selectivity of ethylene with HZSM-5 attributed to its strong acid sites compared to γ-Al_2O_3. Despite high activity of HZSM-5, alumina is a widely used catalyst in chemical industries owing to its high stability, low price, and less coking.

Chen et al. (2007) employed TiO_2/γ-Al_2O_3 catalyst for ethylene production from catalytic dehydration of ethanol and studied effect of metal loading, ethanol concentration, and reaction temperature. As the water concentration increased from 5% to 90% in the feed, a declining conversion of ethanol from 86% to 65% was observed. In contrast, ethylene selectivity increased from 86% to 94% upon increased water content. However, this impact of water on ethanol conversion was nullified at higher temperature. It can be said that for alumina-based catalyst, higher reaction temperature is required to achieve higher selectivity towards ethylene (Chen et al. 2007; Zhang et al. 2008).

Catalytic activity of Si–Al–phosphate (SAPO-11 and SAPO-34) and their modified forms with Mn and Zn were compared at 340 °C (Chen et al. 2010). The results revealed that bare SAPO-11 exhibited higher ethanol conversion (97%) and ethylene selectivity (90%) when compared to Zn^{2+} and Mn^{2+} modified SAPO-11. However, completely different catalytic performance was observed with modified SAPO-34 as it exhibited higher ethanol conversion and high selectivity towards ethylene when compared to bare SAPO-34. At 340 °C, 99.4% ethanol conversion and 98.4% ethylene selectivity were obtained with Mn–SAPO-34 catalyst. This can be correlated to its moderate acidic strength and high surface area.

Zhang et al. (2008) employed γ-Al_2O_3, HZSM-5 (Si/Al = 25), SAPO-34, and Ni-SAPO-34 catalyst and evaluated their catalytic activity and stability in the dehydration of ethanol to ethylene. Ethanol conversion and selectivity towards ethylene in decreasing order as HZSM-5 > NiAPSO-34 > SAPO-34 > γ-Al_2O_3 at the reaction condition: 300 °C for HZSM-5, 350 °C for SAPO-34 and NiAPSO-34, and 450 °C for the γ-Al_2O_3 catalyst at LHSV 3 h^{-1}. It was found that both HZSM-5 and Ni-SAPO-34 achieved more than 90% yield of ethylene. SAPO-34 and Ni-SAPO-34 were found to have better stability (100 h). Hence, considering all aspect, Ni-APSO-34 is evolved as a suitable catalyst in conversion of ethanol to ethylene.

Recently, Masih et al. (2019) used Rho–zeolite catalyst for the formation of ethylene from ethanol in temperature range of 200–400 °C and compared the catalytic activity against three different samples of γ-Al_2O_3 and ZSM-5. All γ-Al_2O_3 catalysts showed poor catalytic activity to that of zeolites (Rho and ZSM-5). At 250 °C, Rho-zeolite was found to be superior than ZSM-5 in terms of ethylene selectivity. Moreover, Rho-zeolite exhibited excellent performance in terms of ethanol conversion (100%) and selectivity towards ethylene (99%) at 350 °C with 65 h on stream reaction. Herein, the concentration and strength of medium–strong acidic sites of Rho-zeolite played vital roles in catalytic performance.

3.3.2 Propylene

Propylene is also considered as an important chemical in the petroleum industries. It is used for the production of polypropylene and propylene oxide. Owing to important features such as comparable thermal and hydrothermal stability, acidity, pore structure, and reactivity of HZSM-5, it has been widely used as a catalyst in various reactions. The industrial transformation of methanol to olefins occurs in the presence of HZSM-5. In addition, it can be used as a catalyst in the conversion of ethanol to ethylene. Ethanol can be converted into a mixture of hydrocarbons containing ethylene, light olefins (C_3–C_4), and C_{5+} long-chain hydrocarbons over zeolite catalyst. The product distribution mainly depends on temperature, Si/Al ratio, surface acidity, and water content in the feed. Surface acidity played crucial role in the formation of propylene with H-ZSM-5 (Si/Al = 30, 80, and 280) (Song et al. 2009).

Lower acidity (Si/Al = 280) of HZSM-5 favored ethylene production, whereas C_{5+} hydrocarbons were dominant with higher acidity (Si/Al = 30), while moderate acid strength (Si/Al = 80) was found to be optimum for propylene (Song et al. 2009).

Metal containing (Zr, P, and Sr) HZSM-5 catalysts exhibited superior yield of propylene when compared to HZSM-5 catalyst at high temperatures. These additives were found to greatly reduce strong acid density while maintaining moderate acid strength and improved propylene selectivity (31–32%) (Goto et al. 2010; Song et al. 2009, 2010). Zr and P effectively prevented de-alumination of catalyst, thus resulting in better stability. Similar activity in terms of stability was obtained by tungsten-modified HZSM-11 (Si/Al = 120) catalyst. W-HZSM-11 exhibited highest selectivity (25%) towards propylene 25% at 550 °C (Inoue et al. 2010). Gayubo et al. (2010, 2011) conducted experimental and kinetic simulations studies and concluded that 1 wt% Ni loading significantly suppressed acidity strength of ZSM-5 from 135 kJ to 125 kJ (mol of NH_3)$^{-1}$, which ultimately resulted in improved selectivity of C_3–C_4.

Apart from zeolites, many transition metals and metal oxides have also been reported for the production of propylene. Various catalytic systems for the conversion of ethanol to propylene such as Ni ion-loaded mesoporous silica MCM-41 (Ni–M41) (Iwamoto et al. 2011), Sc-modified indium oxides (Sc/In_2O_3) (Mizuno et al. 2012), yttrium-loaded zirconia (Y/ZrO_2) (Xia et al. 2017), and many others have been reported. Complete ethanol conversion and 16% yield of propylene have been reported with Ni–MCM-41 (Si/Ni = 23) at 400 °C.

There are two different reaction pathways involved in the dehydration of ethanol to ethylene, especially diethyl ether, as well as the formation of acetaldehyde and ethyl acetate. Furthermore, dimerization, isomerization, and metathesis take place to convert ethylene to propylene (Iwamoto et al. 2011). In contrast, reaction pathways recommended by Mizuno et al. (2012) were quite different. The reaction steps were as follows:

1. Conversion of ethanol to acetaldehyde via dehydrogenation
2. Generation of acetone via condensation/ketonization of acetaldehyde
3. Hydrogenation–dehydration of intermediate acetone to propylene

Mizuno et al. (2012) also demonstrated that Sc/In_2O_3 catalyst exhibited higher propylene yield (60% based on carbon balance) and 100% ethanol conversion at 550 °C in the presence of water and hydrogen. Herein, the role of water was to reduce the deactivation of catalysts and hydrogen to promote the conversion of acetone to propylene. Recently, Y/ZrO_2 prepared by co-precipitation was employed for this reaction. Maximum yield of propylene (44%) with 98% ethanol conversion at 450 °C was achieved with 3 wt% Y/ZrO_2. It was observed that the conversion of ethanol to ethylene takes place on acidic sites, while the transformation of ethanol to propylene occurs via acetaldehyde and acetone on basic sites (Xia et al. 2017).

3.3.3 Butanol

Butanol is a versatile commodity chemical with applications as a feedstock for the production of acrylic esters and acrylic acid. It is also used as a solvent and as an additive in gasoline. Owing to its higher energy density, higher air-to-fuel ratio, lower water solubility, and lower heat of vaporization, it is considered as a biofuel (Ndou 2003; Nanda et al. 2017; Wu et al. 2018). Butanol is commercially produced via oxo process in which the following steps are (a) hydroformylation of propylene to generate butyraldehyde and (b) hydrogenation of butyraldehyde to yield butanol (Angelici et al. 2013; Wu et al. 2018). There are few drawbacks of this process such as difficulty in product separation, costly preparation of catalyst, and adverse environmental impacts. Therefore, transformation of ethanol to butanol seems to be more attractive. For the production of butanol, two different reaction pathways have been proposed such as (a) direct dimerization of two molecules of ethanol and (b) a multistep Guerbet process. In Guerbet process, ethanol is dehydrogenated to acetaldehyde followed by the formation of crotonaldehyde via self-aldol condensation of acetaldehyde and subsequent hydrogenation of crotonaldehyde to butanol (Ndou 2003).

Tsuchida et al. (2006) investigated the influence of reaction temperature on catalytic performance of hydroxyapatite (HAP) catalyst with Ca/P (molar ratio = 1.64). At 300 °C, nearly 14.7% ethanol conversion and 76.3% selectivity to butanol were obtained. Ethanol conversion increased from 14.7% to 95.3% by increasing the temperature from 300 °C to 450 °C. On the other hand, the selectivity of *n*-butanol drastically reduced to 6% from 76.3%. This indicated that lower temperature (≤300 °C) was optimal to produce *n*-butanol.

Homogeneous Ru-bis(diphenylphosphanyl) methane catalyst exhibited exclusive butanol selectivity (94%) and ethanol conversion (20%) at mild reaction temperature of 150 °C. This was ascertained to the catalyst's effective control over aldol reaction of acetaldehyde, which mainly produced butanol rather than higher hydrocarbons. It was observed that the catalytic activity decreased with time on stream. This was due to the decomposition of catalyst over the time period and probably due to weak water tolerance of ligands (Dowson et al. 2013). Wingad et al. (2015) also reported Ru phosphine–amine (P–N) complex, which displayed 93% *n*-butanol selectivity at 31% ethanol conversion with turnover number (TON) above 300. In contrast to bis(diphenylphosphanyl) methane catalyst (Dowson et al. 2013), mixed donor P–N ligands (Wingad et al. 2015) had improved water tolerance and assisted deprotonation in hydrogen transfer reactions.

Marcu et al. (2013) demonstrated that acidic/basic site of catalyst plays a vital role in product selectivity. They employed mixed oxides of M–Mg–Al–O (M = Cu, Pd, Ag, Mn, Fe, Sm, and Yb) for ethanol conversion to butanol. Among the used catalysts, Pd- and Sm-based catalysts exhibited higher selectivity towards butanol such as 72.7% and 66.3%, respectively, at 200 °C for 5 h. In case of Cu-based catalyst, an increase in Cu loading decreased the density of basic sites, which subsequently led to a decrease in the selectivity of *n*-butanol. This indicated that *n*-butanol formation correlates with basic site of catalyst. In addition, the decreased basic sites may raise acid sites strength, which could enhance the production of 1,1-diethoxyethane as the by-product.

Total thirteen different heterogeneous Al_2O_3-supported metal catalysts were examined for liquid phase in a one-pot ethanol conversion to butanol (Riittonen et al. 2012). Nearly 20 wt% Ni/Al_2O_3 (commercial catalyst) evolved as the best catalyst in terms of ethanol conversion (25%) and butanol selectivity (80%) at 250 °C and 70 bar. The trend of metals in terms of butanol selectivity was Ni > Pt > Au~Rh > Ru > > Ag. Furthermore, an increased hydrogen pressure resulted in the decreased selectivity of butanol. This can be due to the presence of hydrogen, which suppressed the dimerization of ethanol.

A similar research group carried out the same reaction with different metals (Cu, Co, and Ni) supported on Al_2O_3 in a continuous fixed-bed reactor at 250 °C, 70 bar, and LHSV = 4.3 h^{-1} (Riittonen et al. 2015). Ni and Cu catalysts exhibited better selectivity towards *n*-butanol and Co catalyst favored ethyl acetate as the main product. This was due to the position of metal cations in the crystal structure. In the inverse spinel structure of metal aluminate, Ni and Cu were octahedrally coordinated, while Co was tetrahedrally occupied. Moreover, the acidic/basic sites also played an important role in the product distributions. High acid strength of catalyst resulted in higher butanol selectivity, while the basic site favored ethyl acetate formation.

The effect of Cu and Ni dopant on porous metal oxides (PMOs) derived from synthetic hydrotalcites was studied for the production of 1-butanol from ethanol via Guerbet coupling (Sun et al. 2017). The results demonstrated that the incorporation of Cu and Ni dopants not only enhanced the conversion of ethanol but also increased the selectivity towards 1-butanol. An equimolar amount of Cu and Ni yielded 56% ethanol conversion and 22% butanol at 320 °C for 6 h with space–time yield reaching up to 704.6 kg^{cat-1} g^{-1}. The structural characterization before and after the reaction revealed that Ni–Cu alloy phase remained intact, which played a significant role during the reaction.

3.3.4 1,3-Butadiene

1,3-Butadiene (butadiene or C_4H_6) is another important molecule that can be produced from ethanol. It is used in the production of a variety of synthetic rubbers, resin, elastomers, and other polymeric materials. 1,3-Butadiene is a by-product of steam crackers that used to produce ethylene. Ethanol conversion to 1,3-butadiene was successfully implemented during World War II. Two different processes are reported for 1,3-butadiene from ethanol, such as (a) a one-step catalytic process known as Lebedev process and (b) a two-step process based on Ostromislenskiy reaction. Both the processes are recognized to follow the same reaction pathway. The following steps are believed to proceed in the reaction mechanism (Angelici et al. 2013; Sun and Wang 2014; Pomalaza et al. 2016):

1. Ethanol dehydrogenated to produce acetaldehyde

2. Acetaldol (3-hydroxybutanal) formation by aldol condensation of acetaldehyde
3. Crotonaldehyde generated via dehydration of 3-hydroxybutanal
4. Crotyl alcohol and acetaldehyde produced through Meerwein–Ponndorf–Verley (MPV) reaction of ethanol and crotonaldehyde
5. Crotyl alcohol dehydrated to 1,3-butadiene

The conversion of ethanol to 1,3-butadiene is a multistep reaction. Hence, it was observed that strength of acidic and basic site or redox properties of catalysts played important role in the reaction. As mentioned above, this reaction proceeds via dehydrogenation, condensation, and dehydration. The dehydrogenation step may occur on the redox and basic sites, while condensation and dehydration may take place on the acidic and basic sites (Sun and Wang 2014; Pomalaza et al. 2016). Considering the complexity of reactions, catalyst with optimal functionality is desired to achieve high productivity.

Niiyama et al. (1972) used Si–Mg oxide catalyst with varying ratio of MgO for the conversion of ethanol to butadiene. They optimized that 85% MgO showed high activity towards butadiene at the optimal reaction conditions and proved that balance and control over acid and basic density were essential to gain the best catalytic activity in terms of conversion and selectivity. Similar observation was reported by Kvisle et al. (1988) in which high content of MgO in silica favored high butadiene selectivity.

Jones et al. (2011) investigated several silica-impregnated bimetallic and trimetallic catalysts in ethanol transformation to butadiene. Zr–Zn catalyst with weight ratio of 1.5/0.5 has emerged as the most efficient catalyst, which displayed highest (66%) selectivity of butadiene when acetaldehyde and ethanol (feed ratio of 8:2) are co-fed together in the reaction mixture. Improved catalytic activity was due to Lewis acid sites of metals, while the presence of acetaldehyde in the co-fed enhanced 1,3-butadiene selectivity via aldol condensation reaction.

Various metal oxide dopants have been studied over MgO/SiO_2 and the oxides bearing Ag and Cu exhibited almost full ethanol conversion and 58.2% and 56.3% yield of butadiene, respectively, at 350 °C with low ethanol concentration. However, when the reaction was performed at higher temperature (400 °C) and at higher ethanol concentration, lower ethanol conversion and reduced butadiene yield were found. Moreover, it was observed that the acidic and basic component ratio and redox property of promoters played crucial roles for improving butadiene yield (Makshina et al. 2012).

Sushkevich et al. (2014) employed a multicomponent catalyst ($M/MO_x/SiO_2$) for the synthesis of butadiene from ethanol in a one-step process. Herein, the metals (M = Ag, Cu, and Ni) advocate the ethanol dehydrogenation, whereas the oxides ($MO_x = ZrO_2$, MgO, Al_2O_3, Nb_2O_5, and TiO_2) were active in aldol condensation and Meerwein–Ponndorf–Verley reduction. Ag- and Cu-containing zirconia supported on silica exhibited similar conversion and selectivity to butadiene at 320 °C. However, $Ag/ZrO_2/SiO_2$ catalyst demonstrated higher stability with time on stream.

Sushkevich et al. (2015) also investigated the catalytic performance of Ag-promoted Zr–BEA, Zr–MCM-41, and ZrO_2/SiO_2 in this reaction. Ag/Zr–BEA

(Si/Zr = 200) and Ag/Zr–MCM-41 (Si/Zr = 200) were found to be more active with respect to $Ag/ZrO_2/SiO_2$ (Si/Zr = 200) at equal amounts of Ag and Zr. This activity was ascertained to the isolation of highly concentrated Zr sites in Zr–BEA and Zr–MCM-41 catalysts in comparison to SiO_2 catalyst. In addition, all the catalyst displayed almost same the selectivity (about 65–66%) towards butadiene. This was ascribed to Zr^{4+} sites located in the framework of molecular sieves and incorporated in amorphous SiO_2 resulting into similar selectivity. It was also observed that butadiene yield correlated with the amount of Lewis acid strength. The variation in by-product distribution was mainly due to the nature and properties of the supports.

Angelici et al. (2014) investigated the effects of the addition of CuO to magnesia–silica materials. They synthesized the catalysts by different methods and found that the synthesis method had effected morphology, acidity, and basicity. MgO–SiO_2 and 1 wt% CuO supported on MgO–SiO_2 prepared using wet-kneaded process exhibited 17% and 53% butadiene yield, respectively, under similar reaction condition. This increased activity in the later was attributed to the redox property of Cu, which catalyzed ethanol dehydrogenation to acetaldehyde.

Catalytic performance of Na doping over the mixed metal oxides, i.e., $Zn_xZr_yO_z$, was investigated (Baylon et al. 2016). It was observed that doping Na to $Zn_xZr_yO_z$ mixed oxide not only suppressed strong acidic sites but also minimized ethylene selectivity and maximized acetaldehyde and butadiene selectivity. At 2000 ppm of Na doping in $Zn_1Zr_{10}O_z$-H catalyst at 350 °C with the weight hourly space velocity (WHSV) of 0.2 h^{-1}, 97% conversion of ethanol with 47% butadiene selectivity and 15.9% ethylene selectivity were observed. At WHSV of 6.2 h^{-1}, the catalyst remained active and produced butadiene at a rate of 0.49 $g_{butadiene}/g_{cat}/h$ with 28% butadiene selectivity and 54.4% ethanol conversion. Over a 60 h of time on stream, a 20% drop in the ethanol conversion along with a minor decrease in butadiene selectivity were observed.

Gao et al. (2018) examined ZrO_2 supported on various silica oxides (nano-SiO_2, SBA-15, MCM-41, and SiO_2) for the production of butadiene from feed mixture of ethanol and acetaldehyde. The characterization results revealed that impregnated Zr on silica support decreases the surface area. However, the average pore size for Zr/nano-SiO_2 and Zr/SiO_2 increased, while that of Zr/SBA-15 and Zr/MCM-41 decreased. More importantly, larger pore size turned into increased butadiene selectivity. The large pores of catalyst make it convenient for the in and out movement of both reactant and product, thus providing more contact time. These results indicated that pore size is also an important factor for determining the activity when catalyst contains a balance of acidic and basic sites. With ZrO_2/nano-SiO_2 catalyst, about 91.4% selectivity to butadiene and 52.4% ethanol conversion at 320 °C with WHSV of 1.8 h^{-1} were achieved. The catalyst also exhibited long-term stability (30 h) without a decline in the ethanol conversion and butadiene selectivity, which indicated better coke tolerance of catalyst.

3.4 Conclusions

Ethanol emerged as a prominent platform molecule to produce a variety of industrially important chemicals. Substantial work has been done for the synthesis of various value-added chemicals from ethanol in the past decades. It is observed that the formation of small oxygenates such as acetaldehyde can take place at basic sites of catalyst, while ethylene is generated on the acidic sites. Moderate acid strength is required for propylene production, whereas bifunctional catalyst is required to produce acetone and multifunctional catalyst is desired to produce 1,3-butadiene effectively.

The reaction mechanism for small oxygenates from ethanol is also well understood. For instance, ethanol is dehydrogenated to produce acetaldehyde via breaking of C–H or O–H bonds, which subsequently oxidize to produce acetic acid. The breaking of bonds generally depends on the nature of catalyst support. In the near future, C_2 or C_3 products derived from ethanol may face economical challenge due to the higher availability of low-cost shell gas in the USA. This has driven the interest towards C–C bond-forming reactions to yield butanol or 1,3-butadiene. The acidic, basic, and redox properties of catalysts are found to play significant roles in the reaction mechanisms. Condensation and dehydrogenation reactions occur on the basic sites, while acidic sites favor dehydration reaction.

Although extensive study has been carried out on the transformation of ethanol to value-added compounds, still some challenges are there which are needed to be addressed such as lower time on stream reported for the majority of catalytic systems. In the case of hydrocarbons, there is still some space to apply engineered catalysts, which could enhance high carbon yield. For instance, reaction in which product is highly active such as ethylene requires proper process design, instead of catalyst. In case of unstable intermediates, the principles of catalyst design can be applied to improve product yields. The development of optimal catalyst should emphasize on the desired product selectivity. Finally, the usage of reaction chemistry can lead to prominent alternatives of fossil fuel resources, thus making the products more robust and cleaner, which are currently not accessible from conventional resources.

References

Angelici C, Weckhuysen BM, Bruijnincx PCA (2013) Chemocatalytic conversion of ethanol into butadiene and other bulk chemicals. ChemSusChem 6:1595–1614

Angelici C, Velthoen MEZ, Weckhuysen BM, Bruijnincx PCA (2014) Effect of preparation method and CuO promotion in the conversion of ethanol into 1,3-butadiene over SiO_2-MgO catalysts. ChemSusChem 7:2505–2515

Baghiyev VL, Bagirova NN, Mirzai JI (2011) Conversion of ethanol into acetone over zinc-calcium oxide catalysts. Process Petrochem Oil-ref 12:101–105

Baylon RAL, Sun J, Wang Y (2016) Conversion of ethanol to 1,3-butadiene over Na doped $Zn_xZr_yO_z$ mixed metal oxides. Catal Today 259:446–452

Beck B, Harth M, Hamilton NG, Carrero C, Uhlrich JJ, Trunschke A, Shaikhutdinov S, Schubert H, Freund H-J, Schlögl R, Sauer J, Schomäcker R (2012) Partial oxidation of ethanol on vanadia catalysts on supporting oxides with different redox properties compared to propane. J Catal 296:120–131

Bharathiraja B, Jayamuthunagai J, Sudharsanaa T, Bharghavi A, Praveenkumar R, Chakravarthy M, Yuvaraj D (2017) Biobutanol—An impending biofuel for future: a review on upstream and downstream processing techniques. Renew Sust Energ Rev 68:788–807

Bussi J, Parodi S, Irigaray B, Kieffer R (1998) Catalytic transformation of ethanol into acetone using copper–pyrochlore catalysts. Appl Catal A Gen 172:117–129

Carotenuto G, Tesser R, Di Serio M, Santacesaria E (2013) Bioethanol as feedstock for chemicals such as acetaldehyde, ethyl acetate and pure hydrogen. Biomass Convers Biorefin 3:55–67

Chen G, Li S, Jiao F, Yuan Q (2007) Catalytic dehydration of bioethanol to ethylene over TiO_2/γ-Al_2O_3 catalysts in microchannel reactors. Catal Today 125:111–119

Chen Y, Wu Y, Tao L, Dai B, Yang M, Chen Z, Zhu X (2010) Dehydration reaction of bio-ethanol to ethylene over modified SAPO catalysts. J Ind Eng Chem 16:717–722

Christensen CH, Jørgensen B, Rass-Hansen J, Egeblad K, Madsen R, Klitgaard SK, Hansen SM, Hansen MR, Andersen HC, Riisager A (2006) Formation of acetic acid by aqueous-phase oxidation of ethanol with air in the presence of a heterogeneous gold catalyst. Angew Chem Int Ed 45:4648–4651

Čičmanec P, Ganjkhanlou Y, Kotera J, Hidalgo JM, Tišler Z, Bulánek R (2018) The effect of vanadium content and speciation on the activity of VO_x/ZrO_2 catalysts in the conversion of ethanol to acetaldehyde. Appl Catal A Gen 564:208–217

de Lima AFF, Zonetti PC, Rodrigues CP, Appel LG (2017) The first step of the propylene generation from renewable raw material: acetone from ethanol employing CeO_2 doped by Ag. Catal Today 279:252–259

del Silva-Calpa L, Zonetti PC, de Oliveira DC, de Avillez RR, Appel LG (2017) Acetone from ethanol employing $Zn_xZr_{1-x}O_{2-y}$. Catal Today 289:264–272

Dowson GRM, Haddow MF, Lee J, Wingad RL, Wass DF (2013) Catalytic conversion of ethanol into an advanced biofuel: unprecedented selectivity for *n*-butanol. Angew Chem Int Ed 52:9005–9008

Freitas IC, Gallo JMR, Bueno JMC, Marques CMP (2016) The effect of Ag in the Cu/ZrO_2 performance for the ethanol conversion. Top Catal 59:357–365

Gao M, Zhang M, Jiang H (2018) 1,3-Butadiene production from bioethanol and acetaldehyde over zirconium oxide supported on series silica catalysts. Catal Surv Jpn 22:118–122

Garbarino G, Riani P, Villa García M, Finocchio E, Escribano VS, Busca G (2019) A study of ethanol dehydrogenation to acetaldehyde over copper/zinc aluminate catalysts. Catal Today. https://doi.org/10.1016/j.cattod.2019.01.002

Gaspar AB, Esteves AML, Mendes FMT, Barbosa FG, Appel LG (2009) Chemicals from ethanol—The ethyl acetate one-pot synthesis. Appl Catal A Gen 363:109–114

Gaspar AB, Barbosa FG, Letichevsky S, Appel L (2010) The one-pot ethyl acetate syntheses: the role of the support in the oxidative and the dehydrogenative routes. Appl Catal A 380:113–117

Gayubo AG, Alonso A, Valle B, Aguayo AT, Olazar M, Bilbao J (2010) Hydrothermal stability of HZSM-5 catalysts modified with Ni for the transformation of bioethanol into hydrocarbons. Fuel 89:3365–3372

Gayubo AG, Alonso A, Valle B, Aguayo AT, Olazar M, Bilbao J (2011) Kinetic modelling for the transformation of bioethanol into olefins on a hydrothermally stable Ni–HZSM-5 catalyst considering the deactivation by coke. Chem Eng J 167:262–277

Gazsi A, Koós A, Bánsági T, Solymosi F (2011) Adsorption and decomposition of ethanol on supported Au catalysts. Catal Today 160:70–78

Gorbanev YY, Kegnæs S, Hanning CW, Hansen TW, Riisager A (2012) Acetic acid formation by selective aerobic oxidation of aqueous ethanol over heterogeneous ruthenium catalysts. ACS Catal 2:604–612

Goto D, Harada Y, Furumoto Y, Takahashi A, Fujitani T, Oumi Y, Sadakane M, Sano T (2010) Conversion of ethanol to propylene over HZSM-5 type zeolites containing alkaline earth metals. Appl Catal A Gen 383:89–95

Guan Y, Hensen EJM (2009) Ethanol dehydrogenation by gold catalysts: the effect of the gold particle size and the presence of oxygen. Appl Catal A Gen 361:49–56

Hidalgo JM, Tišler Z, Kubička D, Raabova K, Bulanek R (2016) (V)/Hydrotalcite, (V)/Al_2O_3, (V)/TiO_2 and (V)/SBA-15 catalysts for the partial oxidation of ethanol to acetaldehyde. J Mol Catal A Chem 420:178–189

Holz MC, Tölle K, Muhler M (2014) Gas-phase oxidation of ethanol over Au/TiO_2 catalysts to probe metal–support interactions. Catal Sci Technol 4:3495–3504

Hu W, Li D, Yang Y, Li T, Chen H, Liu P (2018) Copper ferrite supported gold nanoparticles as efficient and recyclable catalyst for liquid-phase ethanol oxidation. J Catal 357:108–117

Huber GW, Iborra S, Corma A (2006) Synthesis of transportation fuels from biomass: chemistry, catalysts, and engineering. Chem Rev 106:4044–4098

Hulea V (2018) Toward platform chemicals from bio-based ethylene: heterogeneous catalysts and processes. ACS Catal 8:3263–3279

Inoue K, Okabe K, Inaba M, Takahara I, Murata K (2010) Catalytic conversion of ethanol to propylene by H-ZSM-11. React Kinet Mech Catal 101:227–235

Inui K, Kurabayashi T, Sato S, Ichikawa N (2004) Effective formation of ethyl acetate from ethanol over Cu-Zn-Zr-Al-O catalyst. J Mol Catal A Chem 216:147–156

Iwamoto M (2015) Selective catalytic conversion of bio-ethanol to propene: a review of catalysts and reaction pathways. Catal Today 242:243–248

Iwamoto M, Kasai K, Haishi T (2011) Conversion of ethanol into polyolefin building blocks: reaction pathways on nickel ion-loaded mesoporous silica. ChemSusChem 4:1055–1058

Jones MD, Keir CG, Iulio CD, Robertson RAM, Williams CV, Apperley DC (2011) Investigations into the conversion of ethanol into 1,3-butadiene. Catal Sci Technol 1:267–272

Jørgensen B, Christiansen SE, Thomsen MLD, Christensen CH (2007) Aerobic oxidation of aqueous ethanol using heterogeneous gold catalysts: efficient routes to acetic acid and ethyl acetate. J Catal 251:332–337

Jørgensen B, Kristensen S, Kunov-Kruse A, Fehrmann R, Christensen C, Riisager A (2009) Gas-phase oxidation of aqueous ethanol by nanoparticle Vanadia/Anatase catalysts. Top Catal 52:253–257

Kaichev VV, Chesalov YA, Saraev AA, Klyushin AY, Knop-Gericke A, Andrushkevich TV, Bukhtiyarov VI (2016) Redox mechanism for selective oxidation of ethanol over monolayer V_2O_5/TiO_2 catalysts. J Catal 338:82–93

Kvisle S, Aguero A, Sneeden RPA (1988) Transformation of ethanol into 1,3-butadiene over magnesium oxide/silica catalysts. Appl Catal 43:117–131

Letichevsky S, Zonetti PC, Reis PPP, Celnik J, Rabello CRK, Gaspar AB, Appel LG (2015) The role of m-ZrO_2 in the selective oxidation of ethanol to acetic acid employing PdO/m-ZrO_2. J Mol Catal A Chem 410:177–183

Li X, Iglesia E (2007) Selective catalytic oxidation of ethanol to acetic acid on dispersed Mo-V-Nb mixed oxides. Chem Eur J 13:9324–9330

Liu C, Sun J, Smith C, Wang Y (2013) A study of $Zn_xZr_yO_z$ mixed oxides for direct conversion of ethanol to isobutene. Appl Catal A Gen 467:91–97

Liu P, Zhu X, Yang S, Li T, Hensen EJM (2015) On the metal–support synergy for selective gas-phase ethanol oxidation over $MgCuCr_2O_4$ supported metal nanoparticle catalysts. J Catal 331:138–146

Makshina EV, Janssens W, Sels BF, Jacobs PA (2012) Catalytic study of the conversion of ethanol into 1,3-butadiene. Catal Today 198:338–344

Marcu I-C, Tanchoux N, Fajula F, Tichit D (2013) Catalytic conversion of ethanol into butanol over M–Mg–Al mixed oxide catalysts (M = Pd, Ag, Mn, Fe, Cu, Sm, Yb) obtained from LDH precursors. Catal Lett 143:23–30

Masih D, Rohani S, Kondo JN, Tatsumi T (2019) catalytic dehydration of ethanol-to-ethylene over Rho zeolite under mild reaction conditions. Microporous Mesoporous Mater 282:91–99

Men'shchikov VA, Gol'dshtein LK, Semenov IP (2014) Kinetics of ethanol dehydrogenation into ethyl acetate. Kinet Catal 55:12–17

Mielby J, Abildstrøm JO, Wang F, Kasama T, Weidenthaler C, Kegnæs S (2014) Oxidation of bioethanol using zeolite-encapsulated gold nanoparticles. Angew Chem Int Ed 53:12513–12516

Mika LT, Cséfalvay E, Németh Á (2018) Catalytic conversion of carbohydrates to initial platform chemicals: chemistry and sustainability. Chem Rev 118:505–613

Mizuno S, Kurosawa M, Tanaka M, Iwamoto M (2012) One-path and selective conversion of ethanol to propene on scandium-modified indium oxide catalysts. Chem Lett 41:892–894

Mostrou S, Sipőcz T, Nagl A, Fődi B, Darvas F, Föttinger K, van Bokhoven JA (2018) Catalytic oxidation of aqueous bioethanol: an efficient upgrade from batch to flow. React Chem Eng 3:781–789

Murthy R (1988) Conversion of ethanol to acetone over promoted iron oxide catalysis. J Catal 109:298–302

Nakajima T, Nameta H, Mishima S, Matsuzaki I, Tanabe K (1994) A highly active and highly selective oxide catalyst for the conversion of ethanol to acetone in the presence of water vapour. J Mater Chem 4:853–858

Nanda S, Azargohar R, Dalai AK, Kozinski JA (2015) An assessment on the sustainability of lignocellulosic biomass for biorefining. Renew Sust Energ Rev 50:925–941

Nanda S, Golemi-Kotra D, McDermott JC, Dalai AK, Gökalp I, Kozinski JA (2017) Fermentative production of butanol: perspectives on synthetic biology. New Biotechnol 37:210–221

Ndou A (2003) Dimerisation of ethanol to butanol over solid-base catalysts. Appl Catal A Gen 251:337–345

Niiyama H, Morii S, Echigoya E (1972) Butadiene formation from ethanol over silica-magnesia catalysts. Bull Chem Soc Jpn 45:655–659

Ob-eye J, Praserthdam P, Jongsomjit B (2019) Dehydrogenation of ethanol to acetaldehyde over different metals supported on carbon catalysts. Catalysts 9:66

Petrolini DD, Cassinelli WH, Pereira CA, Urquieta-González EA, Santilli CV, Martins L (2019) Ethanol dehydrogenative reactions catalyzed by copper supported on porous Al–Mg mixed oxides. RSC Adv 9:3294–3302

Phillips CB, Datta R (1997) Production of ethylene from hydrous ethanol on H-ZSM-5 under mild conditions. Ind Eng Chem Res 36:4466–4475

Pomalaza G, Capron M, Ordomsky V, Dumeignil F (2016) Recent breakthroughs in the conversion of ethanol to butadiene. Catalysts 6:203

Rahman MM, Davidson SD, Sun J, Wang Y (2016) Effect of water on ethanol conversion over ZnO. Top Catal 59:37–45

Rana PH, Parikh PA (2017a) Bioethanol valorization via its gas phase oxidation over Au &/or Ag supported on various oxides. J Ind Eng Chem 47:228–235

Rana PH, Parikh PA (2017b) Bioethanol selective oxidation to acetaldehyde over Ag-CeO_2: role of metal-support interactions. New J Chem 41:2636–2641

Rana PH, Parikh PA (2018) Role of oxygen storage/supply capacity of mixed oxides of Ce and Zr in ethanol oxidation. Chem Eng Res Des 137:478–487

Rass-Hansen J, Falsig H, Jørgensen B, Christensen CH (2007) Bioethanol: fuel or feedstock? J Chem Technol Biotechnol 82:329–333

Riittonen T, Toukoniitty E, Madnani DK, Leino A-R, Kordas K, Szabo M, Sapi A, Arve K, Wärnå J, Mikkola J-P (2012) One-pot liquid-phase catalytic conversion of ethanol to 1-butanol over aluminium oxide—the effect of the active metal on the selectivity. Catalysts 2:68–84

Riittonen T, Eränen K, Mäki-Arvela P, Shchukarev A, Rautio A-R, Kordas K, Kumar N, Salmi T, Mikkola J-P (2015) Continuous liquid-phase valorization of bio-ethanol towards bio-butanol over metal modified alumina. Renew Energy 74:369–378

Rodrigues CP, Zonetti PC, Silva CG, Gaspar AB, Appel LG (2013) Chemicals from ethanol—The acetone one-pot synthesis. Appl Catal A Gen 458:111–118

Rodrigues CP, Zonetti PDC, Appel LG (2017) Chemicals from ethanol: the acetone synthesis from ethanol employing $Ce_{0.75}Zr_{0.25}O_2$, ZrO_2 and $Cu/ZnO/Al_2O_3$. Chem Cent J 11:30

Sánchez AB, Homs N, Fierro JLG, de la Piscina PR (2005) New supported Pd catalysts for the direct transformation of ethanol to ethyl acetate under medium pressure conditions. Catal Today 107–108:431–435

Sánchez AB, Homs N, Miachon S, Dalmon J-A, Fierro JLG, de la Piscina PR (2011) Direct transformation of ethanol into ethyl acetate through catalytic membranes containing Pd or Pd-Zn: comparison with conventional supported catalysts. Green Chem 13:2569–2575

Sheng P-Y, Bowmaker GA, Idriss H (2004) The reactions of ethanol over Au/CeO_2. Appl Catal A Gen 261:171–181

Simakova OA, Sobolev VI, Koltunov KY, Campo B, Leino A-R, Kordás K, Murzin DY (2010) "Double-peak" catalytic activity of nanosized gold supported on titania in gas-phase selective oxidation of ethanol. ChemCatChem 2:1535–1538

Sobolev VI, Koltunov KY (2011) MoVNbTe mixed oxides as efficient catalyst for selective oxidation of ethanol to acetic acid. ChemCatChem 3:1143–1145

Sobolev VI, Koltunov KY (2015) Gas-phase oxidation of alcohols with dioxygen over an Au/TiO_2 catalyst: the role of reactive oxygen species. Kinet Catal 56:343–346

Sobolev VI, Koltunov KY, Simakova OA, Leino A-R, Murzin DY (2012) Low temperature gas-phase oxidation of ethanol over Au/TiO_2. Appl Catal A Gen 433–434:88–95

Sobolev VI, Danilevich EV, Koltunov KY (2013) Role of vanadium species in the selective oxidation of ethanol on V_2O_5/TiO_2 catalysts. Kinet Catal 54:730–734

Song Z, Takahashi A, Mimura N, Fujitani T (2009) Production of propylene from ethanol over ZSM-5 zeolites. Catal Lett 131:364–369

Song Z, Takahashi A, Nakamura I, Fujitani T (2010) Phosphorus-modified ZSM-5 for conversion of ethanol to propylene. Appl Catal A Gen 384:201–205

Sun J, Wang Y (2014) Recent advances in catalytic conversion of ethanol to chemicals. ACS Catal 4:1078–1090

Sun K-Q, Luo S-W, Xu N, Xu B-Q (2008) Gold nano-size effect in Au/SiO_2 for selective ethanol oxidation in aqueous solution. Catal Lett 124:238–242

Sun Z, Vasconcelos AC, Bottari G, Stuart MCA, Bonura G, Cannilla C, Frusteri F, Barta K (2017) Efficient catalytic conversion of ethanol to 1-butanol via the Guerbet reaction over copper- and nickel-doped porous. ACS Sustain Chem Eng 5:1738–1746

Sushkevich VL, Ivanova II, Ordomsky VV, Taarning E (2014) Design of a metal-promoted oxide catalyst for the selective synthesis of butadiene from ethanol. ChemSusChem 7:2527–2536

Sushkevich VL, Ivanova II, Taarning E (2015) Ethanol conversion into butadiene over Zr-containing molecular sieves doped with silver. Green Chem 17:2552–2559

Takei T, Iguchi N, Haruta M (2011) Support effect in the gas phase oxidation of ethanol over nanoparticulate gold catalysts. New J Chem 35:2227–2233

Takei T, Suenaga J, Ishida T, Haruta M (2015) Ethanol oxidation in water catalyzed by gold nanoparticles supported on NiO doped with Cu. Top Catal 58:295–301

Tembe S, Patrick G, Scurrell M (2009) Acetic acid production by selective oxidation of ethanol using Au catalysts supported on various metal oxide. Gold Bull 42:321–327

Tsuchida T, Sakuma S, Takeguchi T, Ueda W (2006) Direct synthesis of n-butanol from ethanol over nonstoichiometric hydroxyapatite. Ind Eng Chem Res 45:8634–8642

Weusthuis RA, Aarts JMMJG, Sanders JPM (2011) From biofuel to bioproduct: is bioethanol a suitable fermentation feedstock for synthesis of bulk chemicals? Biofuels Bioprod Biorefin 5:486–494

Wingad RL, Gates PJ, Street STG, Wass DF (2015) Catalytic conversion of ethanol to n-butanol using ruthenium P–N ligand complexes. ACS Catal 5:5822–5826

Wu X, Fang G, Tong Y, Jiang D, Liang Z, Leng W, Liu L, Tu P, Wang H, Ni J, Li X (2018) Catalytic upgrading of ethanol to n-butanol: progress in catalyst development. ChemSusChem 11:71–85

Xia W, Wang F, Mu X, Chen K, Takahashi A, Nakamura I, Fujitani T (2017) Highly selective catalytic conversion of ethanol to propylene over yttrium-modified zirconia catalyst. Catal Commun 90:10–13

Xiang N, Xu P, Ran N, Ye T (2017) Production of acetic acid from ethanol over CuCr catalysts via dehydrogenation-(aldehyde–water shift) reaction. RSC Adv 7:38586–38593

Zhang X, Wang R, Yang X, Zhang F (2008) Comparison of four catalysts in the catalytic dehydration of ethanol to ethylene. Microporous Mesoporous Mater 116:210–215

Zhang J, Wang L, Zhu L, Wu Q, Chen C, Wang X, Ji Y, Meng X, Xiao F-S (2015) Solvent-free synthesis of zeolite crystals encapsulating gold-palladium nanoparticles for the selective oxidation of bioethanol. ChemSusChem 8:2867–2871

Zheng N, Stucky GD (2006) A general synthetic strategy for oxide-supported metal nanoparticle catalysts. J Am Chem Soc 128:14278–14280

Zonetti PC, Celnik J, Letichevsky S, Gaspar AB, Appel LG (2011) Chemicals from ethanol – The dehydrogenative route of the ethyl acetate one-pot synthesis. J Mol Catal A Chem 334:29–34

Selective Bioethanol Conversion to Chemicals and Fuels via Advanced Catalytic Approaches

4

Thanh Khoa Phung and Guido Busca

Abstract

Ethanol is one of the most important organic solvents used as a fuel, chemical, and as an intermediate for organic chemical synthesis. However, its use as a fuel has several drawbacks including corroding the engine, producing lower energy than conventional fuel, and adsorbing water. Hence, it needs to upgrade into a highly efficient fuel such as *n*-butanol. In addition, the demand for raw materials increases due to the development of current industries such as plastic, solvent, and surfactant industries. However, the fossil resources, the main sources for current raw materials production, are limited and continuously depleted. As a result, the conversion of low-cost biomaterials like bioethanol into high-value raw materials for current industries is a promising perspective. In this chapter, some potential reaction processes in bioethanol conversion including dehydration, oxidation, steam reforming, dehydrogenation, and Guerbet reaction are summarized. These catalytic reactions and ethanol-derived products are discussed to elucidate the current picture of ethanol transformation.

Keywords

Ethanol · Dehydration · Oxidation · Dehydrogenation · Reforming · Guerbet reaction

T. K. Phung (✉)
School of Biotechnology, International University, Vietnam National University, Hồ Chí Minh City, Vietnam
e-mail: ptkhoa@hcmiu.edu.vn

G. Busca
Department of Civil, Chemical and Environmental Engineering, University of Genova, Genova, Italy

S. Nanda et al. (eds.), *Biorefinery of Alternative Resources: Targeting Green Fuels and Platform Chemicals*, https://doi.org/10.1007/978-981-15-1804-1_4

4.1 Introduction

The greenhouse effect and global warming are increasing in recent years, likely due to the emissions associated with the consumption of fossil fuel. Additionally, the demand for raw materials and current production from fossil resources are increasing due to the development of industry and human being requirement. However, the continuous depletion of fossil resources makes the exploitation of alternative and renewable resources (Huber et al. 2006). Bioethanol generated from lignocellulosic biomass (Fig. 4.1) is the most common biofuel, which can substitute gasoline or blended with gasoline to be used as fuel of spark ignition engines (Dowson et al. 2013; Patzek 2004). However, the use of ethanol as a fuel results in a number of drawbacks including corroding the engine, producing lower energy than conventional fuel, and adsorbing water (Chakraborty et al. 2015; Dziugan et al. 2015; Earley et al. 2015; Aitchison et al. 2016; Nanda et al. 2017a). However, ethanol could become a key primary intermediate in the frame of a new organic chemical industry and fuels based on renewables (Phung et al. 2015c). Therefore, the transformation of ethanol into value-added chemicals and advanced fuels is highly necessary.

Bioethanol is currently produced from biomass via the conventional fermentation process (Bulawayo et al. 1996; Olsson and Hahn-Hägerdal 1996; Birol et al. 1998; Huang et al. 2009). The global ethanol production is increasing rapidly and over 117 million m^3 in 2017 (Statista 2019). More than 90% of ethanol is used as fuel and the rest is applied in beverages and industrial applications. However, the ethanol demand is now lower than its world production. Therefore, the production of raw chemicals and fuels from the available excess bioethanol could be necessary in the biorefinery industry in the near future (Sun and Wang 2014).

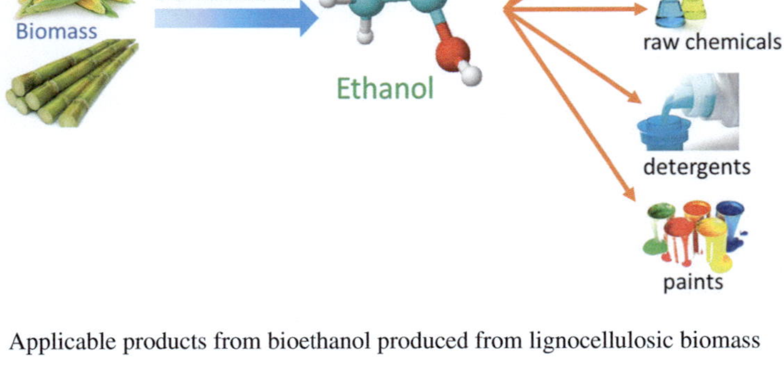

Fig. 4.1 Applicable products from bioethanol produced from lignocellulosic biomass

Bioethanol can be used as a pure fuel or for blending with gasoline, mostly with 5–10 vol% (E5–E10) (Sun and Wang 2014; Nanda et al. 2014b). In Europe, the maximum ethanol content in commercial motor gasoline is 10% (E10). Besides, ethanol can be transformed into many value-added chemicals (Fig. 4.2). Ethanol can be converted to hydrogen through steam reforming (Vaidya and Rodrigues 2006; Fatsikostas et al. 2001; Garbarino et al. 2013; Busca et al. 2010; Nanda et al. 2017b) or dehydrogenation (Autthanit et al. 2018; Campisano et al. 2018; Čičmanec et al. 2018; Clarizia et al. 2019; Garbarino et al. 2019; Guan and Hensen 2013; Inui et al. 2002a, b, 2004; Santacesaria et al. 2003; Shan et al. 2017; Tayrabekova et al. 2018; Tu and Chen 2001; Tu et al. 1994a, b; Weinstein et al. 2011). Dehydration of ethanol generates diethyl ether and ethylene (Phung and Busca 2015a, b; Phung et al. 2014a, b, 2015a, b) and other products such as aromatic and propylene (Bi et al. 2011; Calsavara et al. 2008; Furumoto et al. 2011; Gayubo et al. 2010b; Inaba et al. 2006, 2012; Phung et al. 2015c).

Ethylene oxide and acetic acid originate from the oxidation of ethanol or from ethanol-derived ethylene and acetaldehyde, respectively (Chen et al. 2015; Christensen et al. 2006; Kaichev et al. 2016; Li and Iglesia 2007; Lippits and Nieuwenhuys 2010a, b; Snider et al. 1991). Ethyl acetate is produced from ethanol through the esterification reaction with acetic acid, which can be made through aldol condensation reaction. Additionally, acetaldehyde can be also produced from

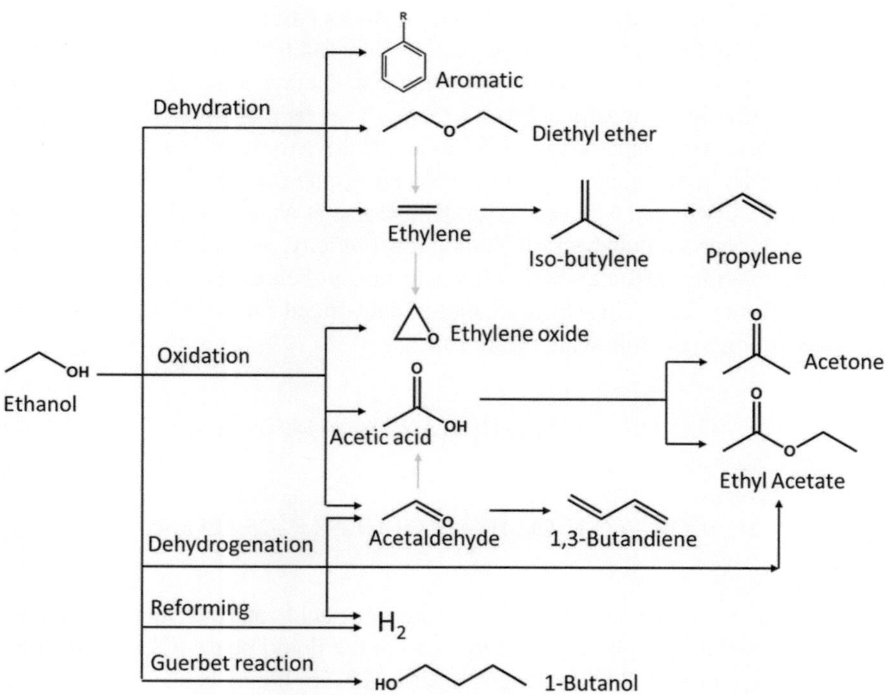

Fig. 4.2 Possible chemicals generated from ethanol

ethanol followed by the catalytic conversion to acetic acid or 1,3-butadiene (Angelici et al. 2014; Makshina et al. 2012). Besides, a high-value fuel, *n*-butanol, originating from ethanol via Guerbet reaction could become a great biofuel in the near future (Aitchison et al. 2016; Carvalho et al. 2012; Koda et al. 2009; Kozlowski and Davis 2013; Ogo et al. 2012; Nanda et al. 2014a).

These abovementioned ethanol-derived chemicals can be used as the intermediates in the biorefinery industry (Sun and Wang 2014). Ethylene and propylene are the widely used compounds in the polymer industry where 1,3-butadiene is also largely applied (Makshina et al. 2012). Ethylene oxide, diethyl ether, acetone, ethyl acetate, and acetaldehyde are industrial solvents and chemical intermediates (Fleischmann et al. 2000; Rebsdat and Mayer 2011; Riemenschneider and Bolt 2005; Sakuth et al. 2010; Sifniades et al. 2011). Moreover, ethanol-derived aromatics and *n*-butanol can be applied as fuels or blended with gasoline (Hahn et al. 2010; Sun and Wang 2014). In this chapter, all the current catalytic processes of ethanol conversion to chemicals and fuels are summarized. An all-inclusive scenario of catalysts used in the ethanol conversion processes is also discussed.

4.2 Dehydration of Ethanol

4.2.1 Diethyl Ether Production

Diethyl ether ($C_2H_5OC_2H_5$) is a simple and colorless ether and a flammable liquid. It has a low solubility in water and low boiling point (34.6 °C); also its melting point is extremely low (−116.3 °C) (Sakuth et al. 2010). Diethyl ether has many applications in the chemical industry such as a solvent in several chemical synthesis processes. Diethyl ether can also be used as a fuel additive for both diesel and gasoline fuels due to its high cetane number (>125) and octane number (>110) (Doğu and Varişli 2007; Erwin and Moulton 1996; Kito-Borsa et al. 1998). However, diethyl ether also has certain drawbacks including low lubricity, easy peroxidation in storage, high volatility, and anesthetic effects to human beings. Diethyl ether can be produced from ethanol, in which ethanol is dehydrated into ethylene and diethyl ether according to the following reactions:

$$C_2H_5OH \rightarrow C_2H_4 + H_2O\left(\Delta H298 = +44.9\,\text{kJ/mol}\right) \tag{4.1}$$

$$2C_2H_5OH \rightarrow C_2H_5OC_2H_5 + H_2O\left(\Delta H298 = -25.1\,\text{kJ/mol}\right) \tag{4.2}$$

Ethanol dehydration is well studied using most acid catalysts such as alumina and zeolites in the gas phase or sulfuric acid in the liquid phase (Chaichana et al. 2019; de Oliveira et al. 2018; Phung and Busca 2015b; Phung et al. 2015b; Sakuth et al. 2010). It is also a byproduct of the ethylene hydration process into ethanol

(Kosaric et al. 2011). For the dehydration of ethanol, acid catalysts favor this reaction than the basic catalysts (Phung et al. 2015a). Several oxides including alumina, titania, and zirconia are the most catalysts used as well as ZSM-5 which is a common catalyst for ethanol dehydration (Chaichana et al. 2019; de Oliveira et al. 2018; Phung and Busca 2015b; Phung et al. 2014a, b, 2015a, b). Additionally, the polyacid catalysts are also used for this reaction (Lee et al. 1992; Varisli et al. 2007).

In parallel with experimental studies, some computational studies and mechanism are also well investigated. The reaction represented in Eq. (4.1) occurs with an E2 concerted elimination mechanism and the mechanism of reaction represented in Eq. (4.2) is an SN2 bimolecular from a protonated and non-protonated form of ethanol (Christiansen et al. 2013; Brown et al. 2014). The mechanism of ethanol dehydration is also confirmed by the density functional theory (DFT) computational studies. However, Eqs. (4.1) and (4.2) are parallel reactions, in which the bimolecular and unimolecular mechanisms are the mechanism in the production of diethyl ether and ethylene, respectively (DeWilde et al. 2013; Roca et al. 1969). Recently, several authors (Phung and Busca 2015a, b; Garbarino et al. 2018) concluded that the parallel-successive path is very active. In fact, diethyl ether can be produced from the reaction of ethoxide group on the surface of catalyst with an adsorbed ethanol molecule (Knözinger 1968) or the reaction of two ethoxide groups (Arai et al. 1967). Moreover, diethyl ether is possibly as an intermediate in the ethylene production from ethanol dehydration (Bokade and Yadav 2011; Christiansen et al. 2013; Ciftci et al. 2012; Kagyrmanova et al. 2011; Yasunaga et al. 2010).

4.2.2 Ethylene Production

Ethylene (C_2H_4) is the simplest alkene and a colorless flammable gas. It has a faint "sweet and musky" odor. Ethylene is soluble in diethyl ether and poorly soluble in water (2.9 mg/L) and in ethanol (4.22 mg/L). Its boiling point is −103.7 °C and melting point is −169.2 °C (Zimmermann and Walzl 2009). Ethylene is one of the largely produced materials in the petrochemical industry with its annual global production of about 150 million metric tons (Lewandowski 2016). North America, Western Europe, Middle East and North East Asia are the main ethylene production regions (Lewandowski 2016). Ethylene has many applications in the production of chemicals and polymers such as polyethylene, ethylene dichloride, and ethylene oxide production (Fan et al. 2013). It was first produced from coke oven gas and Line was the first company commercializing ethylene production plant in the 1930s with the thermal cracking process patent of Standard Oil in 1913. Later in 1941, the first steam cracker was developed by Standard Jersey, and in the 1950s, ethylene became a largest intermediate for the synthesis of organic compounds as the replacement for acetylene as an intermediate (Zimmermann and Walzl 2009).

Nowadays, ethylene is produced commercially from crude oil through steam cracking of paraffinic hydrocarbons process and as a byproduct of fluid catalytic cracking of heavy oils. However, those processes consume high amount of energy as well as emit enormous amounts of greenhouse gas. To reduce the greenhouse gas emissions, bioethylene is produced from bioethanol via dehydration. Ethylene production from ethanol through dehydration has been industrially applied until the 1960s using alumina-based catalysts (Hu 1993). Le Van Mao et al. (1989) developed the bioethanol-to-ethylene (BETE) pathway using zeolites as the catalyst in the 1980s.

The literature indicates high catalytic activities over H-ZSM-5, HBEA, HFAU, HFER, and HMOR zeolites for ethanol dehydration (De las Pozas et al. 1993; Oudejans et al. 1982; Ouyang et al. 2009; Phillips and Datta 1997; Ramesh et al. 2009; Takahara et al. 2005, 2007; Zhang et al. 2008). Those zeolites can be applied to produce ethylene from bioethanol industrially (Fan et al. 2013), but Zhang and Yu (2013) reported that these might be unstable for the dehydration of ethanol. Moreover, the mechanism of this reaction over zeolites has been published with some disagreements between the respective conclusions (Chiang and Bhan 2010; Kondo et al. 2005; Lee et al. 1997; Nguyen et al. 2010; Wang et al. 2005; Zecchina et al. 1996). Currently, bioethylene is commercially produced from ethanol using Syndol catalyst based on $MgO-Al_2O_3/SiO_2$ catalysts (Syndol catalyst, Halcon SD) through the process of Chematur (Fan et al. 2013). The process is performed at a higher temperature (450 °C) giving 99% ethanol conversion with high selectivity of ethylene (96.8%).

4.2.3 Propylene Production

Propylene (C_3H_6) is the second simplest alkene and has one double bond. Propylene has poor solubility in water (0.61 g/m^3) and its boiling point and melting point are −47.6 °C and −185.2 °C, respectively (Eisele and Killpack 2011). Propylene is also the second largest petrochemical after ethylene with many applications in plastics and solvent industries. Due to its wide usage, the global demand for propylene has increased significantly with the global increase forecast of roughly 4.5% (52 billion lbs) by 2020 (Wattanakarunwong 2015).

Propylene is a byproduct from the petrochemical processes involving ethylene and gasoline production (Thinnes 2010; Wattanakarunwong 2015). It is also produced using technologies such as propane dehydrogenation (PDH) and ethylene/butene metathesis. New chemical processes for on-purpose propylene production using low-cost and nonpetroleum source materials such as methanol-to-olefin (MTO) (Chen et al. 2005; Pujadó and Andersen 2006) and methanol-to-propylene (MTP) (Jasper and El-Halwagi 2015; Liebner 2005) have been commercialized. The methanol-to-olefin (MTO) process (UOP/Hydro MTO) uses the MTO-100 (silicoaluminophosphate)-based catalyst, while the MTPROP (ZSM-5-based catalyst, Süd-Chemie) is used for Lurgi's MTP (methanol-to-propylene) process (Koempel and Liebner 2007). Both the processes are based on the methanol obtained in large-scale quantities from natural gas (steam reforming and synthesis) or coal (gasification and synthesis).

Bioethanol is also a suitable choice as a starting chemical for producing biopropylene (Posada et al. 2013). A number of studies on the conversion of ethanol to propylene have been published using zeolite-based catalysts (Aguayo et al. 2002a, b; Arias et al. 1997; Brandão et al. 2002; Gayubo et al. 2010a, 2011; Inaba et al. 2007; Oikawa et al. 2006; Phillips and Datta 1997; Phung et al. 2015c; Song et al. 2009, 2010; Takahara et al. 2005, 2007) and metal oxide catalysts (Bakoyannakis et al. 2001; Carrasco-Marín et al. 1998; Doheim and El-Shobaky 2002; Golay et al. 1999; Phung and Busca 2015b; Phung et al. 2014b, 2015a; Tsuchida et al. 2008; Varisli et al. 2007; Zaki 2005).

Currently, the Brazilian chemical company Braskem, who is the leader in the manufacturing of biopolyethylene, announced to use bioethanol to produce biopropylene (Tullo 2010) for further converting it into biopolypropylene. In this technology, the first step is the fermentation process converting sugars into ethanol followed by the conversion of ethanol into ethylene on one route and dimerized to butene on another route. In the next step, biopropylene is produced by metathesis of bioethylene and biobutene (Eq. 4.3) (Gotro 2013).

$$CH_2 = CH_2 + CH_3CH = CHCH_3 \rightarrow 2CH_3CH = CH_2 \tag{4.3}$$

The mechanism of transformation from ethanol to propylene is still debated. There are many reaction pathways to generate propylene via ethylene (Inaba et al. 2006; Iwamoto 2015; Murata et al. 2008; Takahashi et al. 2012) or acetaldehyde (Hayashi et al. 2014). Hayashi et al. (2014) suggested that ethanol can dehydrogenate to acetaldehyde and transform to acetone by reacting with water on the surface of yttrium-modified ceria catalysts, finally converting into propylene by the hydrogenation reaction (Fig. 4.3a). Moreover, ethanol can also transform directly to propylene.

A few other authors also indicate that propylene can be produced from ethanol via ethylene (Bhadra et al. 2018; Ghashghaee and Shirvani 2019; Khan et al. 2019; Kim et al. 2017; Sarker et al. 2019). However, the reaction mechanism to generate propylene from ethylene can occur via different pathways. Ethanol dehydrates to form ethylene, which trimerizes to give hexane followed by the production of propylene by β-fission (Murata et al. 2008; Oikawa et al. 2006) (Fig. 4.3b). Iwamoto (2015) suggested that propylene production could take place via ethylene and diethyl ether as the immediate. However, ethylene dimerizes to form 1-butene,

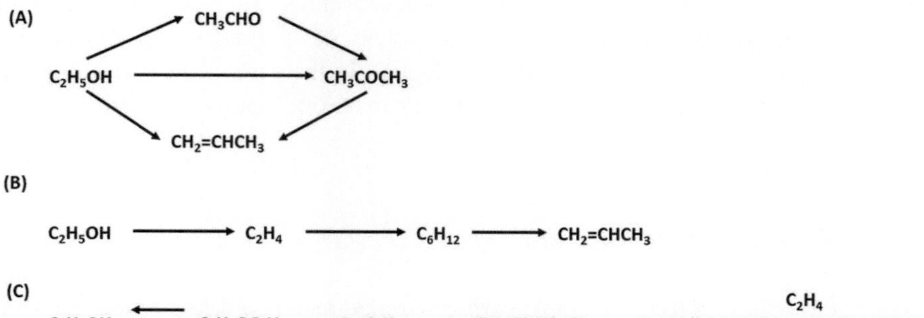

Fig. 4.3 Reaction pathways of ethanol decomposition into propylene

which transforms to 2-butene followed by the reaction of 2-butene with ethylene to produce propylene (Fig. 4.3c).

4.3 Oxidation of Ethanol

4.3.1 Ethylene Oxide Production

Ethylene oxide (C_2H_4O) is a flammable gas and the simplest epoxide. It is a three-membered ring ether with two carbon atoms and one oxygen atom. Therefore, it is easy to open the rings in the addition reactions. Ethylene oxide is miscible with water and has a low boiling point (10.4 °C) and melting point (−112.5 °C) (Rebsdat and Mayer 2011). Ethylene oxide is an important intermediate for many organic syntheses (Salmi et al. 2012). It is commonly utilized for the production of ethylene glycol and polyesters in the fabricating beverage bottles (Bac and Avci 2018; Liu et al. 2017). Additionally, glycols from ethylene oxide can be used in the production of coolants, detergents, polyester fibers, heat transfer fluids, surfactants, polyurethanes, plasticizers, and solvents (Bac and Avci 2018; Liu et al. 2017; Mendes et al. 2007). Polyethylene glycol is used as an important component in the cosmetic, pharmaceutical, lubricant, plasticizer, and solvent industries (Bac and Avci 2018; Liu et al. 2017; Mendes et al. 2007). Ethylene oxide is also utilized to produce ethylene glycol ethers, ethanolamine, and ethoxylates (Bac and Avci 2018; Liu et al. 2017; Mendes et al. 2007).

Ethylene oxide is mainly produced from ethylene as a product from dehydration of ethanol. The studies on the direct oxidation of ethanol into ethylene oxide are relatively scarce. The process mostly involves two reactions of dehydration of ethanol to ethylene and oxidizing the ethylene into ethylene oxide (Coupard and Plennevaux 2014). Copper-, silver-, and gold-based catalysts are the main active sites for the oxidation of ethanol to ethylene oxide (Chen et al. 2015; Lippits and Nieuwenhuys 2010a, b). Catalyst support is also important to enhance the catalytic activity. ZSM-5 (Chen et al. 2015) and Li_2O/Al_2O_3 (Lippits and Nieuwenhuys 2010a, b) showed the best activities as supports of catalysts in terms of the selectivity to ethylene oxide. The literature also reported that the presence of oxygen allows to prevent carbon deposition (Lippits and Nieuwenhuys 2010b). Tuning oxygen concentration and Li_2O doping boost the selectivity of ethylene oxide over silver-based catalysts (Lippits and Nieuwenhuys 2010a). In the case of the ceria support, it can make extra oxygen available to the catalytic reaction (Lippits and Nieuwenhuys 2010b).

4.3.2 Acetic Acid Production

Acetic acid (CH_3COOH) is a simple carboxylic acid and a colorless liquid at room temperature. Its melting point and boiling points are 16–17 °C and 118–119 °C, respectively. In addition, it is a main component of vinegar, an additive used in food and salads for its sour taste (Le Berre et al. 2014). Vinegar is one of the most ancient applications of acetic acid in the world. Acetic acid also has many applications including vinyl acetate monomer, ester production, acetic anhydride, solvents, and in medical use (Le Berre et al. 2014). Vinyl acetate monomer is used as wood glue as well as synthetic fibers and fabrics (Roscher 2000). Other esters arising from acetic acid such as ethyl acetate, propyl acetate, isobutyl acetate, and *n*-butyl acetate are commonly applied in paints, coatings, and inks as solvents (Riemenschneider and Bolt 2005). Acetic anhydride is an acetylation agent, which is used to produce cellulose acetate (Held et al. 2000). Moreover, acetic acid can be involved as a solvent in Friedel-Crafts alkylation reaction or terephthalic acid production. It is also used in medical research such as for the treatment for otitis externa and part of cervical cancer screening (Le Berre et al. 2014).

$$C_2H_5OH + O_2 \rightarrow CH_3COOH + H_2O \quad (4.4)$$

Oxidation of ethanol to acetic acid (Eq. 4.4) was performed in both gas phase (Kaichev et al. 2016; Li and Iglesia 2007) and liquid phase (Christensen et al. 2006; Snider et al. 1991). Palladium-, platinum-, and gold-based catalysts have received attention as suitable catalysts for this reaction (Bianchini et al. 2009; Christensen et al. 2006; Tarnowski and Korzeniewski 1997). In the liquid phase, $Au/MgAl_2O_4$ was selective for ethanol oxidation to acetic acid with a selectivity of 86% at 180 °C (Christensen et al. 2006). Gold-based catalysts with other supports including TiO_2, SiO_2, ZnO, Al_2O_3, and NiO were also optimal catalysts giving high conversion and high acetic acid selectivity (Jørgensen et al. 2007; Sun et al. 2008; Takei et al. 2011). Li and Iglesia (2007) reported that $Mo_{0.61}V_{0.31}Nb_{0.08}O_x/TiO_2$ showed high acetic acid selectivity of 95% with complete ethanol conversion at 237 °C.

The selectivity to acetic acid decreased at high temperature (>200 °C) in the gas phase due to the formation of other products (Kaichev et al. 2016). However, at low temperature (100–150 °C), the main compound was acetaldehyde, which shifted the reaction toward acetic acid at higher temperatures (Kaichev et al. 2016). In monolayer V_2O_5/TiO_2 catalysts, the oxidation of ethanol reaction involved reversible reduction of V^{5+} cation, whereas titanium cations remained in the Ti^{4+} state (Kaichev et al. 2016). In this case, acetic acid selectivity reached 60% at 200 °C, while it achieved 68% selectivity with 60% ethanol conversion at 150 °C using Ce-meso TiO_2 (Eguchi et al. 2008). According to some authors, ethanol is oxidized into acetaldehyde intermediate and transformed into acetic acid in the second step in both gas and liquid phase (Takei et al. 2011). Instead, other authors conclude that the selectivity of acetic acid increases in the presence of water because the oxidation to acetic acid is favored than acetaldehyde production (Li and Iglesia 2007).

4.4 Steam Reforming of Ethanol and Dehydrogenation

4.4.1 Steam Reforming of Ethanol to Hydrogen

Hydrogen has wide applications in energy and chemical industry including manufacture of chemicals (e.g., ammonia and methanol production, upgrading heavy oils), as a coolant, an energy carrier, and in the semiconductor sectors. Hydrogen is considered as a clean energy and mostly used for fuel cell systems to generate electricity. Hydrogen is a highly efficient energy carrier (143 MJ/kg) with respect to that of liquid hydrocarbon-based fuels (Häussinger et al. 2000). Hydrogen is currently produced by several processes including hydrocarbon steam reforming, water electrolysis, and photocatalytic methods (Häussinger et al. 2000; Marone et al. 2014; Ni et al. 2007b; Taboada et al. 2014). Among them, steam reforming of organic compounds is the most common process to produce hydrogen (Faria et al. 2014; Nahar and Dupont 2014). However, using natural gas or oil-derived hydrocarbons, it is not a clean process and causes greenhouse gas emissions. Therefore, hydrogen production through steam reforming of ethanol, arising from renewable biomass, is a clean process to reduce the environmental problems (Nanda et al. 2017b).

Ethanol steam reforming may produce up to six hydrogen molecules per ethanol molecule converted (Eq. 4.5). However, the amount of hydrogen produced is lowered if CO is produced instead of CO_2 (Eq. 4.6) and other side reactions producing methane as a byproduct. Other reactions include dehydration of ethanol to ethylene (Eq. 4.1), hydrogenation of ethanol (Eq. 4.7), dehydrogenation (Eq. 4.8), acetaldehyde decomposition (Eq. 4.9), steam reforming of acetaldehyde (Eq. 4.10), and water-gas shift reaction (Eq. 4.11).

$$C_2H_5OH + 3H_2O \rightarrow 2CO_2 + 6H_2 \tag{4.5}$$

$$C_2H_5OH + H_2O \rightarrow 2CO + 4H_2 \tag{4.6}$$

$$C_2H_5OH + 2H_2 \rightarrow 2CH_4 + H_2O \tag{4.7}$$

$$C_2H_5OH \rightarrow CH_3CHO + H_2 \tag{4.8}$$

$$CH_3CHO \rightarrow CH_4 + CO \tag{4.9}$$

$$CH_3CHO + H_2O \rightarrow 2CO + 3H_2 \tag{4.10}$$

$$CO + H_2O \rightarrow CO_2 + H_2 \tag{4.11}$$

Steam reforming of ethanol may be realized on both noble metal catalysts and non-noble catalysts (Table 4.1), which has an important role in obtaining maximum hydrogen yield and complete ethanol conversion. However, the highest hydrogen yield is also controlled by the reaction temperature, feed composition, and residence time (Hou et al. 2015). Being an endothermic reaction, high reaction temperature, high stream-to-ethanol ratio, and long residence time are required to enhance

Table 4.1 Summary of catalysts used for ethanol steam reforming

Catalyst	Metal loading (wt%)	Temperature (°C)	H_2O/C_2H_5OH (molar ratio)	Ethanol conversion (%)	Hydrogen selectivity (%)	References
Rh/CeO_2	2	300	8/1	58.5	102.1	Diagne et al. (2004)
		400		100	66.3	
		450		100	69.1	
Rh/ZrO_2		300		100	59.6	
		450		100	71.7	
Ru/CeO_2	1	450	–	57	35 (20 min)	Erdohelyi et al. (2006)
				25	23 (100 min)	
Rh/CeO_2	1			80	84 (20 min)	
				56	72 (80 min)	
Ru/Al_2O_3	6	630	1/1	100	67	Chiu et al. (2013)
Rh/Al_2O_3	6			100	72	
Pd/Al_2O_3	6			100	60	
Pt/Al_2O_3	6			100	58	
Rh/MgO	3	650	2.1/1	99 (10 h)	91	Frusteri et al. (2004)
Pd/MgO	3			10 (10 h)	70	
Ni/MgO	21			42 (10 h)	97	
Co/MgO	21			55 (10 h)	92	
Ru/Al_2O_3	1	800	3/1	42	55	Liguras et al. (2003)
	5			100	96	
Rh/Al_2O_3	1			100	95	
	2			100	96	
Pd/Al_2O_3	1			55	50	
Pt/Al_2O_3	1			60	65	
Co/ZnO	10	350	4/1	100 (75 h)	73.4	Llorca et al. (2003)
Co/ZnO (0.06 wt% Na)	10	400	13/1	100	72.1	Llorca et al. (2004)
Co/ZnO (0.23 wt% Na)	10			100	73.4	

(continued)

Table 4.1 (continued)

Catalyst	Metal loading (wt%)	Temperature (°C)	H_2O/C_2H_5OH (molar ratio)	Ethanol conversion (%)	Hydrogen selectivity (%)	References
Co/ZnO (0.78 wt% Na)	10			100	74.2	
Co/Al_2O_3	8	400	3/1	74	60–70	Batista et al. (2004)
	18			99	63–70	
Co/SiO_2	8			89	62–70	
	18			97	69–72	
Co/CeO_2-nano	10	420	21/1	100	93	Machocki et al. (2010)
Co/CeO_2-micro		420		100	76	
Co/ZrO_2-nano		420		100	85	
Co/ZrO_2-micro		420		100	77	
Co/ZnO	20	600	13/1	95	71	Da Costa-Serra et al. (2010)
Ni/ZnO	20	600	12/1	100	67	
Ni/Y_2O_3	20.6	250	3/1	81.9	43.1	Sun et al. (2005)
Ni/Al_2O_3	16.1			76	44	
Ni/La_2O_3	15.3			80.7	49.5	
Ni/La_2O_3	10	350	12/1	38	55	Liu et al. (2010)
	24	350		84	64	
	45	350		100	70	
Ni/$Ce_{0.6}Zr_{0.4}O_2$	10	400	10/1	100	83	Furtado et al. (2009)
Cu/Al_2O_3	1	400	10/1	100	71	
Ni/γ-Al_2O_3	35	500	6/1	100	91	Comas et al. (2004)
Ni/Al_2O_3 (pretreatment at 550 °C)	3.8	450	3/1	96.6	61.5	
		650		100	89.0	
Ni/Al_2O_3 (pretreatment at 700 °C)		450		100	0	
		550		99.2	67.3	
		650		100	87.4	

(continued)

Table 4.1 (continued)

Catalyst	Metal loading (wt%)	Temperature (°C)	H_2O/C_2H_5OH (molar ratio)	Ethanol conversion (%)	Hydrogen selectivity (%)	References
Ni/CeO_2-ZrO_2	10	600	3/1	100	75	Ebiad et al. (2012)
	24	600	3/1	100	70	
Ni/ $ZnAl_2O_4$-CeO_2	7	650	8/1	95	70	Galetti et al. (2011)
Ni/γ-Al_2O_3	10	650	8/1	100	78.2	Akande et al. (2005)
Ni/MgO				100	82.2	
Ni/La_2O_3				100	89.3	
Ni/ZnO				100	89.1	
Ni/La_2O_3	20	500	3/1	35	70	Fatsikostas and Verykios (2004)
		800		~100	95	
Ni/Al_2O_3		700		77	87	
		800		100	96	

hydrogen yields (Piscina and Homs 2008). Besides, high reaction temperature also prevents methane formation (Piscina and Homs 2008), while an increase of space-time favors hydrogen selectivity causing a decrease of intermediates (da Silva et al. 2010).

It is well-known that noble metal catalysts are very active for catalytic ethanol steam reforming. A number of studies reported that Ru, Rh, Pd, and Pt are excellent catalysts for ethanol steam reforming (Auprêtre et al. 2002; Chiu et al. 2013; Diagne et al. 2004; Erdohelyi et al. 2006; Frusteri et al. 2004; Liguras et al. 2003). Rh and Ru are competitive catalysts depending on the catalyst loading (Chiu et al. 2013; Diagne et al. 2004; Frusteri et al. 2004). Rhodium has very high activity at low catalyst loading (0–5 wt%), while Ru gave high catalytic activity but with greater catalyst loading (Chiu et al. 2013; Diagne et al. 2004; Frusteri et al. 2004). The supports and dispersion of metal catalysts on support also play important roles for this reaction (Hou et al. 2015).

Lewis acid γ-Al_2O_3 is remarkably used for steam reforming of ethanol as a support (Auprêtre et al. 2002; Chiu et al. 2013; Liguras et al. 2003), while basic oxides such as La_2O_3, MgO, ZnO, and CeO_2 are also frequently employed as the supports (Diagne et al. 2004; Erdohelyi et al. 2006; Frusteri et al. 2004; Liu et al. 2010; Llorca et al. 2004). The support γ-Al_2O_3 is reported to enhance the ethanol conversion with respect to that of basic supports (Hou et al. 2015). However, acid supports favor ethanol dehydration into ethylene, which can be transformed into coke at high temperatures (Hou et al. 2015; Phung et al. 2015a, b). In fact, 1.0 wt% Rh/γ-Al_2O_3 showed complete ethanol conversion with approximately 95% of hydrogen selectivity at 800 °C with steam:ethanol molar ratio of 3:1 (Liguras et al. 2003).

With high loading of Rh (5 wt%), ethanol conversion was complete at the beginning of the experiment at 650 °C but decreased at the longer times on stream without oxygen addition (43% conversion at 100 h) due to coke formation on the surface of catalysts (Cavallaro et al. 2003). The coke formation also reduces the selectivity of hydrogen (Cavallaro et al. 2003). In contrast, basic supports such as MgO and Mg/Al-based spinel oxide reduce the coke formation rate due to the lower surface acidity, thereby enhancing the stability of the catalysts (Frusteri et al. 2004). However, the deactivation was found mainly due to the sintering of metal (Frusteri et al. 2004). Interestingly, basic MgO support also showed complete ethanol conversion and high stability of catalyst over 3 wt% Rh at 650 °C (Frusteri et al. 2004). In the same support, higher catalyst loading (21 wt%) of Ni and Co gave high hydrogen selectivity (Table 4.1), but they lowered ethanol conversion after 10 h of reaction time (42% conversion for Ni and 55% conversion for Co) (Frusteri et al. 2004).

As mentioned earlier, the catalyst activity increases with an increase in the dispersion of metals on the support. In fact, well dispersion of Ni-Rh bimetallic catalyst on smaller crystals of CeO_2 enhanced the ethanol conversion (Kugai et al. 2006), while the formation of Pd-Zn alloy from the co-deposits of Pd and Zn on ZnO support favored side reactions instead of sufficient steam supply reaction (Casanovas et al. 2006), resulting in a decrease in hydrogen yield. Besides, the reaction mechanism of ethanol steam reforming is less studied. The surface chemistry requires intensive studies to better understand the interactions between the chemical species and the catalyst surface (Hou et al. 2015; Ni et al. 2007a).

Steam reforming normally employs noble metals as the catalysts, but it also utilizes low-cost non-noble metal catalysts (Table 4.1). Nickel is one of the most metal catalysts used with different supports such as Al_2O_3, La_2O_3, and CeO_2-ZrO_2 for steam reforming reaction (Akande et al. 2005; Comas et al. 2004; Ebiad et al. 2012; Fatsikostas and Verykios 2004; Furtado et al. 2009; Galetti et al. 2011; Garbarino et al. 2015; Liu et al. 2010; Sun et al. 2005). Cobalt and copper are also employed for this reaction. Co/CeO_2-nano (X_{C2H5OH}, 100%; S_{H2}, 93%) is comparable with Ni/Al_2O_3 (X_{C2H5OH}, 100%; S_{H2}, 95%) (Machocki et al. 2010; Fatsikostas and Verykios 2004). However, Co achieved high catalytic activity and hydrogen selectivity at lower temperature (420 °C) than that of Ni (800 °C).

Acid support such as Al_2O_3 is still an excellent support for ethanol steam reforming (Comas et al. 2004; Fatsikostas and Verykios 2004). Moreover, La_2O_3, ZnO, and CeO_2-nano are also good support candidates (Akande et al. 2005; Fatsikostas and Verykios 2004; Machocki et al. 2010). In the same support and metal catalysts, the reaction temperature enhanced both ethanol conversion and hydrogen selectivity (Comas et al. 2004; Fatsikostas and Verykios 2004). In fact, higher temperature improved ethanol decomposition by both catalyst and thermal conversions. Additionally, it prevented the formation of methane, resulting in an increase in hydrogen production (Ni et al. 2007a). Interestingly, cobalt metal nanoparticles are also active as a catalyst for steam reforming even when nonsupported (Riani et al. 2016).

4.4.2 Dehydrogenation of Ethanol

4.4.2.1 Acetaldehyde

Acetaldehyde (CH_3CHO) is the second simplest of aldehyde group and a colorless liquid. It is miscible with water and common solvents, including ethanol, ether, benzene, toluene, xylene, turpentine, and acetone. It is a flammable material (flash point of −39 °C) and has a low boiling point (20.2 °C). Acetaldehyde is the main hazard listed in a group of potential occupational carcinogens (Fleischmann et al. 2000). Acetaldehyde can be used in nonalcoholic drinks and natural fruits juices (Fleischmann et al. 2000). Acetaldehyde has a wide application in food industries and plastic industry as well as in the manufacturing of acetic acid and paint binders in alkyd paints (Caro et al. 2012; Fleischmann et al. 2000; Liu and Hensen 2013; Tang et al. 2016; Xu et al. 2016). Additionally, acetaldehyde is also used as a component or in the production of materials in civil, pharmaceutical, and cosmetic industries (Caro et al. 2012; Fleischmann et al. 2000; Liu and Hensen 2013; Tang et al. 2016; Xu et al. 2016). In 2016, the market of acetaldehyde was US $1.26 billion, which is expected to reach up to US $1.80 billion by 2022 and US $1.6 million tons by 2024.

Acetaldehyde can be produced from ethanol through dehydrogenation or partial oxidation (Autthanit et al. 2018; Campisano et al. 2018; Čičmanec et al. 2018; Clarizia et al. 2019; Garbarino et al. 2019; Guan and Hensen 2013; Santacesaria et al. 2003; Shan et al. 2017; Tayrabekova et al. 2018; Tu and Chen 2001; Weinstein et al. 2011). A number of studies have focused on the catalytic reaction (Autthanit et al. 2018; Campisano et al. 2018; Čičmanec et al. 2018; Clarizia et al. 2019; Garbarino et al. 2019; Guan and Hensen 2013; Santacesaria et al. 2003; Shan et al. 2017; Tayrabekova et al. 2018; Tu and Chen 2001; Weinstein et al. 2011) and kinetic and surface study (Cassinelli et al. 2015; Gazsi et al. 2011; Guan and Hensen 2009; Shan et al. 2018). Ethanol dehydrogenation has been studied on mainly copper-based catalysts (Table 4.2). Ethanol conversion increases with an increase of reaction temperature, but acetaldehyde selectivity can decrease meanwhile (Table 4.2). In other words, acetaldehyde selectivity can drop at higher ethanol conversion. It can reach 100% selectivity at ethanol conversion lower than 15% at 250 °C over Ni-Cu alloys, while it is constant at near 90% selectivity with near-complete ethanol conversion on the 19.3 wt% $Cu/ZnAl_2O_4$ at 400 °C (Table 4.2). Additionally, Cassinelli et al. (2015) reported that the addition of Cu^+ enhanced the ethanol dehydrogenation than that of using only Cu^0.

Platinum- and gold-based catalysts are also good candidates for ethanol dehydrogenation (Table 4.2). They showed the same characteristics with Cu-based catalysts in the acetaldehyde selectivity with respect to ethanol conversion and reaction temperature. Gold is the best catalyst with high acetaldehyde selectivity (>90%) at low temperature, i.e., 180 °C (Table 4.2), and at high temperature, i.e., 400 °C

Table 4.2 Summary of ethanol dehydrogenation to acetaldehyde

Catalyst	Support	Temperature (°C)	Ethanol conversion (%)	Acetaldehyde selectivity (%)	References
NiCu alloys	–	250–350	<15	100	Shan et al. (2017)
VO_x	ZrO_2	200	60	90	Čičmanec et al. (2018)
V-Zr-La	SBA-15-HT	400	80	48	Autthanit et al. (2018)
5Cu	ZrO_2	200	27.9	21.0[a]	Freitas et al. (2014)
10Cu	ZrO_2		33.3	16.6[a]	
20Cu	ZrO_2		31.9	16.8[a]	
30Cu	ZrO_2		30.9	20.1[a]	
5Cu	ZrO_2	250	64.3	16.8[a]	
10Cu	ZrO_2		68.4	12.5[a]	
20Cu	ZrO_2		69.3	12.6[a]	
30Cu	ZrO_2		68.8	14.5[a]	
Cu	$ZnAl_2O_4$	400	100	90	Garbarino et al. (2019)
CuO-Cr_2O_3	Al_2O_3	300	55	50	Tayrabekova et al. (2018)
Au(26) Ir(74)	SiO_2	180	56	92[b]	Guan and Hensen (2013)
Pt	CeO_2(rod)	200	4.3	83	Ciftci et al. (2013)
Pt	CeO_2(cube)		2.0	80	
Au	CeO_2(rod)		1.3	100	
Au	CeO_2(cube)		0.2	100	
Pt	CeO_2(rod)	300	30	34	
Pt	CeO_2(cube)		9.0	32	
Au	CeO_2(rod)		4.8	97	
Au	CeO_2(cube)		3.3	100	
Pt	CeO_2(rod)	400	99	0.6	
Pt	CeO_2(cube)		14	27	
Au	CeO_2(rod)		31	66	
Au	CeO_2(cube)		16	93	
Au (5.8)	SBA-16(3.1)	400	90	95	Guan and Hensen (2009)

[a]Mainly ethyl acetate (~70–80%)
[b]Trace amounts of acetic acid and ethyl acetate

(Table 4.2). Au (5.8)/SBA-16 (3.1) showed the best catalytic activity with 95% acetaldehyde selectivity at 90% ethanol conversion. Besides, in the case of zirconia support, the addition of gold reduced ethanol dehydration reaction by passivation of the acid sites of the support and improved dehydrogenation reaction by introducing new sites for this reaction (Wang et al. 2016).

4.4.2.2 Ethyl Acetate

Ethyl acetate ($CH_3COOC_2H_5$) is the ester of ethanol and acetic acid and a colorless liquid at room temperature with a pleasant "fruity" smell. It is a flammable liquid (flash point: – 4 °C) with boiling point and melting point of 77.1 °C and – 83.6 °C, respectively. Ethyl acetate is miscible with common organic solvents (e.g., ethanol, acetone, diethyl ether, and benzene), but poorly soluble in water (83 g/L at 20 °C) (Riemenschneider and Bolt 2005). Ethyl acetate is commonly used as a solvent and has many applications in food, cosmetic, coating, electronics, and printing industries (Cheng et al. 2016; Inui et al. 2002b; Petek et al. 2018; Shen et al. 2017; Singha and Ray 2016; Zonetti et al. 2011). As a solvent, ethyl acetate is used for cleaning, paint removal, and coatings depending on its purity. Its high purity can be applied in cleaning electric circuit boards as well as a nail polish remover, while its low purity can be applied in pharmaceuticals, perfumes, printing, decaffeination of tea or coffee, food, and a carrier solvent for herbicides (Riemenschneider and Bolt 2005).

Traditionally, ethyl acetate is produced from ethanol and acetic acid through esterification reaction (Eq. 4.12). Moreover, it can be produced through the reaction of ethylene and acetic acid (Eq. 4.13); Tishchenko reaction, i.e., the dimerization of acetaldehyde (Eq. 4.14); or direct dehydrogenation of ethanol (Eq. 4.15). The process illustrated in Eq. (4.15) opens a new route to produce ethyl acetate from biomass-derived ethanol. In parallel with Eq. (4.15), the side reaction of dehydrogenation of ethanol to acetaldehyde can also occur.

$$C_2H_5OH + CH_3COOH \rightarrow CH_3COOC_2H_5 + H_2O \tag{4.12}$$

$$C_2H_4 + CH_3COOH \rightarrow CH_3COOC_2H_5 \tag{4.13}$$

$$2CH_3CHO \rightarrow CH_3COOC_2H_5 \tag{4.14}$$

$$2C_2H_5OH \rightarrow CH_3COOC_2H_5 + 2H_2 \tag{4.15}$$

Copper-based materials are the main catalysts used in ethanol dehydrogenation into ethyl acetate (Carotenuto et al. 2013; Colley et al. 2005; Gaspar et al. 2010; Inui et al. 2002a, b, 2004; Santacesaria et al. 2012; Tu et al. 1994a, b; Zonetti et al. 2011). Two families of copper-based catalysts used are copper/copper chromite (Carotenuto et al. 2013; Colley et al. 2005; Santacesaria et al. 2012; Tu et al. 1994a, b) and copper metal supported and/or promoted by different oxides (Gaspar et al. 2010; Inui et al. 2002a, b, 2004; Santacesaria et al. 2012; Zonetti et al. 2011). Copper/copper chromite showed higher ethyl acetate selectivity (98–99%) at 65% of ethanol conversion at 220–240 °C (Santacesaria et al. 2012). Moreover, using K_2CO_3 in the treatment of Cu-Zn-Zr-Al-O catalyst also enhanced ethanol conversion (66%) and ethyl acetate selectivity (85%) at 220 °C (Inui et al. 2004). This reaction is still less studied to produce "green" ethyl acetate from renewable ethanol. Other catalysts such as PdO/SiO_2 (Gaspar et al. 2009), Pd/ZnO, and Pd/SnO_2 (Sánchez et al. 2005) are also employed for this process.

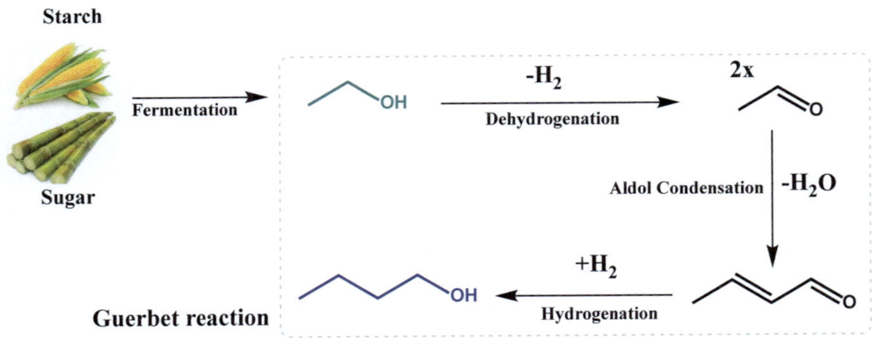

Fig. 4.4 Biobutanol production from biomass-derived bioethanol via Guerbet reaction

4.5 Guerbet Reaction of Ethanol

n-Butanol (C_4H_9OH) is a primary and colorless alcohol and a refractive organic liquid. It has low solubility in water (73 g/L at 25 °C), but it has high solubility in acetone, ethyl ether, and ethanol. It has a high boiling point of 117.7 °C and low melting point of −89.8 °C (Hahn et al. 2010). *n*-Butanol can be produced from ethanol by Guerbet reaction (Fig. 4.4) (Aitchison et al. 2016; Chakraborty et al. 2015; Dowson et al. 2013; Earley et al. 2015; Ho et al. 2016; Koda et al. 2009; Kozlowski and Davis 2013; Tsuchida et al. 2008). The reaction includes ethanol dehydrogenation to acetaldehyde followed by aldol condensation of acetaldehyde to create crotonaldehyde. Butanol is produced via the crotonaldehyde hydrogenation route (Earley et al. 2015).

n-Butanol has many applications in foods and beverages as well as using it as a fuel (Carvalho et al. 2012; Hahn et al. 2010). As a fuel, butanol has many advantages in comparison to ethanol. For example, during combustion, butanol produces higher energy than that of ethanol and it also absorbs lower water and is miscible with gasoline better than ethanol (Carvalho et al. 2012; Dziugan et al. 2015; Earley et al. 2015). Moreover, using *n*-butanol can take advantages of the available infrastructure for the use of gasoline, e.g., pipeline for transportation and direct use in the gasoline engine without modification (Nanda et al. 2017a).

Ethanol can be converted into butanol using both homogeneous and heterogeneous catalysts. There are many homogeneous catalysts including Ir- and Ru-based catalysts (Aitchison et al. 2016; Chakraborty et al. 2015; Dowson et al. 2013; Tseng et al. 2016; Wingad et al. 2016; Xie et al. 2016). Ruthenium pincer complexes catalyst showed 68.3% of butanol selectivity at 66.9% ethanol conversion (Xie et al. 2016). However, the use of a homogeneous catalyst poses a catalyst recycle challenge. Solid-base metal oxides (Ndou 2003), acid-base bifunctional metal oxides (Carvalho et al. 2012, 2013; Ramasamy et al. 2016), hydroxyapatite (Ogo et al. 2011, 2012; Tsuchida et al. 2006, 2008), basic zeolites (Yang and Meng 1993), and transition metal compounds (Gines and Iglesia 1998) have been used for this process.

For Guerbet reaction, effective catalysts are required to integrate multifunctional active sites for hydrogenation, dehydrogenation, and aldol condensation. Alcohol coupling systems commonly need basic materials such as alkali metal, hydroxide, and solid base (Kozlowski and Davis 2013). Among them, alkali metals, which are considered as highly active catalysts of alcohol coupling systems and their metal doping, e.g., Cu or Ni, are capable of catalyzing hydrogenation and dehydrogenation (Kozlowski and Davis 2013).

Hydrogenation step needs hydrogen from dehydrogenation step adsorbed on the surface of catalysts and, likely, the hydrogen adsorbed from dehydration of ethanol over acid sites. In addition, acid sites need to stabilize the reactive ethoxide and enolate intermediates formed (Ho et al. 2016). Thus, the relationship between the acidity and basicity of the catalyst is a key feature of this reaction. Currently, K-$CuMg_5CeO_x$ and Ca/P contain both acid and base properties and show high activity and butanol selectivity (>50%) in Guerbet reaction (Gines and Iglesia 1998; Kozlowski and Davis 2013; Tsuchida et al. 2006, 2008).

4.6 Conclusions

This chapter provided an overview of all the current catalytic bioethanol conversion processes into chemicals and fuels. Some of the processes have already been developed and commercialized such as ethylene and propylene productions from bioethanol. Other potential processes are under research and development to optimize economical and scientific catalysis. These processes provide a route in the production of raw materials from bioethanol, a green chemical from biomass, in order to reduce the dependency on fossil fuels. Additionally, the use of green bioethanol-derived products in the petrochemical industries can mitigate the environmental pollution and improve the quality of life.

References

Aguayo AT, Gayubo AG, Atutxa A, Olazar M, Bilbao J (2002a) Catalyst deactivation by coke in the transformation of aqueous ethanol into hydrocarbons. Kinetic modeling and acidity deterioration of the catalyst. Ind Eng Chem Res 41:4216–4224

Aguayo AT, Gayubo AG, Tarrío AM, Atutxa A, Bilbao J (2002b) Study of operating variables in the transformation of aqueous ethanol into hydrocarbons on an HZSM-5 zeolite. J Chem Technol Biotechnol 77:211–216

Aitchison H, Wingad RL, Wass DF (2016) Homogeneous ethanol to butanol catalysis-guerbet renewed. ACS Catal 6(10):7125–7132

Akande AJ, Idem RO, Dalai AK (2005) Synthesis, characterization and performance evaluation of Ni/Al_2O_3 catalysts for reforming of crude ethanol for hydrogen production. Appl Catal A 287:159–175

Angelici C, Velthoen MEZ, Weckhuysen BM, Bruijnincx PCA (2014) Effect of preparation method and CuO promotion in the conversion of ethanol into 1,3-butadiene over SiO_2–MgO catalysts. ChemSusChem 7:2505–2515

Arai H, Take J-I, Saito Y, Yoneda Y (1967) Ethanol dehydration on alumina catalysts: I. The thermal desorption of surface compounds. J Catal 9:146–153

Arias D, Colmenares A, Cubeiro ML et al (1997) The transformation of ethanol over $AlPO_4$ and SAPO molecular sieves with AEL and AFI topology. Kinetic and thermodynamic approach. Catal Lett 45:51–58

Auprêtre F, Descorme C, Duprez D (2002) Bio-ethanol catalytic steam reforming over supported metal catalysts. Catal Commun 3(6):263–267

Autthanit C, Praserthdam P, Jongsomjit B (2018) Oxidative and non-oxidative dehydrogenation of ethanol to acetaldehyde over different VO_x/SBA-15 catalysts. J Environ Chem Eng 6:6516–6529

Bac S, Avci AK (2018) Ethylene oxide synthesis in a wall-coated microchannel reactor with integrated cooling. Chem Eng J. https://doi.org/10.1016/j.cej.2018.10.041

Bakoyannakis DN, Zamboulis D, Stalidis GA, Deliyanni EA (2001) The effect of preparation method on the catalytic activity of amorphous aluminas in ethanol dehydration. J Chem Technol Biotechnol 76:1159–1164

Batista MS, Santos RKS, Assaf EM, Assaf JM, Ticianelli EA (2004) High efficiency steam reforming of ethanol by cobalt-based catalysts. J Power Sources 134:27–32

Bhadra BN, Song JY, Khan NA, Jun JW, Kim TW, Kim CU, Jhung SH (2018) Conversion of ethylene into propylene with the siliceous SSZ-13 zeolite prepared without an organic structure-directing agent. J Catal 365:94–104

Bi J, Liu M, Song C, Wang X, Guo X (2011) C_2–C_4 light olefins from bioethanol catalyzed by Ce-modified nanocrystalline HZSM-5 zeolite catalysts. Appl Catal B Environ 107:68–76

Bianchini C, Bambagioni V, Filippi J, Marchionnia A, Vizza F, Bert P, Tampucci A (2009) Selective oxidation of ethanol to acetic acid in highly efficient polymer electrolyte membrane-direct ethanol fuel cells. Electrochem Commun 11:1077–1080

Birol G, Önsan Zİ, Kırdar B, Oliver SG (1998) Ethanol production and fermentation characteristics of recombinant *Saccharomyces cerevisiae* strains grown on starch. Enzyme Micro Technol 22:672–677

Bokade VV, Yadav GD (2011) Heteropolyacid supported on montmorillonite catalyst for dehydration of dilute bio-ethanol. Appl Clay Sci 53:263–271

Brandão P, Philippou A, Rocha J, Anderson MW (2002) Dehydration of alcohols by microporous niobium silicate AM-11. Catal Lett 80:99–102

Brown W, Iverson B, Anslyn E, Foote C (2014) Organic chemistry. Cengage Learning, Belmont, CA, USA

Bulawayo B, Bvochora JM, Muzondo MI, Zvauya R (1996) Ethanol production by fermentation of sweet-stem sorghum juice using various yeast strains. World J Microbiol Biotechnol 12:357–360

Busca G, Costantino U, Montanari T, Ramis G, Resini C, Sisani M (2010) Nickel versus cobalt catalysts for hydrogen production by ethanol steam reforming: Ni–Co–Zn–Al catalysts from hydrotalcite-like precursors. Int J Hydrog Energy 35:5356–5366

Calsavara V, Baesso ML, Fernandes-Machado NRC (2008) Transformation of ethanol into hydrocarbons on ZSM-5 zeolites modified with iron in different ways. Fuel 87:1628–1636

Campisano ISP, Rodella CB, Sousa ZSB, Henriques CA, Teixeira da Silva V (2018) Influence of thermal treatment conditions on the characteristics of Cu-based metal oxides derived from hydrotalcite-like compounds and their performance in bio-ethanol dehydrogenation to acetaldehyde. Catal Today 306:111–120

Caro C, Thirunavukkarasu K, Anilkumar M, Shiju NR, Rothenberg G (2012) Selective autooxidation of ethanol over titania-supported molybdenum oxide catalysts: Structure and reactivity. Adv Synth Catal 354:1327–1336

Carotenuto G, Tesser R, Di Serio M, Santacesaria E (2013) Kinetic study of ethanol dehydrogenation to ethyl acetate promoted by a copper/copper-chromite based catalyst. Catal Today 203:202–210

Carrasco-Marín F, Mueden A, Moreno-Castilla C (1998) Surface-treated activated carbons as catalysts for the dehydration and dehydrogenation reactions of ethanol. J Phys Chem B 102:9239–9244

Carvalho DL, de Avillez RR, Rodrigues MT, Borges LEP, Appel LG (2012) Mg and Al mixed oxides and the synthesis of *n*-butanol from ethanol. Appl Catal A 415–416:96–100

Carvalho DL, Borges LEP, Appel LG, Ramírez de la Piscina P, Homs N (2013) *In situ* infrared spectroscopic study of the reaction pathway of the direct synthesis of *n*-butanol from ethanol over MgAl mixed-oxide catalysts. Catal Today 213:115–121

Casanovas A, Llorca J, Homs N, Fierro JLG, Ramírez de la Piscina P (2006) Ethanol reforming processes over ZnO-supported palladium catalysts: effect of alloy formation. J Mol Catal A 250:44–49

Cassinelli WH, Martins L, Passos AR et al (2015) Correlation between structural and catalytic properties of copper supported on porous alumina for the ethanol dehydrogenation reaction. ChemCatChem 7:1668–1677

Cavallaro S, Chiodo V, Freni S, Mondello N, Frusteri F (2003) Performance of Rh/Al_2O_3 catalyst in the steam reforming of ethanol: H_2 production for MCFC. Appl Catal A 249:119–128

Chaichana E, Boonsinvarothai N, Chitpong N, Jongsomjit B (2019) Catalytic dehydration of ethanol to ethylene and diethyl ether over alumina catalysts containing different phases with boron modification. J Porous Mater 26:599–610

Chakraborty S, Piszel PE, Hayes CE, Baker RT, Jones WD (2015) Highly selective formation of *n*-butanol from ethanol through the Guerbet process: a tandem catalytic approach. J Am Chem Soc 137:14264–14267

Chen JQ, Bozzano A, Glover B, Fuglerud T, Kvisle S (2005) Recent advancements in ethylene and propylene production using the UOP/Hydro MTO process. Catal Today 106:103–107

Chen H, Jia X, Li Y, Liu C, Yang Y (2015) Controlled surface properties of Au/ZSM5 catalysts and their effects in the selective oxidation of ethanol. Catal Today 256:153–160

Cheng C, Cong Y, Du C, wang J, Yao G, Zhao H (2016) Solubility determination and thermodynamic models for dehydroepiandrosterone acetate in mixed solvents of (ethyl acetate+methanol), (ethyl acetate+ethanol) and (ethyl acetate+isopropanol). J Chem Thermodyn 101:372–379

Chiang H, Bhan A (2010) Catalytic consequences of hydroxyl group location on the rate and mechanism of parallel dehydration reactions of ethanol over acidic zeolites. J Catal 271:251–261

Chiu W-C, Horng R-F, Chou H-M (2013) Hydrogen production from an ethanol reformer with energy saving approaches over various catalysts. Int J Hydrog Energy 38:2760–2769

Christensen CH, Jørgensen B, Rass-Hansen J, Egeblad K, Madsen R, Klitgaard SK, Hansen SM, Hansen MR, Andersen HC, Riisager A (2006) Formation of acetic acid by aqueous-phase oxidation of ethanol with air in the presence of a heterogeneous gold catalyst. Angew Chem 118:4764–4767

Christiansen MA, Mpourmpakis G, Vlachos DG (2013) Density functional theory-computed mechanisms of ethylene and diethyl ether formation from ethanol on γ-Al_2O_3 (100). ACS Catal 3:1965–1975

Čičmanec P, Ganjkhanlou Y, Kotera J, Hidalgo JM, Tišler Z, Bulánek R (2018) The effect of vanadium content and speciation on the activity of VO_x/ZrO_2 catalysts in the conversion of ethanol to acetaldehyde. Appl Catal A 564:208–217

Ciftci A, Varisli D, Tokay KC, Sezgi NA, Dogu T (2012) Dimethyl ether, diethyl ether & ethylene from alcohols over tungstophosphoric acid based mesoporous catalysts. Chem Eng J 207:85–93

Ciftci A, Ligthart DAJM, Pastorino P, Hensen EJM (2013) Nanostructured ceria supported Pt and Au catalysts for the reactions of ethanol and formic acid. Appl Catal B 130–131:325–335

Clarizia L, Apuzzo J, Di Somma I, Marotta R, Andreozzi R (2019) Selective photo-oxidation of ethanol to acetaldehyde and acetic acid in water in presence of TiO_2 and cupric ions under UV–simulated solar radiation. Chem Eng J 361:1524–1534

Colley SW, Tabatabaei J, Waugh KC, Wood MA (2005) The detailed kinetics and mechanism of ethyl ethanoate synthesis over a Cu/Cr_2O_3 catalyst. J Catal 236:21–33

Comas J, Mariño F, Laborde M, Amadeo N (2004) Bio-ethanol steam reforming on Ni/Al_2O_3 catalyst. Chem Eng J 98:61–68

Coupard V, Plennevaux T (2014) Method for producing ethylene oxide from a thermally integrated ethanol stream. France Patent.
Da Costa-Serra JF, Guil-López R, Chica A (2010) Co/ZnO and Ni/ZnO catalysts for hydrogen production by bioethanol steam reforming. Influence of ZnO support morphology on the catalytic properties of Co and Ni active phases. Int J Hydrog Energy 35:6709–6716
da Silva AM, da Costa LOO, Souza KR, Mattos LV, Noronha FB (2010) The effect of space time on Co/CeO_2 catalyst deactivation during oxidative steam reforming of ethanol. Catal Commun 11:736–740
De las Pozas C, Lopez-Cordero R, Gonzalez-Morales J, Travieso N, Roque-Malherbe R (1993) Effect of pore diameter and acid strength in ethanol dehydration on molecular sieves. J Mol Catal 83:145–156
de Oliveira TKR, Rosset M, Perez-Lopez OW (2018) Ethanol dehydration to diethyl ether over Cu-Fe/ZSM-5 catalysts. Catal Commun 104:32–36
DeWilde JF, Chiang H, Hickman DA, Ho CR, Bhan A (2013) Kinetics and mechanism of ethanol dehydration on γ-Al2O3: the critical role of dimer inhibition. ACS Catal 3:798–807
Diagne C, Idriss H, Pearson K, Gómez-García MA, Kiennemann A (2004) Efficient hydrogen production by ethanol reforming over Rh catalysts. Effect of addition of Zr on CeO_2 for the oxidation of CO to CO_2. Comptes Rendus Chimie 7:617–622
Doğu T, Varişli D (2007) Alcohols as alternatives to petroleum for environmentally clean fuels and petrochemicals. Turk J Chem 31:551–567
Doheim M, El-Shobaky H (2002) Catalytic conversion of ethanol and iso-propanol over ZnO-treated Co_3O_4/Al_2O_3 solids. Colloids Surf A Physicochem Eng Asp 204:169–174
Dowson GR, Haddow MF, Lee J, Wingad RL, Wass DF (2013) Catalytic conversion of ethanol into an advanced biofuel: unprecedented selectivity for *n*-butanol. Angew Chem Int Ed Eng 52:9005–9008
Dziugan P, Jastrzabek KG, Binczarski M, Karski S, Witonska IA, Kolesinska B, Kaminski ZJ (2015) Continuous catalytic coupling of raw bioethanol into butanol and higher homologues. Fuel 158:81–90
Earley JH, Bourne RA, Watson MJ, Poliakoff M (2015) Continuous catalytic upgrading of ethanol to *n*-butanol and $>C_4$ products over Cu/CeO_2 catalysts in supercritical CO_2. Green Chem 17:3018–3025
Ebiad MA, El-Hafiz DRA, Elsalamony RA, Mohamed LS (2012) Ni supported high surface area CeO_2–ZrO_2 catalysts for hydrogen production from ethanol steam reforming. RSC Adv 2:8145–8156
Eguchi Y, Abe D, Yoshitake H (2008) Oxidation state of Ce and ethanol–oxygen reaction of mesoporous titania-supported cerium oxide. Microporous Mesoporous Mater 116:44–50
Eisele P, Killpack R (2011) Propene Ullmann's encyclopedia of industrial chemistry. Wiley, Weinheim
Erdohelyi A, Raskó J, Kecskés T, Tóth M, Dömök M, Baán K (2006) Hydrogen formation in ethanol reforming on supported noble metal catalysts. Catal Today 116:367–376
Erwin J, Moulton S (1996) Maintenance and Operation of the U.S. DOE Alternative Fuel Center. Southwest Research Institute, San Antonio, USA
Fan D, Dai D-J, Wu H-S (2013) Ethylene formation by catalytic dehydration of ethanol with industrial considerations. Mater 6:101–115
Faria EC, Neto RCR, Colman RC, Noronha FB (2014) Hydrogen production through CO_2 reforming of methane over Ni/$CeZrO_2$/Al_2O_3 catalysts. Catal Today 228:138–144
Fatsikostas AN, Verykios XE (2004) Reaction network of steam reforming of ethanol over Ni-based catalysts. J Catal 225:439–452
Fatsikostas AN, Kondarides DI, Verykios XE (2001) Steam reforming of biomass-derived ethanol for the production of hydrogen for fuel cell applications. Chem Commun:851–852
Fleischmann G, Jira R, Bolt HM, Golka K (2000) Acetaldehyde Ullmann's encyclopedia of industrial chemistry. Wiley, Weinheim

Freitas IC, Damyanova S, Oliveira DC, Marques CMP, Bueno JMC (2014) Effect of Cu content on the surface and catalytic properties of Cu/ZrO_2 catalyst for ethanol dehydrogenation. J Mol Catal A Chem 381:26–37

Frusteri F, Freni S, Spadaro L, Chiodo V, Bonura G, Donato S, Cavallaro S (2004) H_2 production for MC fuel cell by steam reforming of ethanol over MgO supported Pd, Rh, Ni and Co catalysts. Catal Commun 5:611–615

Furtado AC, Alonso CG, Cantão MP, Fernandes-Machado NRC (2009) Bimetallic catalysts performance during ethanol steam reforming: Influence of support materials. Int J Hydrog Energy 34:7189–7196

Furumoto Y, Harada Y, Tsunoji N, Takahashi A, Fujitani T, Ide Y, Sadakane M, Sano T (2011) Effect of acidity of ZSM-5 zeolite on conversion of ethanol to propylene. Appl Catal A 399:262–267

Galetti AE, Gomez MF, Arrua LA, Abello MC (2011) Ethanol steam reforming over $Ni/ZnAl_2O_4$-CeO_2. Influence of calcination atmosphere and nature of catalytic precursor. Appl Catal A 408:78–86

Garbarino G, Riani P, Lucchini MA, Canepa F, Kawale S, Busca G (2013) Cobalt-based nanoparticles as catalysts for low temperature hydrogen production by ethanol steam reforming. Int J Hydrog Energy 38:82–91

Garbarino G, Wang C, Valsamakis I, Chitsazan S, Riani P, Finocchio E, Flytzani-Stephanopoulos M, Busca G (2015) A study of Ni/Al_2O_3 and $Ni–La/Al_2O_3$ catalysts for the steam reforming of ethanol and phenol. Appl Catal B 174–175:21–34

Garbarino G, Prasath Parameswari Vijayakumar R, Riani P, Finocchio E, Busca G (2018) Ethanol and diethyl ether catalytic conversion over commercial alumina and lanthanum-doped alumina: Reaction paths, catalyst structure and coking. Appl Catal B 236:490–500

Garbarino G, Riani P, Villa García M, Finocchio E, Sanchez Escribano V, Busca G (2019) A study of ethanol dehydrogenation to acetaldehyde over copper/zinc aluminate catalysts. Catal Today. https://doi.org/10.1016/j.cattod.2019.01.002

Gaspar AB, Esteves AML, Mendes FMT, Barbosa FG, Appel LG (2009) Chemicals from ethanol—The ethyl acetate one-pot synthesis. Appl Catal A 363:109–114

Gaspar AB, Barbosa FG, Letichevsky S, Appel LG (2010) The one-pot ethyl acetate syntheses: The role of the support in the oxidative and the dehydrogenative routes. Appl Catal A 380:113–117

Gayubo AG, Alonso A, Valle B, Aguayo AT, Bilbao J (2010a) Kinetic model for the transformation of bioethanol into olefins over a HZSM-5 zeolite treated with alkali. Ind Eng Chem Res 49:10836–10844

Gayubo AG, Alonso A, Valle B, Aguayo AT, Bilbao J (2010b) Selective production of olefins from bioethanol on HZSM-5 zeolite catalysts treated with NaOH. Appl Catal B Environ 97:299–306

Gayubo AG, Alonso A, Valle B, Aguayo AT, Olazar M, Bilbao J (2011) Kinetic modelling for the transformation of bioethanol into olefins on a hydrothermally stable Ni–HZSM-5 catalyst considering the deactivation by coke. Chem Eng J 167(1):262–277

Gazsi A, Koós A, Bánsági T, Solymosi F (2011) Adsorption and decomposition of ethanol on supported Au catalysts. Catal Today 160:70–78

Ghashghaee M, Shirvani S (2019) Catalytic transformation of ethylene to propylene and butene over an acidic Ca-incorporated composite nanocatalyst. Appl Catal A 569:20–27

Gines MJL, Iglesia E (1998) Bifunctional condensation reactions of alcohols on basic oxides modified by copper and potassium. J Catal 176:155–172

Golay S, Doepper R, Renken A (1999) Reactor performance enhancement under periodic operation for the ethanol dehydration over gamma-alumina, a reaction with a stop-effect. Chem Eng Sci 54:4469–4474

Gotro J (2013) Bio-Based Polypropylene; Multiple Synthetic Routes Under Investigation. https://polymerinnovationblog.com/bio-based-polypropylene-multiple-synthetic-routes-under-investigation/. Accessed 15 Apr 2019

Guan Y, Hensen EJM (2009) Ethanol dehydrogenation by gold catalysts: The effect of the gold particle size and the presence of oxygen. Appl Catal A 361:49–56

Guan Y, Hensen EJM (2013) Selective oxidation of ethanol to acetaldehyde by Au–Ir catalysts. J Catal 305:135–145

Hahn H, Dämbkes G, Rupprich N, Bahl H (2010) Butanols, Ullmann's encyclopedia of industrial chemistry. Wiley, Weinheim

Häussinger P, Lohmüller R, Watson AM (2000) Hydrogen. Ullmann's Encyclopedia of Industrial Chemistry. Wiley, Weinheim

Hayashi F, Tanaka M, Lin D, Iwamoto M (2014) Surface structure of yttrium-modified ceria catalysts and reaction pathways from ethanol to propene. J Catal 316:112–120

Held H, Rengstl A, Mayer D (2000) Acetic anhydride and mixed fatty acid anhydrides. Ullmann's Encyclopedia of Industrial Chemistry. Wiley, Weinheim

Ho CR, Shylesh S, Bell AT (2016) Mechanism and kinetics of ethanol coupling to butanol over hydroxyapatite. ACS Catal 6:939–948

Hou T, Zhang S, Chen Y, Wang D, Cai W (2015) Hydrogen production from ethanol reforming: catalysts and reaction mechanism. Renew Sust Energ Rev 44:132–148

Hu YC (1993) In: McKetta JJ (ed) Chemical processing handbook. Dekker, New York, USA

Huang C-F, Lin T-H, Guo G-L, Hwang W-S (2009) Enhanced ethanol production by fermentation of rice straw hydrolysate without detoxification using a newly adapted strain of *Pichia stipitis*. Bioresour Technol 100:3914–3920

Huber GW, Iborra S, Corma A (2006) Synthesis of transportation fuels from biomass: chemistry, catalysts, and engineering. Chem Rev 106:4044–4098

Inaba M, Murata K, Saito M, Takahara I (2006) Ethanol conversion to aromatic hydrocarbons over several zeolite catalysts. React Kinet Catal Lett 88:135–141

Inaba M, Murata K, Saito M, Takahara I (2007) Production of olefins from ethanol by Fe-supported zeolite catalysts. Green Chem 9:638–646

Inaba M, Murata K, Takahara I, Inoue KI (2012) Production of olefins and propylene from ethanol by Zr-modified H-ZSM-5 zeolite catalysts. Adv Mater Sci Eng. https://doi.org/10.1155/2012/293485

Inui K, Kurabayashi T, Sato S (2002a) Direct synthesis of ethyl acetate from ethanol carried out under pressure. J Catal 212:207–215

Inui K, Kurabayashi T, Sato S (2002b) Direct synthesis of ethyl acetate from ethanol over Cu-Zn-Zr-Al-O catalyst. Appl Catal A 237:53–61

Inui K, Kurabayashi T, Sato S, Ichikawa N (2004) Effective formation of ethyl acetate from ethanol over Cu-Zn-Zr-Al-O catalyst. J Mol Catal A 216:147–156

Iwamoto M (2015) Selective catalytic conversion of bio-ethanol to propene: A review of catalysts and reaction pathways. Catal Today 242:243–248

Jasper S, El-Halwagi M (2015) A techno-economic comparison between two methanol-to-propylene processes. PRO 3:684–698

Jørgensen B, Christiansen SE, Thomsen MLD, Christensen CH (2007) Aerobic oxidation of aqueous ethanol using heterogeneous gold catalysts: Efficient routes to acetic acid and ethyl acetate. J Catal 251:332–337

Kagyrmanova A, Chumachenko V, Korotkikh V, Kashkin V, Noskov A (2011) Catalytic dehydration of bioethanol to ethylene: pilot-scale studies and process simulation. Chem Eng J 176:188–194

Kaichev VV, Chesalov YA, Saraev AA et al (2016) Redox mechanism for selective oxidation of ethanol over monolayer V_2O_5/TiO_2 catalysts. J Catal 338:82–93

Khan NA, Yoo DK, Bhadra BN, Kim TW, Kim CU, Jhung SH (2019) Preparation of SSZ-13 zeolites from beta zeolite and their application in the conversion of ethylene to propylene. Chem Eng J 377:119546

Kim HJ, Kim J-W, Kim N, Kim T-W, Jhung SH, Kim C-U (2017) Controlling size and acidity of SAPO-34 catalyst for efficient ethylene to propylene transformation. Mol Catal 438:86–92

Kito-Borsa T, Pacas DA, Selim S, Cowley SW (1998) Properties of an ethanol–diethyl ether–water fuel mixture for cold-start assistance of an ethanol-fueled vehicle. Ind Eng Chem Res 37:3366–3374

Knözinger H (1968) Dehydration of alcohols on aluminum oxide. Angew Chem Int Ed 7:791–805

Koda K, Matsu-ura T, Obora Y, Ishii Y (2009) Guerbet reaction of ethanol to *n*-butanol catalyzed by iridium complexes. Chem Lett 38:838–839

Koempel H, Liebner W (2007) Lurgi's methanol to propylene (MTP®): report on a successful commercialisation. In: Bellot Noronha F, Schmal M, Falabella Sousa-Aguiar E (eds) Studies in surface science and catalysis, vol 167. Elsevier, Amsterdam, pp 261–267

Kondo JN, Ito K, Yoda E, Wakabayashi F, Domen K (2005) An ethoxy intermediate in ethanol dehydration on Brønsted acid sites in zeolite. J Phys Chem B 109:10969–10972

Kosaric N, Duvnjak Z, Farkas A et al (2011) Ethanol Ullmann's encyclopedia of industrial chemistry. Wiley, Weinheim, pp 4–5

Kozlowski JT, Davis RJ (2013) Heterogeneous catalysts for the guerbet coupling of alcohols. ACS Catal 3:1588–1600

Kugai J, Subramani V, Song C, Engelhard MH, Chin Y-H (2006) Effects of nanocrystalline CeO_2 supports on the properties and performance of Ni–Rh bimetallic catalyst for oxidative steam reforming of ethanol. J Catal 238:430–440

Le Berre C, Serp P, Kalck P, Torrence GP (2014) Acetic acid. Ullmann's encyclopedia of industrial chemistry. Wiley, Weinheim

Le Van Mao R, Nguyen TM, McLaughlin GP (1989) The bioethanol-to-ethylene (BETE) process. Appl Catal 48:265–277

Lee KY, Arai T, Nakata S, Asaoka S, Okuhara T, Misono M (1992) Catalysis by heteropoly compounds. 20. An NMR study of ethanol dehydration in the pseudoliquid phase of 12-tungstophosphoric acid. J Am Chem Soc 114:2836–2842

Lee C-C, Gorte R, Farneth W (1997) Calorimetric study of alcohol and nitrile adsorption complexes in H-ZSM-5. J Phys Chem B 101:3811–3817

Lewandowski S (2016) Ethylene - Global, IHS Markit. https://cdn.ihs.com/www/pdf/Steve-Lewandowski-Big-Changes-Ahead-for-Ethylene-Implications-for-Asia.pdf. Accessed 26 Dec 2019

Li X, Iglesia E (2007) Selective catalytic oxidation of ethanol to acetic acid on dispersed Mo-V-Nb mixed oxides. Chem Eur J 13:9324–9330

Liebner W (2005) Handbook of petrochemicals production processes. McGraw Hill, New York, NY

Liguras DK, Kondarides DI, Verykios XE (2003) Production of hydrogen for fuel cells by steam reforming of ethanol over supported noble metal catalysts. Appl Catal B 43:345–354

Lippits MJ, Nieuwenhuys BE (2010a) Direct conversion of ethanol into ethylene oxide on copper and silver nanoparticles: effect of addition of CeO_x and Li_2O. Catal Today 154:127–132

Lippits MJ, Nieuwenhuys BE (2010b) Direct conversion of ethanol into ethylene oxide on gold-based catalysts: effect of CeO_x and Li_2O addition on the selectivity. J Catal 274:142–149

Liu P, Hensen EJM (2013) Highly efficient and robust $Au/MgCuCr_2O_4$ catalyst for gas-phase oxidation of ethanol to acetaldehyde. J Am Chem Soc 135:14032–14035

Liu J-Y, Lee C-C, Wang C-H, Yeh C-T, Wang C-B (2010) Application of nickel–lanthanum composite oxide on the steam reforming of ethanol to produce hydrogen. Int J Hydrog Energy 35:4069–4075

Liu J, Zhang F, Xu W, Shi N, Mu S (2017) Thermal reactivity of ethylene oxide in contact with contaminants: a review. Thermochim Acta 652:85–96

Llorca J, de la Piscina PR, Dalmon J-A, Sales J, Homs NS (2003) CO-free hydrogen from steam-reforming of bioethanol over ZnO-supported cobalt catalysts: effect of the metallic precursor. Appl Catal B 43:355–369

Llorca J, Homs NS, Sales J, Fierro J-LG, Ramírez de la Piscina P (2004) Effect of sodium addition on the performance of Co–ZnO-based catalysts for hydrogen production from bioethanol. J Catal 222:470–480

Machocki A, Denis A, Grzegorczyk W, Gac W (2010) Nano- and micro-powder of zirconia and ceria-supported cobalt catalysts for the steam reforming of bio-ethanol. Appl Surf Sci 256:5551–5558

Makshina EV, Janssens W, Sels BF, Jacobs PA (2012) Catalytic study of the conversion of ethanol into 1,3-butadiene. Catal Today 198:338–344

Marone A, Izzo G, Mentuccia L, Massini G, Paganin P, Rosa S, Varrone C, Signorini A (2014) Vegetable waste as substrate and source of suitable microflora for bio-hydrogen production. Renew Energy 68:6–13

Mendes GCC, Brandão TRS, Silva CLM (2007) Ethylene oxide sterilization of medical devices: A review. Am J Infect Control 35:574–581

Murata K, Inaba M, Takahara I (2008) Effects of surface modification of H-ZSM-5 catalysts on direct transformation of ethanol into lower olefins. J Jpn Petrol Inst 51:234–239

Nahar G, Dupont V (2014) Hydrogen production from simple alkanes and oxygenated hydrocarbons over ceria–zirconia supported catalysts: Review. Renew Sust Energ Rev 32:777–796

Nanda S, Dalai AK, Kozinski JA (2014a) Butanol and ethanol production from lignocellulosic feedstock: biomass pretreatment and bioconversion. Energ Sci Eng 2:138–148

Nanda S, Mohammad J, Reddy SN, Kozinski JA, Dalai AK (2014b) Pathways of lignocellulosic biomass conversion to renewable fuels. Biomass Conv Bioref 4:157–191

Nanda S, Golemi-Kotra D, McDermott JC, Dalai AK, Gökalp I, Kozinski JA (2017a) Fermentative production of butanol: perspectives on synthetic biology. New Biotechnol 37:210–221

Nanda S, Rana R, Zheng Y, Kozinski JA, Dalai AK (2017b) Insights on pathways for hydrogen generation from ethanol. Sustain Energ Fuel 1:1232–1245

Ndou A (2003) Dimerisation of ethanol to butanol over solid-base catalysts. Appl Catal A 251:337–345

Nguyen CM, Reyniers M-F, Marin GB (2010) Theoretical study of the adsorption of C_1–C_4 primary alcohols in H-ZSM-5. Phys Chem Chem Phys 12:9481–9493

Ni M, Leung DYC, Leung MKH (2007a) A review on reforming bio-ethanol for hydrogen production. Int J Hydrog Energy 32:3238–3247

Ni M, Leung MKH, Leung DYC, Sumathy K (2007b) A review and recent developments in photocatalytic water-splitting using TiO_2 for hydrogen production. Renew Sust Energ Rev 11:401–425

Ogo S, Onda A, Yanagisawa K (2011) Selective synthesis of 1-butanol from ethanol over strontium phosphate hydroxyapatite catalysts. Appl Catal A 402:188–195

Ogo S, Onda A, Iwasa Y, Hara K, Fukuoka A, Yanagisawa K (2012) 1-Butanol synthesis from ethanol over strontium phosphate hydroxyapatite catalysts with various Sr/P ratios. J Catal 296:24–30

Oikawa H, Shibata Y, Inazu K, Iwase Y, Murai K, Hyodo S, Kobayashi G, Baba T (2006) Highly selective conversion of ethene to propene over SAPO-34 as a solid acid catalyst. Appl Catal A 312:181–185

Olsson L, Hahn-Hägerdal B (1996) Fermentation of lignocellulosic hydrolysates for ethanol production. Enzym Microb Technol 18:312–331

Oudejans J, Van Den Oosterkamp P, Van Bekkum H (1982) Conversion of ethanol over zeolite H-ZSM-5 in the presence of water. Appl Catal 3:109–115

Ouyang J, Kong F, Su G, Hu Y, Song Q (2009) Catalytic conversion of bio-ethanol to ethylene over La-modified HZSM-5 catalysts in a bioreactor. Catal Lett 132:64–74

Patzek TW (2004) Thermodynamics of the corn-ethanol biofuel cycle. Crit Rev Plant Sci 23:519–567

Petek A, Krajnc M, Petek A (2018) Saponification of ethyl acetate in the presence of β-cyclodextrin. J Mol Liq 272:313–318

Phillips CB, Datta R (1997) Production of ethylene from hydrous ethanol on H-ZSM-5 under mild conditions. Ind Eng Chem Res 36:4466–4475

Phung TK, Busca G (2015a) Diethyl ether cracking and ethanol dehydration: acid catalysis and reaction paths. Chem Eng J 272:92–101

Phung TK, Busca G (2015b) Ethanol dehydration on silica-aluminas: active sites and ethylene/diethyl ether selectivities. Catal Commun 68:110–115

Phung TK, Herrera C, Larrubia MÁ et al (2014a) Surface and catalytic properties of some γ-Al_2O_3 powders. App Catal A 483:41–51

Phung TK, Lagazzo A, Rivero Crespo MÁ, Sánchez Escribano V, Busca G (2014b) A study of commercial transition aluminas and of their catalytic activity in the dehydration of ethanol. J Catal 311:102–113

Phung TK, Proietti Hernández L, Busca G (2015a) Conversion of ethanol over transition metal oxide catalysts: effect of tungsta addition on catalytic behaviour of titania and zirconia. Appl Catal A 489:180–187

Phung TK, Proietti Hernández L, Lagazzo A, Busca G (2015b) Dehydration of ethanol over zeolites, silica alumina and alumina: Lewis acidity, Brønsted acidity and confinement effects. Appl Catal A 493:77–89

Phung TK, Radikapratama R, Garbarino G, Lagazzo A, Riani P, Busca G (2015c) Tuning of product selectivity in the conversion of ethanol to hydrocarbons over H-ZSM-5 based zeolite catalysts. Fuel Process Technol 137:290–297

Piscina PR, Homs N (2008) Use of biofuels to produce hydrogen (reformation processes). Chem Soc Rev 37:2459–2467

Posada JA, Patel AD, Roes A, Blok K, Faaij AP, Patel MK (2013) Potential of bioethanol as a chemical building block for biorefineries: preliminary sustainability assessment of 12 bioethanol-based products. Bioresour Technol 135:490–499

Pujadó PR, Andersen JM (2006) Handbook of petroleum refinery processes. McGraw Hill, New York, NY

Ramasamy KK, Gray M, Job H, Smith C, Wang Y (2016) Tunable catalytic properties of bi-functional mixed oxides in ethanol conversion to high value compounds. Catal Today 269:82–87

Ramesh K, Hui LM, Han Y-F, Borgna A (2009) Structure and reactivity of phosphorous modified H-ZSM-5 catalysts for ethanol dehydration. Catal Commun 10:567–571

Rebsdat S, Mayer D (2011) Ethylene oxide. Ullmann's encyclopedia of industrial chemistry. Wiley, New York, NY

Riani P, Garbarino G, Infantes-Molina A, Rodríguez-Castellón E, Canepa F, Busca G (2016) Hydrogen from steam reforming of ethanol over cobalt nanoparticles: effect of boron impurities. Appl Catal A 518:67–77

Riemenschneider W, Bolt HM (2005) Esters. Organic Ullmann's encyclopedia of industrial chemistry. Wiley, New York, NY

Roca FF, De Mourgues L, Trambouze Y (1969) Catalytic dehydration of ethanol over silica-alumina. J Catal 14:107–113

Roscher G (2000) Vinyl esters. Ullmann's encyclopedia of industrial chemistry. Wiley, New York, NY

Sakuth M, Mensing T, Schuler J, Heitmann W, Strehlke G, Mayer D (2010) Ethers, Aliphatic Ullmann's encyclopedia of industrial chemistry. Wiley, Weinheim, pp 437–438

Salmi T, Roche M, Hernández Carucci J, Eränen K, Murzin D (2012) Ethylene oxide—kinetics and mechanism. Curr Opin Chem Eng 1:321–327

Sánchez AB, Homs N, Fierro JLG, Piscina PR (2005) New supported Pd catalysts for the direct transformation of ethanol to ethyl acetate under medium pressure conditions. Catal Today 107–108:431–435

Santacesaria E, Sorrentino A, Tesser R, Di Serio M, Ruggiero A (2003) Oxidative dehydrogenation of ethanol to acetaldehyde on V_2O_5/TiO_2-SiO_2 catalysts obtained by grafting vanadium and titanium alkoxides on silica. J Mol Catal A Chem 204–205:617–627.

Santacesaria E, Carotenuto G, Tesser R, Di Serio M (2012) Ethanol dehydrogenation to ethyl acetate by using copper and copper chromite catalysts. Chem Eng J 179:209–220

Sarker M, Khan NA, Yoo DK, Bhadra BN, Jun JW, Kim TW, Kim CU, Jhung SH (2019) Synthesis of SSZ-13 zeolite in the presence of dimethylethylcyclohexyl ammonium ion and direct conversion of ethylene to propylene with the SSZ-13. Chem Eng J 377:120116

Shan J, Janvelyan N, Li H, Liu J, Egle TM, Ye J, Biener MM, Biener J, Friend CM, Flytzani-Stephanopoulos M (2017) Selective non-oxidative dehydrogenation of ethanol to acetaldehyde and hydrogen on highly dilute NiCu alloys. Appl Catal B 205:541–550

Shan J, Liu J, Li M, Lustig S, Lee S, Flytzani-Stephanopoulos M (2018) NiCu single atom alloys catalyze the CH bond activation in the selective non-oxidative ethanol dehydrogenation reaction. Appl Catal B 226:534–543

Shen J, Li F, Yin B, Sun L, Chen C, Wen S, Chen Y, Ruan S (2017) Enhanced ethyl acetate sensing performance of Al-doped In_2O_3 microcubes. Sensors Actuators B Chem 253:461–469

Sifniades S, Levy AB, Bahl H (2011) Acetone. Ullmann's encyclopedia of industrial chemistry. Wiley, New York, NY

Singha R, Ray JK (2016) Selective acetylation of primary alcohols by ethyl acetate. Tetrahedron Lett 57:5395–5398

Snider BB, Merritt JE, Dombroski MA, Buckman BO (1991) Solvent effects on manganese(III)-based oxidative free-radical cyclizations: ethanol and acetic acid. J Organomet Chem 56:5544–5553

Song Z, Takahashi A, Mimura N, Fujitani T (2009) Production of propylene from ethanol over ZSM-5 zeolites. Catal Lett 131:364–369

Song Z, Takahashi A, Nakamura I, Fujitani T (2010) Phosphorus-modified ZSM-5 for conversion of ethanol to propylene. Appl Catal A 384:201–205

Statista (2019) Global ethanol production for fuel use from 2000 to 2017 (in million cubic meters). https://www.statista.com/statistics/274142/global-ethanol-production-since-2000/. Accessed 26 Feb 2019

Sun J, Wang Y (2014) Recent advances in catalytic conversion of ethanol to chemicals. ACS Catal 4:1078–1090

Sun J, Qiu X-P, Wu F, Zhu W-T (2005) H_2 from steam reforming of ethanol at low temperature over Ni/Y_2O_3, Ni/La_2O_3 and Ni/Al_2O_3 catalysts for fuel-cell application. Int J Hydrog Energy 30:437–445

Sun K-Q, Luo S-W, Xu N, Xu B-Q (2008) Gold nano-size effect in Au/SiO_2 for selective ethanol oxidation in aqueous solution. Catal Lett 124:238–242

Taboada E, Angurell I, Llorca J (2014) Dynamic photocatalytic hydrogen production from ethanol–water mixtures in an optical fiber honeycomb reactor loaded with Au/TiO_2. J Catal 309:460–467

Takahara I, Saito M, Inaba M, Murata K (2005) Dehydration of ethanol into ethylene over solid acid catalysts. Catal Lett 105:249–252

Takahara I, Saito M, Matsuhashi H, Inaba M, Murata K (2007) Increase in the number of acid sites of a H-ZSM5 zeolite during the dehydration of ethanol. Catal Lett 113:82–85

Takahashi A, Xia W, Nakamura I, Shimada H, Fujitani T (2012) Effects of added phosphorus on conversion of ethanol to propylene over ZSM-5 catalysts. Appl Catal A 423:162–167

Takei T, Iguchi N, Haruta M (2011) Synthesis of acetoaldehyde, acetic acid, and others by the dehydrogenation and oxidation of ethanol. Catal Surv Jpn 15:80–88

Tang C, Zhai Z, Li X, Sun L, Bai W (2016) Sustainable production of acetaldehyde from lactic acid over the magnesium aluminate spinel. J Taiwan Inst Chem Eng 58:97–106

Tarnowski DJ, Korzeniewski C (1997) Effects of surface step density on the electrochemical oxidation of ethanol to acetic acid. J Phys Chem B 101:253–258

Tayrabekova S, Mäki-Arvela P, Peurla M, Paturi P, Eränen K, Ergazieva GE, Aho A, Murzin DY, Dossumov K (2018) Catalytic dehydrogenation of ethanol into acetaldehyde and isobutanol using mono- and multicomponent copper catalysts. Comptes Rendus Chimie 21:194–209

Thinnes B (2010) 'On-purpose' propylene production. Hydrocarb Process 89:19

Tseng KN, Lin S, Kampf JW, Szymczak NK (2016) Upgrading ethanol to 1-butanol with a homogeneous air-stable ruthenium catalyst. Chem Commun 52:2901–2904

Tsuchida T, Sakuma S, Takeguchi T, Ueda W (2006) Direct synthesis of *n*-butanol from ethanol over nonstoichiometric hydroxyapatite. Ind Eng Chem Res 45:8634–8642

Tsuchida T, Kubo J, Yoshioka T, Sakuma S, Takeguchi T, Ueda W (2008) Reaction of ethanol over hydroxyapatite affected by Ca/P ratio of catalyst. J Catal 259:183–189

Tu Y-J, Chen Y-W (2001) Effects of alkali metal oxide additives on Cu/SiO_2 catalyst in the dehydrogenation of ethanol. Ind Eng Chem Res 40:5889–5893

Tu Y-J, Chen Y-W, Li C (1994a) Characterization of unsupported copper—chromium catalysts for ethanol dehydrogenation. J Mol Catal 89:179–189

Tu Y-J, Li C, Chen Y-W (1994b) Effect of chromium promoter on copper catalysts in ethanol dehydrogenation. J Chem Technol Biotechnol 59:141–147

Tullo A (2010) Braskem to make propylene from ethanol Braskem to make propylene from ethanol. Chemical & Engineering News. http://cenblog.org/the-chemical-notebook/2010/10/brakem-to-make-propylene-from-ethanol/. Accessed 15 Apr 2019

Vaidya PD, Rodrigues AE (2006) Insight into steam reforming of ethanol to produce hydrogen for fuel cells. Chem Eng J 117:39–49

Varisli D, Dogu T, Dogu G (2007) Ethylene and diethyl-ether production by dehydration reaction of ethanol over different heteropolyacid catalysts. Chem Eng Sci 62:5349–5352

Wang W, Jiao J, Jiang Y, Ray SS, Hunger M (2005) Formation and decomposition of surface ethoxy species on acidic zeolite Y. ChemPhysChem 6:1467–1469

Wang C, Garbarino G, Allard LF, Wilson F, Busca G, Flytzani-Stephanopoulos M (2016) Low-temperature dehydrogenation of ethanol on atomically dispersed gold supported on $ZnZrO_x$. ACS Catal 6:210–218

Wattanakarunwong P (2015) IRPC KM: Propylene Market. http://irpc.listedcompany.com/misc/PRESN/20150515-propylene-market-analyst.pdf. Accessed 7 Oct 2018

Weinstein RD, Ferens AR, Orange RJ, Lemaire P (2011) Oxidative dehydrogenation of ethanol to acetaldehyde and ethyl acetate by graphite nanofibers. Carbon 49:701–707

Wingad RL, Bergstrom EJ, Everett M, Pellow KJ, Wass DF (2016) Catalytic conversion of methanol/ethanol to isobutanol—a highly selective route to an advanced biofuel. Chem Commun 52:5202–5204

Xie Y, Ben-David Y, Shimon LJ, Milstein D (2016) Highly efficient process for production of biofuel from ethanol catalyzed by ruthenium pincer complexes. J Am Chem Soc 138:9077–9080

Xu J, Xu X-C, Yang X-J, Han Y-F (2016) Silver/hydroxyapatite foam as a highly selective catalyst for acetaldehyde production via ethanol oxidation. Catal Today 276:19–27

Yang C, Meng ZY (1993) Bimolecular condensation of ethanol to 1-butanol catalyzed by alkali cation zeolites. J Catal 142:37–44

Yasunaga K, Gillespie F, Simmie J, Curran HJ, Kuraguchi Y, Hoshikawa H, Yamane M, Hidaka Y (2010) A multiple shock tube and chemical kinetic modeling study of diethyl ether pyrolysis and oxidation. J Phys Chem A 114:9098–9109

Zaki T (2005) Catalytic dehydration of ethanol using transition metal oxide catalysts. J Colloid Interface Sci 284:606–613

Zecchina A, Bordiga S, Spoto G, Scarano D, Spanò G, Geobaldo F (1996) IR spectroscopy of neutral and ionic hydrogen-bonded complexes formed upon interaction of CH_3OH, C_2H_5OH, $(CH_3)_2O$, $(C_2H_5)_2O$ and C_4H_8O with HY, H-ZSM-5 and H-mordenite: Comparison with analogous adducts formed on the H-Nafion superacidic membrane. J Chem Soc Faraday Trans 92:4863–4875

Zhang M, Yu Y (2013) Dehydration of ethanol to ethylene. Ind Eng Chem Res 52:9505–9514

Zhang X, Wang R, Yang X, Zhang F (2008) Comparison of four catalysts in the catalytic dehydration of ethanol to ethylene. Microporous Mesoporous Mater 116:210–215

Zimmermann H, Walzl R (2009) Ethylene. Ullmann's encyclopedia of industrial chemistry. Wiley, New York, NY

Zonetti PC, Celnik J, Letichevsky S, Gaspar AB, Appel LG (2011) Chemicals from ethanol—The dehydrogenative route of the ethyl acetate one-pot synthesis. J Mol Catal A 334:29–34

A Spotlight on Butanol and Propanol as Next-Generation Synthetic Fuels

5

Sonil Nanda, Rachita Rana, Dai-Viet N. Vo, Prakash K. Sarangi, Trinh Duy Nguyen, Ajay K. Dalai, and Janusz A. Kozinski

Abstract

The exhaustive extraction and prodigious utilization of fossil fuels have led to the large-scale increase in the emissions of greenhouse gases. In addition, the per capita demand of petrochemical resources is escalating due to rapid industrialization and the rising number of vehicles in the transportation sector. There is a growing interest in the development of alternative fuels to reduce the carbon footprint and air pollution caused by the fossil fuels. Biofuels produced from plant residues are carbon neutral and can be produced through biomass-to-liquid and biomass-to-gas conversion technologies. Bioethanol, biopropanol, and biobutanol are some alcohol-based fuels and chemicals that have found multifarious industrial applications. Although bioethanol is blended with gasoline for use as a transportation fuel, it is often criticized over food-versus-fuel debate because of its raw materials being food crops such as corn, sugarcane, and other grains. In addition, butanol and propanol have high potentials over ethanol in replacing gasoline partially or completely due to their advanced fuel properties. This chapter throws light on butanol and propanol as the next-generation synthetic fuels. The aspects discussed in this chapter include their fuel chemistry as

S. Nanda (✉) · R. Rana · A. K. Dalai
Department of Chemical and Biological Engineering, University of Saskatchewan, Saskatoon, Saskatchewan, Canada
e-mail: sonil.nanda@usask.ca

D.-V. N. Vo · T. D. Nguyen
Center of Excellence for Green Energy and Environmental Nanomaterials, Nguyễn Tất Thành University, Hồ Chí Minh City, Vietnam

P. K. Sarangi
Directorate of Research, Central Agricultural University, Imphal, Manipur, India

J. A. Kozinski
Department of Chemical Engineering, University of Waterloo, Waterloo, Ontario, Canada

S. Nanda et al. (eds.), *Biorefinery of Alternative Resources: Targeting Green Fuels and Platform Chemicals*, https://doi.org/10.1007/978-981-15-1804-1_5

well as production technologies from petrochemicals and bio-based feedstocks. The biotechnological developments in the fermentation of lignocellulosic biomass to produce butanol and propanol are provided. The chapter concludes with a note on industrial challenges and future prospects in employing butanol and propanol as commercial biofuels and biochemicals.

Keywords
Biomass · Biofuels · Ethanol · Butanol · Propanol · Fermentation

5.1 Introduction

The worldwide consumption of petroleum and other liquid fossil fuels is expected to escalate from 86 million BPD (barrels per day) in 2008 to 98 million BPD by 2020 and to 112 million BPD by 2035 (USEIA 2011). Figure 5.1 illustrates the global trend for the consumption of fossil fuels in addition to the resulting CO_2 emissions. The steady search for alternative fuels to replace gasoline has brought two biofuels into the limelight, namely ethanol and butanol. The initial focus has been on developing processes to extract the chemical energy stored in waste plant residues and other organic wastes. Lignocellulosic feedstocks (e.g., agricultural biomass and forestry refuse), dedicated energy crops, food waste, waste cooking oil, municipal solid wastes, industrial effluents, cattle manure, and sewage sludge contain renewable carbon in the form of carbohydrates, fats, lipids,

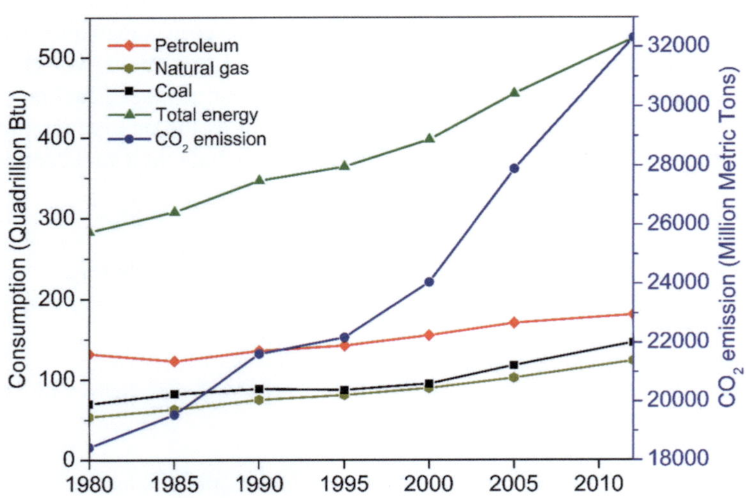

Fig. 5.1 Worldwide fossil fuel consumption and CO_2 emission over the years (Data Source: USEIA 2016)

proteins, lignin, sugars, etc., that can be converted to biofuels through thermochemical and biochemical technologies (Nanda et al. 2013, 2015b, c, 2016b, c, d, 2017b, 2018b, 2019; Gong et al. 2017a, b).

Lignocellulosic biomass can potentially lead to a scenario secured with sustainable energy supply, reduced greenhouse gas emissions, and carbon-neutral footprint (Nanda et al. 2016a, f; Okolie et al. 2019). These materials incorporate agricultural refuse (e.g., straw, bagasse, and husk), energy crops (e.g., temperate grasses and short rotation coppice), and forest residues (e.g., sawdust and infested wood). These wastes are inexpensive resources found abundantly worldwide and have the potential to supply continually alternative biofuels through a suitable biomass-to-liquid and biomass-to-gas conversion technology. Furthermore, lignocellulosic materials are nonedible and pose a negligible threat to agricultural lands and diversion of food crops (e.g., barley, cassava, corn, potato, wheat, etc.) for biorefining (Nanda et al. 2015a). As lignocellulosic biomass does not complete with the food supply chain, they have broad potential in bioethanol and biobutanol refineries. Figure 5.2 shows the recent trend for the generation of widely accepted biofuels such as biodiesel and bioethanol.

The amount of lignocellulosic wastes generated globally every year is in the order of billions of tons. The annual agricultural residues in the USA and Canada are approximately 1147 and 18 million dry tons, respectively (Jones et al. 2007; Gronowska et al. 2009; Mabee and Saddler 2010). The respective estimates for forest residues in the USA and Canada are more than 100 and 46 million dry tons per annum (Gronowska et al. 2009; Mabee and Saddler 2010). The energy crop harvests in the USA and Canada are around 3383 and 433 million dry tons,

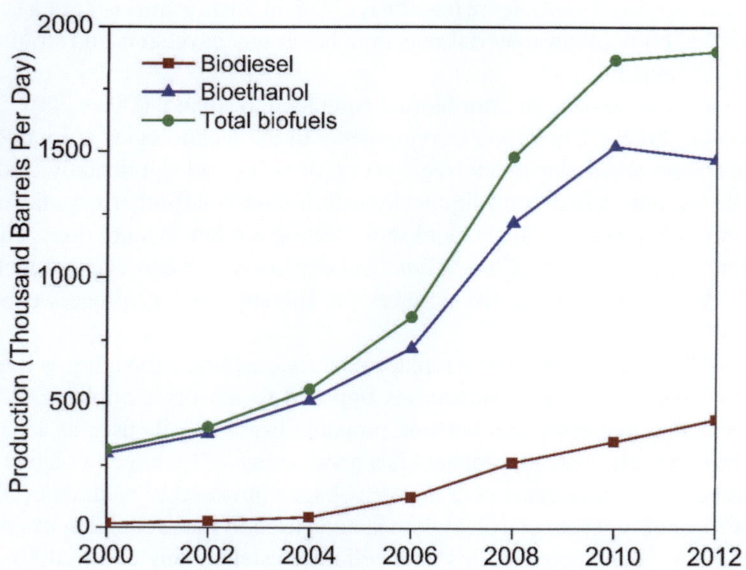

Fig. 5.2 Worldwide biofuel production over the years (Data Source: USEIA 2016)

respectively (Jones et al. 2007; Yemshanov and McKenney 2008). From its annually available residues from agriculture, forests, and energy crops, Canada has a potential to produce around 4.9, 13.8, and 117 billion liters of bioethanol, respectively (Mabee and Saddler 2010). In the UK, almost 300 million tons of wastes are generated every year, which includes nearly 100 million tons of carbonaceous materials and 14 million tons of crop and forest residues (KTN 2016). It is suggested that at least 25 million tons of carbonaceous wastes could be extracted and converted to five million tons of bioethanol with a value in the proximity of £2.4 billion.

Considering the fuel chemistry, aliphatic alcohols such as methanol, ethanol, propanol, and butanol have the potential to be implemented as synthetic fuels in the internal combustion engines (ICE). Currently, the dominant sources for bioethanol are corn and sugarcane. The USA and Brazil together contribute to over 90% of the total bioethanol produced globally (Nanda et al. 2014b). Owing to the fact that corn and sugarcane are food crops, bioethanol is often surmounted by the food-versus-fuel debate (Graham-Rowe 2011). Although bioethanol production is well established at metabolic and commercial scales, its increased dependency on food crops and grains has raised concerns toward compromised food security. Owing to this exploitation of food crops for fuel production, the resulting inflation in the food prices has been directing the global interest to utilizing waste biomass for fuel production.

Lignocellulosic biomasses are being deployed as food crop surrogates and second-generation feedstocks for bioethanol production. However, the development of innovative methods for their efficient bioconversion is still a significant challenge as lignocellulosic biomass requires size reduction, acid pretreatment, enzymatic hydrolysis, delignification, and fermentation (Azargohar et al. 2013, 2018, 2019; Nanda et al. 2018a). Hence, there has always been inquisitiveness to seek a superior fuel with flexibility in raw material selection, better bioconversion, and striking fuel features than ethanol.

Butanol seems to be a superior biofuel compared to ethanol (Dürre 2007; Nanda et al. 2014a, 2017a). Therefore, developments in the technologies at its upstream (bioconversion) and downstream (recovery) processing can significantly lead to its cost-effective production from lignocellulosic biomass and other organic wastes. The chief confinements in the butanol fermentation are low butanol titers, compromised butanol tolerance by *Clostridium*, and expensive solvent recovery technologies. The low butanol productivity makes the solvent recovery process expensive and energy-intensive.

Propanol can also be used as a replacement for gasoline although it is traditionally used as an industrial chemical. As opposed to the other aliphatic alcohols, namely ethanol, methanol, and butanol, propanol is not usually used as a synthetic transportation fuels. Instead, propanol is a good source of hydrogen in direct liquid fuel cell because it can generate a higher voltage compared to methanol (Qi et al. 2002). Propanol in its supercritical fluid state is used in the production of composites of carbon fibers, epoxy resins, and cellulose esters (Jiang et al. 2009). In the automotive sector, 2-propanol is used as "gas dryer" to prevent the freezing of gasoline in the pipelines. Moreover, 2-propanol is used to deice the windshields of vehicles in cold weather conditions. Propanol can also be produced using petrochemical and biological routes like ethanol and butanol. The beneficial fuel and solventogenic properties of butanol and propanol are discussed in this chapter. In addition, the production pathways of these advanced solvents through synthetic and biochemical pathways are also comprehensively discussed.

5.2 Divergent Fuel Properties of Ethanol, Propanol, and Butanol

Table 5.1 makes a comparison of the chemical properties and fuel attributes of ethanol, propanol, and butanol in contrast to gasoline. Butanol is an excellent fuel to replace gasoline. Butanol has a greater energy content (29.2 MJ/L) than that of ethanol (21.2 MJ/L) and propanol (24 MJ/L). In addition, the heating value of butanol is only 10% lower than that of gasoline (32.5 MJ/L). Ethanol has a higher oxygen content of 35% compared to butanol, which contains 22% oxygen (Szulczyk 2010). Moreover, the octane ratings of butanol and propanol are near to those of gasoline indicating their suitability for engine fuel usage. A high octane number suggests better compression of the fuel before its detonation. Motor octane number (MON) and research octane number (RON) are measured by burning the fuel inside an ICE at variable compression ratios as well as with 600 and 900 rpm engine speeds, respectively (Dabelstein et al. 2007). The anti-knock index (AKI) or posted octane

Table 5.1 Comparison of the fuel properties of ethanol, butanol, and propanol with gasoline

Property	Ethanol	Propanol	Butanol	Gasoline
Chemical formula	C_2H_5OH	C_3H_7OH	C_4H_9OH	H, C_4 to C_{12}
Molecular weight (g/mol)	46.07	60.09	74.12	114.23
Density at 20 °C (g/m^3)	0.789	0.803	0.81	0.7
Viscosity at 25 °C (mPa·s)	1.074	1.959	2.573	0.6
Flash point (°C)	17.2	22	35	−43
Autoignition temperature (°C)	365	371	343	280
Boiling point (°C)	78.4	97	117.4	125
Melting point (°C)	−114	−126	−89.8	−56.6[a]
Specific gravity at 15.6 °C	0.79	0.82	0.81	0.72–0.78
Acidity (pK_a)	15.9	16	16.1	–
Reid vapor pressure (psi)	2.3	0.29	0.3	8–15
Calorific value (MJ/L)	21.2	24	29.2	32.5
Lower heating value (MJ/kg)	26.7	30.6	33.1	43.45
Higher heating value (MJ/kg)	29.7	33.6	37.3	46.54
Research octane number (RON)	108.6	118	92	91–99
Motor octane number (MON)	89.7	98	71	81–89
Anti-knock index (AKI)	99.5	108	97	85–96
Air-to-fuel ratio	9	–	11.2	14.7
Solubility in water (%) at 25 °C	100	100	7.3	–

References: Nanda et al. (2017a), Sarangi and Nanda (2018), Biofuel.org.uk (2018)

number (PON) is calculated using the formula (RON + MON)/2. On the other hand, air-to-fuel ratio, a factor to measure antipollution and fuel performance, is the mass ratio of air and the fuel inside an ICE. The air-to-fuel ratio for fuel oil-fired furnaces and natural gas-fired furnaces are in mass units and volume (or mole) units, respectively. As opposed to ethanol, butanol has its air-to-fuel ratio of 11.2, which is much closer to that of gasoline (14.7) (MacLean and Lave 2003).

Ethanol can potentially substitute 32% of global gasoline usage when used in E85 (i.e., 15% gasoline and 85% ethanol) (Kim and Dale 2004). The ethanol-gasoline blended fuel (known as gasohol) is necessary, as the current vehicular engines are mechanical restrictions in using pure ethanol. Moreover, gasohol fuel is mostly compactible for use in flexible-fuel vehicles or dual-fuel vehicles that have ICE designed to operate on more than one fuel or gasoline-blended fuels. Butanol has a tendency to be blended with gasoline at higher levels providing twice the carbon content in every gallon of the blended fuel. The regulations in the USA allow butanol to be blended at 16% v/v with gasoline compared to 10% v/v blending for ethanol (Butamax™ 2018). Furthermore, butanol can be mixed with gasoline in flexible proportions allowing for implementation in current vehicular engines either in pure or blended forms without the requirement of intensive vehicular engine alteration (Qureshi et al. 2007).

Ethanol has a lower viscosity than butanol, which causes progressive wearing of the ICE engine parts. It is often reported that higher alcohol additives provide enhanced lubrication to reduce the chances of wear and tear of the mechanical components inside the vehicular engine. The presence of greater levels of lower alcohols (e.g., methanol and ethanol) in the blended fuel may lead to the corrosion to the ICE engine components due to accumulated moisture content in the fuel and the organic acids generated as the oxygenated compounds (Surisetty et al. 2011). Although methanol and ethanol are soluble in water, they exhibit poor miscibility in gasoline. Hence, phase separation problems are evitable when these alcohols are blended with gasoline at higher concentrations. On the contrary, blending higher alcohols such as propanol, butanol, and decanol with gasoline do not pose any phase separation problems due to their hygroscopic nature. As the length of the hydrocarbon chain in higher alcohols increases, the solubility in water decreases. The decrease in solubility becomes evident with four or higher carbons in the hydrocarbon chain. The reason for lower solubility in the case of longer hydrocarbon chain is more energy requirement to overcome the tightly packed hydrogen bonds between alcohol molecules as their size and mass increase. It is noteworthy to mention that the hydrocarbon chain length also has biological effects.

Butanol has many attributes that are divergent than ethanol, a few of which includes reduced solubility in water, less volatility, low flammability, low toxicity, and less corrosiveness (Nanda et al. 2017a). Moreover, butanol is considered as an excellent fuel extender because it contains only 22% oxygen compared to ethanol's oxygen content of 35%. In contrast to methanol, both propanol and butanol are less toxic with low volatility. Another disadvantage of ethanol and propanol is vapor lock of the engine due to their vapor pressures and low boiling points compared to butanol and gasoline. The fuel chemistry of butanol is comparable with gasoline, which has led to an increased global interest in its fermentative production from renewable feedstocks and eventual end use as a gasoline-blended fuel or substitute.

5.3 Petrochemical Routes for Butanol Production

The petrochemical production of butanol involves (Fig. 5.3) (a) oxo synthesis starting from propylene, (b) Reppe synthesis, (c) aldol process or crotonaldehyde hydrogenation, and (d) Guerbet reaction. Straight chain alcohols and branched alcohols such as ethanol, propanol, and butanol can also be synthesized from syngas (i.e., H_2 and CO) as the precursor through Fischer-Tropsch catalysis using nickel and iron-based catalysts (Zhao et al. 2002). Oxo process (or hydroformylation) is an industrial process where CO and H_2 are supplemented to a C=C compound with catalyst-substituted hydrocarbonyls (e.g., cobalt, rhodium, and ruthenium) (Lee et al. 2008). Because of hydroformylation of propene, aldehydes (e.g., *n*-butanol and isobutyraldehyde) are formed which then undergo catalytic hydrogenation to produce butanol. Butanol is generated in variable isomeric ratios depending on operating temperature, pressure, and catalyst. The oxo process operated at 1–5 MPa with rhodium catalyst yields up to 95% *n*-butanol and 5% isobutanol (or 2-methyl-1-propanol) (Uyttebroek et al. 2015). The conventional oxo process also leads to the generation of *n*-butyraldehyde from propylene. *n*-Butyraldehyde is a precursor in the production of 2-ethyl hexanol via aldol condensation as well as *n*-butanol. Deoxygenation of butyraldehyde by ethanol results in butylenes, whereas its reduction without dehydration produces butanol. Acetone and butanol are conventional solvents to synthesize isoprene rubber monomers, butadiene, and dimethyl butadiene (Jones and Woods 1986).

Reppe synthesis deals with catalytic carbonylation of propene (or propylene) at 100 °C under 0.5–2 MPa pressure involving the reaction between propene, water, and CO to produce butanol (Lee et al. 2008). The catalyst used in the process is usually a polynuclear iron carbonyl hydride or a tertiary ammonium salt. *n*-Butanol and isobutanol are obtained in a ratio of 86:14 (Uyttebroek et al. 2015). Although Reppe process results in preferable ratio of *n*-butanol-to-isobutanol under mild reaction conditions, it is usually more successful than the oxo process from the economic point of view.

Crotonaldehyde hydrogenation of acetaldehyde is another process to produce butanol, which is comprised of several reactions such as aldol condensation, dehydration, and hydrogenation (Lee et al. 2008). Although the process involves alkali catalysts and petrochemical resources, it usually requires ambient temperature and pressure. The dehydration step is prompted by acidification using acetic acid or phosphoric acid followed by distillation. Copper catalysts are required in the hydrogenation step that is executed in the gas or liquid phase. Approximately 1350 kg of acetaldehyde can result in 1000 kg of *n*-butanol from this process (Uyttebroek et al. 2015).

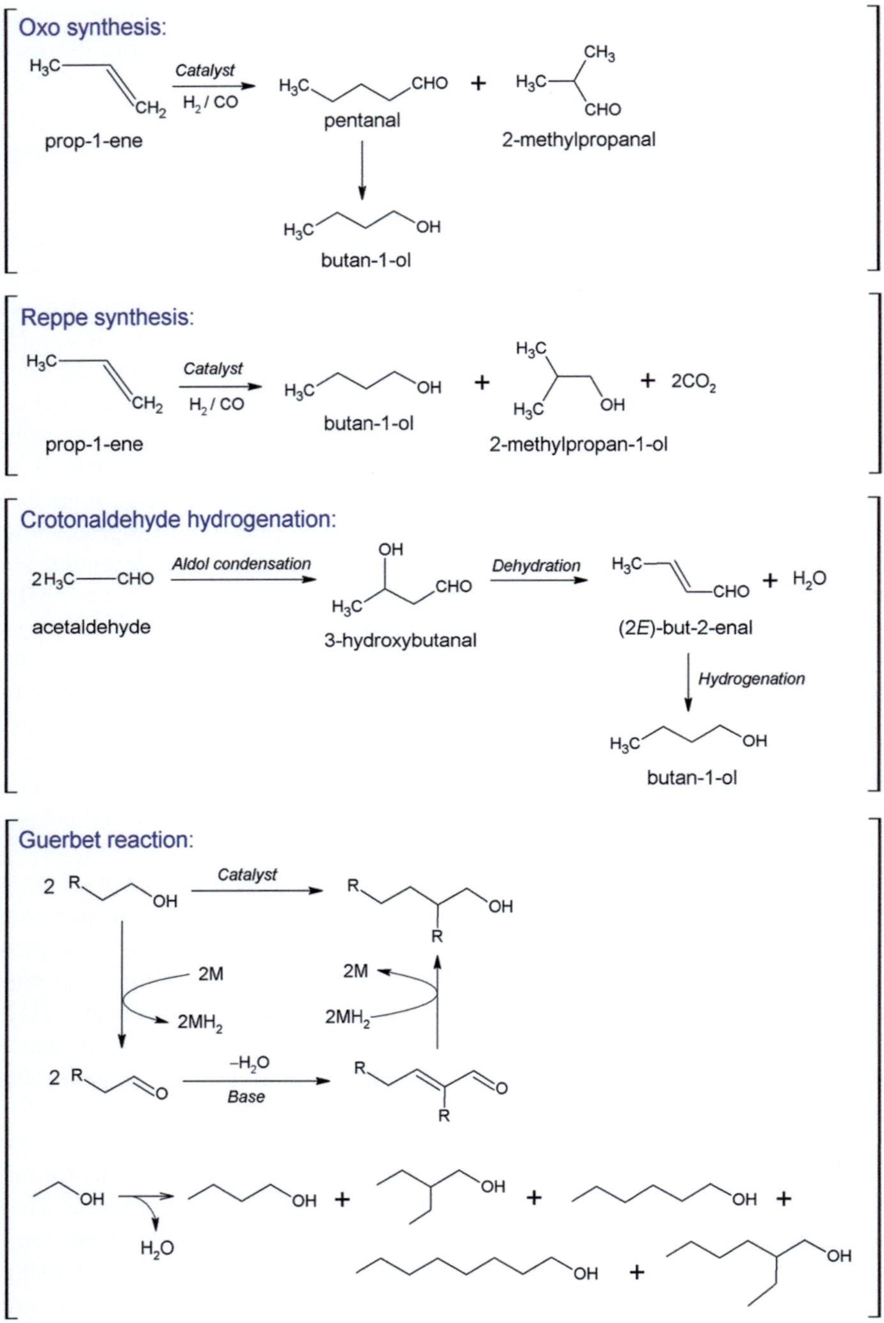

Fig. 5.3 Petrochemical routes for butanol production

The industrial Guerbet process involves the following main steps: (a) dehydrogenation of a primary alcohol, (b) base-catalyzed aldol-coupling reaction, and (c) hydrogenation of α,β-unsaturated aldehyde (Chakraborty et al. 2015). The final step leads to the uniting of the two primary alcohols to form a long-chain alcohol via the "borrowed hydrogen" phenomenon that has no net loss of gaseous hydrogen. Guerbet reaction allows facile C-C bonding with unreactive alcohols using ruthenium or iridium catalysts (Aitchison et al. 2016). Guerbet process is performed usually with long-chain primary alcohols, but utilizing ethanol as a substrate in the reaction to produce butanol remains a challenge. These main limitations include the dehydrogenation of ethanol, which is a thermodynamic challenge, as well as the generation of undesired by-products resulting from the uncontrolled aldol reaction due to the highly reactive acetaldehyde (Carlini et al. 2003).

5.4 Biological Route for Butanol Production

The microorganisms involved in ethanol or butanol production can be assessed in terms of feedstock utilization, fermentation parameters (e.g., oxygen, pH, light, temperature, feed concentration, residence time, etc.), nutritional requirements, and bioreactor design. An ideal ethanol-producing microorganism should retain the following features (Dien et al. 2003):

1. Ethanol yields >90% of theoretical estimation
2. Ethanol productivity >1 g/L/h
3. Ethanol tolerance >40 g/L
4. Ability to grow in inexpensive and simple growth media
5. Growth requirements (e.g., pH, temperature, oxygen necessity, etc.) to retard the occurrence of contaminants
6. Metabolic tolerance to inhibitors
7. Resistance to high temperatures and acidic pH

The preferred microorganisms for ethanol production include fungi and yeast. *Saccharomyces cerevisiae* is the most popular industrial ethanol-producing fungi because it demonstrates a high tolerance to ethanol and sustained growth in oxygen-limiting fermentation conditions. However, *S. cerevisiae* lacks the natural potency to ferment hemicellulose (pentose or C_5 sugars), especially xylose (Ha et al. 2011). Nevertheless, the enzyme xylose isomerase facilitates the conversion of xylose to xylulose, which can be fermented to ethanol by *S. cerevisiae*. A few fungi having the ability to ferment directly cellulose to ethanol include *Aspergillus*, *Fusarium*, *Monilia*, *Neurospora*, *Paecilomyces*, *Penicillium*, *Phanerochaete*, *Schizophyllumi*, *Sclerotium*, and *Trichoderma* (Duff and Murray 1996). A few fungi are exempted from their inefficiency in xylose fermentation such as *Aspergillus oryzae* and *Trichoderma reesei*. While *Aspergillus oryzae* degrades xylan by secreting β-xylosidase, *T. reesei* degrades hemicellulose sugars by producing two endo-β-xylanase enzymes, viz., xylanase I and xylanase II (Nanda et al. 2014a).

Some bacteria and actinomycetes investigated for the secretion of cellulase enzymes for ethanol production include *Bacillus*, *Bacteroides*, *Clostridium*, *Cellulomonas*, *Ruminococcus*, *Thermomonospora*, *Streptomyces*, *Acetivibrio*, *Erwinia*, and *Microbispora* (Bisaria 1991; Prasad et al. 2007). The following features should be present in bacteria to make them more effective compared to fungi for ethanol production: (a) ethanol production reliably in large-scale bioreactors, (b) less requirement of saccharifying enzymes, and (c) viability and stability of cells during bulk fermentation. Anaerobic and thermophilic bacteria are beneficial over the traditional ethanol-producing yeasts for their tendency to endure extreme temperatures and utilize a broad spectrum of low-cost feedstocks. However, the low ethanol resistance (<2% v/v) by most bacteria is a key constraint in the industrial fermentation (Georgieva et al. 2007). Unlike ethanol production that is dominated by aerobic spore-forming fungi (i.e., *S. cerevisiae*), butanol production is performed almost exclusively by a consortium of bacteria that are obligate anaerobes, spore-forming, and Gram-positive such as *Clostridium* spp. On the other hand, *Clostridium* spp. are also robust in utilizing both hemicellulose (pentose or C_5 sugars) and cellulose (hexose or C_6 sugars) unlike *S. cerevisiae* that flourishes only in utilizing hexose sugars (Qureshi et al. 2007).

Butanol production is accomplished through acetone–butanol–ethanol (ABE) fermentation by using a few responsible *Clostridium* spp., namely *C. acetobutylicum*, *C. beijerinckii*, *C. butylicum*, *C. saccharobutylicum*, *C. aurantibutyricum*, *C.*

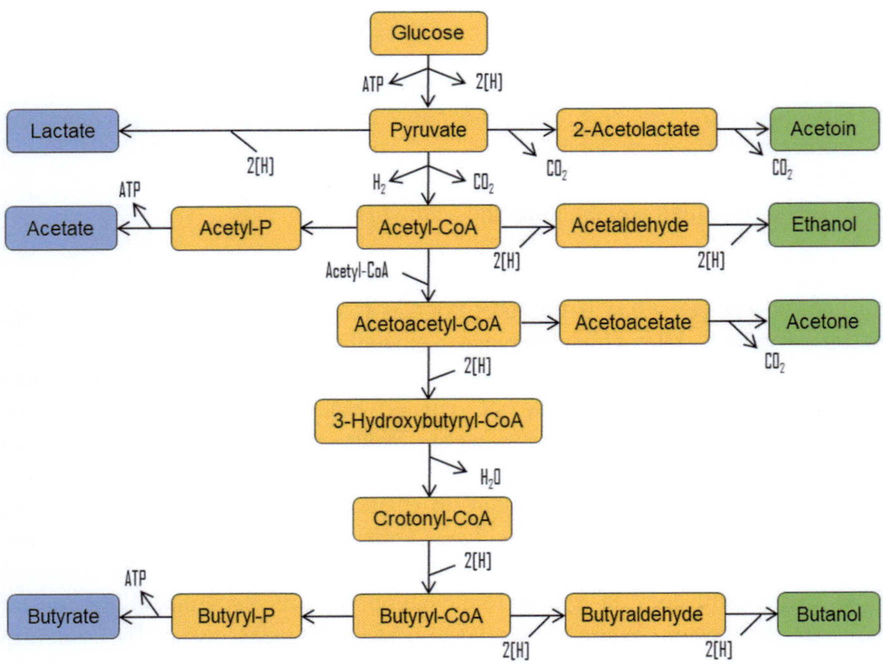

Fig. 5.4 Biological route for butanol production. (Adapted from Nanda et al. (2017a))

saccharoperbutylacetonicum, etc. Figure 5.4 represents the metabolic steps involved in butanol fermentation involving *Clostridium* spp. Typically, ABE fermentation is carried out under anaerobic conditions at 35 °C for 36–72 h to produce a total solvent (acetone, butanol, and ethanol) level up to 20–25 g/L (Qureshi and Ezeji 2008). The ABE fermentation involves acidogenic (or acid producing) and solventogenic (or solvent producing) phases. The acidogenic phase results in the production of acids such as acetic acid and butyric acid, whereas solventogenic phase leads to the generation of acetone, butanol, and ethanol. The ABE fermentation progresses with the involvement of biosynthetic enzymes, especially aldehyde/alcohol dehydrogenase, acetoacetyl-CoA thiolase, 3-hydroxybutyryl-CoA dehydrogenase, butyryl-CoA dehydrogenase, crotonase, and crotonyl-CoA. The characteristic fractions of acetone, butanol, and ethanol obtained through ABE fermentation are in the ratio of 3:6:1 (Jones and Woods 1986).

The current butanol recovery technologies include distillation, adsorption, liquid-liquid extraction, pervaporation, perstraction, gas stripping, and supercritical fluid extraction (Ezeji et al. 2007). While adsorption uses molecular sieves and resins to separate butanol, the gas stripping recycles the fermentation gases (i.e., CO_2 and H_2) to create butanol vapors for subsequent purification by condensation. Gas stripping lowers cell toxicity and results in high substrate utilization. Liquid-liquid extraction uses an extractant to separate the products based on their variable distribution coefficients. Perstraction upgrades liquid-liquid extraction by implementing permeable membrane to separate the cell culture from the extracting solvent, thereby preventing the issues of toxicity and emulsion formation. Pervaporation uses the principle of membrane permeation using nonporous membrane followed by evaporation to dehydrate the organic solvent and separate the products. Further improvements in butanol recovery technologies could expedite its large-scale production as a flexible fuel or drop-in fuel for use in the current vehicular engines without any significant modification.

5.5 Petrochemical Routes for Propanol Production

Apart from being the main source of vehicular fuels, petrochemical industries are also the source of several organic solvents. Petrochemical resources such as natural gas and crude oil are the main source of these organic solvents. Various organic fractions and aromatic compounds that are produced during the reforming or catalytic cracking of naphtha finally result in the formation of aliphatic hydrocarbons. To produce oxygenated solvents, certain key ingredients are required such as ethylene, propylene, butylene, and methane (Wypych 2001). Capello et al. (2009) performed life cycle analysis and environmental risk assessment for the technologies relating to the production of significant organic solvents including propanol from petrochemical feedstocks. The synthesis of industrial solvents (including propanol) from petrochemical resources can be represented through the following four routes as reported in the literature (Chauvel and Lefebre 1989a, b; Wells 1999). These routes differ based on their key process, unit operation, and precursor materials. The four major solvent production pathways can be characterized as follows:

1. Methanol pathway: This pathway uses methanol as the precursor that is produced from natural gas through catalytic gas-to-liquid conversion technology. Acetic acid, formaldehyde, and tetrahydrofuran are some of the chief solvents obtained through this route.
2. Steam cracking or naphtha pathway: This pathway involves steam cracking of naphtha obtained from crude oil to produce a broad spectrum of alkanes and alkenes. Aliphatic alcohols including ethanol, propanol, and butanol are obtained along with certain intermediate by-products, which can be processed to derive acetates.
3. Benzene-toluene-xylene (BTX) pathway: In this pathway, naphtha is derived from BTX reformates and further extracted through molecular sieves. After extraction, the recovered naphtha and its derivatives are used to produce alkanes including hexanes and heptanes.
4. BTX splitting pathway: This pathway involves thermochemical and catalytic splitting of BTX reformate and pyrolyzed gasoline to produce aromatic compounds and organic solvents such as benzene, xylene, and toluene. The resulting benzene, toluene, and intermediate by-products are treated to produce aldehydes and organic solvents such as benzaldehyde, cyclohexane, and acetone.

The naphtha or steam cracking route is the most diversified. It accommodates a wide variety of precursors as the raw materials to generate essential chemicals and solvents including propene and ethane. The worldwide productions of propane and ethane are 30×10^6 tons (Eisele and Killpack 2002) and 80×10^6 tons (Zimmermann and Walzl 2002), respectively. Certain side reactions mark a low increase in the cumulative energy demand (CED) per kg of the product because the subordinates with CED values are needed for these reactions. A few of such reactions are (a) hydrogenation to propanol, (b) methanol supplement to produce methyl tertiary butyl ether (MTBE), and (c) esterification to generate isoamyl acetate, butyl acetate, and isobutyl acetate. The processes resulting in greater fluctuations in CED are hydroformylations for isoamyl alcohol, isobutanol, pentanol, and dioxane. For hydroformylation process, the challenge in CED is due to the requirement of high energy input and large quantities of precursors and syngas (Chauvel and Lefebre 1989a, b).

5.6 Biological Routes for Propanol Production

Propanol is commercially produced from petrochemical sources. The techniques employed in propanol production from petrochemical sources are expensive and environmentally unfriendly. In addition, with increasing stress on the petroleum resources, the emphasis is now on the biological production of propanol from renewable biomass. There is lack of information on responsible microorganisms

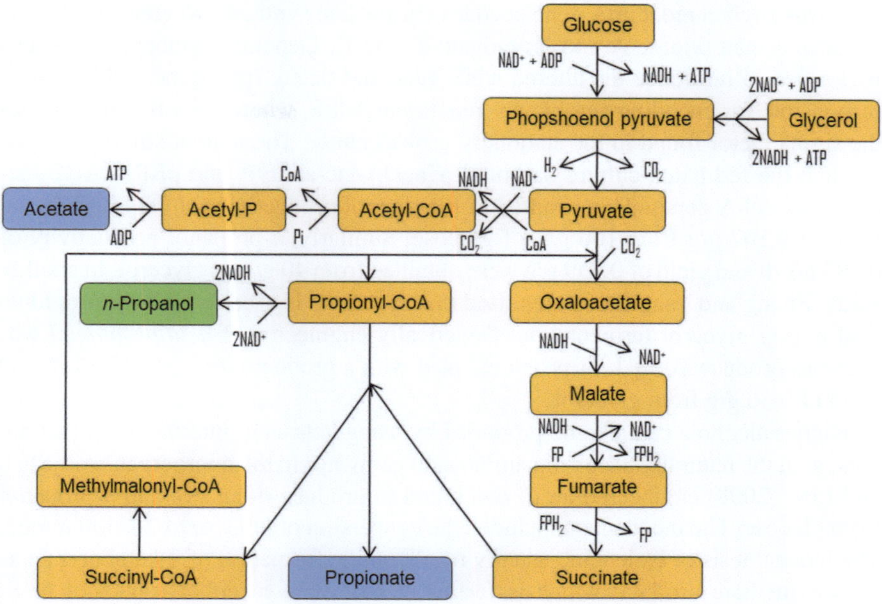

Fig. 5.5 Biological route for propanol production. (Adapted from Ammar et al. (2013))

that can be employed in the production of propanol in significant amounts (Janssen 2004). *Clostridium* spp. are found to produce smaller amounts of propanol from threonine catabolism. Jang et al. (2012) have compiled propanol production from carboxylic acids (propionic acid), hydroxy acids (3-hydroxypropionic acid and lactic acid), diols (1,2-propandiol and 1,3-propandiol). and alcohols.

Propanol can be obtained through propionic acid fermentation using two main *Propionibacterium* spp. such as *P. acidipropionici* and *P. freudenreichii* ssp. *shermanii* (Himmi et al. 2000; Eden et al. 2001; Liu et al. 2011). Reports suggest that an overexpressed bifunctional aldehyde or alcohol dehydrogenase (adhE) can respectively transform butyryl-CoA and acetyl-CoA to butanol and ethanol (Nair et al. 1994; Yu et al. 2011). Ammar et al. (2013) studied the possibility for synthesizing *n*-propanol using propionyl-CoA as represented in Fig. 5.5. They investigated the pairing of CoA with propionyl to produce *n*-propanol. Hence, a bifunctional adhE gene cloned from *Escherichia coli* was expressed in *P. freudenreichii*. The production of *n*-propanol from both glucose and glycerol was found along with the higher rate of consumption of propionic acid by the adhE gene-expressing mutant. This study was the first effort in the biosynthesis of *n*-propanol via genetically engineered *Propionibacterium*.

Choi et al. (2012) engineered an *E. coli* strain to produce *n*-propanol under anaerobic conditions. L-Threonine was found to be overproduced by the mutated *E. coli*. The first step was to introduce feedback-resistant ilvA gene that encoded L-threonine dehydratase with the deletion of all other genes that could compete for the metabolic pathways. Additionally, the overexpression of cimA gene and ackA gene was engineered. cimA gene encodes citramalate synthase, whereas ackA gene encodes acetate kinase A and propionate kinase II. Genetic engineering was also performed to introduce the altered adhE gene and delete rpoS gene. adhE gene is responsible for encoding aerobically functional AdhE, whereas rpoS gene encodes the sigma factor found in the stationary growth phase. The mutant strain produced through the fed-batch culture harbored pTacDA-tac-adhE^{mut} and pBRthrABC-tac-cimA-tac-ackA genes. This resulted in *n*-propanol productivity of 0.144 g/L/h and yield of 0.107 g/g from 100 g/L of glucose. Similarly, *n*-propanol productivity of 0.083 g/L/h and yield of 0.259 g/g were obtained from 40 g/L of glycerol. In another study, Zhang and Yang (2009) reported that succinate is used to produce propionic acid during glycerol fermentation. Genetically engineered *P. acidipropionici* was found to produce 106 g/L of propionic acid with a productivity of 0.035 g/L/h and yield of 0.56 g/g from glycerol.

Higher alcohols can also be produced by using keto-acid intermediates that are present in the naturally occurring amino acid pathways in the microorganisms. Shen and Liao (2008) engineered an *E. coli* strain to produce *n*-butanol and *n*-propanol from glucose. The mechanism included the conversion of glucose to 2-ketobutyrate. The biosynthesis of isoleucine usually results in the formation of 2-ketobutyrate as an intermediate product, which is further transformed to either *n*-butanol or *n*-propanol. The reaction for propanol is catalyzed using heterologous decarboxylase and dehydrogenase. In contrast, butanol is catalyzed to produce amino acid norvaline. The competing pathways were eliminated, and the amino acid biosynthesis was deregulated to improve the yields of *n*-butanol and *n*-propanol. The final strain showed the production of 2 g/L of total solvents with a ratio of 1:1 for propanol and butanol.

There can be several metabolic pathways for synthesizing the desired solvent product. Both natural evolution and synthetic biology pathways exhibit distinct properties that are specific for the wild-type and engineered microorganisms. Shen and Liao (2013) illustrated the principle of synergy between two varying methods of *n*-propanol synthesis by *E. coli.* The studies on the metabolic pathways by the host gene removal showed their discrete traits with the relevance of tricarboxylic acid cycle for threonine pathways. In contrast, citramalate pathway was found independent of the tricarboxylic acid cycle. Both the pathways were found to be complementary in driving the energy requirements in the host microorganisms. The experiments on solvent production strengthened the synergistic impacts predicted by the yield model. Thus, this study successfully incorporated the synergy into model design principles of engineered metabolic pathways of the microorganisms to enhance the product yield.

5.7 Challenges and Future Perspectives

Butanol can be produced both through petrochemical and biological means. Uyttebroek et al. (2015) performed a sustainability metric analysis of butanol derived from petrochemical and biological routes. Petrochemical routes (92%) for butanol synthesis demonstrate a higher material efficiency compared to fermentative route (27–42%). The material efficiency in this study was calculated as the total weight of valuable products divided by the total weight of valuable products and waste residues (Sheldon and Sanders 2015). The E-factor is a critical concept in estimating the material efficiency. About 0.6 kg of propene is consumed to produce nearly 1.0 kg of butanol, referring to 95% butanol yield in pure form (Uyttebroek et al. 2015). This resulted in an E-factor of 0.1 with a greater material efficiency of 92% considering the exclusion of water used in the petrochemical process. On the contrary, an E-factor of 3.7 and a lower material efficiency of 21% were reported for the bio-based process considering butanol as the only product of interest. However, when butanol was the main product along with acetone and ethanol as the by-products, the E-factor decreased to 2.7 and the material efficiency slightly increased to 27%. The E-factor further decreased to 1.4 and the material efficiency increased to 42% when biomass fiber and residual proteins were considered as the useful products along with butanol, ethanol, and acetone. Petrochemical route for butanol production typically results in more butanol yields and less by-product formation leading to higher material efficiency. On the contrary, fermentative butanol production results in lower butanol yields due to the by-products such as acetone, ethanol, acetic acid, and butyric acid. Therefore, finding value for the by-products of ABE fermentation can decrease the E-factor and increase the material efficiency in the biological butanol production route.

Petrochemical routes (69 GJ/ton) are also less energy-intensive than fermentative route (116 GJ/ton) (Uyttebroek et al. 2015). This makes the fermentative route (1041 €/ton) relatively expensive than petrochemical routes (915 €/ton). In contrast to a fermentative route that has a land use requirement of 0.29 ha/ton, petrochemical routes predominantly rely on petroleum supply. Regardless of the lower energy and cost requirement by petrochemical routes, the interest in biological butanol production is gaining momentum because of the volatility of gasoline and other fossil fuel supplies, their increasing prices, and legislations restricting their future usage.

As discussed earlier, butanol production through ABE fermentation has many limitations such as butanol toxicity, low butanol yields, and challenges in butanol recovery and bacteriophage contamination. The challenges and potential solutions during fermentative butanol production are summarized in Table 5.2. Butanol-producing *Clostridium* spp. can barely tolerate butanol concentrations of 2% in the fermentation medium, which results in solvent toxicity and ceasing of microbial fermentation (Jones and Woods 1986). Although ethanol initiation begins at 24 g/L, *Saccharomyces cerevisiae* can tolerate up to 90 g/L of ethanol during fermentation (Ghosh and Tyagi 1979). According to reports, naturally occurring *Clostridium* spp. can tolerate 12–13 g/L of butanol (Dürre 2008; García et al. 2011). However, genetically modified *Clostridium beijerinckii* BA101 has been reported to withstand 19.6 g/L of butanol (Qureshi and Blaschek 1999). Butanol inhibits the clostridial growth in the fermentation medium by dissolving the phospholipid layers present in the cell membrane of the bacteria. The solubility of the alcohol in the bacterial cell membrane affects its membrane fluidity, which increases with the growing chain length of the hydrocarbon (Jones and Woods 1986). The low butanol titer levels in

Table 5.2 Challenges and possible solution for butanol biosynthesis

Biorefining stage	Challenge	Solution
Feedstock selection	High feedstock (starch-based) cost and ethical issues of food versus fuel	Transition toward cheaper, renewable, and abundantly available lignocellulosic feedstocks (e.g., straw, wood and grasses)
Biomass pretreatment (upstream processing)	Culture degeneration by inhibitory intermediary by-products generated during biomass pretreatment	Removal of inhibitory by-products from the hydrolysate can make the process expensive. Detoxifying or neutralizing the process using techniques such as over-liming can reduce the concentration of inhibitory compounds. Developing engineered microorganisms that can resist the effects of inhibitory compounds
Fermentation	Low butanol yield and incomplete sugar conversion	Developing engineered microorganisms with improved butanol production. Process designing for continuous fermentation and in situ product removal to enhance butanol tolerance, better sugar conversion, and volumetric productivity
	Formation of numerous fermentation by-products	Modification of knockout genes to delete the trait for by-product formation in recombinant microorganisms. Shifting the carbon flow and energy flux exclusively toward butanol production pathway
	Bacteriophage contamination	Good factory hygiene by using nonionic detergents and antibiotics. Developing strains less vulnerable to bacteriophage infection
Product recovery (downstream processing)	Expensive and energy-intensive butanol recovery technologies	Retrofitting or developing new recovery technologies for enhanced solvent recovery and purification. Lowering of recovery cost and energy consumption by increasing butanol titer

References: Green (2011), Dürre (2011)

the fermentation also make the butanol recovery process challenging, expensive, and energy-intensive.

Similar to butanol production, fermentative propanol production is also challenging and expensive, which restricts from making it a common fuel. Since the energy gains of propanol over ethanol are minimal, its industrial production and commercial uses are demanding. The challenge remains in improving both butanol and propanol yields from the fermentative processes. Therefore, the genetic engineering manipulations in *E. coli* and *Saccharomyces cerevisiae* have been investigated to enhance bioproduction of propanol (Shen and Liao 2008; Atsumi and Liao 2008; Matsuda et al. 2011; Choi et al. 2012). Propanol yields were reported to increase up to 4 g/L and 10 g/L from glucose and glycerol, respectively, by employing metabolically engineering *E. coli* (Atsumi and Liao 2008; Shen and Liao 2008; Choi et al. 2012).

Syngas fermentation is a hybrid technology that connects both the petrochemical and biochemical solvent production processes (Nanda et al. 2016e). In this technique, suitable microorganisms are used to convert syngas (a mixture of H_2 and CO) produced from biomass gasification to yield solvents such as ethanol, butanol, and traces of propanol (Wilkins and Atiyeh 2011). Syngas fermentation is at the verge of commercialization to produce fuels and chemicals from syngas (Liu et al. 2014). It is necessary to select a low-cost fermentation medium to make the process feasible. For example, replacement of corn steep liquor with yeast extract as the fermentation medium using *Alkalibaculum bacchi* strain CP15 decreased the cost of fermentation by 27% and resulted in 78% more ethanol. When a 7-L fermenter was used for continuous fermentation, the ethanol yield was 6 g/L in yeast extract and yeast extract-free media. When yeast extract was replaced with corn steep liquor, the maximum yields of ethanol (8 g/L), *n*-propanol (6 g/L), and *n*-butanol (1 g/L) were noticed. *A. bacchi* strain CP15 was not typical for *n*-butanol and *n*-propanol production. Another 16S rRNA gene-based study showed that a mixture of *Alkalibaculum bacchi* strain CP15 (56%) and *Clostridium propionicum* (34%) gave better yields of higher alcohols such as propanol and butanol from syngas (Liu et al. 2014).

Recently, DuPont Tate & Lyle BioProducts has announced the expansion of their biopropanol facility in Tennessee, USA, by increasing the production of 1,3-propandiol up to 35 million pounds per annum (DuPont Tate & Lyle Bio Products, 2018). DuPont and British Petroleum have also collaborated to form Butamax™ Advanced Biofuels, which intends to make biobutanol an inexpensive, versatile, and flexible fuel for the motor vehicles (Butamax™ 2018). Furthermore, recombinant DNA technology and metabolic engineering of butanol-producing *Clostridium* have resulted in significant developments in ABE fermentation. A few process improvements include (a) high cell growth, stability, and viability, (b) lower chances of endospore formation by *Clostridium*, (c) high butanol selectivity and tolerance, (d) tolerance to lower levels of oxygen during anaerobic fermentation, and (e) utilization of a wide variety of feedstocks (Papoutsakis 2008).

5.8 Conclusions

There is a mounting interest worldwide to efficiently produce butanol and propanol from inexpensive feedstocks as newer biofuels for use in the transportation sector. Butanol overshadows ethanol with its progressive fuel properties and chemistry, which are comparable to gasoline. Butanol production through ABE fermentation involves the solventogenic activity of *Clostridium* spp., which yields three solvents, namely acetone, butanol, and ethanol in the characteristic ratio of 3:6:1. However, by-products such as acetic acid and butyric acid are also obtained from acidogenic stage of ABE fermentation. As opposed to ethanol fermentation by yeast *Saccharomyces cerevisiae* which can only ferment hexose sugars, ABE fermentation by *Clostridium* spp. is beneficial because the latter can ferment both pentose and hexose sugars. However, ethanol fermentation is commercially viable due to high yields and greater resistance of yeast to acidic pH and high levels of ethanol in the fermentation. Furthermore, ABE fermentation has many shortcomings such as low butanol yields due to clostridial inhibition to butanol even at low titer levels, uncontrolled bacteriophage infection, premature seizure of the fermentation, and expensive butanol recovery techniques.

On the other hand, propanol is also emerging as a new candidate in the biofuel spectrum. Similar to ethanol and butanol, propanol can also be produced from petrochemicals and bio-based feedstocks. However, the literature on propanol fermentation is nascent at the current stage. Several factors such as research, development, and innovation in the bioconversion technologies, upstream processing of feedstocks, effective pretreatment technology to recover maximum sugars, as well as metabolic engineering of fermenting microorganisms can help determine the sustainable production of butanol and propanol from waste biomass. The principal focus should be on developing efficient biochemical technologies including genetic and metabolic engineering of fermenting microorganisms for converting nonedible plant residues and biogenic wastes to butanol and propanol, which can potentially outshine the quintessential fossil fuels.

References

Aitchison H, Wingad RL, Wass DF (2016) Homogeneous ethanol to butanol catalysis—Guerbet renewed. ACS Catal 6:7125–7132

Ammar EM, Wang Z, Yang ST (2013) Metabolic engineering of *Propionibacterium freudenreichii* for n-propanol production. Appl Microbiol Biotechnol 97:4677–4690

Atsumi S, Liao JC (2008) Directed evolution of *Methanococcus jannaschii* citramalate synthase for biosynthesis of 1-propanol and 1-butanol by *Escherichia coli*. Appl Environ Microbiol 74:7802–7808

Azargohar R, Nanda S, Rao BVSK, Dalai AK (2013) Slow pyrolysis of deoiled canola meal: product yields and characterization. Energy Fuel 27:5268–5279

Azargohar R, Nanda S, Dalai AK (2018) Densification of agricultural wastes and forest residues: a review on influential parameters and treatments. In: Sarangi PK, Nanda S, Mohanty P (eds) Recent advancements in biofuels and bioenergy utilization. Springer Nature, Singapore, pp 27–51

Azargohar R, Nanda S, Kang K, Bond T, Karunakaran C, Dalai AK, Kozinski JA (2019) Effects of bio-additives on the physicochemical properties and mechanical behavior of canola hull fuel pellets. Renew Energy 132:296–307

Biofuel.org.uk (2018). http://biofuel.org.uk/bioalcohols.html. Accessed 29 Dec 2018

Bisaria VS (1991) Bioprocessing of agro-residue to glucose and chemicals. In: Martin AM (ed) Bioconversion of waste materials to industrial products. Elsevier, London, pp 187–223

Butamax™ (2018) A joint venture between BP and DuPont. http://www.butamax.com. Accessed 29 Dec 2018

Capello C, Wernet G, Sutter J, Hellweg S, Hungerbühler K (2009) A comprehensive environmental assessment of petrochemical solvent production. Int J Life Cycle Assess 14:467–479

Carlini C, Di Girolamo M, Macinai A, Marchionna M, Noviello M, Galletti AMRM, Sbrana G (2003) Selective synthesis of isobutanol by means of the Guerbet reaction: Part 2. Reaction of methanol/ethanol and methanol/ethanol/*n*-propanol mixtures over copper based/MeONa catalytic systems. Mol Catal A: Chem 200:137–146

Chakraborty S, Piszel PE, Hayes CE, Baker RT, Jones WD (2015) Highly selective formation of *n*-butanol from ethanol through the Guerbet process: a tandem catalytic approach. J Am Chem Soc 137:14264–14267

Chauvel A, Lefebre G (1989a) Petrochemical processes: technical and economic characteristics, Synthesis-gas derivatives and major hydrocarbons, vol 1. Editions Technip, Paris

Chauvel A, Lefebre G (1989b) Petrochemical processes: Technical and economic characteristics, Major oxygenated, chlorinated and nitrated derivatives, vol 2. Editions Technip, Paris

Choi YJ, Park JH, Kim TY, Lee SY (2012) Metabolic engineering of *Escherichia coli* for the production of 1-propanol. Metab Eng 14:477–486

Dabelstein W, Reglitzky A, Schutze A, Reders K (2007) Automotive fuels. Ullmann's Encycl Ind Chem 4:425–458

Dien BS, Cotta MA, Jeffries TW (2003) Bacteria engineered for fuel ethanol production: current status. Appl Microbiol Biotechnol 63:258–266

Duff SJB, Murray WD (1996) Bioconversion of forest products industry waste cellulosics to fuel ethanol: a review. Bioresour Technol 55:1–33

DuPont Tate & Lyle Bio Products (2018) Expanding bio-based propanediol production in Tennessee. http://www.duponttateandlyle.com/DuPont-Tate-Lyle-Bio-Products-Expanding-Bio-based-Propanediol-Production-in-Tennessee. Accessed 29 Dec 2018

Dürre P (2007) Biobutanol: an attractive biofuel. Biotechnol J 2:1525–1534

Dürre P (2008) Fermentative butanol production bulk chemical and biofuel. Ann N Y Acad Sci 1125:353–362

Dürre P (2011) Fermentative production of butanol—the academic perspective. Curr Opin Biotechnol 22:331–336

Eden A, Van Nedervelde L, Drukker M, Benvenisty N, Debourg A (2001) Involvement of branched-chain amino acid aminotransferases in the production of fusel alcohols during fermentation in yeast. Appl Microbiol Biotechnol 55:296–300

Eisele P, Killpack R (2002) Propene. In: Wiley-VCH (ed) Ullmann's encyclopedia of industrial chemistry. Wiley-VCH, Weinheim

Ezeji TC, Qureshi N, Blaschek HP (2007) Bioproduction of butanol from biomass: from genes to bioreactors. Curr Opin Biotechnol 18:220–227

García V, Päkkilä J, Ojamo H, Muurinen E, Keiski RL (2011) Challenges in biobutanol production: how to improve the efficiency? Renew Sust Energ Rev 15:964–980

Georgieva TI, Skiadas IV, Ahring BK (2007) Effect of temperature on ethanol tolerance of a thermophilic anaerobic ethanol producer *Thermoanaerobacter* A10: modeling and simulation. Biotechnol Bioeng 98:1161–1170

Ghosh TK, Tyagi RD (1979) Rapid ethanol fermentation of cellulose hydrolysate. II. Product and substrate inhibition and optimization of fermentor design. Biotechnol Bioeng 21:1401–1420

Gong M, Nanda S, Hunter HN, Zhu W, Dalai AK, Kozinski JA (2017a) Lewis acid catalyzed gasification of humic acid in supercritical water. Catal Today 291:13–23

Gong M, Nanda S, Romero MJ, Zhu W, Kozinski JA (2017b) Subcritical and supercritical water gasification of humic acid as a model compound of humic substances in sewage sludge. J Supercrit Fluids 119:130–138

Graham-Rowe D (2011) Agriculture: beyond food versus fuel. Nature 474:S6–S8

Green EM (2011) Fermentative production of butanol—the industrial perspective. Curr Opin Biotechnol 22:337–343

Gronowska M, Joshi S, MacLean HL (2009) A review of U.S. and Canadian biomass supply studies. Bioresources 4:341–369

Ha SJ, Galazka JM, Kim SR, Choi JH, Yang X, Seo JH, Glass NL, Cate JHD, Jin YS (2011) Engineered *Saccharomyces cerevisiae* capable of simultaneous cellobiose and xylose fermentation. PNAS 108:504–509

Himmi EH, Bories A, Boussaid A, Hassani L (2000) Propionic acid fermentation of glycerol and glucose by *Propionibacterium acidipropionici* and *Propionibacterium freudenreichii* ssp. *shermanii*. Appl Microbiol Biotechnol 53:435–440

Jang YS, Kim B, Shin JH, Choi YJ, Choi S, Song CW, Lee J, Park HG, Lee SY (2012) Bio-based production of C2–C6 platform chemicals. Biotechnol Bioeng 109:2437–2459

Janssen PH (2004) Propanol as an end product of threonine fermentation. Arch Microbiol 182:482–486

Jiang G, Pickering SJ, Lester EH, Turner TA, Wong KH, Warrior NA (2009) Characterisation of carbon fibres recycled from carbon fibre/epoxy resin composites using supercritical *n*-propanol. Compos Sci Technol 69:192–198
Jones DT, Woods DR (1986) Acetone-butanol fermentation revisited. Microbiol Rev 50:484–524
Jones A, O'Hare M, Farrell A (2007) Biofuel boundaries: estimating the medium-term supply potential of domestic biofuels. Research report UCB-ITS-TSRC-RR-2007-4. University of California, Berkeley, CA
Kim S, Dale BE (2004) Global potential bioethanol production from wasted crops and crop residues. Biomass Bioenergy 26:361–375
Lee SY, Park JH, Jang SH, Nielsen LK, Kim J, Jung KS (2008) Fermentative butanol production by clostridia. Biotechnol Bioeng 101:209–228
Liu Y, Zhang YG, Zhang RB, Zhang F, Zhu J (2011) Glycerol/glucose co-fermentation: one more proficient process to produce propionic acid by *Propionibacterium acidipropionici*. Curr Microbiol 62:152–158
Liu K, Atiyeh HK, Stevenson BS, Tanner RS, Wilkins MR, Huhnke RL (2014) Continuous syngas fermentation for the production of ethanol, *n*-propanol and *n*-butanol. Bioresour Technol 151:69–77
Mabee WE, Saddler JN (2010) Bioethanol from lignocellulosics: status and perspectives in Canada. Bioresour Technol 101:4806–4813
MacLean HL, Lave LB (2003) Evaluating automobile fuel/propulsion system technologies. Prog Energy Combust Sci 29:1–69
Matsuda F, Furusawa C, Kondo T, Ishii J, Shimizu H, Kondo A (2011) Engineering strategy of yeast metabolism for higher alcohol production. Microb Cell Factories 10:70
Nair RV, Bennett GN, Papoutsakis ET (1994) Molecular characterization of an aldehyde/alcohol dehydrogenase gene from *Clostridium acetobutylicum* ATCC 824. J Bacteriol 176:871–885
Nanda S, Mohanty P, Pant KK, Naik S, Kozinski JA, Dalai AK (2013) Characterization of North American lignocellulosic biomass and biochars in terms of their candidacy for alternate renewable fuels. Bioenergy Res 6:663–677
Nanda S, Dalai AK, Kozinski JA (2014a) Butanol and ethanol production from lignocellulosic feedstock: biomass pretreatment and bioconversion. Energ Sci Eng 2:138–148
Nanda S, Mohammad J, Reddy SN, Kozinski JA, Dalai AK (2014b) Pathways of lignocellulosic biomass conversion to renewable fuels. Biomass Convers Bioref 4:157–191
Nanda S, Azargohar R, Dalai AK, Kozinski JA (2015a) An assessment on the sustainability of lignocellulosic biomass for biorefining. Renew Sust Energ Rev 50:925–941
Nanda S, Reddy SN, Hunter HN, Butler IS, Kozinski JA (2015b) Supercritical water gasification of lactose as a model compound for valorization of dairy industry effluents. Ind Eng Chem Res 54:9296–9306
Nanda S, Reddy SN, Hunter HN, Dalai AK, Kozinski JA (2015c) Supercritical water gasification of fructose as a model compound for waste fruits and vegetables. J Supercrit Fluids 104:112–121
Nanda S, Dalai AK, Berruti F, Kozinski JA (2016a) Biochar as an exceptional bioresource for energy, agronomy, carbon sequestration, activated carbon and specialty materials. Waste Biomass Valor 7:201–235
Nanda S, Dalai AK, Gökalp I, Kozinski JA (2016b) Valorization of horse manure through catalytic supercritical water gasification. Waste Manag 52:147–158
Nanda S, Dalai AK, Kozinski JA (2016c) Supercritical water gasification of timothy grass as an energy crop in the presence of alkali carbonate and hydroxide catalysts. Biomass Bioenergy 95:378–387
Nanda S, Isen J, Dalai AK, Kozinski JA (2016d) Gasification of fruit wastes and agro-food residues in supercritical water. Energ Convers Manage 110:296–306
Nanda S, Kozinski JA, Dalai AK (2016e) Lignocellulosic biomass: a review of conversion technologies and fuel products. Curr Biochem Eng 3:24–36
Nanda S, Reddy SN, Mitra SK, Kozinski JA (2016f) The progressive routes for carbon capture and sequestration. Energ Sci Eng 4:99–122

Nanda S, Golemi-Kotra D, McDermott JC, Dalai AK, Gökalp I, Kozinski JA (2017a) Fermentative production of butanol: perspectives on synthetic biology. New Biotechnol 37:210–221

Nanda S, Gong M, Hunter HN, Dalai AK, Gökalp I, Kozinski JA (2017b) An assessment of pinecone gasification in subcritical, near-critical and supercritical water. Fuel Process Technol 168:84–96

Nanda S, Rana R, Sarangi PK, Dalai AK, Kozinski JA (2018a) A broad introduction to first, second and third generation biofuels. In: Sarangi PK, Nanda S, Mohanty P (eds) Recent advancements in biofuels and bioenergy utilization. Springer Nature, Singapore, pp 1–25

Nanda S, Reddy SN, Vo DVN, Sahoo BN, Kozinski JA (2018b) Catalytic gasification of wheat straw in hot compressed (subcritical and supercritical) water for hydrogen production. Energ Sci Eng 6:448–459

Nanda S, Rana R, Hunter HN, Fang Z, Dalai AK, Kozinski JA (2019) Hydrothermal catalytic processing of waste cooking oil for hydrogen-rich syngas production. Chem Eng Sci 195:935–945

Okolie JA, Rana R, Nanda S, Dalai AK, Kozinski JA (2019) Supercritical water gasification of biomass: a state-of-the-art review of process parameters, reaction mechanisms and catalysis. Sustain Energ Fuel 3:578–598

Papoutsakis ET (2008) Engineering solventogenic clostridia. Curr Opin Biotechnol 19:420–429

Prasad S, Singh A, Joshi HC (2007) Ethanol as an alternative fuel from agricultural, industrial and urban residues. Resour Conserv Recycl 50:1–39

Qi Z, Hollett M, Attia A, Kaufman A (2002) Low temperature direct 2-propanol fuel cells. Electrochem Solid-State Lett 5:A129–A130

Qureshi N, Blaschek HP (1999) Production of acetone butanol ethanol (ABE) by a hyperproducing mutant strain of *Clostridium beijerinckii* BA101 and recovery by pervaporation. Biotechnol Prog 15:594–602

Qureshi N, Ezeji TC (2008) Butanol, 'a superior biofuel' production from agricultural residues (renewable biomass): recent progress in technology. Biofuels Bioprod Biorefin 2:319–330

Qureshi N, Saha BC, Cotta MA (2007) Butanol production from wheat straw hydrolysate using *Clostridium beijerinckii*. Bioprocess Biosyst Eng 30:419–427

Sarangi PK, Nanda S (2018) Recent developments and challenges of acetone-butanol-ethanol fermentation. In: Sarangi PK, Nanda S, Mohanty P (eds) Recent advancements in biofuels and bioenergy utilization. Springer Nature, Singapore, pp 111–123

Sheldon RA, Sanders JPM (2015) Toward concise metrics for the production of chemicals from renewable biomass. Catal Today 239:3–6

Shen CR, Liao JC (2008) Metabolic engineering of *Escherichia coli* for 1-butanol and 1-propanol production via the keto-acid pathways. Metab Eng 10:312–320

Shen CR, Liao JC (2013) Synergy as design principle for metabolic engineering of 1-propanol production in *Escherichia coli*. Metab Eng 17:12–22

Surisetty VR, Dalai AK, Kozinski J (2011) Alcohols as alternative fuels: an overview. Appl Catal A Gen 404:1–11

Szulczyk KR (2010) Which is a better transportation fuel—butanol or ethanol? Int J Energ Environ 1:501–512

The Knowledge Transfer Network (KTN) (2016) From shale gas to biomass: the future of chemical feedstocks. Horsham, UK

USEIA, U.S. Energy Information Administration (2011) International Energy Outlook 2011. http://www.worldenergyoutlook.org. Accessed 15 Dec 2011

USEIA, U.S. Energy Information Administration (2016) International Energy Statistics. https://www.eia.gov/. Accessed 1 June 2017

Uyttebroek M, Hecke WV, Vanbroekhoven K (2015) Sustainability metrics of 1-butanol. Catal Today 239:7–10

Wells M (1999) Handbook of petrochemicals and processes. Ashgate Publishing Co, Brookfield, VT

Wilkins MR, Atiyeh HK (2011) Microbial production of ethanol from carbon monoxide. Curr Opin Biotechnol 22:326–330

Wypych G (2001) Handbook of solvents, vol 1, 2nd edn. ChemTec Publishing, Ontario, p 870

Yemshanov D, McKenney D (2008) Fast-growing poplar plantations as a bioenergy supply source for Canada. Biomass Bioenergy 32:185–197

Yu M, Zhang Y, Tang IC, Yang ST (2011) Metabolic engineering of *Clostridium tyrobutyricum* for *n*-butanol production. Metab Eng 13:373–382

Zhang A, Yang ST (2009) Engineering *Propionibacterium acidipropionici* for enhanced propionic acid tolerance and fermentation. Biotechnol Bioeng 104:766–773

Zhao N, Xu R, Wei W, Sun Y (2002) Cu/Mn/ZrO_2 catalyst for alcohol synthesis by Fischer-Tropsch modified elements. React Kinet Catal Lett 75:297–304

Zimmermann H, Walzl R (2002) Ethylene. In: Wiley-VCH (ed) Ullmann's encyclopedia of industrial chemistry. Wiley-VCH, Weinheim

Technological Advancements in the Production and Application of Biomethanol

6

Prakash K. Sarangi, Sonil Nanda, and Dai-Viet N. Vo

Abstract

There is a worldwide concern for the emission of greenhouse gases from the overdependence and exploitation of fossil fuels. Several technological and industrial infrastructures heavily rely on fossil fuel-based energy systems. There is a paradigm shift in replacing fossil fuels with cleaner fuels derived from renewable biomass due to their carbon neutrality, sustainability, and global acceptance. Therefore, renewable energy sources are considered as alternative energy pool for sustainable purposes. In this context, biomass is seen as a promising option with great potential for the conversion to biofuels and biochemicals. Biomethanol is one of such biochemicals, which can be produced from biomass and biogenic wastes through thermochemical and biological routes. In this chapter, the industrial applications and production pathways of biomethanol is comprehensively discussed. The challenges and perspectives of several technologies used in biomethanol production are summarized.

Keywords

Biomethanol · Biomass · Biofuels · Biochemical · Methane · Hydrogen

P. K. Sarangi (✉)
Directorate of Research, Central Agricultural University, Imphal, Manipur, India
e-mail: sarangi77@yahoo.co.in

S. Nanda
Department of Chemical and Biological Engineering, University of Saskatchewan, Saskatoon, Saskatchewan, Canada

D.-V. N. Vo
Center of Excellence for Green Energy and Environmental Nanomaterials, Nguyễn Tất Thành University, Hồ Chí Minh City, Vietnam

S. Nanda et al. (eds.), *Biorefinery of Alternative Resources: Targeting Green Fuels and Platform Chemicals*, https://doi.org/10.1007/978-981-15-1804-1_6

6.1 Introduction

The rapid urbanization and infrastructural developments in automobile and industrial sectors are continually increasing worldwide energy demands, which mostly rely on fossil fuels. Excessive use of fossil fuels emit toxic pollutants and greenhouse gases, thus causing global warming and ecological imbalance (Nanda et al. 2016c). Several environmental concerns such as pollution, energy shortages, and global warming have led the researchers around the world to seek alternative energy sources that are clean and carbon-neutral (Nanda et al. 2015).

Many biomass resources such as lignocellulosic biomass, municipal solid wastes, industrial effluents, and cattle manure are utilized for bioenergy, biofuel, biochemical, and biomaterial production to have both economic and ecological benefits (Nanda et al. 2013, 2018). Waste biomass and biogenic wastes can be converted to biofuels and biochemicals through thermochemical (e.g., pyrolysis, gasification, liquefaction, and carbonization) and biological (e.g., fermentation and anaerobic digestion) conversion technologies (Nanda et al. 2014b, 2016b, 2017d). Specific to the conversion technologies, different fuel products can be produced from waste biomass, which include bio-oil, biodiesel, syngas, biohydrogen, bioethanol, biobutanol, biomethanol, and biopropanol (Reddy et al. 2014, 2018; Nanda et al. 2014a, 2017a; Sarangi and Nanda 2018).

Methanol is a simple alcohol otherwise called as wood spirits, wood alcohol, or methyl alcohol. It is a promising alcohol traditionally used as a platform chemical, solvent, and recently as an alternative fuel. Different thermochemical and biochemical processes can be exploited for its production. Having myriad applications in day-to-day life, the value chain of methanol has gained prominent dimensions toward production of high-value chemicals and materials. Like ethanol, methanol is colorless, flammable, and liquid at room temperature with a distinctive odor. Some other characteristics of methanol are polar at room temperature, miscible with water, petroleum, and organic compounds.

Methanol is obtained as a by-product of biodiesel refineries along with free fatty acids, crude glycerol, wastewater, and alkali catalysts (Ortiz et al. 2011). Methanol can undergo reforming reaction in supercritical water to produce hydrogen-rich syngas (Van Bennekom et al. 2011 Reddy et al. 2016). There is wide array of feedstocks for the production of methanol, which includes natural gas, coal, and waste biomass. The production of methanol from renewable biomass is regarded as one of the renewable methods when compared to its production from natural gas. Presently, one of the main routes of biomethanol production is through Fischer-Tropsch catalysis from synthesis gas (Venvik and Yang 2017). In this chapter, different applications of biomethanol are extensively discussed along with alternative production technologies from biomass and methane. The potential of different biomasses for biomethanol production is also summarized.

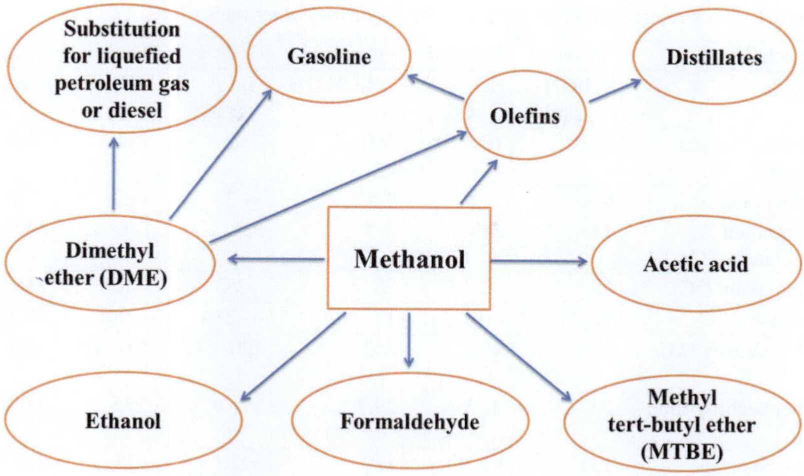

Fig. 6.1 Applications of methanol in fuel and chemical industries

6.2 Industrial Applications of Methanol

Methanol has wide range of applications as far as energy and chemical industries are concerned (Fig. 6.1). Methanol had received a great attention for supplementing fuel oil in motor vehicles in the USA during the global fuel crisis in the 1970s. Methanol is widely used in chemical industries, plastics, paints, and explosives. In winter, methanol can be used as an antifreeze agent in pipelines and windshield washer fluids in vehicles. Thus, a wide range of applications of methanol has been detected from simple chemical industries to novel applications like energy, transportation, and fuel cells.

Methanol has several important fuel applications as mentioned below:

1. Used as transportation fuel
2. Blending with petroleum
3. Converted into dimethyl ether (DME) for use as diesel alternative
4. Fuel cell application to generate electricity
5. Used in transesterification to produces biodiesel

Table 6.1 makes a comparison of the fuel properties of methanol along with other alcohol and hydrocarbon fuels. The major application of methanol has been recognized in transportation sector due to its sustainability, carbon neutrality, and cost-effectiveness compared to other fuels. The favorable combustion characteristics of methanol can help in increasing the engine performance and efficiencies when compared to gasoline (Bromberg and Cheng 2010). Another advantage of methanol over petroleum-based fuels is its environmentally benign nature and biodegradability (Malcolm 1999).

Table 6.1 Comparison of fuel properties of methanol with other fuels

Characteristic	Methanol	Gasoline	Ethanol	Hydrogen	Methane	Butanol
Formula	CH_3OH	H, C_4-C_{12}	CH_3CH_2OH	H_2	CH_4	C_4H_9OH
Molecular weight (kg/kmol)	32.04	100–105	46.1	2.02	16.04	74.1
Air/fuel ratio	6.5	14.6	9.0	34.3	17.2	11.2
Autoignition temperature (°C)	435	280	365	585	540	343
Boiling point (°C)	65	32–210	78	−253	−162	118
Flash point (°C)	11	−43	13	<−253	−188	28. 9
Energy density (MJ/kg)	19.6	44.5	26.9	120–142	50–55.5	33.1
Higher heating value (kJ/g)	20	47.5	29.7	141.9	55.5	37.3
Research octane number	136	91–99	129	>130	120	96
Viscosity at 25 °C (mPa·s)	0.54	0.6	1.07	0.009	0.011	2.6

References: MacLean and Lave (2003), Dürre (2007), Lee et al. (2008), Szulczyk (2010), Surisetty et al. (2011), Sarangi and Nanda (2018)

As methanol burns at a low temperature, it is superior to that of gasoline as far as transportation sector is concerned (Adamson and Pearson 2000). There is low risk of explosion due to low volatility nature of methanol. Furthermore, transportation of methanol is very easy in its liquid form, which is superior to other gaseous fuels such as natural gas and hydrogen. Its volumetric energy content is less than that of gasoline, thereby requiring nominal changes for fuel distribution purposes. As far as fuel parameters are concerned, it has greater octane number than gasoline (Fatih et al. 2011). Being an attractive fuel substitute, it is used as a blending fuel with gasoline in M85 (85% methanol and 15% unleaded gasoline), which does not need any kind of technical modifications to the engines (Yanju et al. 2008). Thus, methanol is a promising alternative for high-compression engine applications with low cost and increased power.

Biomethanol has the potential for power generation (Suntana et al. 2009). Methanol is regarded as a clean and efficient substitute for gas turbines and can be used in power industries (Galindo and Badr 2007). Methanol shows a wide range of downstream applications like fuel cell-powered vehicles due to easily degradable to CO_2 and H_2 in the presence of steam. Methanol is also focused as a liquid hydrogen carrier and as a hydrogen storage compound. In addition to chemicals and automobile fuels, methanol acts as the substrate for production of biodiesel through transesterification. A wide variety of chemicals is also produced from methanol like acetic acid formaldehyde, methyl tertiary butyl ether (MTBE), and dimethyl ether (DME). In addition, methanol acts as an anti-frost agent and organic solvent (Ptasinski et al. 2002).

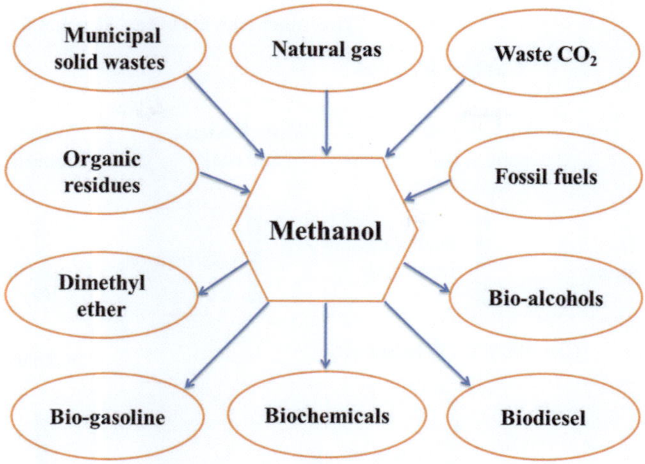

Fig. 6.2 Major sources of production and applications of methanol

About 0.19 tons of biomethanol was produced per ton of fossil fuel, resulting in a reduction of 0.42 MT of CO_2 emissions annually (Ptasinski et al. 2002). Although methanol, ethanol, and dimethyl ether are alternative fuels, major advantages of methanol have been detected in higher CO_2 reduction (Leduc et al. 2010). As far as environmental threats are concerned, CO_2 recycling is necessary for its reduction in atmospheric concentration. Hence, the production of methanol from waste CO_2 is another way of reducing greenhouse gases from the atmosphere. Methanol is produced using CO_2 from flue gas and H_2 through CO_2 reforming or Fischer-Tropsch synthesis (Treacy and Ross 2004; Singh et al. 2018). Therefore, methanol can be considered as a suitable substrate for the production of fuel, chemical, and energy for future generations. A wide array of sources from which methanol can be produced and applied in various sectors as shown in Fig. 6.2.

6.3 Thermochemical Production of Biomethanol

The production of methanol from biomass can be possible by two important methods such as thermochemical and biological process (Fig. 6.3). The thermochemical route involves high temperatures and pressures to produce synthesis gas (syngas) from biomass through gasification, which includes CO, CO_2, H_2, and CH_4 (Nanda et al. 2017b; Azargohar et al. 2019; Okolie et al. 2019). The conditioning of syngas can help to remove many impurities like tars and methane, thereby adjusting the ratio of H_2 to CO up to a level of 2:1. Fischer-Tropsch catalysis or reforming process involves catalytic reaction of syngas over a suitable metal catalyst at high temperatures and pressure to produce methanol.

Depending on the gasifier, biomass is crushed and dried to reduce the moisture content to less than 10 wt%. Biomass is dried using hot flue gas or steam. Biomass

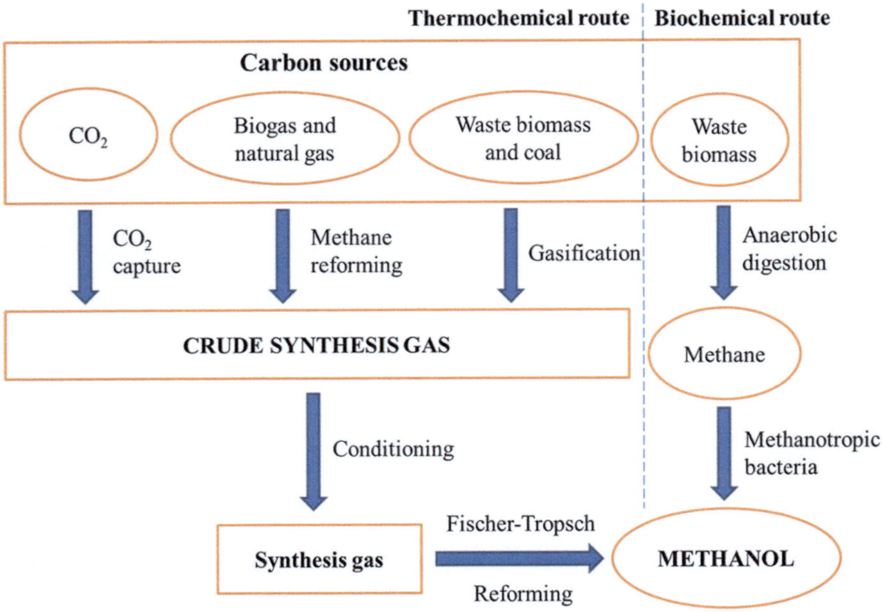

Fig. 6.3 Production of methanol by thermochemical and biochemical routes

is converted to produce a gaseous mixture of CO, CO_2, H_2, H_2O, and light hydrocarbons through gasification and steam reforming. Steam reforming is a process for the production of synthesis gas from natural gas to produce CO and H_2 through nickel or other suitable metal catalysts at high temperatures (Katofsky 1993). The synthesis gas containing H_2, CH_4, and other hydrocarbons accounts for a vital part to determine the heating value of the gas. Autothermal reforming (ATR) facilitates partial oxidation followed by steam reforming (Nanda et al. 2017c). The shift of energy value of CO to H_2 can be performed by water-gas shift reaction (Eq. 6.4).

$$CH_4 + H_2O \rightarrow CO + 3H_2 \tag{6.1}$$

$$C_2H_4 + 2H_2O \rightarrow 2CO + 4H_2 \tag{6.2}$$

$$C_2H_6 + 2H_2O \rightarrow 2CO + 5H_2 \tag{6.3}$$

$$CO + H_2O \leftrightarrow CO_2 + H_2 \tag{6.4}$$

The presence of considerable amount of CO_2 in the synthesis gas can be removed, which is increased after reforming method. To achieve the proper value in the ratio $(H_2 + CO_2)/(CO + CO_2)$ for methanol production, some amount of CO_2 can be removed. There are various types of techniques for removal of CO_2 such as physical and chemical processes. The most common method commercially proven is the chemical absorption using amines. Methanol synthesis takes places through the hydrogenation of CO and CO_2 over a suitable catalyst such as copper oxide, zinc oxide, or chromium oxide.

$$CO + 2H_2 \leftrightarrow CH_3OH \tag{6.5}$$

$$CO_2 + 3H_2 \leftrightarrow CH_3OH + H_2O \tag{6.6}$$

Methane, which is a major component of natural gas, is considered as a next-generation carbon feedstock (Hwang et al. 2014). Moreover, biomethane can be produced through anaerobic digestion of organic wastes by using methanogenic bacteria. Methane is produced from agricultural activities like enteric fermentation and rice cultivation (Environmental Protection Agency 2013). Nevertheless, CH_4 is a potent greenhouse gas, which has global warming potential nearly 25 times greater than CO_2 (Nanda et al. 2016c). The anthropogenic share of CH_4 emission was estimated to be 481 billion m^3 in 2010, which is almost equivalent to the global warming potential of 6867 Mt of CO_2 (Environmental Protection Agency 2013).

There is much potential for the conversion of methane to methanol. Due to the inert and saturated nature of methane, high temperature and high pressure are employed for its conversion into methanol. Different products such as formaldehyde, formic acid, CO, and CO_2 are also formed during the process of controlled oxidative conversion of methane to methanol (Gesser et al. 1985). A single-step process involving the oxidation of methane to methanol has shown as the direct way of conversion. This process is an exothermic reaction, thereby avoiding the syngas formation step (Bjorck et al. 2018).

The dissociation energy of C–H bond in methane (440 kJ/mol) is greater than that of methanol (393 kJ/mol). There is also a two-step process in the conversion of methane to methanol through which methane is first converted to syngas through steam reforming and the second step involves catalytic synthesis of methanol. The overall process for methanol production from natural gas or syngas is an endothermic reaction. The first step toward the conversion of CO and H_2 is mostly energy-intensive step that operates at very high temperature about 900 °C and pressures of 3 MPa (Wood et al. 2012). Naqvi et al. (2012) demonstrated methanol production through gasification of black liquor, thereby proving CO_2 as the potential substrate for biomethanol industries by reduction process.

According to Hamelinck and Faaij (2002), the production of methanol is achieved from biomass with a range between 54% and 58% of the net higher heating value (HHV) energy efficiency. During their investigations, they detected the HHV energy efficiency is about 55%. On the other hand, Williams et al. (1995) reported the range for HHV energy efficiencies to be between 56.6% and 67.7% depending on different gasification technologies. Isaksson et al. (2012) found the net HHV energy efficiencies of 51%. The thermal efficiency is the ratio between the energy content of the methanol on HHV basis to the sum of energy content of all primary energy input sources including biomass, electricity, and heat.

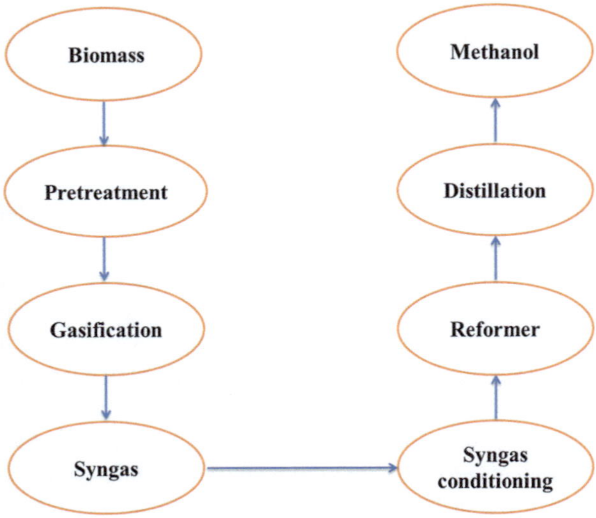

Fig. 6.4 A simplified route for biomethanol production from biomass

6.4 Biological Production of Biomethanol

The production of biomethanol is achieved via anaerobic metabolism by using several methanotrophic bacteria. During ethanol fermentation process, methanol is also formed as a by-product. The lignocellulosic biomass containing cellulose, hemicelluloses, and lignin can be used for the conversion to a wide range of chemicals and fuels (Nanda et al. 2016a). Lignocellulosic biomass including agricultural crop residues and forestry biomass contain hexose and pentose sugars, which can be extracted through a suitable physical, chemical, or biological pretreatment for conversion to alcohol-based biofuels (Sarangi and Nanda 2019). By avoiding the final step (i.e., distillation), crude methanol with about 80% purity is obtained. Being an intermediate grade product, the major application of such crude methanol is found as a marine fuel (Brynolf et al. 2014; Winnes et al. 2015). A simplified thermochemical version of biomass conversion into biomethanol is shown in Fig. 6.4.

Nakagawa et al. (2007) reported methanol yields of 55%, 36%, and 39% from rice bran, rice straw, and rice husks, respectively. According to the experiments conducted by Chia (2011), the productions of biomethanol from sugarcane bagasse and corncob were 5.93% and 0.67%, respectively. Other organic residues that have demonstrated the production of methanol are animal and livestock manure (Weimer et al. 1996), industrial effluents and sewage sludge (Caballero et al. 1997), and municipal solid wastes (Bhattacharyya et al. 2008). The production of methanol is possible by the degradation of methyl esters. The conversion of pectin releases uronic acid, which helps in the formation of methanol by the combination of methoxyl group of uronic acid with ether (Bhattacharyya et al. 2008).

The potential of methanotrophic bacteria such as *Bacillus methanicus* has been focused for the conversion of methane as a sole source of carbon and energy to

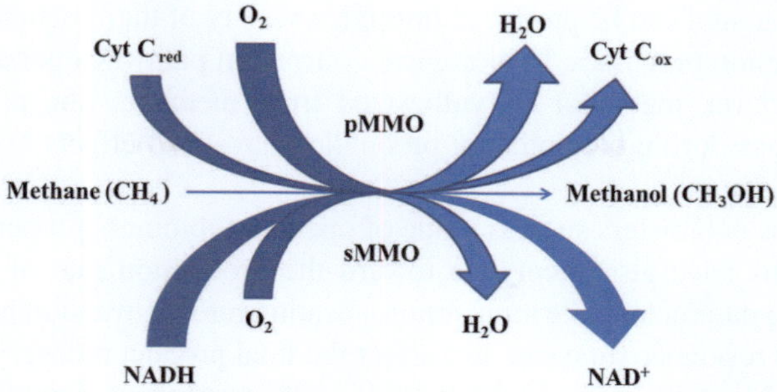

Fig. 6.5 Conversion of methane to methanol by methanotrophic bacteria

methanol by partial oxidation process with efficient and high conversion rate (Bender and Conrad 1992; Hanson and Hanson 1996). The biocatalytic action of methanotrophic bacteria helps in conversion of methane to methanol. These methane-oxidizing bacteria are naturally present in soil, swamps, rivers, oceans, ponds, and sewage sludge representing nearly about up to 8% of the total heterotrophic population (Higgins et al. 1981). *Methylosinus trichosporium* is one of the notable microorganisms responsible for biomethanol production (Mehta et al. 1987; Sugimori et al. 1995; Takeguchi et al. 1997; Xin et al. 2009a, b).

There are two categories of methanotrophic microorganisms, especially low affinity and high affinity. The low-affinity methanotrophic bacteria have the potential to degrade methane at a higher concentrations level (i.e., >40 ppm). These are generally found in soils having high methane concentrations. On the other hand, the high-affinity methanotrophs use low concentration of methane (i.e., ~2 ppm). The biocatalytic function of methane monooxygenase (MMO) enzymes helps in the conversion of methane into methanol through oxidation process (Fig. 6.5). Two forms of MMO have been detected in methanotrophs such as soluble cytoplasmic form (sMMO) and particulate membrane-bound form (pMMO). The conversion processes of methanol take place as follows (Hanson and Hanson 1996; Xin et al. 2009a, b):

1. Oxidation of methanol to formaldehyde by methanol dehydrogenase (MDH)
2. Conversion of formaldehyde to formic acid by formaldehyde dehydrogenase (FADH)
3. Conversion of formic acid to CO_2 by formate dehydrogenase (FDH)

6.5 Conclusions

As the energy security and environmental sustainability are concerned, waste biomass is the potential alternative for biofuel and biochemical production. The production of biomethanol from renewable waste materials has potential applications in automobile and industrial sectors supplying fuel and energy. Depending on the feedstock, methanol can be produced through a variety of thermochemical and biological conversion processes. In biological conversion pathway operated by methanotrophic bacteria, methanol is synthesized from methane. The production and purity of syngas during biomethanol production have also effects toward the final product recovery.

The process parameters such as mode of operation, biomass properties, and process conditions have also great role toward the production rate of biomethanol. Other reaction parameters like temperature, heating rate, conversion processes, particle size, and residence time can also affect the final product recovery. Overall, the economics of biomethanol not only depends on the production parameters and conditions but also on the purity grade of the final product. Significant improvements in the conversion process can result in the large-scale production of methanol for applications in fuel and chemical industries.

References

Adamson KA, Pearson P (2000) Hydrogen and methanol: a comparison of safety, economics, efficiencies and emissions. J Power Sources 86:548–555

Azargohar R, Nanda S, Dalai AK, Kozinski JA (2019) Physico-chemistry of biochars produced through steam gasification and hydro-thermal gasification of canola hull and canola meal pellets. Biomass Bioenergy 120:458–470

Bender M, Conrad R (1992) Kinetics of CH_4 oxidation in oxic soils exposed to ambient air or high CH_4 mixing ratios. FEMS Microbiol Ecol 101:261–270

Bhattacharyya JK, Kumar S, Devotta S (2008) Studies on acidification in two-phase biomethanation process of municipal solid waste. Waste Manag 28:164–169

Bjorck CE, Dobson PD, Pandhal J (2018) Biotechnological conversion of methane to methanol: evaluation of progress and potential. AIMS Bioeng 5:1–38

Bromberg L, Cheng WK (2010) Methanol as an alternative transportation fuel in the US: Options for sustainable and/or energy-secure transportation. Sloan Automotive Laboratory, MIT, Cambridge, MA, Report No.: 4000096701:78

Brynolf S, Fridell E, Andersson K (2014) Environmental assessment of marine fuels: liquefied natural gas, liquefied biogas, methanol and bio-methanol. J Clean Prod 74:86–95

Caballero JA, Front R, Marcilla A, Coma JA (1997) Characterization of sewage sludges by primary and secondary pyrolysis. J Anal Appl Pyrolysis 40–41:433–450

Chia SK (2011) Production of bio-methanol from agricultural waste by pyrolysis method, Thesis. Department of Chemical and Process Engineering, National University of Malaysia, Bandar Baru Bangi

Dürre P (2007) Biobutanol: an attractive biofuel. Biotechnol J 2:1525–1534

Environmental Protection Agency (US) (2013) Global mitigation of non-CO_2 greenhouse gases, Washington, DC. Report No.: EPA 430-R-13-011:410

Fatih DM, Balat M, Balat H (2011) Biowastes to biofuels. Energy Convers Manag 52:1815–1828

Galindo CP, Badr O (2007) Renewable hydrogen utilisation for the production of methanol. Energ Convers Manage 48:519–527

Gesser H, Hunter N, Prakash C (1985) The direct conversion of methane to methanol by controlled oxidation. Chem Rev 85:235–244

Hamelinck CN, Faaij APC (2002) Future prospects for production of methanol and hydrogen from biomass. J Power Sources 111:1–22

Hanson RS, Hanson TE (1996) Methanotrophic bacteria. Microbiol Rev 60:439–471

Higgins IJ, Best DJ, Hammond RC, Scott D (1981) Methane-oxidizing microorganisms. Microbiol Rev 45:556–590

Hwang IY, Lee SH, Choi YS, Park SJ, Na JG, Chang IS, Kim C, Kim HC, Kim YH, Lee JW, Lee EY (2014) Biocatalytic conversion of methane to methanol as a key step for development of methane-based biorefineries. J Microbiol Biotechnol 24:1597–1605

Isaksson J, Pettersson K, Mahmoudkhani M, Åsblad A, Berntsson T (2012) Integration of biomass gasification with a Scandinavian mechanical pulp and paper mill - consequences for mass and energy balances and global CO_2 emissions. Energy 44:420–428

Katofsky RE (1993) The production of fluid fuels from biomass. Princeton University, Center for Energy and Environmental Studies, Princeton, NJ

Leduc S, Schmid E, Obersteiner M, Riahib K (2010) Methanol production by gasification using a geographically explicit model. Appl Energy 87:68–75

Lee SY, Park JH, Jang SH, Nielsen LK, Kim J, Jung KS (2008) Fermentative butanol production by clostridia. Biotechnol Bioeng 101:209–228

MacLean HL, Lave LB (2003) Evaluating automobile fuel/propulsion system technologies. Prog Energy Combust Sci 29:1–69

Malcolm PI (1999) Evaluation of the fate and transport of methanol in the environment. American Methanol Institute, Washington DC Report No: 3522-002:69

Mehta P, Mishra S, Ghose T (1987) Methanol accumulation by resting cells of *Methylosinus trichosporium* (I). J Gen Appl Microbiol 33:221–229

Nakagawa H, Harada T, Ichinose T, Takeno K, Matsumoto S, Kobayashi M, Sakai M (2007) Bio-methanol production and CO_2 emission reduction from forage grasses, trees, and crop residues. Jpn Agric Res Q 41:173–180

Nanda S, Mohanty P, Pant KK, Naik S, Kozinski JA, Dalai AK (2013) Characterization of North American lignocellulosic biomass and biochars in terms of their candidacy for alternate renewable fuels. Bioenergy Res 6:663–677

Nanda S, Dalai AK, Kozinski JA (2014a) Butanol and ethanol production from lignocellulosic feedstock: biomass pretreatment and bioconversion. Energ Sci Eng 2:138–148

Nanda S, Mohammad J, Reddy SN, Kozinski JA, Dalai AK (2014b) Pathways of lignocellulosic biomass conversion to renewable fuels. Biomass Conv Bioref 4:157–191

Nanda S, Azargohar R, Dalai AK, Kozinski JA (2015) An assessment on the sustainability of lignocellulosic biomass for biorefining. Renew Sustain Energy Rev 50:925–941

Nanda S, Dalai AK, Kozinski JA (2016a) Supercritical water gasification of timothy grass as an energy crop in the presence of alkali carbonate and hydroxide catalysts. Biomass Bioenergy 95:378–387

Nanda S, Kozinski JA, Dalai AK (2016b) Lignocellulosic biomass: a review of conversion technologies and fuel products. Curr Biochem Eng 3:24–36

Nanda S, Reddy SN, Mitra SK, Kozinski JA (2016c) The progressive routes for carbon capture and sequestration. Energ Sci Eng 4:99–122

Nanda S, Golemi-Kotra D, McDermott JC, Dalai AK, Gökalp I, Kozinski JA (2017a) Fermentative production of butanol: perspectives on synthetic biology. New Biotechnol 37:210–221

Nanda S, Gong M, Hunter HN, Dalai AK, Gökalp I, Kozinski JA (2017b) An assessment of pinecone gasification in subcritical, near-critical and supercritical water. Fuel Process Technol 168:84–96

Nanda S, Li K, Abatzoglou N, Dalai AK, Kozinski JA (2017c) Advancements and confinements in hydrogen production technologies. In: Dalena F, Basile A, Rossi C (eds) Bioenergy systems for the future. Woodhead Publishing, Elsevier, pp 373–418

Nanda S, Rana R, Zheng Y, Kozinski JA, Dalai AK (2017d) Insights on pathways for hydrogen generation from ethanol. Sustain Energ Fuel 1:1232–1245

Nanda S, Rana R, Sarangi PK, Dalai AK, Kozinski JA (2018) A broad introduction to first, second and third generation biofuels. In: Sarangi PK, Nanda S, Mohanty P (eds) Recent advancements in biofuels and bioenergy utilization. Springer Nature, Singapore, pp 1–25

Naqvi M, Yan J, Dahlquist E (2012) Bio-refinery system in a pulp mill for methanol production with comparison of pressurized black liquor gasification and dry gasification using direct causticization. Appl Energy 90:24–31
Okolie JA, Rana R, Nanda S, Dalai AK, Kozinski JA (2019) Supercritical water gasification of biomass: a state-of-the-art review of process parameters, reaction mechanisms and catalysis. Sustain Energ Fuel 3:578–598
Ortiz FG, Ollero P, Serrera A, Sanz A (2011) Thermodynamic study of the supercritical water reforming of glycerol. Int J Hydrogen Energ 36:8994–9013
Ptasinski KJ, Hamelinck C, Kerkhof PJAM (2002) Exergy analysis of methanol from the sewage sludge process. Energ Convers Manage 43:1445–1457
Reddy SN, Nanda S, Dalai AK, Kozinski JA (2014) Supercritical water gasification of biomass for hydrogen production. Int J Hydrogen Energ 39:6912–6926
Reddy SN, Nanda S, Kozinski JA (2016) Supercritical water gasification of glycerol and methanol mixtures as model waste residues from biodiesel refinery. Chem Eng Res Des 113:17–27
Reddy SN, Nanda S, Sarangi PK (2018) Applications of supercritical fluids for biodiesel production. In: Sarangi PK, Nanda S, Mohanty P (eds) Recent advancements in biofuels and bioenergy utilization. Springer Nature, Singapore, pp 261–284
Sarangi PK, Nanda S (2018) Recent developments and challenges of acetone-butanol-ethanol fermentation. In: Sarangi PK, Nanda S, Mohanty P (eds) Recent advancements in biofuels and bioenergy utilization. Springer Nature, Singapore, pp 111–123
Sarangi PK, Nanda S (2019) Recent advances in consolidated bioprocessing for microbe-assisted biofuel production. In: Nanda S, Sarangi PK, Vo DVN (eds) Fuel processing and energy utilization. CRC Press, Boca Raton, FL, pp 141–157
Singh S, Kumar R, Setiabudi HD, Nanda S, Vo DVN (2018) Advanced synthesis strategies of mesoporous SBA-15 supported catalysts for catalytic reforming applications: a state-of-the-art review. Appl Catal A Gen 559:57–74
Sugimori D, Takeguchi M, Okura I (1995) Biocatalytic methanol production from methane with *Methylosinus trichosporium* OB3b: an approach to improve methanol accumulation. Biotechnol Lett 17:783–784
Suntana AS, Vogt KA, Turnblom EC, Ravi U (2009) Bio-methanol potential in Indonesia: forest biomass as a source of bio-energy that reduces carbon emissions. Appl Energy 86:215–221
Surisetty VR, Dalai AK, Kozinski J (2011) Alcohols as alternative fuels: an overview. Appl Catal A Gen 404:1–11
Szulczyk KR (2010) Which is a better transportation fuel—butanol or ethanol? Int J Energ Environ 1:501–512
Takeguchi M, Furuto T, Sugimori D, Okura I (1997) Optimization of methanol biosynthesis by *Methylosinus trichosporium* OB3b: an approach to improve methanol accumulation. Appl Biochem Biotechnol 68:143–152
Treacy D, Ross JRH (2004) The potential of the CO_2 reforming of CH_4 as a method of CO_2 mitigation. A thermodynamic study. Prepr Pap-Am Chem Soc Div Fuel Chem 49:126–127
Van Bennekom JG, Venderbosch RH, Assink D, Heeres HJ (2011) Reforming of methanol and glycerol in supercritical water. J Supercrit Fluids 58:99–113
Venvik HJ, Yang J (2017) Catalysis in microstructured reactors: short review on small-scale syngas production and further conversion into methanol, DME and Fischer-Tropsch products. Catal Today 285:135–146
Weimer T, Schaber K, Specht M, Bandi A (1996) Methanol from atmospheric carbon dioxide: a liquid zero emission fuel for the future. Energ Convers Manage 37:1351–1356
Williams R, Larson E, Katofsky R, Chen J (1995) Methanol and hydrogen from biomass for transportation. Energy Sustain Dev 1:18–34
Winnes H, Styhre L, Fridell E (2015) Reducing GHG emissions from ships in port areas. Res Transp Bus Manag 17:73–82
Wood DA, Nwaoha C, Towler BF (2012) Gas-to-liquids (GTL): a review of an industry offering several routes for monetizing natural gas. J Nat Gas Sci Eng 9:196–208

Xin JY, Cui JR, Niu JZ, Hua SF, Li SB, Zhu LM (2009a) Production of methanol from methane by methanotrophic bacteria. Biocatal Biotransformation 22:225–229

Xin JY, Cui JR, Niu JZ, Hua SF, Xia CG, Li SB, Zhu LM (2009b) Production of methanol from methane by methanotrophic bacteria. Biocatal Biotransformation 22:225–229

Yanju W, Shenghua L, Hongsong L, Rui Y, Jie L, Ying W (2008) Effects of methanol/gasoline blends on a spark ignition engine performance and emissions. Energy Fuel 222:1254–1259

Biorefinery Approaches for the Production of Fuels and Chemicals from Lignocellulosic and Algal Feedstocks

7

Venkateswara R. Naira, R. Mahesh, Suraj K. Panda, and Soumen K. Maiti

Abstract

The fuels and chemicals produced from the fossil-derived petroleum are on the verge of completion. As a result, petro-refinery-based commodity prices are rising despite their increasing demand. Moreover, the greenhouse gas emissions from petro-based fuels are devastating to the environment. Thus, the concerns over the search for alternative, renewable and green sources of fuels and platform chemicals have been raised. Among the explored bio-based sources, the biorefinery potential of lignocellulosic and algal feedstocks has been found to be the most promising due to their sustainability and versatility of multi-product generation. This chapter reviews the biorefinery approaches in various thermochemical and biological conversion routes of lignocellulosic and algal feedstocks for co-producing biofuels and biochemicals that can be widely applied in the industrial sectors like energy, food, pharmaceuticals, cosmetics and textiles. From lignocellulosic biomass, the co-production technologies of chemicals and fuels via lignin and sugar (C_5 and C_6) platforms have been focused. On the other hand, the generation of bioactive components such as lipids, carbohydrates and proteins from algae along with co-production of biofuels and biochemicals is also discussed. The utilization of recycle streams like crude glycerol and spent biomass residues is emphasized in the algal biorefineries. Finally, the current industrial scenarios of both lignocellulosic and algal biorefineries are outlined.

Keywords

Biorefinery · Lignocellulosic biomass · Algae · Biofuels · Biochemicals

V. R. Naira · R. Mahesh · S. K. Panda · S. K. Maiti (✉)
Department of Biosciences and Bioengineering, Indian Institute of Technology Guwahati, Guwahati, Assam, India
e-mail: soumenkm@iitg.ac.in

S. Nanda et al. (eds.), *Biorefinery of Alternative Resources: Targeting Green Fuels and Platform Chemicals*, https://doi.org/10.1007/978-981-15-1804-1_7

7.1 Introduction

Energy is one of the driving forces to meet the various demands of human population and technological advancements. Among various forms of energy, fossil fuels, especially crude petroleum and diesel, fulfil the needs of automobile industry and transportation. The rising prices, global energy crisis, depletion of fossil fuels, air and water pollution, greenhouse gas emissions, ozone layer depletion and global warming are some of the environmental concerns related to the exploiting usage of fossil fuels (Nanda et al. 2016b). In such a scenario, the energy security and collective demands for the use of renewable and alternative sources of energy such as biofuels (bioethanol, biodiesel and bio-oil) can be met by using lignocellulosic and algal biomasses (Nanda et al. 2015). Apart from biofuel, these renewable feedstocks can also be utilized for the production of biochemicals in an eco-friendly way via several biorefinery approaches.

The major obstacle in lignocellulosic and algal biomass biorefinery is the difficulty in the separation of value-added products without causing damage to the formation of other products simultaneously. Concomitant synthesis of different intracellular and extracellular product fractions in microbial cell factories would be an excellent biorefinery outlook to reduce the upstream and downstream processing costs. The synthesis of high-value products concurrently from biomass feedstocks reduces the overall cost of their production and yields huge profits. Lignocellulosic biomass and algal biomass can be subjected to different thermochemical (pyrolysis, liquefaction and gasification) and biochemical processes (saccharification, fermentation and anaerobic digestion) for the production of alternative fuels and high-value by-products (Nanda et al. 2014).

Biorefinery is the sustainable processing of biomass into a spectrum of bio-based products (e.g. food, feed, chemicals and materials) and bioenergy (e.g. biofuels, power and/or heat) (Jong et al. 2012). The biorefinery concept can be comparable with traditional crude petroleum oil refinery (Perez et al. 2017), except that the biorefinery involves the processing of biomass for biofuels and platform chemicals unlike the crude petroleum in fossil fuels. Firstly, biofuels from the second-generation (agricultural and woody biomass) and third-generation feedstocks (microalgae) have been developed for replacing the fossil fuels partially or completely. However, the final price of biofuel could not compete with the existing prices of fossil fuels due to the expensive upstream and downstream processing involved in biomass conversion technologies.

The current cost of biodiesel production from 60% oil containing microalgae (third-generation feedstock) is US$3.96–10.56 per L, which is nearly tenfold expensive than soybean-derived biodiesel (a first-generation feedstock) (Kapoore et al. 2018). Hence, biorefinery concept is introduced into both second- and third-generation biofuel platforms to produce value-added chemicals along with the respective biofuel to compensate the high costs of biomass processing. For example, when a lignocellulosic feedstock contains two-thirds of sugars (cellulose and hemicellulose) and one-third of lignin, 75% of total sugars for ethanol and 25% of sugars for platform chemicals are more economically viable than using all the sugars for ethanol (Yamakawa et al. 2018). Alongside the enhancement of product scope, researchers are also working extensively on process design and engineering for enhancing the net energy ratios. However, this chapter elucidates the recent trends in the production of platform chemicals and biofuels from the biorefinery of second- and third-generation feedstocks.

In this chapter, lignocellulosic wastes (second-generation feedstocks) and algae (third-generation feedstocks) are discussed due to their high potential and versatility in biorefinery towards the generation of biofuels and biochemicals. The conventional biofuel production from high-cost starch-based food crops such as sugarcane molasses and corn has raised global concerns for food security and economic viability. Hence, the search for a cheaper alternative, that is agro-based feedstocks like rice straw, wheat straw and sugarcane bagasse, has paved the way to explore lignocellulosic biomass for second-generation biofuels production. Such feedstocks are available in ample amounts and could be the sustainable resources for biofuels and biochemicals production. The lignocellulosic biomass also constitutes bamboo, wood from forestry wastes, municipal wastes, etc.

On the other hand, algal biomass production has also received a great interest as the third-generation feedstock due to its high growth rates, usability of non-arable lands and less intensive cultivation requirements (Chisti 2007). However, the biofuel production has not yet fully commercialized from this feedstock as the net energy ratios (NER) are relatively lower (Slade and Bauen 2013). Apart from solving the aspects of biomass production technologies, simultaneous generation of biofuels and value-added co-products from microalgae via biorefinery concept can greatly increases the net energy ratios by adding value to the final product ranges (Chia et al. 2018).

7.2 Biofuels and Platform Chemicals from Lignocellulosic and Algal Biomasses

Lignocellulosic biomass can be valorized into a wide range of fuels and platform chemicals that are extracted either from cellulose, hemicellulose or lignin fraction of biomass (Fig. 7.1). Lignin can be transformed into value-added chemicals like lignosulphonates, ferulic acid and vanillin which are applied as cement water reducer, binders dye dispersants, cosmetics, flavouring agents, etc. The cellulose/hemicellulose part can be hydrolysed into sugars such as glucose and xylose, which are biochemically converted into a broad range of liquid biofuels (e.g. bioethanol and biobutanol), solvents (e.g. acetone), organic acids (e.g. acetic acid, butyric acid and lactic acid for pharmaceutical and cosmetic applications) and other alcoholic compounds (e.g. isopropanol, isobutanol, 1,3-propane diol, xylitol for food and precursors of high-value industrial products). The hydrolysed sugars can also be converted into furfural, 5-hydroxymethylfurfural (HMF), levulinic acid, formic acid and 2,5-dimethyl furan (DMF) via chemical synthesis for use as precursors for pesticides, textile dyes and resin production. The residual bacterial and fungal biomasses used for saccharification and fermentation of lignocellulosic biomasses can also be used for producing

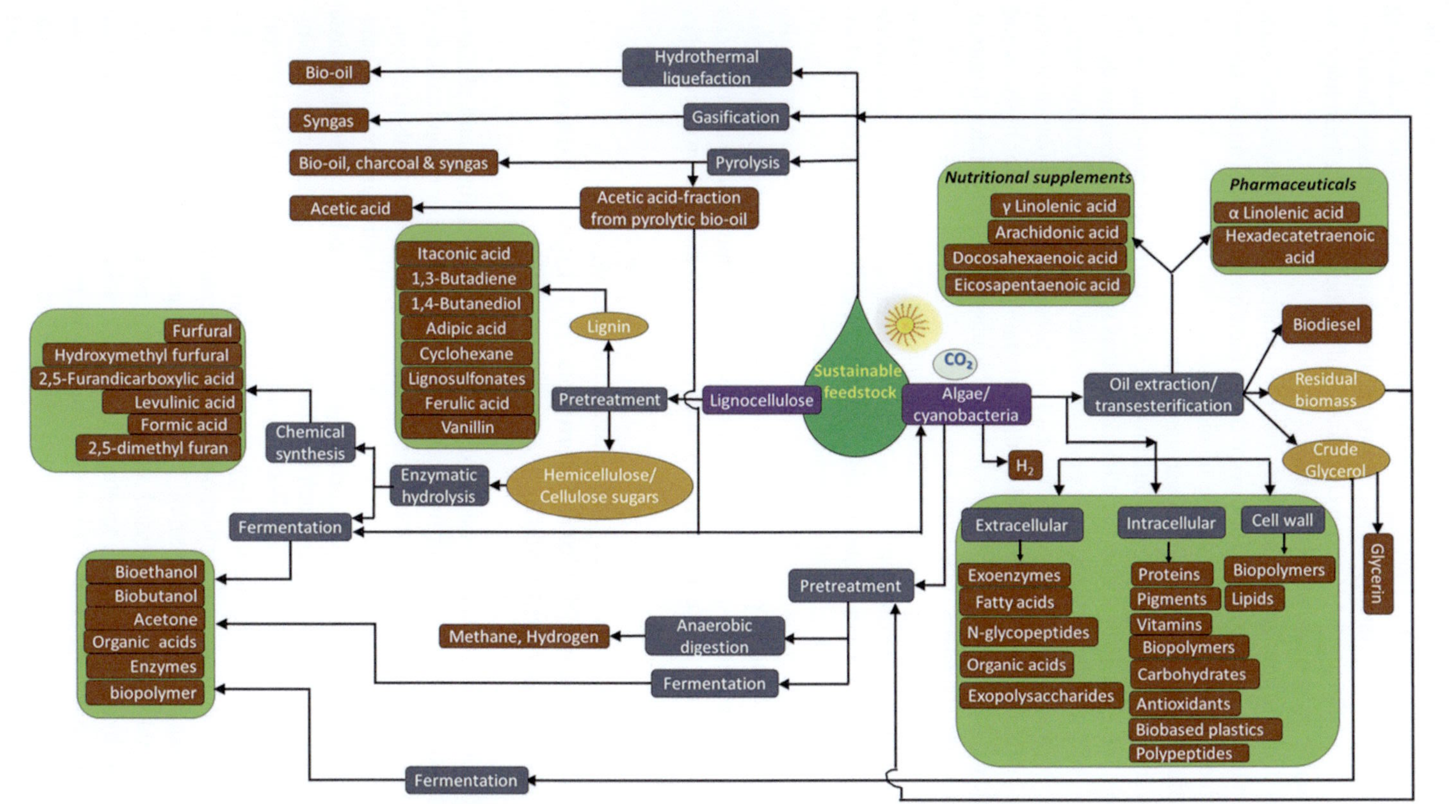

Fig. 7.1 The scope for production of biofuels and biochemicals from lignocellulosic biomass and algae

biopolymers like chitin, polyhydroxyalkanoates (PHA) and polyhydroxybutyrate (PHB), which are used for the manufacturing of bioplastics.

Microalgae mainly offer a wide range of bioproducts that are derived from their biomass directly by various downstream processing techniques. The scope of intracellular product generation includes carbohydrates, proteins, vitamins, lipids, specific polypeptides, pigments, biopolymers and high-value fatty acids. Lipids are the precursors for biodiesel production from the algae via transesterification process. The proteins like phycoerythrin, phycocyanin, allophycocyanin and phycobiliproteins have been extracted from algal sources extensively for producing fluorescent dyes and food pigments. Most valuable algae-derived pigments are carotenoids (β-carotene), fucoxanthin and astaxanthin, widely reported as dyes in food industries. Vitamins of classes B, C and E are also identified in algae (Kapoore et al. 2018).

Some specific strains of cyanobacteria and green algae are capable of photobiological evolution of hydrogen (a clean burning fuel) via biophotolysis of water. Some industrial algal species have been reported to produce extracellular products such as exopolysaccharides, extracellular proteins (e.g. exoenzymes, protease inhibitors and phycoerythrin-like proteins), organic acids, extracellular lipids, extracellular phytohormones and allelopathic chemicals which have a wide range of industrial applications in food, pharmaceuticals, cosmetics, textiles and detergent sectors.

7.3 Lignocellulosic Biorefineries

Lignocellulosic biomass is comprised of cellulose (30–35%) which is shielded by a matrix of hemicellulose (25–30%) and embedded by branched structure of lignin (15–20%) (Nanda et al. 2013). Cellulose is composed of microfibrils, which are homopolysaccharides made up of glucose units linked by β-1,4-glycosidic bonds. Hemicellulose is a branched structure mainly composed of C_5 (pentose) sugars such as xylose and arabinose and C_6 (hexose) sugars such as glucose, mannose, galactose, fructose and rhamnose. Lignin provides stiffness and rigidity to plant cell walls. Lignin is mainly composed of phenolic compounds.

Lignocellulosic biomass is very rigid and compact in nature due to its complex structure of cellulosic and hemicellulosic network embedded by the recalcitrant lignin. Moreover, the biomass has to be refined into small particle size for enhancing the homogeneity of subsequent biochemical and thermochemical conversion processes. Hence, the mechanical methods such as chipping, milling and grinding are used for reducing the particle size, cellulose crystallinity and degree of polymerization while increasing the flow properties, surface area and porosity of biomass. The commonly used milling methods are ball milling, knife milling, centrifugal milling and vibratory milling.

The conversion of pre-processed lignocellulosic biomass into fuels and value-added products in the biorefinery needs fractionation of biomass into component lignin, hemicellulose and cellulose. The recalcitrance of lignin and shielding effects of hemicellulose around the cellulose hinder the hydrolytic enzymes from penetrating the plant cell wall and producing fermentable sugars. Lignin can be removed by treating lignocellulosic biomass with alkali reagents (e.g. hydroxyl derivatives of sodium, potassium, calcium and ammonium salts) at ambient temperatures via hydrothermal treatment typically at 15 psi and 121 °C for 10–30 min of holding time. This treatment effectively degrades the lignin by breaking the side chains of the glycosides and esters of lignocellulosic biomass (Durot et al. 2003). Some other lignin-removal methods like ozonolysis, ionic liquid treatment (e.g. cholinium ionic liquid), and ammonia fibre explosion (at 60–90 °C and 3 MPa for 30–60 min) have also been reported (Kumar et al. 2009).

The lignin platform offers the production of various chemicals like syringic acid, *p*-coumaric acid, caffeic acid, syringaldehyde, naphthalene, 4-hydroxybenzaldehyde, 4-hydroxybenzoic acid, methylnaphthalene, ferulic acid, methanol, benzoic acid, terephthalic acid, eugenol and vanillin (Isikgor and Becer 2015). These chemicals have a range of applications in pharmaceuticals, nutraceutical and cosmetic industries. For example, ferulic acid is used as antioxidant in pharmaceuticals and cosmetic industries and as preservative in nutraceutical industries for the stabilization of vitamins C and E. Vanillin is also widely used as a flavouring agent and fragrant substances.

The fuel products of fermentation like bioethanol and biobutanol can be blended with gasoline because of their high octane number, complete combustion, less greenhouse gas emission characteristics, etc. The enzymatic products, e.g. cellulase enzymes, can be widely used in detergent industry, textile and paper industries for bleaching and deinking, and cellulosic biofuel production (Ravindran and Jaiswal 2016).

7.3.1 Pretreatment of Lignocellulosic Biomass

The above-mentioned pretreatments of lignocellulosic biomass is effective for lignin removal but not for the extraction of hemicellulose and cellulose. Hence, the delignified biomass has to be treated by various hydrothermal processes like steam explosion, dilute acid pretreatment (e.g. H_2SO_4, HCl and H_3PO_4) and ammonia percolation method. The steam explosion process is usually carried out with supersaturated steam at temperatures of 180–240 °C and pressures of 1–3.5 MPa with a biomass loading of 10–15%. The dilute acid pretreatment is carried out with an acid concentration of 1–4% and a temperature of 120 °C for 10–60 min with 10–15% of biomass loading. The ammonia percolation method is carried out with liquid ammonia (5–15% v/v) at temperatures of 140–200 °C with biomass loading of 10–30%.

Most of these pretreatment methods generate certain amount of by-products due to sugar degradation which generally inhibit the growth of fermenting microorganism or limit the production of end products. For example, the degradation of pentose sugar such as xylose from hemicellulose fraction forms furfural. Similarly, glucose also degrades into 5-hydroxymethylfurfural (HMF). Under severe pretreatment conditions (e.g. long reaction time, high temperature or high acid concentrations), HMF further degrades into levulinic and formic acid. During pretreatment, some other microbial growth inhibitory compounds like acetic acid from hemicellulose fraction and phenolic compounds such as ferulic acid from lignin fraction are also released. Thus, the detoxification steps like charcoal adsorption, solvent-solvent extraction, nanofiltration have to be applied prior to the fermentative process (Jönsson and Martín 2016). These inhibitors after recovery from hydrolysate can be used as value-added chemicals. Some value-added products such as sugar acid (e.g. galacturonic acid and glucuronic acid) and oligosaccharides are also released during pretreatment. The pretreated lignocellulosic biomass undergoes several conversion steps for the biorefinery production of liquid fuels and platform chemicals (Fig. 7.2).

The pretreated lignocellulosic biomass can be directly utilized for enzymatic saccharification and fermentation. This step has certain disadvantages such as the presence of inhibitors that hinder the fermentation process and the productivity of end products. The pretreated slurry can be separated into solid cellulose fraction and liquid hydrolysate fraction in which the solid part is utilized for end-product generation. This has several advantages such as the availability of high sugar content and absence of inhibitors. The fractionated hemicellulosic hydrolysate having inhibitory compounds has a great potential of generating value-added products via direct fermentation or chemical synthesis as it consists of soluble pentose and hexose sugars.

7.3.2 Enzymatic Saccharification

After biomass pretreatment, the recovered pretreated biomass slurry or solid cellulose fraction is allowed for enzymatic degradation as the high crystallinity of cellulose resists the chemical degradation. Hence, the enzymatic hydrolysis of cellulose into glucose monomeric units can be carried out by cellulase enzymes. Cellulase is an enzyme complex mainly consisting of endoglucanase, exoglucanase and β-glucosidase enzymes. They act synergistically for the degradation of cellulose. Endoglucanases act on the random interior sections of cellulosic chain to break into smaller fragments, while exoglucanases act at the ends of the fragments to catalyse and produce disaccharides like cellobioses. Cellobioses are subsequently degraded into glucose units by β-glucosidase enzyme (Zhang and Lynd 2004). In the whole process, the effective utilization of sugar platform can be established by using suitable fermenting strains, low-cost cellulase enzymes and an efficient process design for the biochemical conversion of both pentose and hexose sugar platforms into fuels and chemicals. The different modes of fermentation process of pretreated lignocellulosic biomass have been discussed below. Different feedstocks, products and microorganisms used for the saccharification of lignocellulosic biomass are described in Table 7.1.

The conventional processes of cellulosic fermentation have been carried out by separate hydrolysis and fermentation (SHF) where initially the enzymatic hydrolysis of cellulose fraction takes place in one reactor vessel using commercial cellulose enzyme followed by fermentation process of the resulted sugars in the second vessel. However, the disadvantages of SHF are requirement of several steps and inhibition of cellulase enzymes by end products, which make the process expensive. By using the SHF process, *Clostridium sporogenes* BE01 produced 5.52 g/L butanol from 4% H_2SO_4 pretreated rice straw (Gottumukkala et al. 2013).

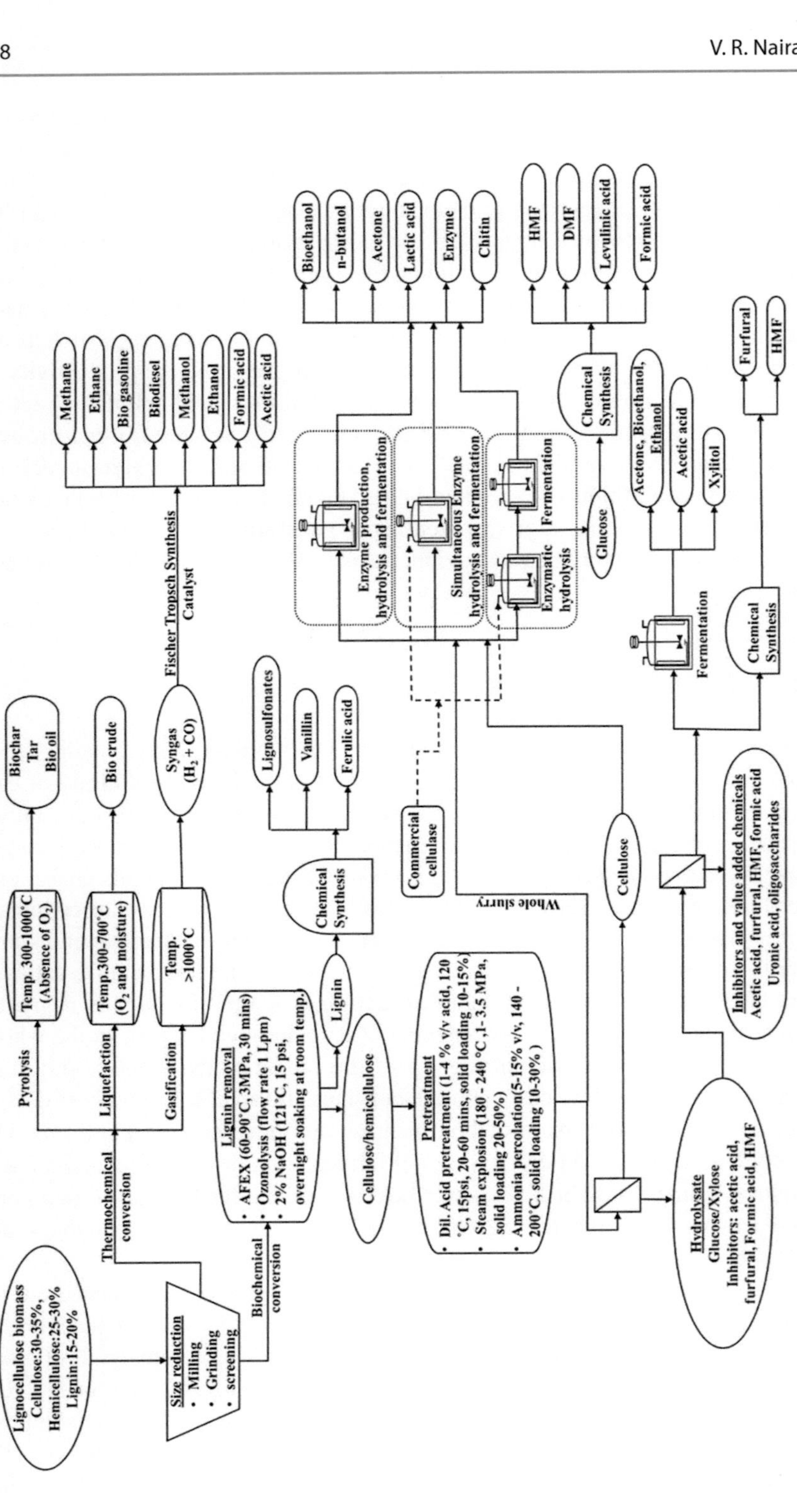

Fig. 7.2 Process network of lignocellulosic biorefinery for the production of biofuels and platform chemicals

Table 7.1 Various saccharification and fermentation processes for lignocellulosic biomass

Pretreatment	Feedstock	Type of fermentation	Microorganism	Product	References
Pretreated solid cellulose fraction					
1% (v/v) H_2SO_4 (100 °C for 40 min)	Sugarcane bagasse	Solid state	*Trichoderma koningii* INCQS 40331	Cellulase (8.2 IU/g substrate)	Salomão et al. (2019)
1 mL of 0.3 N NaOH/g substrate	Sugarcane bagasse	Solid state	*Aspergillus tamarii*	Xylanase (750 IU/g dry weight); β-xylosidase (250 IU/g dry weight)	Ferreira et al. (1999)
4% (w/w) H_2SO_4, solid loading 15%, 121 °C and 60 min	Rice straw	Separate hydrolysis and fermentation	*Clostridium sporogenes* BE01	Butanol (5.52 g/L)	Gottumukkala et al. (2013)
1% NaOH, 121 °C and 60 min	Rice straw	Simultaneous saccharification and fermentation	*Candida acidothermophilum* ATCC 20381	Ethanol (139 g/L) by vacuum condensation	Roychoudhury et al. (1992)
3.8 mg H_2SO_4/g feed, 175 °C, 5 min and solid/liquid ratio 2:1 (w/w)	Corn stover	Simultaneous saccharification and fermentation	*Corynebacterium glutamicu* SIIM B253	L-lysine (33.8 g/L)	Chen et al. (2019)
1% NaOH, solid-to-liquid ratio 1:15 (w/v), 50 °C and 72 h	Rice straw	Consolidated bioprocessing	Mixed culture from cattle and pig manure, compost and corn field soil	Butyric acid (16.2 g/L, productivity 2.69 g/L/d)	Ai et al. (2016)
Pretreated liquid hydrolysate					
3% (w/w) SO_2 and steam explosion (195 °C and 5 min)	Sugarcane straw	Fermentation	*Moorella thermoacetica* ATCC 39073	Acetic acid (17 g/L, 71% of theoretical value)	Ehsanipour et al. (2016)

(continued)

Table 7.1 (continued)

Pretreatment	Feedstock	Type of fermentation	Microorganism	Product	References
12.9 g/L of H_2SO_4, steam explosion (0.8 MPa and 5 min)	Corn stover	Fermentation	*Rhizopus oryzae* AS 3.819	Chitosan (10.96 g/L)	Yang et al. (2017)
Pretreated whole slurry mixture					
Wet explosion, (170 °C, 20 min and 79.8 psi O_2)	Corn stover	Separate hydrolysis and fermentation	*Saccharomyces cerevisiae*	Ethanol (26.8 g/L, 65.3% theoretical yield)	Rana et al. (2014)
Steam explosion (217 °C and 10 min)	Wheat straw	Separate hydrolysis and fermentation	*Saccharomyces cerevisiae*	Ethanol (32 g/L, 98% theoretical yield)	Alfani et al. (2000)
10% (v/v) H_2SO_4, 8% solid loading, 121 °C and 60 min	Wheat straw	Separate hydrolysis and fermentation	*Clostridium beijerinckii* P260	Acetone, butanol, ethanol (ABE: 25 g/L)	Qureshi et al. (2007)
Maleic acid, 190 °C, 0.2% (w/v) and 20 min	Rice straw	Simultaneous saccharification and fermentation	*Saccharomyces cerevisiae*	Ethanol (11.2 g/L, 62.8% theoretical yield)	Jung et al. (2015)
1% NaOH (121 °C and 1 h)	Sunflower straw	Simultaneous saccharification and fermentation	*Pichia stipites* CECT 1922	Ethanol (3.6 g/L, 34.5% theoretical yield)	Antonopoulou et al. (2016)
Dilute H_2SO_4	Wheat straw	Consolidated bioprocessing	*Trichoderma reesei* RUT-C30, *Saccharomyces cerevisiae* and *Scheffersomyces stipitis*	Ethanol (9.8 g/L, 69% theoretical yield)	Brethauer and Studer (2014)
Steam explosion (230 °C and 15 min)	Beech wood	Consolidated bioprocessing	*Trichoderma reesei* RUT-C30, *Lactobacillus pentosus* and *Lactobacillus brevis*	Lactic acid (20 g/L)	Shahab et al. (2018)

To avoid the problems of two-step process and end-product inhibition, simultaneous saccharification and fermentation (SSF) has been developed in which the glucose released through enzymatic saccharification (using commercial enzymes) is simultaneously utilized by the fermenting microorganism in a single vessel. A maximum concentration of 67 g/L of lactic acid was produced from sugarcane bagasse through SSF process by using *Lactobacillus delbrueckii* (Adsul et al. 2007). The highest productivity and yield of lactic acid were 0.93 g/L/h and 0.83 g/g, respectively. The concomitant production of acetone, butanol and ethanol through ABE fermentation using *Clostridium beijerinckii* from sugarcane bagasse through a combination of pretreatment, microwave decomposition, ammonia immersion, liquid hot water, microbial decomposition and enzyme hydrolysis yielded 11.9 g/L of ABE by SSF process (Su et al. 2015).

Consolidated bioprocessing (CB) is another interesting process that involves co-culturing a microbial strain (for cellulase production via enzymatic saccharification) and a fermenting strain (for simultaneous fermentation of released sugar) in a single vessel. This process has several advantages over the other processes, e.g. reduction of enzyme cost, simultaneous biochemical conversion of pentose and hexose by using both pentose-utilizing and hexose-utilizing strains at the same time. For example, the consolidated bioprocessing of dilute H_2SO_4-pretreated wheat straw using *Trichoderma reesei* RUT-C30 (for cellulase production), *Saccharomyces cerevisiae* (for glucose fermentation) and *Scheffersomyces stipitis* (for xylose fermentation) produced 9.8 g/L of bioethanol with 69% theoretical yield (Brethauer and Studer 2014).

In a solid-state (SS) fermentation, high solid loading of pretreated biomass is used at low moisture content for enzymes and metabolites production. The advantages are high concentration of end products, low water requirement, lower energy requirement, lower sterility requirement and easy aeration. The amylase production was carried out by *Aspergillus niger* using sugarcane bagasse in a solid-state fermentation in which the maximum amylase activity obtained was 457.82 IU/gds with yeast extract as the nitrogen source (Rosés and Guerra 2009).

7.3.3 Conversion of Lignocellulosic-Derived Sugars

The C_5 (pentose) and C_6 (hexose) sugars obtained after the enzymatic saccharification of cellulose and hemicellulosic fractions can be chemically transformed into a number of value-added products. The reduction of C_6 (e.g. glucose) and C_5 sugars (e.g. xylose) into useful chemicals via different platforms is shown in Table 7.2 (Isikgor and Becer 2015). Every chemical has its own application in various industrial sectors. For example, 2,5-dimethylfuran (DMF) from HMF platform has higher calorific value than ethanol and can be blended with gasoline (Tian et al. 2011). Similarly, high-value derivatives from levulinic acid can be utilized for agro-based applications like potential herbicide and insecticide production. Furfural has industrial applications for the extraction of aromatics, manufacturing of synthetic rubber, phenolic resins, wax recovery from petroleum and vegetative oils, etc. Apart from

Table 7.2 Scope of chemicals produced from various platforms derived from C_5 and C_6 sugars

Platform	Chemicals
1,4-Diacid	Succinic acid; fumaric acid; 1,4-butanediol; malic acid; 1,4-butanediamine; succinic ester; γ-butyrolactone; tetrahydrofuran; 2-pyrrolidone; *n*-methyl-2-pyrrolidone
3-Hydroxypicolinic acid	3-hydroxypicolinic acid; 3-hydroxypropionic esters; 1,3-propanediol; propene; acrylamide; malonic acid; β-propiolactone; acrylic acid; acrylates; acrolein; acrylonitrile
3-Hydroxy-γ-butyrolactone	3-Hydroxy-γ-butyrolactone; epoxylactone; 2(5H)-furanone; β-methacryloyloxy-γ-butyrolactone; acrylate lactone
5-Hydroxymethylfurfural and 2,5-furandicarboxylic acid	5-hydroxymethylfurfural; 2,5-furandicarboxylic acid; 2,5-diformylfuran; 2,5-dimethylfuran; 1,6-hexanediol; 1,2,6-hexanetriol; 2,5-bis(hydroxymethyl furan); 2,5-bis(aminomethyl furan); 2,5-dihydroxymethyl-tetrahydrofuran; 1,6-hexanediamine
Acetone-butanol-ethanol	Acetone; 2-propanol; propene; 1,2-dichloroethane; vinyl chloride; ethylenediamine; glyoxal; ethanol; butadiene; acetic acid; ethylene; ethylene oxide; ethylene glycol; glycolic acid; butanol; butyric acid; 1-butene; isobutene; α-olefins; vinyl acetate; methyl methacrylate; styrene
Aspartic acid	Aspartic acid; fumaric acid; maleic acid; 2-amino-1,4-butanediol; β-alanine; aspartic anhydride; 3-aminobutyrolactone
Glucaric acid	Gluconic acid; glucaric acid; glucaro-1,4-lactone; α-ketoglucarates; methylglucoside; glucuronic acid; glucarodilactone
Glutamic acid	Glutamic acid; 1,2,5-pentanetriol; glutaminol; 1,5-pentanediamine; 1,5-pentanediol; γ-aminobutyric acid; glutaric acid; 4-amino-1-butanol
Glycerol	Glycerol; diglycerol; mannitol; 1,3-propanediol; propylene glycol; glycerol carbonate; ethylene glycol; glycidol; propanol; dihydroxyacetone; 3-hydroxypropionaldehyde; propene; propane; acrolein; acrylic acid; glyceraldehyde
Itaconic acid	Itaconic acid; itaconic diamide; Itaconic anhydride; 3-methylpyrrolidine; 2-methyl-1,4-butanediol; 2-methyl-1,4-butanediamine; 2- and 3-methyl-γ-butyrolactones; 3-methyl tetrahydrofuran
Lactic acid	Lactic acid; lactide; acetaldehyde; 2,3-pentanedione; pyruvic acid; lactates; oxalic acid; propylene glycol; acrylic acid; propylene oxide; pyruvaldehyde; propanoic acid; acrylates
Levulinic acid	Levulinic acid; diphenolic acid; levulinate ketal; 5-aminolevulinic acid; angelica lactones; 2-methyl tetrahydrofuran; 1,4-pentanediol; γ-valerolactones; hydroxy(amino)amide; butane; adipic acid; adiponitrile; pentanoic acid; pentanol; 5-nonanone; 5-nonanol; ethylpentenoate; dimethyl adipate; methyl valerate; 2-butanone; levulinic esters

(continued)

Table 7.2 (continued)

Platform	Chemicals
Sorbitol	Sorbitol; isosorbide; 1,4-sorbitan; sorbose; 2-ketogulonic acid; vitamin C
Xylose-furfural-arabinitol	Xylose; xylitol; xylaric acid; hydroxyfurans; furoic acid; 2-(5H)-furanone; furfural; furfuryl; furan; tetrahydrofuran, furanacrylic acid; maleic anhydride; maleic acid; 2-methylfuran; difurfuryldiamines; furfuryl alcohol; tetrahydrofurfuryl alcohol; 2-hydroxymethyl-5-vinyl furan
Others	Citric acid; erythrose; 1,3-dihydroxyacetone dimer; 2,3-butanediol; isoprene; hydroxyalkanoic acid; levoglucosenone; limonene; formic acid; glycolaldehyde dimer; isobutanol; erythritol

the monomeric C_5 and C_6 sugar platforms, oligosaccharides released during the degradation of cellulose and hemicellulose can be converted into polymeric compounds like polyesters, polyamides, polycarbonates, polyurea, polyanhydrides and polyurethanes which are used in plastic and packaging industries.

The hemicellulose fraction obtained after the pretreatment of lignocellulosic biomass is usually a liquid hydrolysate that is rich in xylose and a small proportion of glucose in soluble form and other by-products (treated as inhibitors) from sugar degradation. Hence, it can be fermented after inhibitor separation to produce fuels and chemicals. The separated inhibitors from the hydrolysate can be directly used as value-added products or the whole fraction can be bypassed for direct fermentation. The inhibitors such as furfural, HMF, acetic acid and formic acid present in the hydrolysate can be separated by either nanofiltration technique, solvent–solvent extraction with ethyl acetate (Fenske et al. 1998) or trialkylamine (Zhu et al. 2011), or liquid–solid extraction by using activated carbon (Parajo et al. 1997). The use of nanofiltration membranes for maximizing the sugars' yield by separating inhibitors simultaneously from rice straw hydrolysate has been demonstrated. In the study, xylose, glucose, arabinose and cellobiose were concentrated by 100%, 104%, 93% and 151%, respectively, while the final concentration of inhibitors was reduced to 3% of the initial concentration in the sugar stream (Maiti et al. 2012).

By separating or not separating the inhibitors from the hemicellulosic fraction, the liquid hydrolysate can be used for direct fermentation into bioethanol by pentose (xylose)-fermenting yeast such as *Pichia stipites* and *Fusarium oxysporum* (McMillan 1993). *Candida boidinii*, *Candida tropicalis* and *Pachysolen tannophilus* can convert xylose into xylitol. Xylitol has a wide range of industrial applications as an alternative sweetener in foods and syrups and also used in pharmaceutical formulations as an additive for patients with diabetes and dental caries (Mohamad et al. 2015). Chitin production from fermented biomass of fungus grown using corn stover hydrolysate was reported by Yang et al. (2017). Chitin has many medical applications because of its biodegradability, biocompatibility and good adsorbing behaviour within the cellular matrix. Hence, it is widely applied for stitching the surgeries and wounds (Jardine and Sayed 2016). In acetone-butanol-ethanol (ABE) fermentation, acetic acid and butyric acid production can also be achieved from hemicellulosic sugar fermentation (Nanda et al. 2017). The fermentation of hydrolysate obtained from the pretreatment of wheat straw, forest residues, switchgrass and sugarcane straw was carried out using *Moorella thermoacetica* in which sugarcane straw hydrolysate yielded 17 g/L of acetic acid with 71% theoretical conversion yield (Ehsanipour et al. 2016).

The generation of multiple co-products along with the principal product via biorefinery approach from lignocellulosic biomass can solve the constraints for commercialization of a single product. The choice of a suitable pretreatment strategy based on the specificity of product, separation of treated biomass into multiple streams (e.g. lignin, cellulose, hemicellulose and their hydrolysates) for the generation of value-added products at different stages is crucial for lignocellulosic biorefineries. For example, the fractionation of solid and liquid residues of pretreated sugarcane bagasse (lime and alkaline H_2O_2 treatment) where the solid fractions are enzymatically hydrolysed to produce bioethanol and lignin fraction can be streamed to boilers for the generation of heat. The generated waste from liquid fraction and leftover residues after enzymatic hydrolysis are anaerobically digested with mixed culture of thermophilic microorganisms for methane (biogas) production (Rabelo et al. 2011). Recent pretreatment and biorefinery approaches of various lignocellulosic feedstocks are summarized in Table 7.3.

7.4 Algal Biorefineries

7.4.1 Algae Cultivation

The generation of products in algae biorefinery is mainly decided by the compositions of lipids, proteins and carbohydrates of algal species, which are the precursors of biofuels and platform chemicals (Laurens et al. 2017). This composition of these compounds is species-specific and can be easily varied by nutritional supply (e.g. nitrogen starvation for biodiesel production) (Reitan et al. 1994). Algae can also be used for the production of pigments, vitamins, nutraceuticals, pharmaceuticals and antioxidants via selective extraction techniques. They can also secrete a wide range of extracellular products (e.g. exopolysaccharides and exoenzymes). Thus, algal biorefinery offers a broad range of co-products with desirable quantities of main products (Fig. 7.3).

Several cultivation systems are established to achieve large-scale manufacturing of products from algae. Algae are grown in both open (raceway ponds) and closed systems (photobioreactors, PBRs) (Chisti 2007). Mostly practised cultivation systems are artificial open ponds due to their economic feasibility, easy installation and maintenance. Due to their low biomass productivities and contamination issues, PBR systems were developed to obtain high biomass productivities by allowing better control over culture conditions. All PBRs result in higher biomass productivities, but they have their own advantages and disadvantages. The tubular and flat panel PBRs allow higher biomass concentrations with better control over the culture

Table 7.3 Multi-product generation from lignocellulosic feedstocks

Feedstock	Pretreatment	Process	Principal product and co-product	Microorganism	References
Dairy manure and other organic wastes	2% NaOH treatment	Anaerobic digestion, simultaneous saccharification and fermentation, and electrocoagulation	Methane, chitin, reclaimed water and electricity	*Rhizopus oryzae*	Liu et al. (2016b)
Hemp	Dilute H_2SO_4 and alkaline-oxidative method	Simultaneous saccharification and fermentation	Bioethanol and succinic acid	*Saccharomyces cerevisiae* and *Actinobacillus succinogenes*	Kuglarz et al. (2016)
Rice husk and corn stover	Steam explosion	Fast pyrolysis, and separate hydrolysis and fermentation	Bio-oil and succinic acid	*Escherichia coli* MG-PYC	Wang et al. (2013)
Rice straw	Dilute H_2SO_4 pretreatment and sulfomethylation reagent	Lignin fractionation, and simultaneous saccharification and fermentation	Lignosulfonate, xylose recovery, bioethanol	*Saccharomyces cerevisiae*	Zhu et al. (2015)
Softwood	–	Pyrolysis and algae growth using alkali treated bio-oil fraction	Bio-oil and algae biomass	*Chlamydomonas reinhardtii*	Zhao et al. (2013)
Sugarcane bagasse	Lime and alkaline H_2O_2	Separate hydrolysis and fermentation, anaerobic digestion, and combustion	Bioethanol, methane and heat energy	*Saccharomyces cerevisiae*	Rabelo et al. (2011)
Sugarcane bagasse	Milling, H_2SO_4 pretreatment	Simultaneous saccharification and fermentation	Bioethanol, xylitol	*Saccharomyces cerevisiae* and *Candida tropicalis*	Unrean and Ketsub (2018)
Wheat straw	Steam explosion	Simultaneous saccharification and fermentation, and pyrolysis	Bioethanol and bio-oil	*Kluyveromyces marxianus*	Tomas-Pejo et al. (2017)
Wheat straw	Hydrothermal pretreatment	Batch fermentation and anaerobic digestion	Bioethanol, biohydrogen and biogas (methane)	*Saccharomyces cerevisiae* and thermophilic mixed culture	Kaparaju et al. (2009)

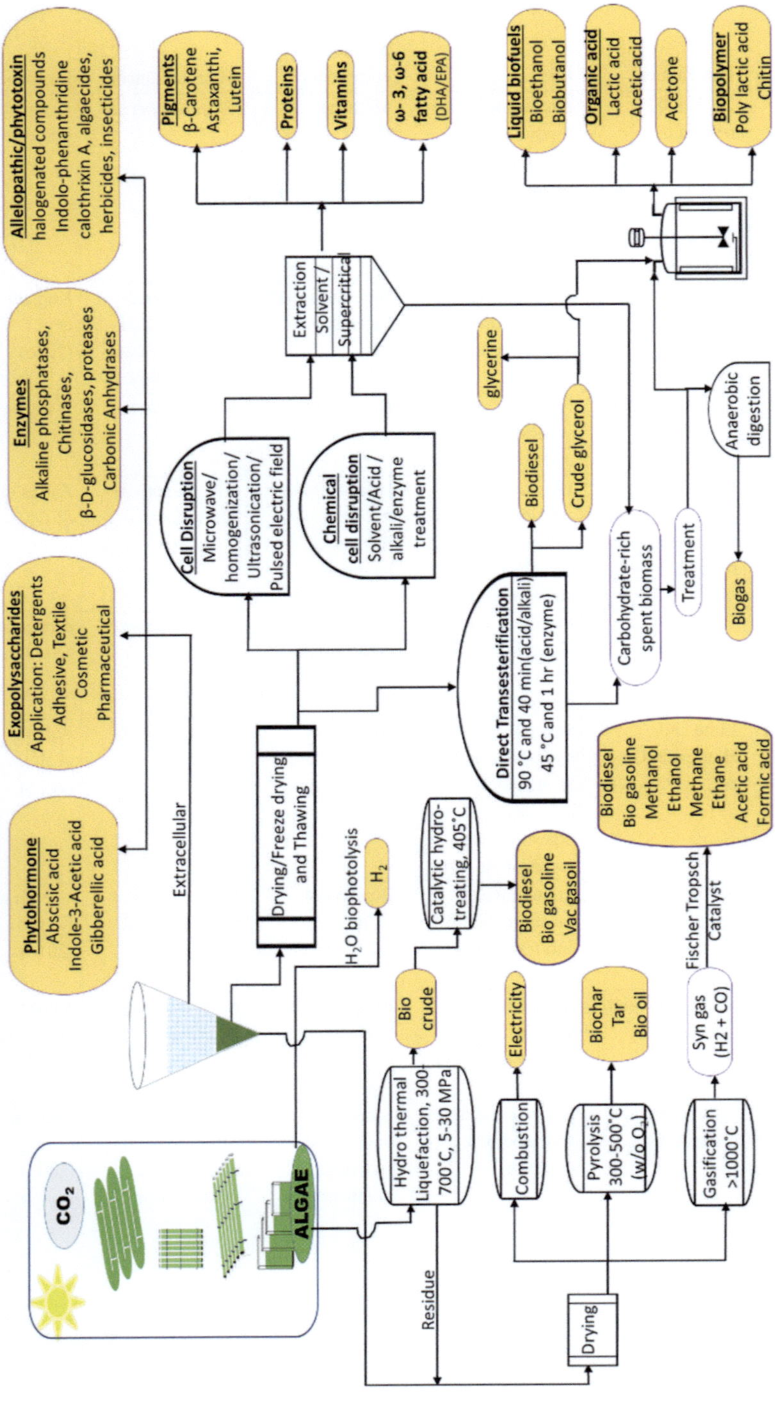

Fig. 7.3 Process network of algal biorefinery for the production of biofuels and platform chemicals

medium, i.e. control of CO_2 supply, pH, temperature and salinity, and they are highly suitable for the commercial-scale production of algae.

Algae can be cultivated in three modes, namely photoautotrophic, heterotrophic and mixotrophic. Photoautotrophic growth using CO_2 (from flue gases) of algae is beneficial as CO_2 is cheaper than any other sugars and organic acids, and it is mostly obtained as a waste greenhouse gas emission from industries. Heterotrophic mode of cultivation includes the utilization of carbon sources such as sugars and organic acids by algae for their growth without light. In mixotrophic mode, both organic source and CO_2 are used for cultivating algae using a light source.

After cultivation, algae are subjected to harvesting processes such as centrifugation, filtration (cake and ultrafiltration), sedimentation, flotation, flocculation (chemical, biological, magnetic, electro and auto-flocculation) to separate the algal biomass and extracellular products. The leftover media after fractionation of the extracellular products can be recycled to the next batch with adequate nutrient re-supplementation. The wet/dried algal biomass after harvesting is extracted and fractionated into platforms of lipids, carbohydrates and proteins via various downstream processing techniques.

7.4.2 Downstream Processing of Algae for Fuels and Chemicals

Algal biorefining products are mainly intact within the cell membrane or reside inside the cells. Hence, downstream processing is very critical for algal biorefinery and its economics is mainly decided by the cell disruption techniques involved. Depending on the cell wall rigidness of specific microalgal species and products of interest, the cell disruption technology has to be decided (Günerken et al. 2015; Lee et al. 2017). Broadly, the cell disruption methods are mainly divided into two types, namely mechanical and non-mechanical.

The mechanical type of cell disruption needs the energy in the form of shear force, wave energy, current and heat. The methods such as bead milling and high-pressure homogenization can be categorized under shear-force-based cell disruption. Bead milling breaks the algal cells by creating compressive and shear forces applied through fast-moving solid beads (Günerken et al. 2015). In high-pressure homogenization, the cells are pressed through a small orifice where cell lysis is promoted by turbulence, shear stress and cavitation mechanisms. Even though shear-force-based cell disruption extracts all the intracellular cellular components, especially lipids/biodiesel, their high-energy demand and heat generation make them unsuitable for the biorefinery approaches. Moreover, the fragile compounds like proteins lose their activity due to heat generation (Doucha and Livansky 2008).

The wave-energy-based cell disruption techniques include ultrasonication, microwaves and pulsed electric field treatment (Günerken et al. 2015). In ultrasonication, the cells are lysed due to shock waves generated by bursting of microbubbles caused by cavitation effect. During microwaves treatment, the algal cell walls are disrupted due to high heating in short time which damages the pectin and cellulose structures of algal cell wall. In the pulsed electric field treatment, the cell disrupts due to an electric potential that generates across the cell wall. Except the microwave treatment due to its high heat, other two techniques can be used for algal biorefineries that involve separation of fragile compounds like proteins. However, scaling up of those techniques is still in early development stages. With regard to the heat-based cell disruption, steam explosion and hydrothermal liquefaction were explored recently. In steam explosion, the wet/dry biomass is subjected to temperatures in the range of 180–240 °C and pressures of 1–3.5 MPa. A sudden depressurizing explodes the algal cells to release lipids, soluble sugars and solid residues (Cheng et al. 2015).

The non-mechanical cell disruption methods include chemical and biological techniques. In chemical methods, chemicals like antibiotics, ionic liquids, special nanoparticles, chelating agents, chaotropes, detergents, solvents, hypochlorites, acids and alkalis are used (Günerken et al. 2015; Lee et al. 2017). The most commonly used chemical methods are conventional lipid extraction followed by transesterification. Both the methods are developed mainly for biodiesel production from lipids, with the by-products such as crude glycerol and residual algal biomass. The direct transesterification decreases the downstream cost and time durations compared to the conventional ones. The direct transesterification process is carried out using methanol and acid/base catalysts (e.g. H_2SO_4 and NaOH) (Mahesh et al. 2019; Naira et al. 2018).

High temperature requirement makes the technique unfavourable for mild algal biorefinery. Hence, mild biological cell disruption techniques like enzyme lysis and algicidal treatment are developed. In enzymatic cell disruption, cellulases, lipases, lysozymes, pectinases and proteases are used for lysing the rigid algal cell wall containing cellulose, membrane lipids and proteins (Wu et al. 2017). The cocktail of the enzymes is more powerful in breaking the cells than the single enzyme (Wu et al. 2017). However, the enzymatic cell disruption is energy efficient as the cost of enzyme, stability and reusability are the challenges that need to be resolved for large-scale applications. In the algicidal treatment, bacteria, cyanobacteria, microalgae and viruses are the microorganisms that are reported to lyse the cell wall of algae by releasing extracellular allelopathic compounds (Chen et al. 2013). Hence, algicidal treatment can be considered for biorefinery applications that can also overcome the challenges of enzyme cost, stability and reusability. The recent advances in the application of available cell disruption methods for biorefinery product generation are tabulated in Table 7.4.

7.4.3 Extracellular Products from Algae and Their Applications

The production of extracellular metabolites during the algal growth or starvation periods provides an excellent biorefinery opportunity for the generation of a wide range of platform chemicals like exopolysaccharides (EPS), proteins, fatty acids, organic acids, phytoharmones and allelopathic chemicals. This extracellular platform is highly beneficial in the perspective of downstream processing economics as they require less energy to recover. Exopolysaccharides are used in food (e.g. jelly, candy, sauce and beverages), pharmaceuticals (e.g. capsule covering and antitumour

Table 7.4 Various cell disruption techniques for multi-product generation

Cell disruption technique	Methodology	Algal species	Products	References
Algicidal treatment	*Flammeovirga yaeyamensis* secreted amylase, cellulase and xylanase	*Chlorella vulgaris* ESP-31	Lipids	Chen et al. (2013)
Enzymatic cell disruption	Cocktail of cellulase, lipase and protease enzymes	*Nannochloropsis*	Lipids, proteins and glucose	Chen et al. (2016)
Enzymatic cell disruption	Aqueous enzymatic extraction using autolysin	*Chlamydomonas reinhardtii*	Lipids and proteins	Sierra et al. (2017)
Hydrothermal liquefaction	Heterogeneous catalysts (Pd/C, Pt/C and Ru/C) under inert (He) atmosphere and high-pressure reducing (H_2) conditions	*Nannochloropsis*	Biocrude oil, H_2, CO_2 and CH_4	Duan and Savage (2011)
Steam explosion	Steam explosion followed by membrane filtration and *n*-hexane extraction	*Nannochloropsis gaditana*	Soluble monomeric sugars and lipids	Lorente et al. (2017)
Ultrasonication	4 N HCL treatment followed by ultrasonication-assisted acetone extraction	*Haematococcus pluvialis*	Lipid and astaxanthin	Dong et al. (2014)
Ultrasonication	Ultrasonication-assisted DMSO, ethanol and water extraction	*Nannochloropsis* spp.	Phenolic compounds and chlorophylls	Parniakov et al. (2015)
Ultrasonication	Ultrasonication-assisted methanol extraction	*Dunaliella salina*	Carotenoids and chlorophylls	Macías-Sánchez et al. (2009)
Ultrasonication	Ultrasonication-assisted acetone, ethanol and hexane extraction	*Chlorella vulgaris*	Carotenoids, chlorophylls, steroids, phytols and fatty acids	Plaza et al. (2012)

drugs), textiles (e.g. printing and dyeing), cosmetics (e.g. body lotion and emulsifier), detergents, adhesives, wastewater treatment, bioremediation, bioflocculants and oil recovery (Liu et al. 2016a).

The exopolysaccharides from various algal groups like rhodophytes, cyanophytes chlorophytes and dinoflagellates have been widely reported to exhibit antibacterial, antiviral, antitumour, anticoagulant and antioxidant activities (Chen et al. 2010; Freire-Nordi et al. 2005). In a recent study, biodiesel (1.5 g/L) and exopolysaccharides (236 mg/L) were concomitantly produced via *Scenedesmus abundans* (Mahesh et al. 2019). These exopolysaccharides were shown to have monosaccharides like glucose, galactose, arabinose and raminose. Exoenzymes such as alkaline phosphatases, chitinases, β-D-glucosidases and proteases were also identified in algal species like *Chlorella*, *Scenedesmus* and *Anabaena*. Extracellular phenoloxidases such as phenoloxidase laccase and laccase-like enzymes were found in the cultures of *Chlorella*, *Scenedesmus* and *Anabaena* (Liu et al. 2016a; Zhang et al. 2008). These phenoloxidases can be used for bioremediation of phenolic pollutants.

Extracellular phytohormones such as abscisic acid (from *Chlorella vulgaris* and *Stichococcus bacillaris*), indole-3-acetic acid (from *Chlorella pyrenoidosa* and *Scenedesmus armatus*) and gibberellic acid (from *Scytonema hofmanni*) are reported to be used in food crops for alleviating stresses (Liu et al. 2016a; Rodriguez et al. 2006). Allelopathic compounds like unsaturated C18 fatty acids (from *Chlorella*), polysaccharides (from *Chlorella vulgaris* and *Anabaena* PCC 7120) and alkaloids (from *Calothrix* sp. and *Anabaena flosaquae*) are reported to inhibit the growth of bacteria and other algae (Ikawa et al. 1984; Liu et al. 2016a).

During the cultivation of some cyanobacterial and green algal species, biohydrogen can also be co-produced by converting light energy with the help of [Fe-Fe]-hydrogenase

Table 7.5 Recent advances in the utilization of recycle streams for expanded algal biorefineries

Recycle stream	Microorganism or process	Products	References
Recycle from transesterification or lipid extraction process			
Crude glycerol	*Lactobacillus diolivorans*	Hydroxypropionaldehyde; 3-hydroxypropanoic acid; and 1,3-propanediol	Lindlbauer and Marx (2017)
Crude glycerol	*Propionibacterium freudenreichii* ssp. *shermanii*	Propionic acid; succinic acid; acetic acid and trehalose	Pawlicka-Kaczorowska and Czaczyk (2017)
Crude glycerol	*Schizochytrium limacinum*	Docosahexaenoic acid; lipids; proteins and carbohydrates	Pyle et al. (2008)
Spent biomass of *Chlamydomonas* sp. after transesterification	Hydrolysis	ε-Polylysine	Sivaramakrishnan et al. (2019)
Spent biomass of *Tetraselmis* sp. after supercritical CO_2 extraction	Anaerobic digestion	Biogas (methane)	Hernandez et al. (2014)
Recycle from hydrothermal liquefaction process			
Hydrothermal liquefaction wastewater of *Desmodesmus* sp.	*Desmodesmus* sp.	Algal biomass	Alba et al. (2013)
Hydrothermal liquefaction wastewater of *Spirulina*	*Chlorella minutissima*	Algal biomass	Jena et al. (2011)

enzyme, that reduces protons to molecular hydrogen using photosynthetically reduced ferredoxin (Fd) as a donor (Greenbaum 1988). The utilization of recycle streams from biorefineries is an innovative step towards the development of green and sustainable environment. Moreover, it adds value to the economy of biorefinery industries. Some recent advances in using crude glycerol, spent biomass and organic-source-rich fraction as recycle streams from various bio-based industries are tabulated in Table 7.5.

7.5 Thermochemical Conversion of Lignocellulosic Biomass and Algae

Conventionally, lignocellulosic biomass and algal feedstock can be converted into fuels and chemicals via various thermochemical conversion routes such as pyrolysis, gasification, direct combustion and hydrothermal liquefaction. Before the thermochemical treatments, the biomass is ground or milled for decreasing their particle size and increasing the surface area.

7.5.1 Pyrolysis

Pyrolysis is the thermochemical decomposition of biomass feedstock at elevated temperatures and non-oxidizing conditions into solid biochar, liquid bio-oil and condensable or incondensable gaseous products. The process of pyrolysis mainly consists of three stages. In stage I, dehydration of algae sample is performed where water is released. In stage II, pyrolytic decomposition is conducted. In stage III, char transformation and mineral matter decomposition are done. Owing to the recalcitrant nature of lignocellulosic biomass, it can be subjected to three different kinds of pyrolysis processes, namely slow pyrolysis (300–700 °C) to yield biochar as the main product as well as fast pyrolysis (450–650 °C) and flash pyrolysis (800–1000 °C) to yield bio-oil as the main product (Demirbas and Arin 2002).

Apart from bio-oil and biochar, the gaseous products like CO_2, CO, H_2, CH_4, C_2H_6, C_2H_4 and a small amount of other gases like C_3H_8, NH_3, NO_x and SO_x are evolved from the degradation of hemicellulose, cellulose and lignin. In the case of algal biomass, the carbohydrates are decomposed into CO, H_2O, CH_4, CH_3Cl, CH_3OH, CH_3SH, C_7H_8, C_2H_5OH and CO_2 at the temperature range of 125–235 °C. Similarly, the proteins are decomposed into CH_4N, C_2H_6, HCN, CO, CH_4 and CO_2 at the temperature range of 295–365 °C. On the other hand, the lipid fraction is decomposed into ketones (C_2H_2O and C_3H_5O), CH_4, HCN, C_2H_4, CH_4N, C_2H_6, C_7H_8, CO and CO_2 at the temperature range of 400–600 °C. Although the product of interest varies with the type of pyrolytic process, the final products are biochar and a fraction of volatile compounds in which condensable and incondensable gases are present. The condensable stream of gases can be further converted into bio-oil (main product of fast and ultrafast pyrolysis), which can be upgraded for use as a transportation fuel. The non-condensable fraction comprising H_2, CH_4, CO and CO_2 can be used as fuel gases for energy generation.

The pyrolytic bio-oil (also known as bio-crude) can be upgraded for the production of transportation liquid fuels via catalytic cracking. It can also be utilized as a precursor for platform chemicals like phenols for resin manufacturing, additives for fertilizers, pharmaceutical purposes and flavouring agents (e.g. glycolaldehyde) for food industries (Balat 2011). Combustible syngas and hydrogen fuels can be generated via gasification and steam reforming. The leftover solid biochar has a wide range of agricultural applications such as heavy metal adsorption, biodegradation of pesticides and increasing water-holding capacity of soil (Nanda et al. 2016a).

7.5.2 Combustion

Combustion is a traditional way of burning a fuel to release heat and gaseous streams in the presence of oxygen. Similar to the pyrolytic process stages, combustion can also be of three stages. However, the temperature range of decomposition stage is 145–500 °C and the biochar transforms into ash due to the involvement of oxygen in the process. Due to the oxidation reaction in all the stages of combustion, CO_2 is seen as the dominant gaseous product. The produced gas stream can be used to run a steam turbine to generate electricity and the ash can be utilized as an additive in fertilizer industries.

7.5.3 Gasification

The process of gasification involves subliming the biomass by feeding air or pure oxygen to decompose lignocellulosic and algal biomasses thermally at higher temperatures (>1000 °C), which results in a wide range of gaseous products. In gasification, the biomass is heated in air, oxygen or steam to generate syngas, which contains H_2, CO, CO_2 and CH_4 (Okolie et al. 2019). The physical conditions and catalysts play important roles in gasification for increasing the carbon yield. For example, *Spirulina* (algae) biomass at temperatures ranging from 800 °C to 1000 °C in the presence of oxygen produced syngas with traces of O_2, N_2 and C_2H_4 (Hirano et al. 1998). Upon increasing the temperature, the H_2 yield and carbon conversion efficiency increased. In another study, the combined gasification and methane (from biogas synthesis) reforming of wood and black liquor (lignocellulosic biomass) increased H_2/CO ratio in the syngas (Åberg et al. 2015).

In another investigation with *Chlorella vulgaris*, the Ni catalyst in gasification reaction increased the content of CH_4 by decreasing H_2 (Minowa and Sawayama 1999). Moreover, the nitrogen present in the feedstock was converted into NH_3 due to the Ni presence, making the product to manufacture a premium fertilizer. This syngas can be utilized directly as a fuel or as precursor for hydrocarbon fuels via Fischer-Tropsch synthesis, H_2 fuel and electricity. For biorefinery routes, syngas has been utilized for ammonia synthesis followed by biohydrogen and biomethanol production along with various other fuels and platform chemicals (Table 7.6) (Quarderer 1986; Van et al. 1995; Mokrani and Scurrell 2009; Molino et al. 2016; Arora et al. 2016).

Table 7.6 Fuels and platform chemicals from bio-based syngas

Source	Transformation route	Bioproduct
Bio-based syngas	Fischer-Tropsch (Fe, Co and Ru)	Gasoline, diesel, waxes and olefins
	Oxosynthesis ($HCo(CO)_4$)	Aldehydes and alcohols
	Alkali-doped (ZnO)	Mixed alcohols
	Isosynthesis (ZrO_2)	*I*-C_4 hydrocarbons
	Water-gas shift reaction (H_2O)	H_2
	Cu/ZnO catalysis	Biomethanol
	Co/Rh catalysis; syngas fermentation (acetogenic bacteria)	Bioethanol
H_2 (product of bio-based syngas)	N_2 over Fe/FeO	Ammonia
Biomethanol (product of bio-based syngas)	Ag catalysis	Formaldehyde
	Acidic ion exchange, isobutylene	Methyl tertiary butyl ether
	Carbonylation; $CH_3OH + CO$, Co, Rh and Ni	Acetic acid
	Zeolite catalysis	Olefins and gasoline
	Homologation (Co)	Bioethanol
	Al_2O_3 catalysis	Biodimethyl ether
	Direct use as fuel	M100 and M85 blending

7.5.4 Hydrothermal Liquefaction

Hydrothermal liquefaction (HTL) is an energy-efficient thermochemical treatment that converts biomass feedstock directly into bio-oil (bio-crude). In the HTL process, the lignocellulosic and algal biomasses are exposed to high pressure (5–30 MPa) and high temperature (200–700 °C) under inert atmosphere (e.g. N_2 or He) or reducing gases (e.g. H_2 or CO) (Tian et al. 2014). The process yields bio-crude as an important product along with some solid residues and gaseous phases. The solid phase can be converted into biochar and gaseous phase into syngas and other gaseous fuels. The aqueous phase is mainly composed of NH_4^+, PO_4^{3-}, K^+, N^+ and Mg^{2+} ions. Hence, bacteria or algae can directly use the post-HTL wastewater for fermentation or carbon utilization.

Other organic compounds from the aqueous phase can also be used as carbon sources for fermentation of bacterial/fungal cultures and even for heterotrophic algal culturing (Pham et al. 2013). The gaseous phase products from the HTL of algae are comprised mostly of CO_2; hence, the gaseous stream can be recycled back for the growth of algae (Amin 2009). The remaining solid residue is composed of a number of inorganic compounds that might be used for the cultivation of algae or synthesis of new catalysts (Liu et al. 2012). A recent novel biorefinery study on hydrothermal co-liquefaction of microalgae (*Chlorella pyrenoidosa*) and lignocellulosic biomass (rice husk) with equal ratios at 300 °C for 60 min yielded highest bio-crude having cyclic oxygenates (20.6%) and esters, ketones and alcohols content (17.2%) (Gai et al. 2015). This co-liquefaction strategy resulted in the decreased acidity and lower nitrogen content of bio-crude oil.

7.6 Industrial Scenario of Lignocellulosic and Algal Biorefineries

In the USA, the companies like Abengoa and POET-DSM have been working on commercial lignocellulosic biorefineries. Abengoa has the capability to produce 25 million gallons of cellulosic ethanol per annum, 21 MW of green electricity, lignin and animal feed from cellulosic biomass, crop residues and plant fibres as the feedstock (Chemicals Technology 2019). In India, Ethanol Blended Programme (EBP) was implemented by the Government of India for boosting the agricultural sector and to reduce imported energy requirements (Best Current Affairs 2019). In support of EBP, 12 lignocellulosic biorefineries were planned by major oil companies (e.g. Bharat Petroleum Corporation Ltd.). The company has planned to use agricultural wastes and crop residues as the feedstock for producing second-generation bioethanol and organic manure for improving soil fertility. Assam Biorefinery Pvt. Ltd. in India was set up as a joint venture between Numaligarh Refinery Ltd. (NRL), Chempolis, Finland, and Fortum 3 B.V., Netherlands, for producing 62 million litres of bioethanol annually along with the co-production of furfural and acetic acid from bamboo (0.5 million tons) as lignocellulosic feedstock by 2021 (Numaligarh Refinery Limited 2017).

Volkswagen research group (Wolfsburg, Germany) has been working on microalgae that could excrete fuel compounds like 1-butanol, isobutanol, octanol or bisabolene (Bippes 2018). Buggypower (Portugal) is also producing high-quality marine microalgae with high nutritional properties such as polyunsaturated fatty acids (PUFAs), pigments, vitamins, essential amino acids and minerals (Real 2018). A paper-based industry from Italy, Favini Srl., has been started extracting platform chemicals like alginate, agar and carrageenan along with paper production from seaweeds (macroalgae) (Monegato 2018).

Global companies like Microsynbiontix, Cellana, Fermentalg, Microphyt, Algenuity, Algae for Future (a4f), Algafarm, Phytoplancton Marino and Parry Nutraceuticals have exploited microalgal products like essential fatty acids, pigments (astaxanthin and ketocarotenoid) and a wide range of platform chemicals for applications in nutraceuticals, aquaculture and animal feed industries (Algal Biomass Organization 2019). In India, Reliance Industries Ltd. and Algenol have collaboratively designed an algal biorefinery project to convert 1 tonne of CO_2 from industrial processes into 144 gallons of fuels like ethanol, gasoline, diesel and jet fuel (Green Car Congress 2015).

7.7 Conclusions

Both lignocellulosic biomass and algal feedstock have a huge potential for establishing multi-product biorefineries for co-production of fuels and value-added platform chemicals. The recent co-production technologies of both the feedstocks for multiple products have been discussed in this chapter. In lignocellulosic biorefinery, fermentation technology can be improved to produce more diversified products as compared to the existing products. For example, while co-culturing different microbial strains for maximizing the utilization of complete lignocellulosic sugar content via consolidated bioprocessing, the microbial strains can be chosen specifically for co-production of different compounds in a single fermentation vessel although the co-culture compatibility is a challenging task.

In the case of algae biorefinery, the advancements in developing an efficient and mild cell disruption technology by the combination of existing techniques have to be continued to reduce the cost of downstream processing as well as to extract fragile molecules like proteins along with the lipids and carbohydrates. The innovation in utilizing recycle streams of both the biorefineries for either microbial (bacterial or fungal) fermentation or algae biomass production can be a promising opportunity for integrating both the industries together. In the present industrial scenario, lignocellulosic biorefineries have already been proving their potential across the world. However, the existing industrial algae biorefineries are limited, though algae have been highly exploited for single or dual products use.

References

Åberg K, Pommer L, Nordin A (2015) Syngas production by combined biomass gasification and in situ biogas reforming. Energy Fuel 29:3725–3731

Adsul MG, Varma AJ, Gokhale DV (2007) Lactic acid production from waste sugarcane bagasse derived cellulose. Green Chem 9:58–62

Ai BL, Chi X, Meng J, Sheng ZW, Zheng LL, Zheng XY, Li JZ (2016) Consolidated bioprocessing for butyric acid production from rice straw with undefined mixed culture. Front Microbiol 7:1648

Alba LG, Torri C, Fabbri D, Kersten SRA, Brilman DWF (2013) Microalgae growth on the aqueous phase from hydrothermal liquefaction of the same microalgae. Chem Eng J 228:214–223

Alfani F, Gallifuoco A, Saporosi A, Spera A, Cantarella M (2000) Comparison of SHF and SSF processes for the bioconversion of steam-exploded wheat straw. J Ind Microbiol Biotechnol 25:184–192

Algal Biomass Organization (2019). www.algaebiomass.org. Accessed 14 Apr 2019

Amin S (2009) Review on biofuel oil and gas production processes from microalgae. Energy Convers Manag 50:1834–1840

Antonopoulou G, Vayenas D, Lyberatos G (2016) Ethanol and hydrogen production from sunflower straw: the effect of pretreatment on the whole slurry fermentation. Biochem Eng J 116:65–74

Arora P, Hoadley AFA, Mahajani SM, Ganesh A (2016) Small-scale ammonia production from biomass: a techno-enviro-economic perspective. Ind Eng Chem Res 55:6422–6434

Balat M (2011) An overview of the properties and applications of biomass pyrolysis oils. Energy Sources A 33:674–689

Best Current Affairs (2019) Second generation ethanol bio-refineries in India. www.bestcurrentaffairs.com/second-generation-ethanol-bio-refineries-india. Accessed 14 Apr 2019

Bippes M (2018) The H2020-project PHOTOFUEL: biocatalytic solar fuels for sustainable mobility in Europe. ALGAEUROPE, Amsterdam, p 37

Brethauer S, Studer MH (2014) Consolidated bioprocessing of lignocellulose by a microbial consortium. Energy Environ Sci 7:1446–1453

Chemicals Technology (2019) Abengoa cellulosic ethanol biorefinery, Kansas. www.chemicals-technology.com/projects/abengoa-cellulosic-ethanol-biorefinery. Accessed 14 Apr 2019

Chen B, You W, Huang J, Yu Y, Chen W (2010) Isolation and antioxidant property of the extracellular polysaccharide from *Rhodella reticulata*. World J Microbiol Biotechnol 26:833–840

Chen CY, Bai MD, Chang JS (2013) Improving microalgal oil collecting efficiency by pretreating the microalgal cell wall with destructive bacteria. Biochem Eng J 81:170–176

Chen L, Li R, Ren X, Liu T (2016) Improved aqueous extraction of microalgal lipid by combined enzymatic and thermal lysis from wet biomass of *Nannochloropsis oceanica*. Bioresour Technol 214:138–143

Chen ZY, Liu G, Zhang J, Bao J (2019) A preliminary study on L-lysine fermentation from lignocellulose feedstock and techno-economic evaluation. Bioresour Technol 271:196–201

Cheng J, Huang R, Li T, Zhou J, Cen K (2015) Physicochemical characterization of wet microalgal cells disrupted with instant catapult steam explosion for lipid extraction. Bioresour Technol 191:66–72

Chia SR, Chew KW, Show PL, Yap YJ, Ong HC, Ling TC, Chang JS (2018) Analysis of economic and environmental aspects of microalgae biorefinery for biofuels production: a review. Biotechnol J 13:1700618

Chisti Y (2007) Biodiesel from microalgae. Biotechnol Adv 25:294–306

Demirbas A, Arin G (2002) An overview of biomass pyrolysis. Energy Sources 24:471–482

Dong S, Huang Y, Zhang R, Wang S, Liu Y (2014) Four different methods comparison for extraction of astaxanthin from green alga *Haematococcus pluvialis*. Sci World J 2014:7

Doucha J, Livansky K (2008) Influence of processing parameters on disintegration of *Chlorella* cells in various types of homogenizers. Appl Microbiol Biotechnol 81:431–440

Duan P, Savage PE (2011) Hydrothermal liquefaction of a microalga with heterogeneous catalysts. Ind Eng Chem Res 50:52–61

Durot N, Gaudard F, Kurek B (2003) The unmasking of lignin structures in wheat straw by alkali. Phytochemistry 63(5):617–623

Ehsanipour M, Suko AV, Bura R (2016) Fermentation of lignocellulosic sugars to acetic acid by *Moorella thermoacetica*. J Ind Microbiol Biotechnol 43:807–816

Fenske JJ, Griffin DA, Penner MH (1998) Comparison of aromatic monomers in lignocellulosic biomass prehydrolysates. J Ind Microbiol Biotechnol 20:364–368

Ferreira GL, Boer CG, Peralta RM (1999) Production of xylanolytic enzymes by *Aspergillus tamarii* in solid state fermentation. FEMS Microbiol Lett 173:335–339

Freire-Nordi CS, Vieira AAH, Nascimento OR (2005) The metal binding capacity of *Anabaena spiroides* extracellular polysaccharide: an EPR study. Process Biochem 40:2215–2224

Gai C, Li Y, Peng N, Fan A, Liu Z (2015) Co-liquefaction of microalgae and lignocellulosic biomass in subcritical water. Bioresour Technol 185:240–245

Gottumukkala LD, Parameswaran B, Valappil SK, Mathiyazhakan K, Pandey A, Sukumaran RK (2013) Biobutanol production from rice straw by a non acetone producing *Clostridium sporogenes* BE01. Bioresour Technol 145:182–187

Green Car Congress (2015) Algenol and Reliance launch algae fuels demonstration project in India. www.greencarcongress.com/2015/01/20150121-algenol.html. Accessed 14 Apr 2019

Greenbaum E (1988) Energetic efficiency of hydrogen photoevolution by algal water splitting. Biophys J 54:365–368

Günerken E, D'Hondt E, Eppink MHM, Garcia-Gonzalez L, Elst K, Wijffels RH (2015) Cell disruption for microalgae biorefineries. Biotechnol Adv 33:243–260

Hernandez D, Solana M, Riano B, Garcia-Gonzalez MC, Bertucco A (2014) Biofuels from microalgae: lipid extraction and methane production from the residual biomass in a biorefinery approach. Bioresour Technol 170:370–378

Hirano A, Hon-Nami K, Kunito S, Hada M, Ogushi Y (1998) Temperature effect on continuous gasification of microalgal biomass: theoretical yield of methanol production and its energy balance. Catal Today 45:399–404

Ikawa M, Hartshorne T, Caron L-A, Iannitelli RC, Barbero LJ, Wegener K (1984) Inhibition of growth of the green alga *Chlorella pyrenoidosa* by unsaturated fatty acids. J Am Oil Chem Soc 61:1877–1878

Isikgor FH, Becer CR (2015) Lignocellulosic biomass: a sustainable platform for the production of bio-based chemicals and polymers. Polym Chem 6:4497–4559

Jardine A, Sayed S (2016) Challenges in the valorisation of chitinous biomass within the biorefinery concept. Curr Opin Green Sustain Chem 2:34–39

Jena U, Vaidyanathan N, Chinnasamy S, Das KC (2011) Evaluation of microalgae cultivation using recovered aqueous co-product from thermochemical liquefaction of algal biomass. Bioresour Technol 102:3380–3387

Jong ED, Higson A, Walsh P, Wellish M (2012) Bio-based chemicals: value-added products from biorefineries. International Energy Agency (IEA) report. Bioenergy Task 42

Jönsson LJ, Martín C (2016) Pretreatment of lignocellulose: formation of inhibitory by-products and strategies for minimizing their effects. Bioresour Technol 199:103–112

Jung YH, Park HM, Kim DH, Park YC, Seo JH, Kim KH (2015) Combination of high solids loading pretreatment and ethanol fermentation of whole slurry of pretreated rice straw to obtain high ethanol titers and yields. Bioresour Technol 198:861–866

Kaparaju P, Serrano M, Thomsen AB, Kongjan P, Angelidaki I (2009) Bioethanol, biohydrogen and biogas production from wheat straw in a biorefinery concept. Bioresour Technol 100:2562–2568

Kapoore RV, Butler TO, Pandhal J, Vaidyanathan S (2018) Microwave-assisted extraction for microalgae: from biofuels to biorefinery. Biology 7:18

Kuglarz M, Alvarado-Morales M, Karakashev D, Angelidaki I (2016) Integrated production of cellulosic bioethanol and succinic acid from industrial hemp in a biorefinery concept. Bioresour Technol 200:639–647

Kumar P, Barrett DM, Delwiche MJ, Stroeve P (2009) Methods for pretreatment of lignocellulosic biomass for efficient hydrolysis and biofuel production. Ind Eng Chem Res 48:3713–3729

Laurens LML, Markham J, Templeton DW, Christensen ED, Van Wychen S, Vadelius EW, Chen-Glasser M, Dong T, Davis R, Pienkos PT (2017) Development of algae biorefinery concepts for biofuels and bioproducts; a perspective on process-compatible products and their impact on cost-reduction. Energy Environ Sci 10(8):1716–1738

Lee SY, Cho JM, Chang YK, Oh Y-K (2017) Cell disruption and lipid extraction for microalgal biorefineries: a review. Bioresour Technol 244:1317–1328

Lindlbauer KA, Marx H (2017) 3-Hydroxypropionaldehyde production from crude glycerol by lactobacillus diolivorans with enhanced glycerol uptake. Biotechnol Biofuels 10:295

Liu X, Bi XT, Liu C, Liu Y (2012) Performance of Fe/AC catalyst prepared from demineralized pine bark particles in a microwave reactor. Chem Eng J 193:187–195

Liu L, Pohnert G, Wei D (2016a) Extracellular metabolites from industrial microalgae and their biotechnological potential. Marine Drug 14:191

Liu Z, Liao W, Liu Y (2016b) A sustainable biorefinery to convert agricultural residues into value-added chemicals. Biotechnol Biofuels 9:197

Lorente E, Hapońska M, Clavero E, Torras C, Salvadó J (2017) Microalgae fractionation using steam explosion, dynamic and tangential cross-flow membrane filtration. Bioresour Technol 237:3–10

Macías-Sánchez MD, Mantell C, Rodríguez M, Martínez de la Ossa E, Lubián LM, Montero O (2009) Comparison of supercritical fluid and ultrasound-assisted extraction of carotenoids and chlorophyll a from *Dunaliella salina*. Talanta 77:948–952

Mahesh R, Naira VR, Maiti SK (2019) Concomitant production of fatty acid methyl ester (biodiesel) and exopolysaccharides using efficient harvesting technology in flat panel photobioreactor with special sparging system via *Scenedesmus abundans*. Bioresour Technol 278:231–241

Maiti SK, Thuyavan YL, Singh S, Oberoi HS, Agarwal GP (2012) Modeling of the separation of inhibitory components from pretreated rice straw hydrolysate by nanofiltration membranes. Bioresour Technol 114:419–427

McMillan J (1993) Xylose fermentation to ethanol. A review. National Renewable Energy Lab, Golden, CO

Minowa T, Sawayama S (1999) A novel microalgal system for energy production with nitrogen cycling. Fuel 78:1213–1215

Mohamad N, Mustapa Kamal S, Mokhtar M (2015) Xylitol biological production: a review of recent studies. Food Rev Int 31:74–89

Mokrani T, Scurrell M (2009) Gas conversion to liquid fuels and chemicals: the methanol route-catalysis and processes development. Catal Rev 51:1–145

Molino A, Chianese S, Musmarra D (2016) Biomass gasification technology: the state of the art overview. J Energy Chem 25:10–25

Monegato A (2018) Use of seaweeds in paper production: the FAVINI experience. ALGAEUROPE, Amsterdam, p 54

Naira VR, Das D, Maiti SK (2018) Designing a CO_2 supply strategy for microalgal biodiesel production under diurnal light in a cylindrical-membrane photobioreactor. Bioresour Technol 250:936–941

Nanda S, Mohanty P, Pant KK, Naik S, Kozinski JA, Dalai AK (2013) Characterization of north American lignocellulosic biomass and biochars in terms of their candidacy for alternate renewable fuels. Bioenergy Res 6:663–677

Nanda S, Mohammad J, Reddy SN, Kozinski JA, Dalai AK (2014) Pathways of lignocellulosic biomass conversion to renewable fuels. Biomass Conv Bioref 4:157–191

Nanda S, Azargohar R, Dalai AK, Kozinski JA (2015) An assessment on the sustainability of lignocellulosic biomass for biorefining. Renew Sust Energy Rev 50:925–941

Nanda S, Dalai AK, Berruti F, Kozinski JA (2016a) Biochar as an exceptional bioresource for energy, agronomy, carbon sequestration, activated carbon and specialty materials. Waste Biomass Valor 7:201–235

Nanda S, Reddy SN, Mitra SK, Kozinski JA (2016b) The progressive routes for carbon capture and sequestration. Energy Sci Eng 4:99–122

Nanda S, Golemi-Kotra D, McDermott JC, Dalai AK, Gökalp I, Kozinski JA (2017) Fermentative production of butanol: perspectives on synthetic biology. New Biotechnol 37:210–221

Numaligarh Refinery Limited (2017) NRL inks framework agreement with M/s Chempolis Oy for implementation of bio refinery. www.nrl.co.in/internal_Default.aspx?id=news&Nid=10253. Accessed 14 Apr 2019

Okolie JA, Rana R, Nanda S, Dalai AK, Kozinski JA (2019) Supercritical water gasification of biomass: a state-of-the-art review of process parameters, reaction mechanisms and catalysis. Sustain Energy Fuel 3:578–598

Parajo JC, Dominguez H, Dominguez JM (1997) Improved xylitol production with *Debaryomyces hansenii* Y-7426 from raw or detoxified wood hydrolysates. Enzym Microb Technol 21:18–24

Parniakov O, Apicella E, Koubaa M, Barba FJ, Grimi N, Lebovka N, Pataro G, Ferrari G, Vorobiev E (2015) Ultrasound-assisted green solvent extraction of high-added value compounds from microalgae *Nannochloropsis* spp. Bioresour Technol 198:262–267

Pawlicka-Kaczorowska J, Czaczyk K (2017) Effect of crude and pure glycerol on biomass production and trehalose accumulation by *Propionibacterium freudenreichii* ssp. *shermanii* 1. Acta Biochim Pol 64:621–629

Perez ATE, Camargo M, Rincon PCN, Marchant MA (2017) Key challenges and requirements for sustainable and industrialized biorefinery supply chain design and management: a bibliographic analysis. Renew Sust Energy Rev 69:350–359

Pham M, Schideman L, Scott J, Rajagopalan N, Plewa MJ (2013) Chemical and biological characterization of wastewater generated from hydrothermal liquefaction of *Spirulina*. Environ Sci Technol 47:2131–2138

Plaza M, Santoyo S, Jaime L, Avalo B, Cifuentes A, Reglero G, García-Blairsy Reina G, Señoráns FJ, Ibáñez E (2012) Comprehensive characterization of the functional activities of pressurized liquid and ultrasound-assisted extracts from *Chlorella vulgaris*. LWT—Food Sci Technol 46:245–253

Pyle DJ, Garcia RA, Wen Z (2008) Producing docosahexaenoic acid (DHA)-rich algae from biodiesel-derived crude glycerol: effects of impurities on DHA production and algal biomass composition. J Agric Food Chem 56:3933–3939

Quarderer GJ (1986) Mixed alcohols from synthesis gas. In 78th National AICHE Meeting, New Orleans, LA, USA

Qureshi N, Saha BC, Cotta MA (2007) Butanol production from wheat straw hydrolysate using *Clostridium beijerinckii*. Bioprocess Biosyst Eng 30:419–427

Rabelo S, Carrere H, Maciel Filho R, Costa A (2011) Production of bioethanol, methane and heat from sugarcane bagasse in a biorefinery concept. Bioresour Technol 102:7887–7895

Rana V, Eckard AD, Ahring BK (2014) Comparison of SHF and SSF of wet exploded corn stover and loblolly pine using in-house enzymes produced from *T. reesei* RUT C30 and *A. saccharolyticus*. Springerplus 3:516

Ravindran R, Jaiswal A (2016) Microbial enzyme production using lignocellulosic food industry wastes as feedstock: a review. Bioengineering 3:30

Real G (2018) Large-scale commercial microalgae production to supply the food, feed and cosmetic markets. ALGAEUROPE, Amsterdam, p 54

Reitan KI, Rainuzzo JR, Olsen Y (1994) Effect of nutrient limitation on fatty-acid and lipid-content of marine microalgae. J Phycol 30:972–979

Rodriguez AA, Stella AM, Storni MM, Zulpa G, Zaccaro MC (2006) Effects of cyanobacterial extracellular products and gibberellic acid on salinity tolerance in *Oryza sativa L.* Saline Syst 2:7

Rosés RP, Guerra NP (2009) Optimization of amylase production by *Aspergillus niger* in solid-state fermentation using sugarcane bagasse as solid support material. World J Microbiol Biotechnol 25:1929–1939

Roychoudhury PK, Ghose TK, Ghosh P (1992) Operational strategies in vacuum-coupled SSF for conversion of lignocellulose to ethanol. Enzym Microb Technol 14:581–585

Salomão GSB, Agnezi JC, Paulino LB, Hencker LB, de Lira TS, Tardioli PW, Pinotti LM (2019) Production of cellulases by solid state fermentation using natural and pretreated sugarcane bagasse with different fungi. Biocatal Agric Biotechnol 17:1–6

Shahab RL, Luterbacher JS, Brethauer S, Studer MH (2018) Consolidated bioprocessing of lignocellulosic biomass to lactic acid by a synthetic fungal-bacterial consortium. Biotechnol Bioeng 115:1207–1215

Sierra LS, Dixon CK, Wilken LR (2017) Enzymatic cell disruption of the microalgae *Chlamydomonas reinhardtii* for lipid and protein extraction. Algal Res 25:149–159

Sivaramakrishnan R, Suresh S, Incharoensakdi A (2019) *Chlamydomonas* sp. as dynamic biorefinery feedstock for the production of methyl ester and ε-polylysine. Bioresour Technol 272:281–287

Slade R, Bauen A (2013) Micro-algae cultivation for biofuels: cost, energy balance, environmental impacts and future prospects. Biomass Bioenergy 53:29–38

Su H, Liu G, He M, Tan F (2015) A biorefining process: sequential, combinational lignocellulose pretreatment procedure for improving biobutanol production from sugarcane bagasse. Bioresour Technol 187:149–160

Tian G, Xu H, Daniel R (2011) DMF-a new biofuel candidate. InTech, Surrey

Tian C, Li B, Liu Z, Zhang Y, Lu H (2014) Hydrothermal liquefaction for algal biorefinery: a critical review. Renew Sust Energy Rev 38:933–950

Tomas-Pejo E, Fermoso J, Herrador E, Hernando H, Jiménez-Sánchez S, Ballesteros M, González-Fernández C, Serrano D (2017) Valorization of steam-exploded wheat straw through a biorefinery approach: bioethanol and bio-oil co-production. Fuel 199:403–412

Unrean P, Ketsub N (2018) Integrated lignocellulosic bioprocess for co-production of ethanol and xylitol from sugarcane bagasse. Ind Crop Prod 123:238–246

Van DK, Van DR, Van EV, Van HH, Schipper W, Stam H (1995) Methanol from natural gas: conceptual design & comparison of processes. Delft University of Technology, Delft

Wang C, Thygesen A, Liu Y, Li Q, Yang M, Dang D, Wang Z, Wan Y, Lin W, Xing J (2013) Bio-oil based biorefinery strategy for the production of succinic acid. Biotechnol Biofuels 6:74

Wu C, Xiao Y, Lin W, Li J, Zhang S, Zhu J, Rong J (2017) Aqueous enzymatic process for cell wall degradation and lipid extraction from *Nannochloropsis* sp. Bioresour Technol 223:312–316

Yamakawa CK, Qin F, Mussatto SI (2018) Advances and opportunities in biomass conversion technologies and biorefineries for the development of a bio-based economy. Biomass Bioenergy 119:54–60

Yang L, Li X, Lai C, Fan Y, Ouyang J, Yong Q (2017) Fungal chitosan production using xylose rich of corn stover prehydrolysate by *Rhizopus oryzae*. Biotechnol Biotechnol Equip 31:1160–1166

Zhang YHP, Lynd LR (2004) Toward an aggregated understanding of enzymatic hydrolysis of cellulose: noncomplexed cellulase systems. Biotechnol Bioeng 88:797–824

Zhang J, Liu X, Xu Z, Chen H, Yang Y (2008) Degradation of chlorophenols catalyzed by laccase. Int Biodeterior Biodegrad 61:351–356

Zhao X, Chi Z, Rover M, Brown R, Jarboe L, Wen Z (2013) Microalgae fermentation of acetic acid-rich pyrolytic bio-oil: reducing bio-oil toxicity by alkali treatment. Environ Prog Sustain Energy 32:955–961

Zhu JJ, Yong QA, Xu Y, Yu SY (2011) Detoxification of corn stover prehydrolyzate by trialkylamine extraction to improve the ethanol production with *Pichia stipitis* CBS 5776. Bioresour Technol 102:1663–1668

Zhu S, Huang W, Huang W, Wang K, Chen Q, Wu Y (2015) Pretreatment of rice straw for ethanol production by a two-step process using dilute sulfuric acid and sulfomethylation reagent. Appl Energy 154:190–196

Conversion of Rice Husk and Nutshells into Gaseous, Liquid, and Solid Biofuels

8

Anton P. Koskin, Inna V. Zibareva, and Aleksey A. Vedyagin

Abstract

Nowadays, biomass is considered as a main renewable energy source with a high contribution value to the worldwide energetics. In general, biomass is represented by solid waste materials from plant and animal farming, horticulture, food processing, etc. All these wastes are nonedible and cannot be used as food or feed. Among the agricultural solid wastes, rice husk and nutshells are of a special attention due to their high production. In this chapter, the main approaches and concepts of the rice husk and nutshells transformation into valuable gaseous, liquid, and solid products are overviewed. In some cases, the initial biomass should be pretreated to enhance the efficiency of its further conversion. The transformation processes can be classified onto direct combustion (solid fuel), pyrolysis and hydrothermal methods (solid and liquid fuels) and gasification (gaseous fuel). The commonly used fixed-bed and fluidized-bed reactors including catalytic ones, auger reactor, and plasma gasifiers are widely applied.

Keywords

Agricultural solid wastes · Rice husk · Nutshells · Biofuels · Lignocellulosic biomass · Waste-to-energy

A. P. Koskin · I. V. Zibareva
Boreskov Institute of Catalysis SB RAS, Novosibirsk, Russia

A. A. Vedyagin (✉)
Boreskov Institute of Catalysis SB RAS, Novosibirsk, Russia

National Research Tomsk Polytechnic University, Tomsk, Russia
e-mail: vedyagin@catalysis.ru

S. Nanda et al. (eds.), *Biorefinery of Alternative Resources: Targeting Green Fuels and Platform Chemicals*, https://doi.org/10.1007/978-981-15-1804-1_8

8.1 Introduction

The global population growth leads to a continuous increase in the total amount of the worldwide agricultural crops. The subsequent processing of the agricultural sources into the food products results in the accumulation of huge quantities of the solid wastes. Most of these solid agricultural wastes originate from farmlands, orchards, vineyards, dairies and feedlots. The total amount of the annual crop waste production exceeds one billion tons. The high volume and low packing density of the crop wastes including rice husk and nutshells complicate their transportation, thereby creating problems of their proper disposal (Rice Market Monitor 2017). For example, rice husk is usually disposed of in the farmlands, mulched or burnt, thus affecting the environment through greenhouse gas emissions. Therefore, the utilization of the agricultural solid wastes is of extreme importance.

The agricultural and wood processing wastes along with crop residues, urban organic or animal wastes are considered as renewable biomass resources. The utilization methods for processing of agricultural waste and other types of biomass have been developed for a long time. In this connection, one of the most perspective directions is their usage as a renewable energy source. Solar energy is accumulated in various biomass forms during the photosynthesis. Therefore, a variety of gaseous, liquid, and solid fuels, or the so-called biofuels, can be derived from the biomass wastes. This concept is known as waste-to-energy or energy-from-waste conversion.

Lignocellulosic biomass is a complex solid material constructed from oxygen-containing organic polymers produced via natural processes. Carbohydrate polymers and oligomers (65–75%) and lignin (18–35%) represent the major part of the structural chemical high-molecular components. In the future, an important advantage of using biofuels in comparison with traditional petrochemical hydrocarbons would be the diminishing gaseous emissions, which affect the environment and climate (Sudiro and Bertucco 2007). Additionally, it allows one to reduce the cost

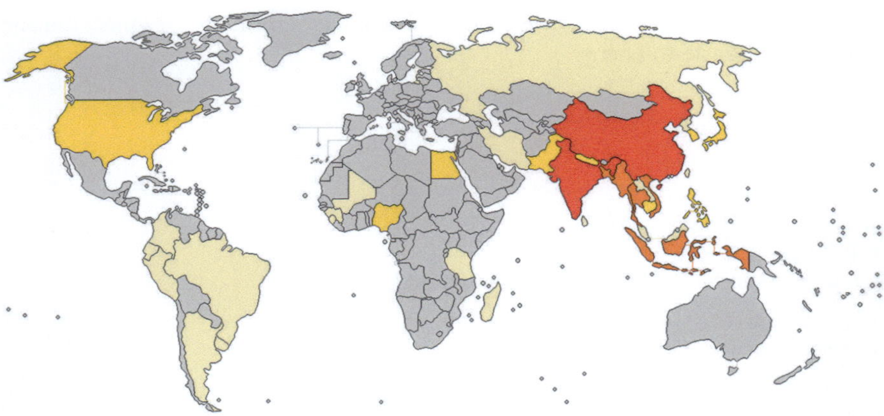

Fig. 8.1 World map with highlighted countries producing rice in million metric tons per annum (MMTPA): >150 MMTPA (red); >25 MMTPA (orange); >5 MMTPA (yellow); >1 MMTPA (light yellow)

of the catalytic posttreatment (Marquit 2009). Most agricultural wastes contain a significantly lower amount of fixed sulfur and nitrogen when compared with the fossil fuels (e.g., bituminous coal). Therefore, such fuels derived from biomass can be considered as renewable with a diminished level of SO_x and NO_x emissions.

In this chapter, the approaches of biofuel production from rice residues (e.g., rice husk) and dry tree nuts wastes (e.g., almond, walnut, cashew, pistachio, and hazelnut nutshells) are reviewed. The attractiveness of rice wastes as a source of biofuels is stipulated by a huge popularity of this crop in the world (747 million tons per year). In terms of a worldwide tonnage, rice is inferior to the totality of the cereal crops. At the same time, it gives the largest amount of the solid wastes (Nakhshiniev et al. 2014). Figure 8.1 presents the rice production areas in the world map. As it can be seen that the top ten countries producing rice are China, India, Indonesia, Vietnam, Thailand, Bangladesh, Myanmar, the Philippines, Brazil, and Pakistan. Therefore, nearly 90% of world's areas of rice cultivation are in the Asian countries. The leading positions in this list are occupied by China (market share of 28%) and India (market share of 28%).

The major rice residues are husk and straw. Both these types of wastes have a great potential to be considered as the source for waste-to-energy concept. The annual volume of rice husk production is estimated to be 149 million tons. On the other hand, it should be mentioned that rice husk is characterized with high ash content, which is equal to 20–25 wt% depending on the geographical area for cultivation. From this point of view, agricultural wastes of nuts cultivation (nutshells) are much more attractive, since the content of ash component irrespective to the nut type does not exceed 5 wt%, which is significantly lower.

The countries producing majority of tree nuts are shown in Fig. 8.2. The world leaders are African countries (cashew), Turkey (hazelnut and pistachio), Iran (pistachio), India (cashew), China (walnut), the USA (almond, walnut, and pistachio), and Vietnam (cashew). It is worth mentioning that the efficient processing of agricultural wastes into

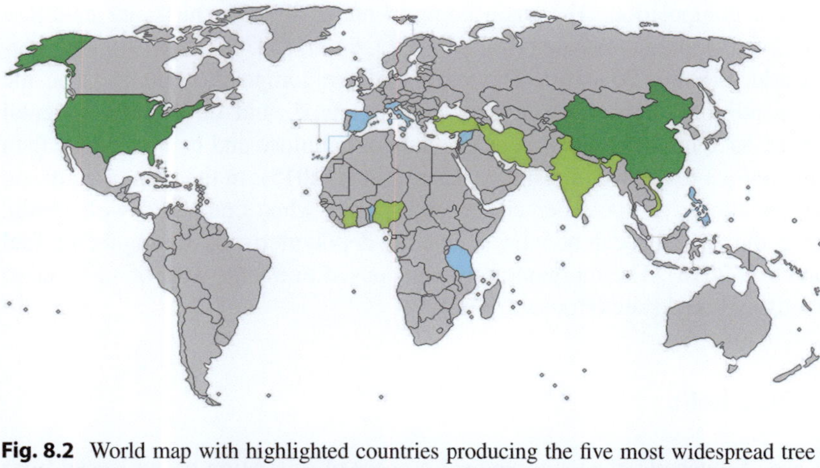

Fig. 8.2 World map with highlighted countries producing the five most widespread tree nuts in million metric tons per annum (MMTPA): >1 MMTPA (green); >0.02 MMTPA (light green); >0.005 MMTPA (blue)

biofuels is more advantageous comparing with the cultivation of biofuel-oriented crops (*Jatropha*, *Miscanthus*, etc.) (Nigam and Singh 2011). In this case, firstly, the technology of agricultural food production can become non-waste and environmentally friendly, and secondly, no re-orientation of the current land tenure is required.

8.2 Physicochemical Characterization of Rice Husk and Nutshells

8.2.1 Rice Husk

The technological scheme to produce alimentary rice from paddy consists of a shelling process with the separation of rice husk, which results in the appearance of the so-called brown rice. Subsequent milling and classification procedures allow one to separate rice bran and broken rice. Therefore, the main solids being formed during the rice production are rice straw, husk, ash, bran, and broken rice (Moraes et al. 2014). The hull formed during the grain growth represents the rice husk, and it is of 8–10 mm in length, 2–3 mm in width, and 0.2 mm in thickness (Fang et al. 2004). The density of the husk is quite low, i.e., 122 kg/m^3. On an average, its volume is estimated to be about 20% of the rice weight.

Since the rice husk contains a low level of nutrients and a high amount of silica, it cannot be considered as a component for the production of fodder (Moraes et al. 2014). Along with a set of building and materials science applications (e.g., incineration with subsequent usage of rice husk ash in concretes (Fapohunda et al. 2017)), other applications of rice harvest residues include the synthesis of the catalyst using rice husk ash (Tang et al. 2018), production of highly porous sorbents (Larichev et al. 2013, 2015; Yeletsky et al. 2009), carbon-based materials for supercapacitors (Lebedeva et al. 2018a, b, 2015), and use as solid fuel and feedstock in biofuel production.

When biomass is considered as a precursor for biofuels, the key characteristic is its chemical composition. The main components of the rice husk are cellulose (28.6–41.5 wt%), hemicellulose (14–28.6 wt%), lignin (20.4–33.7 wt%), and inorganic residues (about 20 wt%). The variety of the composition depends on the weather conditions, agronomical treatment, soil type, and other environmental parameters. Rice husk contains more amounts of cellulose and hemicellulose than lignin in contrast to the wood residue (Goenka et al. 2015). In the case of cellulose and hemicellulose, the polymer chains are shorter when compared with lignin. Therefore, the rice husk can be more efficiently depolymerized forming the biofuel (Nelson et al. 2018). The inorganic residues consist of 95–98 wt% of silica in an amorphous hydrated state (Housten 1972).

8.2.2 Nutshells

In the case of nutshells, the technological process of separation of nut kernel from other parts of nut (soft nutshell or wet nutshell) can be varied significantly depending on the type of nut. Besides the traditional drying, stages such as extraction of the components contained in the soft nutshell can be included. The finally obtained dry nuts undergo further processing such as shelling and separation procedures, giving at the end the nut kernels and the dry nut shell. It should be emphasized that compared to the solid wastes from the milled rice production, the amount of the processing wastes from the nut kernel production is significantly higher. For instance, in the case of cashew, the weight of waste exceeds 70% of the initial unshelled nut by weight.

The range of the compounds present in the nutshell is noticeably wider. The extracts derived from the soft nutshell are efficiently applied in synthetic chemistry, perfumery, cosmetics, and other related sectors (Balachandran et al. 2013; Balgude and Sabnis 2013). In particular, soft cashew nutshell contains about 30–35 wt% of cashew nutshell liquid (CNSL or cashew shell oil), which is commercially available and widely used in biomedicine, materials science, and chemical industry (Hamad and Mubofu 2015; Lomonaco et al. 2017; Tullo 2008). The main difference between rice husk and nutshells is lower ash content of the latter. This makes the nutshells a more suitable type of the agricultural wastes as for direct combustion as for the synthesis of biofuels. Table 8.1 summarizes the data on the composition of rice husk and nutshells under consideration. It should be noted that the processes for nutshells conversion are much less reported in the literature than those for rice husk, which is due to the lower volumes of nuts production.

8.3 Conversion of Rice Husk and Nutshells into Biofuels

The processes of the agricultural wastes utilization with energy production can be divided into two categories. The first category considers direct usage of biomass as a fuel for combustion with the production of heat and electricity. The second category is represented by the processes for preliminary transformation of raw biomass into other forms of energetic products (e.g., depolymerization with elimination of ash-forming components) for their further application as an energy source (Goyal et al. 2008). This category includes biochemical (e.g., anaerobic digestion) and thermochemical (e.g., pyrolysis or gasification) conversions. All these processes facilitate the synthesis of solid, liquid, and gaseous biofuels.

A tendency toward the complexity of the biomass conversion technological schemes is observed (Fig. 8.3). Such schemes include the primary procedures of the thermochemical and biochemical treatments of the raw feedstock, thereby transforming it into an easily transportable precursor of biofuels (or synthetic transportation fuels). Furthermore, the biofuel precursors are responsible for the realization of the final synthesis stage of biofuel with high-energy characteristics. During the last few decades, biofuel upgrading processes have gained wide attention. For example, hydrodeoxygenation process is aimed at reducing the amount of oxygenates in the composition of biomass-derived fuels, thereby diminishing the corrosive properties and increasing the energy characteristics (Bykova et al. 2012; Dundich et al. 2010; Kukushkin et al. 2015).

Table 8.1 Composition and higher heating values of the rice husk and nutshells

Biomass source	Moisture (wt%)	Volatile matter (wt%)	Cellulose (wt%)	Hemicellulose (wt%)	Lignin (wt%)	Fixed carbon (wt%)	Ash (wt%)	Atomic ratios		Higher heating value (MJ/kg)	References
								O/C	H/C		
Rice husk	6.0	52.0	25.1	28.6	24.4	15.0	16.9	0.65	1.55	13.4	Fang et al. (2004), Goenka et al. (2015)
Almond nutshell	8.7	79.7	38.5	28.8	29.5	19.3	2.3	0.70	1.44	18.9	Demirbaş (2010), Savova et al. (2001)
Walnut shell	7.7	77.9	26.9	22. 5	47.7	13.5	0.94	0.61	1.52	17.2	Kar (2011)
Cashew nutshell cake	10.4	69.3	–	–	–	19.3	1.0	0.66	1.71	24.1	Das and Ganesh (2003)
Pistachio nutshell	5.8	84.9	43.0	25.3	16.4	12.9	2.2	0.66	1.7	18.0	Goldfarb et al. (2017), Li et al. (2018)
Hazelnut shell	10.9	69.0	25.9	25.9	42.5	19.4	1.35	0.62	1.27	21.4	Çepelioğullar and Pütün (2013), Demirbaş (1997)

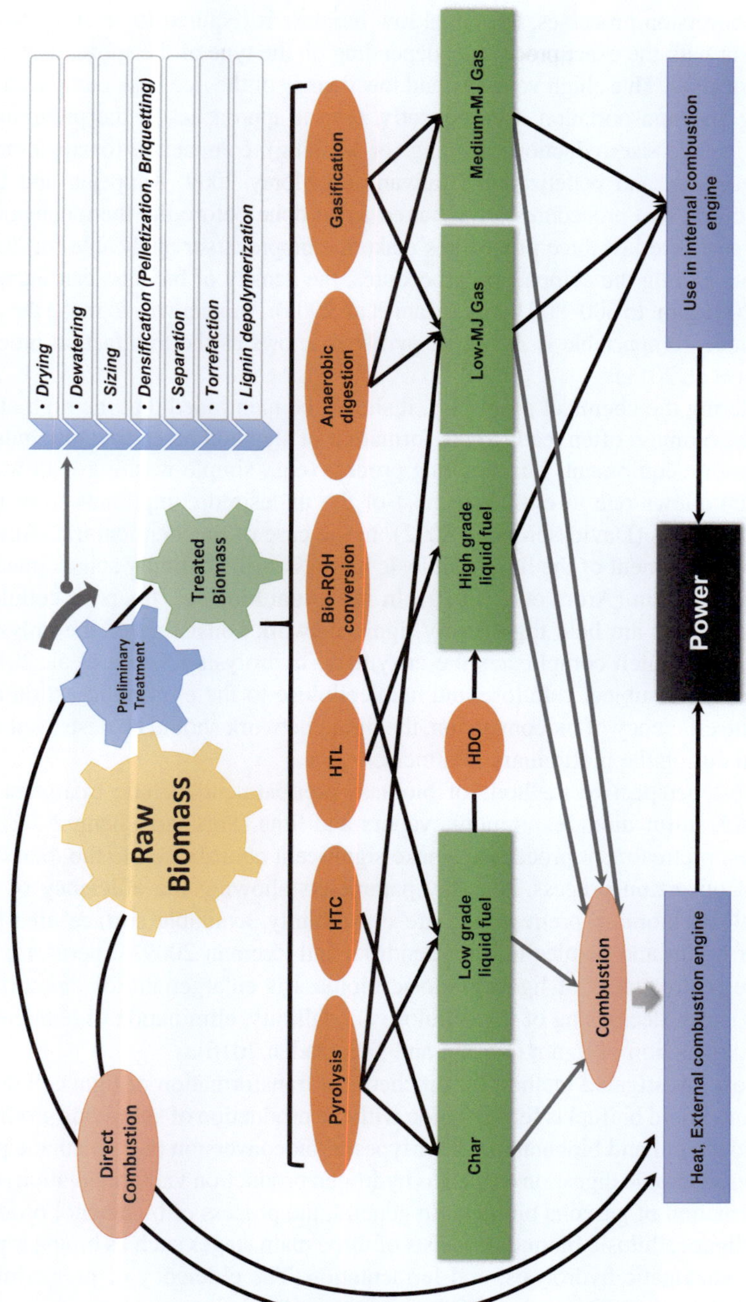

Fig. 8.3 General scheme of biomass conversion (rice husk and nutshells) into solid, liquid, and gaseous biofuels

8.3.1 Pretreatment of Biomass

In some conversion processes, the initial raw biomass is required to be subjected to pretreatment with the exact procedures depending on the type of the subsequent conversion processes. Thus, high volatility and low density of the rice husk complicate its processing and transportation. Consequently, typical approaches to the preliminary treatment include size reduction, leaching (or washing), compaction (or agglomeration), briquetting, and pelletization (Kaliyan and Morey 2009; Koppejan and Loo 2012). The first two procedures are required to be done before the thermochemical processes since the last three procedures make the bioprecursor applicable for direct combustion. During the compaction procedures, the density of biomass can increase from 40–200 kg/m^3 to 600–800 kg/m^3 (Mani et al. 2003). Another way to make the raw biomass more compactable is its partial pyrolysis knows as the torrefaction process (Tumuluru et al. 2011).

Considering the chemical properties, it should be mentioned that the high alkalinity of the biomass often leads to the formation of slag and other contaminants in the combustion equipment. The leaching process (e.g., simple washing with water or solvents) allows one to eliminate most of the undesired compounds from rice husk and nutshells (Davidsson et al. 2002). In the case of biochemical transformations, the pretreatment of the lignocellulosic biomass is the defining stage aimed to depolymerize lignin (Arora et al. 2019). In the initial biomass, fibers of cellulose and hemicellulose are held together by lignin network consisting of the polysaccharide layers, which complicates the enzymatic hydrolysis (Nanda et al. 2014). Consequently, to subject cellulose and hemicellulose to the enzymatic action and increase the efficiency of bioconversion, the lignin network should be destroyed and eliminated during the preliminary treatment.

The most perspective methods of biomass pretreatment include treatment by diluted acid, sulfur dioxide, ammonia vapors and lime (Sun and Cheng 2002). In some cases, pretreatment procedures make significant contribution to the total cost of the bioconversion process. The key parameters showing the efficiency of the lignocellulosic biomass pretreatment are crystallinity, available surface area and content of lignin and hemicellulose (Hendriks and Zeeman 2009). Therefore, the aim of the pretreatment of lignocellulosic biomass is enlargement of the surface area of biomass, decreasing of the cellulose crystallinity, elimination of hemicellulose, and destruction of lignin (Karimi and Taherzadeh 2016).

The most investigated method for biochemical transformation of lignocellulosic biomass into liquid biofuel is fermentation with the production of spirituous products such as bioethanol and biobutanol. Other types of bioconversion (e.g., methane production via anaerobic digestion as well as hydrogen production via fermentation) can lead to formation of gaseous biofuels. In general, the process of bioethanol production from lignocellulosic biomass consists of three main stages such as biomass pretreatment, enzymatic hydrolysis, and fermentation. The efficiency of pretreatment methods defines the effectiveness of the biofuel synthesis process as a whole. The main challenge is to destroy the lignin network, which forms protective covering over cellulose and hemicellulose and making them inaccessible for hydrolysis.

Subsequent processes of enzymatic hydrolysis and fermentation include transformation of cellulose into glucose and hemicellulose into a number of pentose and hexose sugars (Taherzadeh and Niklasson 2004) and production of bio-alcohols using special microorganisms (Arora et al. 2019). The methods of simultaneous saccharification and fermentation (SSF) are also under research (Karimi et al. 2006).

8.3.2 Direct Combustion of Rice Husk and Nutshells

Direct combustion of agricultural wastes is known as one of the oldest and frequently applied processes for energy production (Yin et al. 2008). It is useful for calorification, drying, and generation of steam for the processes of parboiling treatment of the agricultural crops (Kwofie and Ngadi 2017). In an ideal case, all the oxidized components of biomass can be saturated with oxygen in accordance with the following equation:

$$C_xO_yH_zN_wS_v + aO_2 \rightarrow \text{Energy} + bCO_2 + cH_2O + eN_mO_p + dSO_r \qquad (8.1)$$

Incineration of the agricultural wastes can be performed in the stoves of various types such as stoves of opened fire, boilers, fixed-bed combustors, or fluidized-bed combustors. It should be mentioned that in the case of the fixed-bed incineration, only biomass with moisture less than 50% could be used. Among the disadvantages of this technology, incomplete combustion of biomass, high atmospheric emissions, incrustation, and corrosion in the burning chamber and ancillary equipment are reported (Miles et al. 1996; Saidur et al. 2011).

Completeness of combustion can be achieved by using the preliminary treatment described above (e.g., drying, milling, and leaching) or by the application of other costly pretreatment technologies. For example, the highest efficiency of rice husk and nutshells conversion processes including combustion, pyrolysis, and gasification can be attained in a fluidized-bed reactor. Such reactor consists of a chamber containing a bed of inert particles such as sand or catalytically active particles supported by a distributor plate (Natarajan et al. 1998). Extremely high heat transfer and mass transfer characteristics of the fluidized-bed regime allow performing fast and practically complete transformation of biomass into energy. The main advantage of fluidized-bed reactors is that low-grade, less heat content, and low-density biomass sources with irregular size of grains and high moisture content can be efficiently used.

The modern state of the technologies based on the fluidized-bed reactors guarantees the high uniformity of the side combustion products. It gives an opportunity to selectively trap particles of a defined size (e.g., silica-reached particles of rice husk ash) for their further practical application (Fernandes et al. 2016; Pode 2016). However, despite all the disadvantages of the direct biomass combustion in the fixed-bed reactors, this technology is cost effective, which makes its application more attractive in rural areas and especially in the developing countries (Moraes et al. 2014). It should be emphasized that the rice production capacity in the world is higher in the rural regions. Therefore, low-cost fixed-bed reactors could be a suitable alternative for bioenergy generation.

Another approach to enhance the efficiency of the direct combustion processes is a briquetting technology. Rice husk and nutshells briquettes are obtained by pressing or compaction of biomass via compression at elevated temperatures (Ndindeng et al. 2015). To achieve the required compaction degree, various binding compounds can be used. The final briquettes are cylindrical in shape with a varied diameter from 3 to 20 cm and length from 15 to 50 cm. Thus, direct combustion of the briquette made from rice husk emits noticeably lower amount of particulate matter when compared with raw rice husk. Besides, the higher heating value and density of this solid biofuel can be increased up to 15.5–16.3 MJ/kg and 360–600 kg/m^3, correspondingly (Yank et al. 2016).

Comparative analysis of the direct combustion processes with similar technologies is reported in a number of research and review papers (Baetge and Kaltschmitt 2018; Lim et al. 2012; Pode 2016; Quispe et al. 2017; Simonov et al. 2003). Unfortunately, the combustion of nutshells is poorly described in the literature. da Silva et al. (2017) studied the pistachio shell combustion. Other researchers have reported a benefit of an alternative technology considering the usage of joint combustion (or co-firing) of rice husk or nutshells along with other types of biomass such as sugarcane bagasse and coal (Kuprianov et al. 2006; Sathitruangsak and Madhiyanon 2017; Wu et al. 2015). Combining briquetting and co-firing technologies together can be promising. In this case, rice husk or nutshells are pressed together with fossil fuel residues (e.g., lignite) or other types of wastes (Özyuğuran et al. 2017). No additional binding compounds are required to achieve the desired density of the briquettes.

8.4 Rice Husk and Nutshells as Sources of Biochar

During the last decades, the more advanced methods alternative to the direct combustion of the agricultural wastes are under intensive development. Within the concept of these methods, biomass is preliminary converted into high-calorific value product with denser and less-ash-containing materials (Basu 2010). The energy content is usually increased by the elimination of oxygenates from the raw biomass. Compression of the solid fuel materials simplifies the logistics of the processes. The main methods for the synthesis of solid fuels from biomass are predominantly slow and intermediate pyrolysis, inert and hydrothermal carbonization, and torrefaction. In the case of accelerated processes (e.g., fast and flash pyrolysis), liquid and gaseous fuels are being formed (Mohan et al. 2006).

8.4.1 Pyrolysis

Pyrolysis is a thermal process used for the decomposition of organic sources at high temperatures in the absence of oxygen. Such kind of thermal treatment results in irreversible changes in both the chemical composition and physical properties of biomass (Mohan et al. 2006). The pyrolysis processes can be easily used in rural regions dealing with a huge variety of biomass and agricultural byproducts, which could make these processes the most advantageous. Characteristics of the main pyrolysis processes are summarized in Table 8.2. As seen, in the most cases, a composition of liquids (bio-oil), gases (producer gas) and solids (biochar) results from the process. Carbonization process leads to the formation of biochar. The phase composition and energy characteristics of the pyrolysis products depend significantly on the process conditions (Neves et al. 2011). Thus, depending on process conditions such as heating rate, residence time, reaction temperature, and feed rate, the pyrolysis processes can be divided into carbonization and slow, fast and flash pyrolysis (Basu 2010; Mohan et al. 2006; Nanda et al. 2016).

In some investigations, intermediate and ultra-fast pyrolysis are also mentioned. In general, the conditions of the process are selected to maximize the yield of the desired product. Byproducts are often used for the generation of energy, which is required for drying of raw biomass (pretreatment) or considered as a part of energy consumed during the pyrolysis (Dufour et al. 2009). Non-condensable gases (e.g., H_2, CO, CO_2, CH_4, and C_xH_y) formed as pyrolysis byproducts can be reused as gaseous fuel in boilers or as a source for hydrocarbon production using catalytic Fischer-Tropsch process.

In general, the yield of the solid fuel (biochar) can be increased by using the smaller fraction of biomass, lower heating rates, longer duration of the process, and lower maximum temperature (Neves et al. 2011). Carbonization method is optimal in terms of the highest biochar yield. However, in this case, the expenditure of energy is not comparable with energy profit. The efficiency of this process can be improved by performing it in the presence of water steam under high pressure (i.e., hydrothermal carbonization) or methane (i.e., methanopyrolysis) instead of inert atmosphere. The method of methanopyrolysis is quite unique since it allows to obtain the semi-products for synthetic chemistry (Mohan et al. 2006).

Both slow and intermediate pyrolysis processes are carried out at temperatures below 500 °C in a regime of slow heating rate and longer residence time (Park et al. 2014). Another process of biomass conversion into solid biofuel is torrefaction. In this process, raw biomass is partly pyrolyzed within a temperature range of 200–300 °C, which leads to diminishing the moisture content in biomass and depolymerization of hemicellulose (Medic et al. 2012). Therefore, this process allows producing energy dense and easily transportable precursors of solid fuels.

Different generic types of biomass gasifiers and pyrolyzers have been developed and commercialized for the conversion of biomass of different nature. The construction of these reactors can be categorized using the following criteria:

1. Type of the mutual displacement of the layers of biomass and decomposition medium inside the reactor.
2. Type of the process defined by the heating rate and temperature, decomposition medium, and biomass feeding rate (residence time).

According to the abovementioned criteria, the reactors can be divided into batch reactors, moving-bed reactors, reactors with mechanical movement, and fluidized-bed reactors (Basu 2010). In the case of batch reactors, no solid movement through

Table 8.2 Main types of the pyrolysis processes

	Reaction conditions				Biofuel yield (wt%)		
Process	Temperature (°C)	Residence time	Heating rate	Medium	Char	Liquid	Gas
Carbonization	400	Days	Slow	Inert	<40	–	–
Slow pyrolysis	400–600	5–30 min	Slow	Inert	25–35	20–50	20–50
Fast pyrolysis	450–600	1–2 s	Fast	Inert	10–25	50–70	10–20
Flash pyrolysis	<650	<1 s	Fast	Inert	–	70–80	20–30
Vacuum pyrolysis	350–550	2–30 s	Medium	Vacuum	–	60–70	–
Hydrothermal carbonization	180–250	1–2 h	Fast	Water vapors	50–90	–	–

Reference: Mohan et al. (2006), Basu (2010), Nelson et al. (2018)

the reactor takes place during the pyrolysis process. The moving-bed regime is realized in the shaft furnaces. Mechanical forces (e.g., rotary kiln, rotating screw, or auger reactor) result in the movement and mixing of biomass particles. In the case of fluidized-bed reactors, movement of biomass is caused by a fluid flow (e.g., fluidized-bed, spouted-bed, and entrained-bed). The majority of the reactors mentioned above are applied for conversion of rice husk and nutshells into solid biofuels (Chakma et al. 2015; Demirbas 2006; Jeong et al. 2016; Lim et al. 2012; Moreira et al. 2016; Quispe et al. 2017; Taghizadeh-Alisaraei et al. 2017; Unrean et al. 2018). Some of the technologies of rice husk processing into biochar by the pyrolysis methods are realized on a pilot-scale or large-tonnage scale in China, the USA, India, and Brazil (Lim et al. 2012).

8.4.2 Hydrothermal Carbonization

Hydrothermal processes for biomass transformation are of great interest. Similarly to the traditional pyrolysis, hydrothermal carbonization is realized at anaerobic conditions, but water steam at elevated pressure is used instead of the inert medium (Wang et al. 2018). Initial raw biomass can be used without preliminary drying, and this is the advantage of the present approach. Dissolution and crystallization of biomass occur under pre-critical and supercritical conditions. It allows avoiding problems such as agglomeration that occur in the case of traditional pyrolysis. As described above for pyrolysis, the conditions of the hydrothermal process determine the phase composition of the resulting biofuel (Wang et al. 2018).

Hydrothermal biomass conversion processes include hydrothermal carbonization (HTC), hydrothermal liquefaction (HTL), and hydrothermal gasification (HTG). The choice of the exact process depends on the target products. In general, the hydrothermal processes are performed at relatively low temperatures (180–250 °C) and under autogenous pressure (Funke and Ziegler 2010). As a result, the oxygen and hydrogen contents become significantly lower due to dehydration and decarboxylation reactions.

On the other hand, it is worth to note that additional stage of filtration is required to obtain the char in a purified state. Moreover, the necessity of the usage of high pressures and catalysts reduces the attractiveness of this process. Despite the great fundamental interest to the conversion of rice husk and nutshells using this approach (Kalderis et al. 2014; Yang et al. 2015), the continuous process of the hydrothermal carbonization is not yet realized on the industrial scale as most of the achievements are obtained on the laboratory scale (Elliott et al. 2015).

8.5 Rice Husk and Nutshells as Sources of Bio-oil: Pyrolysis and Hydrothermal Liquefaction

In the case of slow pyrolysis, the liquid phase (bio-oil) is considered as an undesired byproduct since it contains a large amount of oxygenates (Mohan et al. 2006; Mohanty et al. 2013). The fast pyrolysis methods give higher portion of liquid biofuels, which are characterized by better energy parameters (Goyal et al. 2008). Typically, fast pyrolysis is carried out at a temperature range of 500–700 °C with fast heating and low contact times (Balat and Balat 2009). For vacuum pyrolysis, the heating rates are noticeably lower when compared to fast pyrolysis. Moreover, an increase in the bio-oil yield in the case of vacuum pyrolysis is achieved due to lower residence time of the pyrolysis vapors and limited secondary gas-phase reactions (Nachenius et al. 2013).

The production of bio-oil via the conversion of rice husk and nutshells of pistachio, cashew, hazelnut, and walnut have been described in detail elsewhere (Chakma et al. 2015; Das and Ganesh 2003; Goldfarb et al. 2017; Lim et al. 2012; Miranda et al. 2012; Patel et al. 2011; Pütün et al. 1999; Taghizadeh-Alisaraei et al. 2017; Unrean et al. 2018; Zhang et al. 2017). The conceptual scheme of the pyrolysis process is shown in Fig. 8.4. According to the literature, rice husk is preferable than any other crop residues to obtain bio-oil. First of all, it is connected with the lower amount of oxygenates formed as a result of rice husk conversion (Lim et al. 2012). Secondly, pyrolysis or hydrothermal conversion of biomass with high content of lignin including rice husk and nutshells gives a large portion of bio-oil. Due to this feature of rice husk and nutshells, they were efficiently applied in various methods of joint pyrolysis (co-pyrolysis) or hydrothermal co-carbonization along with other types of biomass and solid wastes (Çepelioğullar and Pütün 2013; Hu et al. 2018; Kar 2011; Onay 2014). For example, during the co-pyrolysis and co-hydrothermal liquefaction of rice husk with wet algae, a synergetic effect attributed to the improvement in the bio-oil quality was observed (Hu et al. 2018).

The quality of the produced bio-oil can be improved by using catalysts during pyrolysis (Kwofie and Ngadi 2017; Yin et al. 2008). Various zeolite-based catalysts and rice husk ash can be used in the catalytic pyrolysis processes of rice husk and nutshells. Another alternative method is based on the intensified mass transfer between pseudo-fluidized medium containing gaseous nitrogen and gases formed during the pyrolysis process. It allows to increase the bio-oil yield up to 60 wt% at a higher heating value (HHV) of 24.8 MJ/kg (Park et al. 2010).

The selection of hydrothermal conversion conditions gives a possibility to consider the present method as an approach for the production of liquid biofuels (hydrothermal liquefaction). During the process of thermal depolymerization, raw biomass is being converted into bio-crude oil under moderate temperatures and high

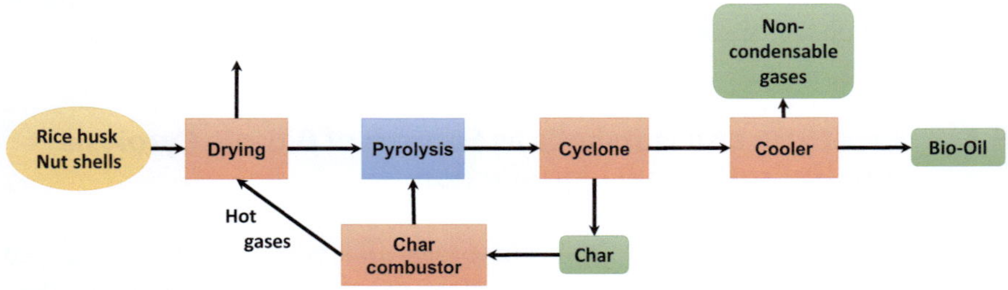

Fig. 8.4 Conceptual scheme of pyrolysis process for rice husk and nutshells

pressures (Shi et al. 2013). Carbon and hydrogen contents in the biomass undergo transformation into hydrophobic compounds with low viscosity and high solubility. Depending on the treatment conditions, the biofuel can be used either directly or with upgrading in the heavy engines including ships and railway transport vehicles. Bio-crude oil can also be upgraded to motor fuels as diesel fuel, gasoline, or jet engine fuel (Quispe et al. 2017). However, it should be noted that hydrothermal liquefaction and hydrothermal carbonization usually require the use of homogeneous or heterogeneous catalyst to improve the yield and quality of bio-oils. Moreover, these processes are also considered to be operated under high pressures. Along with it, aperiodicity of the technological regime worsens their practical value (Elliott et al. 2015).

8.6 Rice Husk and Nutshells as Sources of Gaseous Fuels

8.6.1 Gasification

Gasification of biomass is a process of thermochemical conversion when biomass is being transformed into synthesis gas consisting of CO and H_2 gas mixture with traces of CO_2 and CH_4 (Nelson et al. 2018). A schematic diagram of the gasification process is presented in Fig. 8.5. The process is performed in gasifiers with controlled feeding of the gasification agents such as steam, oxygen, and/or air and CO_2. The gasification process includes several oxidation reactions with the participation of oxygen and steam as mentioned below.

$$C + H_2O \rightarrow CO + H_2 \tag{8.2}$$

$$C + CO_2 \rightarrow 2CO \tag{8.3}$$

The composition of synthesis gas depends on the gasification conditions and the type of the gasifier. Another factor affecting the composition is the type of the biomass source. Thus, the gas produced from rice husk typically contains CO_2 (12.6%), CO (17.9%), N_2 (57.0%), H_2 (8.8%), CH_4 (1.9%), and others gases (1.8%) (Pode 2016).

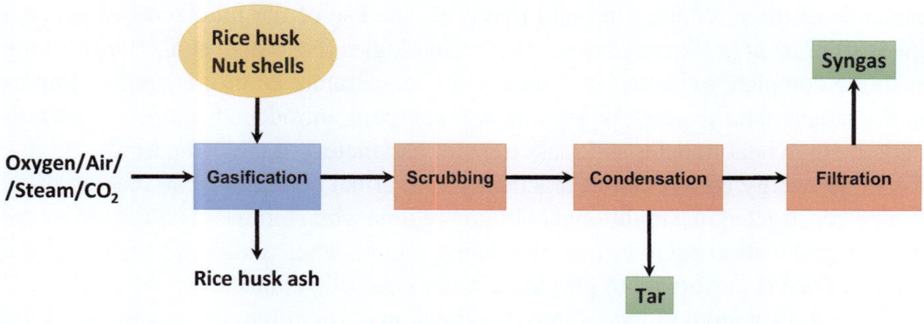

Fig. 8.5 Conceptual scheme of gasification process for rice husk and nutshells

All the characteristics of biomass (i.e., particle size, moisture content, and density) are also of equal importance. Ash formed during the gasification can be eliminated in a dry state or in the form of slag. The slag-forming gasifiers are characterized by the low gasification agent/carbon ratio. Along with it, the temperature of the process is higher than the ash fusion temperature.

All the gasification processes have one general disadvantage, i.e., formation of high-molecular gases, which undergo condensation and form tar (Reddy et al. 2014). It decreases the portion of biomass converted into target gaseous biofuel. The HHV of synthesis gas produced is also reduced with tar formation. There are a few alternative approaches to prevent the tar formation (Nelson et al. 2018). For example, the amount of formed tar can be decreased by thermal cracking in the gasifying reactor working at high temperature (>1000 °C) or by catalytic cracking at relatively lower temperatures (≤750 °C) (Moghtaderi 2007). The produced synthesis gas can be further used for the production of heat, electricity, or other types of fuel. In average, the gasification of biomass allows to achieve an increase in HHV from 12 to 28 MJ/m^3 (Sikarwar et al. 2016).

The gasification of rice husk and nutshells in different reactors is described elsewhere (Farzad et al. 2016; Natarajan et al. 1998; Nelson et al. 2018). In the case of the fixed-bed reactors, the gasification agent (steam, oxygen or air) is purged through the biomass bed in counter-current or co-current flow configuration. In the entrained flow regime, the gasification process (interaction with the gasification agent) takes place in a dense cloud of dry pulverized biomass fine particles in co-current flow. In the case of the moving-bed reactors, the biomass bed is purged with the gasification agent at continuous stirring. Auger technology is based on the tubular reactor of continuous action with biomass feeding by rotating screw, and the heat is transferred along the reactor wall. Another process, which should be mentioned, is the gasification under autothermal conditions. In this case, the heat required for the gasification of biomass is produced directly from the combustion of a part of raw biomass (Milhé et al. 2013). Plasma gasifiers are also widely investigated (Rutberg et al. 2011). Unfortunately, plasma-based processes are still unprofitable.

All the technologies mentioned above were applied for the gasification of rice husk and nutshells (Chakma et al. 2015; Chen and Zhang 2015; Dogru et al. 2002; Jain 2000; Makwana et al. 2019; Nadaleti 2019; Pode 2016; Shen et al. 2015; Sun et al. 2009; Unrean et al. 2018; Wu et al. 2009; Yin et al. 2002). Similarly to the methods of direct combustion and pyrolysis, the use of the fluidized-bed concept gives significant preferences over other technologies (Natarajan et al. 1998). Along with the complete oxidation of biomass, the temperature control is another important characteristic feature. When biomass source is introduced into the fluidized-bed, the high heat transfer and mass transfer parameters of the bed permit the rapid energy conversion at practically isothermal condition. The reactors used can be constructed in a traditional bubbling regime when biomass is fluidized by the flow of gasification agent or in a circulating regime when additional stirring of the fluidized bed is performed to provide a better gas–solid contact.

Since the potential of the biomass gasification is very high, the possibility of the local electricity production using the biomass gasifying reactor system was recently estimated (Abe et al. 2007; Mehrpooya et al. 2018; Pode 2016). It was shown that despite the high-energy potential of rice husk, in the long term, additional biomass resources would be required. The reasonability of such large-scale projects mainly depends on the location of the processing plant, which affects the resource availability and logistics costs of raw biomass supply. Therefore, it can be concluded that to be prospective, the approach for the processing of rice husk and nutshells into gaseous fuels should consider performing the transformation of biomass into easily transportable precursors and their further conversion (e.g., drying, pyrolysis, volatiles oxidation/cracking, and gasification) at pilot-scale plants.

8.6.2 Anaerobic Digestion and Hydrogen Production Via Fermentation

In the process of anaerobic digestion, microorganisms convert biomass in the absence of oxygen into biogas, which is a mixture of CH_4 and CO_2. The gaseous products can be subsequently used as a fuel to produce heat and electricity. This process is well known and was reviewed several times (Angelidaki and Batstone 2010; Saxena et al. 2009). The anaerobic digestion of rice residues and nutshells is widely reported in the literature, but most of these reports indicate the higher prospect of rice straw and soft nutshells compared to rice husk and dry nutshells (Gnansounou and Dauriat 2010; Khoo 2015). In the case of residue with increased lignin content and low moisture content, higher economic resources are required in the biomass pretreatment stage to facilitate the depolymerization of lignin.

Contrary to anaerobic digestion, the process of hydrogen production via fermentation of agricultural wastes is relatively new. During the fermentation, anaerobic bacteria transform carbohydrates into volatile fatty acids, H_2, and CO_2. The fermentation process can be divided into photo fermentation and dark fermentation, when different types of bacteria functionalize at different working conditions (Hallenbeck 2002). The key parameter defining the applicability of biomass in the bioconversion processes is the ratio of the holocellulose (cellulose and hemicellulose together) and lignin (Baetge and Kaltschmitt 2018).

The theoretical (maximum possible) yield of biogas in the case of rice husk was calculated to be 542 m^3/tODM. At the same time, experimentally this value was found to be only 212 m^3/tODM. The observed difference is attributed to the fact mentioned above that lignin forms stable network which blocks the accessibility of cellulose and hemicellulose. Thus, effective conversion of biomass with high lignin content requires additional pretreatment procedures that can also increase the cost of the entire process. Among the considered agricultural wastes, pistachio nutshells only contain relatively lower amount of lignin. However, their usage is also limited due to low moisture content. In this term, bioconversion of pistachio soft nutshells is more economically preferable (Taghizadeh-Alisaraei et al. 2017). The same conclusions can be made for other types of nutshells when comparing them in the soft and dry states.

8.7 Calorific Values of Rice Husk and Nutshells as Biofuels

The most important parameters of the raw biomass including agricultural wastes and the derived biofuels are volatile matter, moisture, ash, and fixed carbon contents. The contents of volatile matters in the solid fuel can be measured experimentally as a part exuded in the gaseous form including water at a temperature range of 25–850 °C in an inert atmosphere. The ash content can be also determined experimentally. The fixed carbon value can be calculated as the difference excluding ash content. The volatile matters and fixed carbon parameters define the value of the chemical energy accumulated in the solid fuel. Therefore, the HHV depends on the biomass composition and is estimated as the amount of energy which can be potentially obtained from the biomass or biofuel.

To compare the biomass sources or the derived biofuels, the elemental composition is proposed as an essential parameter. The significance of O/C and H/C atomic ratios of the biomass or biofuels on their calorific value is illustrated in Fig. 8.6 using the Van Krevelen diagram (Van Krevelen 1950). The values of H/C and O/C atomic ratios depend on the biomass source, operating conditions, moisture content, and the treatment procedures. Figure 8.6 demonstrates these ratios for various biomasses and products.

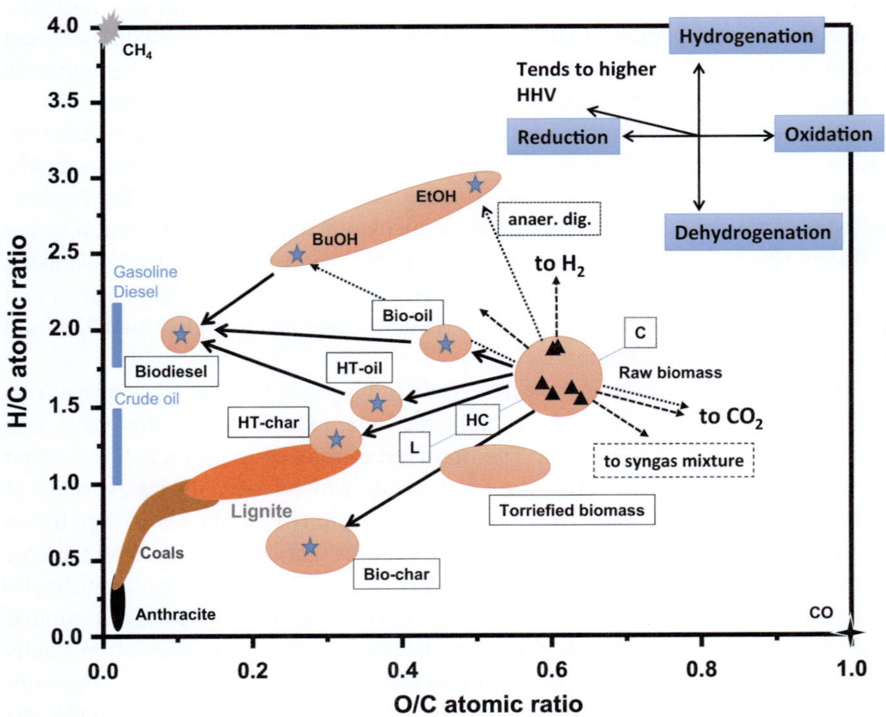

Fig. 8.6 Van Krevelen diagram comparing the heating value of different biomass and fuel products derived via biomass conversion processes

8.8 Conclusions

The growing demands for crops have resulted in an increasing amount of accumulated agricultural solid wastes annually. On the other hand, these wastes can be considered as the bioresources to produce biofuels. In this chapter, the conversion of rice husk and nutshells into solid, liquid, and gaseous biofuels was reviewed. The areas of application and efficiency of the conversion technologies strongly depend on the chemical composition and physical properties of the processed biomass. Thus, direct combustion of rice husk and dry nutshells is the simplest and cheapest method of their usage as a solid fuel, but it is not the highly efficient one. On the contrary, a variety of the solid and liquid biofuels can be achieved by tuning the technological conditions of the pyrolysis processes. Gasification, anaerobic digestion, and fermentation processes allow the production of energy-rich gases (e.g., synthesis gas, methane, and hydrogen), which can be easily used in the existing industrial infrastructures. The final choice of the appropriate process should be done considering the necessity of the additional preliminary treatment procedures, which can make significant contribution to the overall cost of the process.

Acknowledgments The financial support by the Ministry of Science and High Education of Russian Federation (project AAAA-A17-117041710086-6) is acknowledged with gratitude.

References

Abe H, Katayama A, Sah BP, Toriu T, Samy S, Pheach P, Adams MA, Grierson PF (2007) Potential for rural electrification based on biomass gasification in Cambodia. Biomass Bioenergy 31:656–664

Angelidaki I, Batstone DJ (2010) Anaerobic digestion: process. In: Christensen TH (ed) Solid waste technology & management. Wiley, New York, pp 583–600

Arora R, Sharma NK, Kumar S, Sani RK (2019) Lignocellulosic ethanol: feedstocks and bioprocessing. In: Ray RC, Ramachandran S (eds) Bioethanol production from food crops. Elsevier, Amsterdam, pp 165–185

Baetge S, Kaltschmitt M (2018) Rice straw and rice husks as energy sources—comparison of direct combustion and biogas production. Biomass Conv Bioref 8:719–737

Balachandran VS, Jadhav SR, Vemula PK, John G (2013) Recent advances in cardanol chemistry in a nutshell: from a nut to nanomaterials. Chem Soc Rev 42:427–438

Balat M, Balat H (2009) Recent trends in global production and utilization of bio-ethanol fuel. Appl Energy 86:2273–2282

Balgude D, Sabnis AS (2013) CNSL: an environment friendly alternative for the modern coating industry. J Coat Technol Res 11:169–183

Basu P (2010) Biomass gasification and pyrolysis. Elsevier, Burlington, MA

Bykova MV, Ermakov DY, Kaichev VV, Bulavchenko OA, Saraev AA, Lebedev MY, Yakovlev VA (2012) Ni-based sol–gel catalysts as promising systems for crude bio-oil upgrading: Guaiacol hydrodeoxygenation study. Appl Catal B Environ 113-114:296–307

Çepelioğullar Ö, Pütün AE (2013) Thermal and kinetic behaviors of biomass and plastic wastes in co-pyrolysis. Energ Convers Manage 75:263–270

Chakma S, Ranjan A, Choudhury HA, Dikshit PK, Moholkar VS (2015) Bioenergy from rice crop residues: role in developing economies. Clean Technol Environ Pol 18:373–394

Chen Z-M, Zhang L (2015) Catalyst and process parameters for the gasification of rice husk with pure CO_2 to produce CO. Fuel Process Technol 133:227–231
da Silva JCG, Alves JLF, Galdino WVA, Moreira RFPM, José HJ, de Sena RF, Andersen SLF (2017) Combustion of pistachio shell: physicochemical characterization and evaluation of kinetic parameters. Environ Sci Pollut Res 25:21420–21429
Das P, Ganesh A (2003) Bio-oil from pyrolysis of cashew nut shell—a near fuel. Biomass Bioenergy 25:113–117
Davidsson KO, Korsgren JG, Pettersson JBC, Jäglid U (2002) The effects of fuel washing techniques on alkali release from biomass. Fuel 81:137–142
Demirbaş A (1997) Calculation of higher heating values of biomass fuels. Fuel 76:431–434
Demirbas A (2006) Production and characterization of bio-chars from biomass via pyrolysis. Energ Sour Part A 28:413–422
Demirbaş A (2010) Fuel characteristics of olive husk and walnut, hazelnut, sunflower, and almond shells. Energy Sources 24:215–221
Dogru M, Howarth CR, Akay G, Keskinler B, Malik AA (2002) Gasification of hazelnut shells in a downdraft gasifier. Energy 27:415–427
Dufour A, Girods P, Masson E, Rogaume Y, Zoulalian A (2009) Synthesis gas production by biomass pyrolysis: effect of reactor temperature on product distribution. Int J Hydrogen Energ 34:1726–1734
Dundich VO, Khromova SA, Ermakov DY, Lebedev MY, Novopashina VM, Sister VG, Yakimchuk AI, Yakovlev VA (2010) Nickel catalysts for the hydrodeoxygenation of biodiesel. Kinet Catal 51:704–709
Elliott DC, Biller P, Ross AB, Schmidt AJ, Jones SB (2015) Hydrothermal liquefaction of biomass: developments from batch to continuous process. Bioresour Technol 178:147–156
Fang M, Yang L, Chen G, Shi Z, Luo Z, Cen K (2004) Experimental study on rice husk combustion in a circulating fluidized bed. Fuel Process Technol 85:1273–1282
Fapohunda C, Akinbile B, Shittu A (2017) Structure and properties of mortar and concrete with rice husk ash as partial replacement of ordinary Portland cement—a review. Int J Sustain Built Environ 6:675–692
Farzad S, Mandegari MA, Görgens JF (2016) A critical review on biomass gasification, co-gasification, and their environmental assessments. Biofuel Res J 3:483–495
Fernandes IJ, Calheiro D, Kieling AG, Moraes CAM, Rocha TLAC, Brehm FA, Modolo RCE (2016) Characterization of rice husk ash produced using different biomass combustion techniques for energy. Fuel 165:351–359
Funke A, Ziegler F (2010) Hydrothermal carbonization of biomass: a summary and discussion of chemical mechanisms for process engineering. Biofuels Bioprod Biorefin 4:160–177
Gnansounou E, Dauriat A (2010) Techno-economic analysis of lignocellulosic ethanol: a review. Bioresour Technol 101:4980–4991
Goenka R, Parthasarathy P, Gupta NK, Biyahut NK, Narayanan S (2015) Kinetic analysis of biomass and comparison of its chemical compositions by thermogravimetry, wet and experimental furnace methods. Waste Biomass Valor 6:989–1002
Goldfarb JL, Dou G, Salari M, Grinstaff MW (2017) Biomass-based fuels and activated carbon electrode materials: an integrated approach to green energy systems. ACS Sustain Chem Eng 5:3046–3054
Goyal HB, Seal D, Saxena RC (2008) Bio-fuels from thermochemical conversion of renewable resources: a review. Renew Sust Energ Rev 12:504–517
Hallenbeck P (2002) Biological hydrogen production; fundamentals and limiting processes. Int J Hydrogen Energ 27:1185–1193
Hamad F, Mubofu E (2015) Potential biological applications of bio-based anacardic acids and their derivatives. Int J Mol Sci 16:8569–8590
Hendriks ATWM, Zeeman G (2009) Pretreatments to enhance the digestibility of lignocellulosic biomass. Bioresour Technol 100:10–18
Housten D (1972) Rice chemistry and technology. American Association of Cereal Chemists, St. Paul, MN

Hu Y, Wang S, Li J, Wang Q, He Z, Feng Y, Abomohra AE-F, Afonaa-Mensah S, Hui C (2018) Co-pyrolysis and co-hydrothermal liquefaction of seaweeds and rice husk: comparative study towards enhanced biofuel production. J Anal Appl Pyrolysis 129:162–170

Jain BC (2000) Commercialising biomass gasifiers: Indian experience. Energ Sustain Develop 4:72–82

Jeong CY, Dodla SK, Wang JJ (2016) Fundamental and molecular composition characteristics of biochars produced from sugarcane and rice crop residues and by-products. Chemosphere 142:4–13

Kalderis D, Kotti MS, Méndez A, Gascó G (2014) Characterization of hydrochars produced by hydrothermal carbonization of rice husk. Solid Earth Discuss 6:657–677

Kaliyan N, Morey RV (2009) Factors affecting strength and durability of densified biomass products. Biomass Bioenergy 33:337–359

Kar Y (2011) Co-pyrolysis of walnut shell and tar sand in a fixed-bed reactor. Bioresour Technol 102:9800–9805

Karimi K, Taherzadeh MJ (2016) A critical review on analysis in pretreatment of lignocelluloses: degree of polymerization, adsorption/desorption, and accessibility. Bioresour Technol 203:348–356

Karimi K, Emtiazi G, Taherzadeh MJ (2006) Ethanol production from dilute-acid pretreated rice straw by simultaneous saccharification and fermentation with *Mucor indicus*, *Rhizopus oryzae*, and *Saccharomyces cerevisiae*. Enzym Microb Technol 40:138–144

Khoo HH (2015) Review of bio-conversion pathways of lignocellulose-to-ethanol: sustainability assessment based on land footprint projections. Renew Sust Energ Rev 46:100–119

Koppejan J, Loo SV (2012) The handbook of biomass combustion and co-firing. Taylor & Francis Group, London

Kukushkin RG, Bulavchenko OA, Kaichev VV, Yakovlev VA (2015) Influence of Mo on catalytic activity of Ni-based catalysts in hydrodeoxygenation of esters. Appl Catal B Environ 163:531–538

Kuprianov VI, Janvijitsakul K, Permchart W (2006) Co-firing of sugar cane bagasse with rice husk in a conical fluidized-bed combustor. Fuel 85:434–442

Kwofie EM, Ngadi M (2017) A review of rice parboiling systems, energy supply, and consumption. Renew Sust Energ Rev 72:465–472

Larichev YV, Eletskii PM, Tuzikov FV, Yakovlev VA (2013) Porous carbon-silica composites and carbon materials from rice husk: production technology, texture, and dispersity. Catal Ind 5:350–357

Larichev YV, Yeletsky PM, Yakovlev VA (2015) Study of silica templates in the rice husk and the carbon–silica nanocomposites produced from rice husk. J Phys Chem Solids 87:58–63

Lebedeva MV, Yeletsky PM, Ayupov AB, Kuznetsov AN, Yakovlev VA, Parmon VN (2015) Micro–mesoporous carbons from rice husk as active materials for supercapacitors. Mater Renew Sustain Energ 4:20

Lebedeva MV, Yeletsky PM, Ayupov AB, Kuznetsov AN, Gribov EN, Parmon VN (2018a) Rice husk derived activated carbon/polyaniline composites as active materials for supercapacitors. Int J Electrochem Sci 13:3674–3690

Lebedeva MV, Yeletsky PM, Ayupov AB, Kuznetsov AN, Gribov EN, Parmon VN (2018b) Rice husk derived micro-mesoporous carbon materials as active components of supercapacitor electrodes. Catal Ind 10:173–180

Li X, Liu Y, Hao J, Wang W (2018) Study of almond shell characteristics. Materials (Basel) 11:1782

Lim JS, Abdul Manan Z, Wan Alwi SR, Hashim H (2012) A review on utilisation of biomass from rice industry as a source of renewable energy. Renew Sust Energ Rev 16:3084–3094

Lomonaco D, Mele G, Mazzetto SE (2017) Cashew nutshell liquid (CNSL): from an agro-industrial waste to a sustainable alternative to petrochemical resources. In: Anilkumar P (ed) Cashew nut Shell liquid. Springer, New York, pp 19–38

Makwana JP, Pandey J, Mishra G (2019) Improving the properties of producer gas using high temperature gasification of rice husk in a pilot scale fluidized bed gasifier (FBG). Renew Energy 130:943–951

Mani S, Tabil LG, Sokhansanj S (2003) An overview of compaction of biomass grinds. Powder Handl Process 15:160–168

Marquit M (2009) Jatropha helps Air New Zealand cut its CO_2 emissions by more than 60% (17 Jun 2009). https://phys.org/news/2009-06-jatropha-air-zealandco2-emissions.html. Accessed 11 Feb 2019

Medic D, Darr M, Shah A, Rahn S (2012) The effects of particle size, different corn Stover components, and gas residence time on torrefaction of corn Stover. Energies 5:1199–1214

Mehrpooya M, Khalili M, Sharifzadeh MMM (2018) Model development and energy and exergy analysis of the biomass gasification process (based on the various biomass sources). Renew Sust Energ Rev 91:869–887

Miles TR, Miles TR, Baxter LL, Bryers RW, Jenkins BM, Oden LL (1996) Boiler deposits from firing biomass fuels. Biomass Bioenergy 10:125–138

Milhé M, van de Steene L, Haube M, Commandré J-M, Fassinou W-F, Flamant G (2013) Autothermal and allothermal pyrolysis in a continuous fixed bed reactor. J Anal Appl Pyrolysis 103:102–111

Miranda R, Sosa C, Bustos D, Carrillo E, Rodrguez-Cant M (2012) Characterization of pyrolysis products obtained during the preparation of bio-oil and activated carbon. In: Montoya VH, Bonilla-Petriciolet A (eds) Lignocellulosic precursors used in the synthesis of activated carbon: characterization techniques and applications in the wastewater treatment. InTech Open, London, pp 77–92

Moghtaderi B (2007) Effects of controlling parameters on production of hydrogen by catalytic steam gasification of biomass at low temperatures. Fuel 86:2422–2430

Mohan D, Pittman CU, Steele PH (2006) Pyrolysis of wood/biomass for bio-oil: a critical review. Energy Fuel 20:848–889

Mohanty P, Nanda S, Pant KK, Naik S, Kozinski JA, Dalai AK (2013) Evaluation of the physiochemical development of biochars obtained from pyrolysis of wheat straw, timothy grass and pinewood: effects of heating rate. J Anal Appl Pyrolysis 104:485–493

Moraes CAM, Fernandes IJ, Calheiro D, Kieling AG, Brehm FA, Rigon MR, Berwanger Filho JA, Schneider IAH, Osorio E (2014) Review of the rice production cycle: by-products and the main applications focusing on rice husk combustion and ash recycling. Waste Manag Res 32:1034–1048

Moreira R, dos Reis Orsini R, Vaz JM, Penteado JC, Spinacé EV (2016) Production of biochar, bio-oil and synthesis gas from cashew nut shell by slow pyrolysis. Waste Biomass Valor 8:217–224

Nachenius RW, Ronsse F, Venderbosch RH, Prins W (2013) Biomass pyrolysis. In: Murzin DY (ed) Chemical engineering for renewables conversion. Elsevier, Amsterdam, pp 75–139

Nadaleti WC (2019) Utilization of residues from rice parboiling industries in southern Brazil for biogas and hydrogen-syngas generation: heat, electricity and energy planning. Renew Energy 131:55–72

Nakhshiniev B, Biddinika MK, Gonzales HB, Sumida H, Yoshikawa K (2014) Evaluation of hydrothermal treatment in enhancing rice straw compost stability and maturity. Bioresour Technol 151:306–313

Nanda S, Mohammad J, Reddy SN, Kozinski JA, Dalai AK (2014) Pathways of lignocellulosic biomass conversion to renewable fuels. Biomass Conv Bioref 4:157–191

Nanda S, Dalai AK, Berruti F, Kozinski JA (2016) Biochar as an exceptional bioresource for energy, agronomy, carbon sequestration, activated carbon and specialty materials. Waste Biomass Valor 7:201–235

Natarajan E, Nordin A, Rao AN (1998) Overview of combustion and gasification of rice husk in fluidized bed reactors. Biomass Bioenergy 14:533–546

Ndindeng SA, Mbassi JEG, Mbacham WF, Manful J, Graham-Acquaah S, Moreira J, Dossou J, Futakuchi K (2015) Quality optimization in briquettes made from rice milling by-products. Energ Sustain Develop 29:24–31

Nelson L, Park S, Hubbe MA (2018) Thermal depolymerization of biomass with emphasis on gasifier design and best method for catalytic hot gas conditioning. Bioresources 13:4630–4727
Neves D, Thunman H, Matos A, Tarelho L, Gómez-Barea A (2011) Characterization and prediction of biomass pyrolysis products. Prog Energy Combust Sci 37:611–630
Nigam PS, Singh A (2011) Production of liquid biofuels from renewable resources. Prog Energy Combust Sci 37:52–68
Onay Ö (2014) The catalytic co-pyrolysis of waste tires and pistachio seeds. Energ Sour Part A 36:2070–2077
Özyuğuran A, Acma HH, Dahiloğlu E (2017) Production of fuel briquettes from rice husk-lignite blends. Environ Prog Sustain Energy 36:742–748
Park HJ, Heo HS, Jeon J-K, Kim J, Ryoo R, Jeong K-E, Park Y-K (2010) Highly valuable chemicals production from catalytic upgrading of radiata pine sawdust-derived pyrolytic vapors over mesoporous MFI zeolites. Appl Catal B Environ 95:365–373
Park J, Lee Y, Ryu C, Park Y-K (2014) Slow pyrolysis of rice straw: analysis of products properties, carbon and energy yields. Bioresour Technol 155:63–70
Patel RN, Bandyopadhyay S, Ganesh A (2011) Extraction of cardanol and phenol from bio-oils obtained through vacuum pyrolysis of biomass using supercritical fluid extraction. Energy 36:1535–1542
Pode R (2016) Potential applications of rice husk ash waste from rice husk biomass power plant. Renew Sust Energ Rev 53:1468–1485
Pütün AE, Özcan A, Pütün E (1999) Pyrolysis of hazelnut shells in a fixed-bed tubular reactor: yields and structural analysis of bio-oil. J Anal Appl Pyrolysis 52:33–49
Quispe I, Navia R, Kahhat R (2017) Energy potential from rice husk through direct combustion and fast pyrolysis: a review. Waste Manag 59:200–210
Reddy SN, Nanda S, Dalai AK, Kozinski JA (2014) Supercritical water gasification of biomass for hydrogen production. Int J Hydrogen Energ 39:6912–6926
Rice Market Monitor (2017) Food and Agriculture Organization of the United Nations. Statistical database, Oct 2017
Rutberg PG, Bratsev AN, Kuznetsov VA, Popov VE, Ufimtsev AA, Shtengel' SV (2011) On efficiency of plasma gasification of wood residues. Biomass Bioenergy 35:495–504
Saidur R, Abdelaziz EA, Demirbas A, Hossain MS, Mekhilef S (2011) A review on biomass as a fuel for boilers. Renew Sust Energ Rev 15:2262–2289
Sathitruangsak P, Madhiyanon T (2017) Effect of operating conditions on the combustion characteristics of coal, rice husk, and co-firing of coal and rice husk in a circulating fluidized bed combustor. Energy Fuel 31:12741–12755
Savova D, Apak E, Ekinci E, Yardim F, Petrov N, Budinova T, Razvigorova M, Minkova V (2001) Biomass conversion to carbon adsorbents and gas. Biomass Bioenergy 21:133–142
Saxena RC, Adhikari DK, Goyal HB (2009) Biomass-based energy fuel through biochemical routes: a review. Renew Sust Energ Rev 13:167–178
Shen Y, Zhao P, Shao Q, Takahashi F, Yoshikawa K (2015) In situ catalytic conversion of tar using rice husk char/ash supported nickel–iron catalysts for biomass pyrolytic gasification combined with the mixing-simulation in fluidized-bed gasifier. Appl Energy 160:808–819
Shi W, Li S, Jin H, Zhao Y, Yu W (2013) The hydrothermal liquefaction of rice husk to bio-crude using metallic oxide catalysts. Energ Sour Part A 35:2149–2155
Sikarwar VS, Zhao M, Clough P, Yao J, Zhong X, Memon MZ, Shah N, Anthony EJ, Fennell PS (2016) An overview of advances in biomass gasification. Energy Environ Sci 9:2939–2977
Simonov AD, Mishenko TI, Yazykov NA, Parmon VN (2003) Combustion and processing of rice husk in the vibrofluidized bed of catalyst or inert material. Chem Sustain Dev 11:277–283
Sudiro M, Bertucco A (2007) Synthetic fuels by a limited CO_2 emission process which uses both fossil and solar energy. Energy Fuel 21:3668–3675
Sun Y, Cheng J (2002) Hydrolysis of lignocellulosic materials for ethanol production: a review. Bioresour Technol 83:1–11
Sun S, Zhao Y, Ling F, Su F (2009) Experimental research on air staged cyclone gasification of rice husk. Fuel Process Technol 90:465–471

Taghizadeh-Alisaraei A, Assar HA, Ghobadian B, Motevali A (2017) Potential of biofuel production from pistachio waste in Iran. Renew Sust Energ Rev 72:510–522

Taherzadeh MJ, Niklasson C (2004) Ethanol from lignocellulosic materials: pretreatment, acid and enzymatic hydrolyses, and fermentation. In: Saha BC, Hayashi K (eds) Lignocellulose biodegradation, ACS Symp Series. American Chemical Society, Washington, DC, pp 49–68

Tang Z-E, Lim S, Pang Y-L, Ong H-C, Lee K-T (2018) Synthesis of biomass as heterogeneous catalyst for application in biodiesel production: state of the art and fundamental review. Renew Sust Energ Rev 92:235–253

Tullo AH (2008) A nutty chemical. Chem Eng News 86:26–27

Tumuluru JS, Sokhansanj S, Hess JR, Wright CT, Boardman RD (2011) A review on biomass torrefaction process and product properties for energy applications. Ind Biotechnol 7:384–401

Unrean P, Lai Fui BC, Rianawati E, Acda M (2018) Comparative techno-economic assessment and environmental impacts of rice husk-to-fuel conversion technologies. Energy 151:581–593

Van Krevelen DW (1950) Graphical-statistical method for the study of structure and reaction processes of coal. Fuel 29:269–284

Wang T, Zhai Y, Zhu Y, Li C, Zeng G (2018) A review of the hydrothermal carbonization of biomass waste for hydrochar formation: process conditions, fundamentals, and physicochemical properties. Renew Sust Energ Rev 90:223–247

Wu C-Z, Yin X-L, Ma L-L, Zhou Z-Q, Chen H-P (2009) Operational characteristics of a 1.2-MW biomass gasification and power generation plant. Biotechnol Adv 27:588–592

Wu H-C, Ku Y, Tsai H-H, Kuo Y-L, Tseng Y-H (2015) Rice husk as solid fuel for chemical looping combustion in an annular dual-tube moving bed reactor. Chem Eng J 280:82–89

Yang W, Shimanouchi T, Kimura Y (2015) Characterization of the residue and liquid products produced from husks of nuts from *Carya cathayensis* sarg by hydrothermal carbonization. ACS Sustain Chem Eng 3:591–598

Yank A, Ngadi M, Kok R (2016) Physical properties of rice husk and bran briquettes under low pressure densification for rural applications. Biomass Bioenergy 84:22–30

Yeletsky PM, Yakovlev VA, Mel'gunov MS, Parmon VN (2009) Synthesis of mesoporous carbons by leaching out natural silica templates of rice husk. Microporous Mesoporous Mater 121:34–40

Yin XL, Wu CZ, Zheng SP, Chen Y (2002) Design and operation of a CFB gasification and power generation system for rice husk. Biomass Bioenergy 23:181–187

Yin C, Rosendahl LA, Kær SK (2008) Grate-firing of biomass for heat and power production. Prog Energy Combust Sci 34:725–754

Zhang H, Ma Y, Shao S, Xiao R (2017) The effects of potassium on distributions of bio-oils obtained from fast pyrolysis of agricultural and forest biomass in a fluidized bed. Appl Energy 208:867–877

A Review of Thermochemical and Biochemical Conversion of *Miscanthus* to Biofuels

9

Arshdeep Singh, Sonil Nanda, and Franco Berruti

Abstract

Miscanthus is recently being considered as an energy crop for biofuel production because of certain features, such as adaptability to lower temperature, efficient use of water and nutrients, low or no need of nitrogen fertilisers, high biomass yield, fast-growing cycle and less-intensive agricultural cultivation practices than other energy crops. This review is focused on the value-added applications and conversion of *Miscanthus* for bioenergy and biomaterial applications. The thermochemical conversion technologies reviewed in this chapter include pyrolysis, liquefaction, torrefaction and gasification, whereas biochemical conversion technologies include enzymatic saccharification and fermentation. The value-added applications of *Miscanthus* discussed in this chapter include pulp and papermaking, biocomposites and biochemical production. The physicochemical properties of bio-oil and biochar generated from *Miscanthus* have been thoroughly described for fuel and material applications.

Keywords

Miscanthus · Energy crop · Bio-oil · Biochar · Pyrolysis · Liquefaction

A. Singh · F. Berruti
Department of Chemical and Biochemical Engineering, University of Western Ontario, London, Ontario, Canada

S. Nanda (✉)
Department of Chemical and Biological Engineering, University of Saskatchewan, Saskatoon, Saskatchewan, Canada
e-mail: sonil.nanda@usask.ca

S. Nanda et al. (eds.), *Biorefinery of Alternative Resources: Targeting Green Fuels and Platform Chemicals*, https://doi.org/10.1007/978-981-15-1804-1_9

9.1 Introduction

Waste plant biomass has been used as the traditional source for heating and cooking by the rural population worldwide. Owing to their renewable nature, low-cost availability and abundancy, waste plant biomass and organic residues are considered for the production of biofuels that are carbon neutral and generate low net greenhouse gas emissions (Nanda et al. 2016d, 2017d). However, for the sustainable production of bioenergy, it is important that the biomass is non-edible to prevent any food versus fuel controversy with no competition to food supply and arable lands (Nanda et al. 2015a). Lignocellulosic biomass is a collective group of non-edible plant residues containing cellulose, hemicellulose and lignin, examples of which include agricultural biomass, forestry residues, energy crops and invasive plants (Nanda et al. 2013). Lignocellulosic biomass is also known as second-generation feedstock because it is non-edible and includes residues such as wheat straw, corncobs, rice husk, hybrid poplar, switchgrass, *Miscanthus*, etc.

Coal has been traditionally one of the most widely used fossil fuel resources because it has carbon content of 75% to 90% (Jenkins et al. 1998). On the other hand, biomass consists of 50% carbon, along with considerable amounts of oxygen, generating lower heating value than other fossil fuels. The combustion of biomass is also hindered by the presence of alkali and alkaline earth metals, which form ash, resulting in corrosion, plugging, agglomeration, silicate melt-induced slagging and ash fusion in biomass-based power plants (Niu et al. 2016). Hence, suitable thermochemical and biochemical conversion technologies should be implemented to efficiently convert waste biomass to liquid and gaseous fuels (Fig. 9.1). Moreover, lignocellulosic biomass is more suitable for thermochemical conversion to produce alternate fuels than petrochemical resources, such as coal, because of its high volatile components.

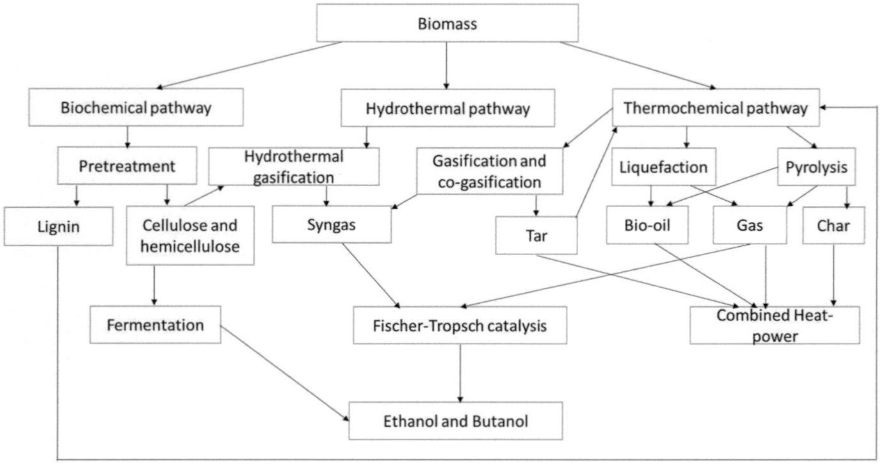

Fig. 9.1 Conversion of biomass to biofuels through thermochemical, hydrothermal and biochemical technologies

Miscanthus is one of the invasive plants, a genus of almost 20 perennial grass species, predominantly found in Asia and the pacific islands but invasive in most other geographical regions. Most of the native species of *Miscanthus* are found in China, Korea, Japan, Taiwan and the Philippines (Clark et al. 2014). *Miscanthus* has been recently targeted as a potential energy crop for biofuel production. As an energy crop, *Miscanthus* has many salient features, such as adaptability to lower temperature, low requirement of nutrient, efficient use of water and nutrients, low or no need of nitrogen fertilisers, high biomass yield, fast-growing cycle and less-intensive agricultural cultivation practices (Nanda et al. 2016b). Some of the commonly found species of *Miscanthus* are *M. giganteus*, *M. sinensis*, *M. sacchatiflorus*, *M. floridulus*, *M. fuscus*, *M. junceus* and *M. changii*.

Miscanthus exhibits better properties than switchgrass, another energy crop, in terms of tolerance to low temperatures, higher biomass yield, higher heating value and lower moisture content (Robbins et al. 2012). For instance, the typical biomass yield from *M. floridulus* (27.8–38 tons/ha/year) is much higher than that from switchgrass, *Panicum virgatum L.* (15 tons/ha/year) (Lee and Kuan 2015). In contrast, another invasive crop, giant reed (*Arundo donax*), has an even higher biomass yield and enhanced tolerance to drought than *Miscanthus* (Ge et al. 2016). However, *Miscanthus* has high tolerance to the flooding pattern than giant reed. When compared with maize, *Miscanthus* can perform photosynthesis and grow at much lower temperatures (Dohleman and Long 2009).

Energy crops, such as switchgrass (Yu et al. 2016), timothy grass (Nanda et al. 2016b), elephant grass (Fontoura et al. 2015), hybrid poplar (Shooshtarian et al. 2018), giant reed (Low et al. 2011) and microalgae (Su et al. 2017), have been investigated for second-generation biofuel production through thermochemical and biological conversion processes. However, there is limited literature available on the biorefining of *Miscanthus* for biofuel, biochemical and biomaterial production. This chapter aims at reviewing the current knowledge on the potential of *Miscanthus* as an energy crop for biorefining. This chapter summarises the thermochemical conversion technologies (e.g. pyrolysis, gasification, liquefaction and torrefaction) and biological conversion technologies (enzymatic hydrolysis and fermentation) of *Miscanthus* to produce value-added products, such as biofuels, biochar, as well as combined heat and power (Fig. 9.1).

9.2 Cultivation of *Miscanthus*

The annual biomass yields from *M. giganteus*, *M. floridulus* and *M. lutarioriparius* is found to be 14.8–33.5, 27.8–38 and 32 tons/ha, respectively (Lee and Kuan 2015). *M. giganteus* is a sterile hybrid between *M. sinensis* and *M. sacchatiflorus*, which is beneficial in generating high biomass yields. Typically, mature *M. giganteus* can grow up to 3–4 m in height. Moreover, it is also resistant to pests and diseases and has extraordinary tolerance to draught and cold temperatures, thus making it a desirable energy crop (Lewandowski et al. 2003, Yu et al. 2013). However, the disadvantage of culturing *M. giganteus* is that it does not produce any seed. Therefore, it can only be propagated through rhizome cutting (Bousiosa and Worrell 2017). Typically, C4 plants can convert solar energy into carbohydrates in their biomass through photosynthesis 40% higher than C3 plants.

The plants can be classified based on the process of photosynthesis, i.e. light reaction and dark reaction. In light reaction, chlorophyll in plants in the presence of sunlight (solar energy) synthesise energy-rich compounds such as adenosine triphosphate (ATP) and some co-enzymes, whereas in dark reactions, ATP and co-enzymes are converted into carbohydrates and CO2. The dark reaction follows C3 and C4 cycles. C3 plants use Calvin cycle (or C3 cycle), while C4 plants use Hatch-Slack pathway (or C4 cycle) for photosynthesis. The names C3 and C4 appears in these photosynthetic pathways because of the first stable carbon products, such as phosphoglyceric acid (a three-carbon compound) and oxaloacetic acid (a four-carbon compound), respectively.

Product quality and quantity mainly depend on the composition of the raw material. The composition of *Miscanthus* genotype varies according to the harvesting season. The contents of hemicellulose and cellulose in *Miscanthus* in the summer harvest are higher than in the winter harvest. For instance, the yield of cultivation of *M. giganteus* in Austria in autumn harvest was 17–30 tDW/ha compared to the winter harvest of 22 tDW/ha (Nsanganwimana et al. 2014). Similarly, biomass yield from *M. giganteus* in Germany during the autumn season (17–30 tDW/ha) was higher than in the winter season (10–20 tDW/ha). Likewise, the autumn harvest from *M. giganteus* cultivated in Portugal was also higher (39 tDW/ha) than in the winter (26–30 tDW/ha) (Nsanganwimana et al. 2014).

In a study, *M. sinensis* genotype showed a significant difference in the production of biogas because of two different harvests, i.e. summer cut versus winter cut (van der Weijde et al. 2016). The biogas yield from *Miscanthus* from its summer harvest (539–591 mL/g dry matter) was found to be greater than its winter harvest (441–520 mL/g dry matter). Fermentation and the ability to release sugar was also higher in the summer harvest than in the winter harvest. An important parameter for combustion quality, i.e. ash content, was higher in the summer harvest (3.3 wt%) than in the winter harvest (1.5 wt%). Therefore, a delay in the spring harvest benefits the combustion of *Miscanthus* quality due to the relatively lower K, Cl and N contents (Brosse et al. 2012).

Miscanthus can be cultivated for up to 25 years. It has two growth phases, namely building phase and adult phase. The biomass yield in the first year of cultivation is around 5.9 ton/ha, whereas in the second and third year of the growth, the biomass yield can be between 8 and 13 ton/ha (Arnoult and Hulmel 2015). The canopy height and stem mass also increase rapidly within three years. The genotype variability is more evident in the initial two to three years. Weather conditions and atmospheric temperature affect the biomass production in the first year of crop establishment. Low atmospheric temperatures after the first winter of crop establishment decreases biomass yield. For instance, *M. giganteus* and *M. sacchariflorus* died after the first winter when cultivated in Sweden and Denmark, whereas *M. sinensis* clones survived. The drop in the temperature explained the cause of the death during the first year of plantation (Arnoult and Hulmel 2015).

Nitrogen application via fertiliser also affects the biomass quality of *M. giganteus*. When the content of nitrogen was increased, the levels of cellulose, hemicellulose and lignin in the above-ground biomass decreased while ash concentration increased. The heating value of biomass is dependent upon the elemental composition (carbon, hydrogen and oxygen) and the variation in the content of cell wall composition and ash (Nanda et al. 2014a). Typically, on a dry basis, *Miscanthus* contains 47.1–49.7 wt% carbon, 5.38–5.92 wt% hydrogen and 41.4–44.6 wt% oxygen. The reported higher heating value of *M. giganteus* ranges between 17 and 20 MJ/kg (Brosse et al. 2012).

Miscanthus is a candidate energy crop within the lignocellulosic biomass family. As the name suggests, lignocellulosic biomass comprises of cellulose (35–55 wt%), hemicellulose (20–40 wt%) and lignin (10–25 wt%) (Nanda et al. 2017c). Cellulose is a repeating polysaccharide of β-D-glucopyranose units, whereas hemicellulose is a matrix polysaccharide containing pentose sugars, hexose sugars and sugar acids. Lignin is a three-dimensional phenylpropane polymer, which binds cellulose and hemicellulose together to provide structural rigidity and integrity to the biomass (Fougere et al. 2016). The relative concentration of lignin acts as the determinant factor during the thermochemical and biochemical conversion of lignocellulosic biomass to biofuels. Since lignin is insoluble in acids and enzymes, except for alkali, its presence creates hurdles for the biological conversion of lignocellulosic biomass to alcohol-based biofuels. However, for thermochemical conversion, such as pyrolysis and hydrothermal liquefaction, the high-content lignin results in bio-oil with satisfactory fuel properties (Hodgson et al. 2011).

The composition of *Miscanthus* depends upon different genotypes, harvesting time, growth seasons, geographical locations and type of fertilisers applied. *Miscanthus* also contains fatty acids, sterols and other aromatic compounds as the extractives (Brosse et al. 2012). These extractives are of high value in terms of precursors for industrial chemicals and materials. The polysaccharides in *Miscanthus* include α-cellulose (50.9 wt%), β-cellulose (11.9 wt%) and γ-cellulose (10.6 wt%), whereas monosaccharides include xylose (14.9 wt%), arabinose (1.1 wt%) and galactose (0.3 wt%) (Villaverde et al. 2010). The compositions of cellulose in *M. giganteus*, *M. floridulus* and *M. lutarioriparius* are, respectively, 33.9, 43.1 and 43.9 wt% (Lee and Kuan 2015). Similarly, lignin contents in *M. giganteus*, *M. floridulus* and *M. lutarioriparius* were found to be 26.9 wt%, 22.3 wt% and 23.2 wt%, respectively. The composition of acid insoluble and acid soluble lignin in *Miscanthus* is 20.8 wt% and 0.9 wt%, respectively (Villaverde et al. 2010). The effect of biomass composition upon the higher heating value (HHV) also depends on the presence of lignin because lignin has nearly 30% higher calorific value than cellulose and hemicellulose together, along with lower oxygen content (Nanda et al. 2015b).

9.3 Value-Added Industrial Applications of *Miscanthus*

Some developing applications of *Miscanthus* in the chemical, pulp and papermaking and biocomposite industries have been reported (Fig. 9.2). The cellulose present in *Miscanthus* is being considered as a promising substrate for ethylene glycol production (Pang et al. 2014). Ethylene glycol is one of the bulk chemicals used

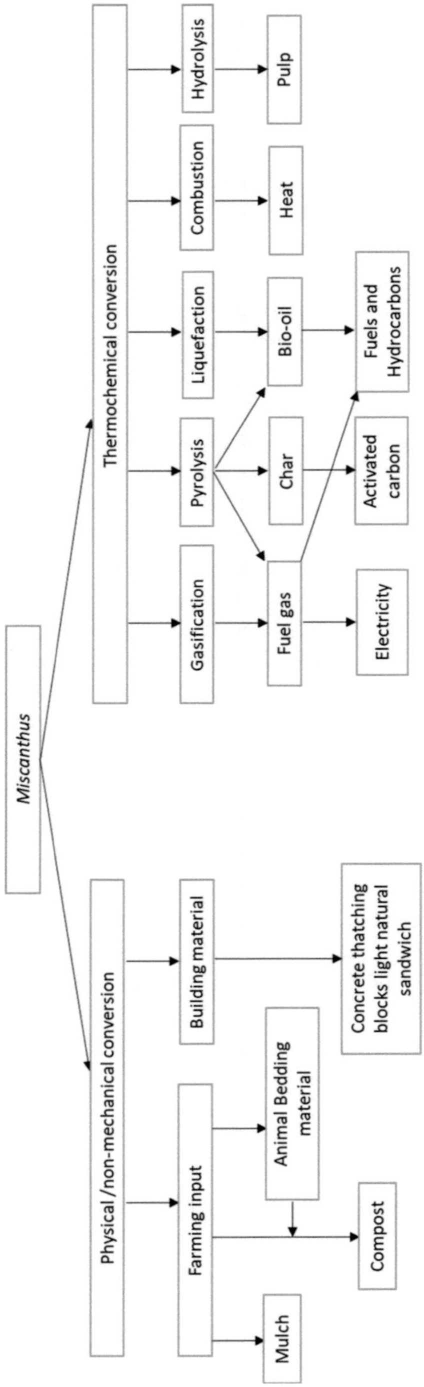

Fig. 9.2 Uses of *Miscanthus* for biofuel and biomaterial production

worldwide with its consumption increasing over the years. Ethylene glycol is mostly used to produce polyester and polyester resins and as a component for anti-freeze solvent and reductive agent. It is traditionally produced from ethylene oxide, which is a petroleum-base material. Another factor that makes *Miscanthus* more attractive for ethylene glycol production is its lower lignin content, compared to other lignocellulosic feedstocks. Moreover, lignin in herbaceous biomass is more easily decomposed than woody biomass-based lignin. The water-soluble components in *Miscanthus* (5.3 wt%) are much lower than those of corn stalk (33.1 wt%) and ligneous woody biomass (15–25 wt%), which makes it a suitable raw material to produce ethylene glycol.

To produce paper and board from lignocellulosic biomass, many methods have been used to reduce pretreatment and logistics cost. Pulp and paper industries at a global scale are now seeking for alternative feedstocks (i.e. agricultural biomass and invasive plants) to reduce the dependency on woody biomass. Invasive plants like *Miscanthus* have shown some promising attributes for use in pulp and paper industries. In China, *M. sacchariflorus* is being used for papermaking because of its fast-growth cycle, high biomass yield and easy pulping (Cappelletto et al. 2000). When compared with oat hull, pulp obtained from *Miscanthus* with sodium hydroxide treatment gave better yield, which could be used as a main component of low-grade paper and cardboard owing to satisfactory structural-dimensional characteristics (Budaeva et al. 2015). On the experiment basis, *Miscanthus* pulp also showed promising results concerning yield and strength than hybrid poplar pulp (Bousiosa and Worrell 2017). Refined *Miscanthus* soda pulp has abilities for use in packaging paper, which also reduces the amount of starch added to the paper for strength enhancement. In the Netherlands, it is already used to produce writing paper (Bousiosa and Worrell 2017).

Nowadays, biopolymers are attracting global attention because of their biodegradable properties, which are retained from their sustainable biomass precursors. However, due to lack of efficient technologies, the large-scale production of biopolymers is costlier than that of polymers derived from petroleum resources. To reduce the cost of biopolymers, filler or binder materials from lignocellulosic biomass are often used. Lignocellulosic polymers seem very effective because they can be added up to 49% as binders without affecting the quality of the biocomposite material. Adding fibres from *Miscanthus* into biopolymeric materials can improve their performance and decreases their cost of biopolymer (Johnson et al. 2005). However, it requires more research to be applied at an industrial level.

Two of the major components present in lignocellulosic biomass, i.e. cellulose and hemicellulose, are used to generate natural fibres and biopolymers for biocomposite applications. On the other hand, polymeric and aromatic lignin has many applications to be used to produce biofuels and platform chemicals. Many heterogeneous metal catalysts can be used to convert lignin into chemicals. Nickel is one of such heterogeneous metal catalysts that can cleave the C–O and C–C bonds of lignin. Different noble metals can also be used to depolymerise lignin, but their high cost restricts large-scale applications (Finch et al. 2012).

In catalytic depolymerisation of lignin, *Miscanthus* was milled to pass through a 40-mesh screen and reacted with Ni/C catalyst in methanol solvent at 225 °C under H_2 pressure (Luo et al. 2016). The liquid phase of lignin (279 mg/g) conversion contained aromatic products rich in phenolics, whereas the solid phase (612 mg/g) was composed of carbohydrates (e.g. glucan, xylan and arabinan). These depolymerised products can be further converted into high-value chemicals. Despite the relatively lower content of lignin in *Miscanthus* (13 wt%) than in hardwood (20–25 wt%), its higher conversion in *Miscanthus* was observed, which makes it a potential feedstock to produce phenolic compounds. Lewis acid catalysts were also used to convert the solid carbohydrate residues to furfurals and levulinic acid (Luo et al. 2016). Regardless, the major application of *Miscanthus* is realised in biofuel production. Table 9.1 summarises the thermochemical and biochemical conversion processes involved in the conversion of *Miscanthus* to solid, liquid and gaseous biofuels.

9.4 Pretreatment and Bioconversion of *Miscanthus*

9.4.1 Pretreatment Technologies

Lignocellulosic biomass requires pretreatment technologies involving chemical, physical and biological agents to depolymerise the cellulose–hemicellulose–lignin matrix and release the fermentable sugars for fermentation to alcoholic biofuels and chemicals (Nanda et al. 2014c). There are many pretreatment methods available, but some of the promising ones include mechanical methods (grinding, milling and crushing of biomass), acid and alkali treatment, liquid hot water, organosolv, wet oxidation, ozonolysis, CO_2 explosion, steam explosion, ammonia fibre explosion (AFEX) and ionic liquids (Menon and Rao 2012).

Liquid hot water has certain advantages over other widely used pretreatment technologies (e.g. acid/alkaline pretreatment, ozonolysis, ammonia fibre explosion, microwave etc.) concerning no chemical involvement, non-corrosiveness and lower production of intermediate components, such as furfural and 5-hydroxymethylfurfural (HMF). Liquid hot water is an environmental-friendly method and an attractive process for the biomass pretreatment. A study by Li et al. (2013) indicated that water washing of *Miscanthus* resulted in approximately 75 w/w% of suspended solids, 18 w/w% of precipitated solids, as well as 6 w/w% sand and salt.

Both furfural and HMF are sugar dehydration products obtained as intermediates of biomass pretreatment. Owing to their wide applications, these are considered among the top ten chemicals derived from bio-based materials (Yi et al. 2015). However, the presence of furfural and HMF in the biomass hydrolysate is inhibitory for the microorganisms to ferment monomeric sugars to alcohol-based biofuels, i.e. bioethanol or biobutanol (Nanda et al. 2014b, 2017a, b; Sarangi and Nanda 2018). Therefore, their recovery and separation from the biomass hydrolysate is highly essential for the bioconversion of biomass to bioethanol and biobutanol. Moreover,

Table 9.1 Comparative yields of solid, liquid and gaseous biofuels from *Miscanthus*

Process	Process conditions	Product yield	References
Slow pyrolysis	Temperature: 500 °C Reaction time: 30 min Feed amount: 400 g Heating rate: 7 °C/min N_2 flow rate: 2 L/min Reactor type: fixed bed batch reactor	Bio-oil: 45–51 wt% Biochar: 30 wt% Gas: 20–25 wt%	Oginni et al. (2017)
Fast pyrolysis	Temperature: 350–550 °C Feeding rate: 2.5 g/min N_2 flow rate: 3.0 and 4.0 L/min Reactor type: fluidised-bed reactor	Bio-oil: 25–69.2 wt% Biochar: 15–30 wt% Gas: 5–55 wt%	Heo et al. (2010)
Fast pyrolysis	Temperature: 350–500 °C Feeding rate: 150 g/h Residence time: 1.29 s, 1.93 s and 3.87 s Run time: 1 h N_2 flow rate: 3.0 and 4.0 L/min Reactor type: fluidised-bed reactor	Bio-oil: 47.7–57.2 wt% Biochar: 17.1–22.0 wt% Gas: 20.9–35.5 wt%	Kim et al. (2014)
Gasification	Temperature: 639–726 °C Biomass flow rate: 3.51–3.15 kg/h Air flow rate: 52.5–53 Ndm^3/min Bed material loaded: 6.6 kg Equivalence ratio (ER): 0.234–0.264 N_2: 5.5 kg/h Reactor type: air-blown bubbling Fluidised-bed gasifier	For lowest ER (0.234) at 639 °C: CO: 23.4 vol% H_2: 16.9 vol% CO_2: 42.5 vol% For highest ER (0.262) at 639 °C: CO: 39.5 vol% CO_2: 33.3 vol% H_2: 17 vol%	Xue et al. (2014a)
Liquefaction	Temperature: 220–280 °C Feed: 10 g Heating rate: 15 °C/min Liquefying agent: water/ethanol Catalysts: formic acid, zinc chloride, trifluoroacetic acid and sodium carbonate Biomass/solvent ratio (w/w): 1:6, 1:8 and 1:10 Water/ethanol ratio (v/v): 100:0, 90:10, 80:20, 70:30, 60:40 and 50:50 Reactor type: high-pressure Parr reactor with stirrer	50% bio-oil yield at 280 °C and 50% water/ethanol ratio; 280 °C was considered as the optimal temperature	Hafez and Hassan (2015)

(continued)

Table 9.1 (continued)

Process	Process conditions	Product yield	References
Torrefaction	Temperature: 250 °C Run time: 30 min Feed: 130 g Heating rate: 10 °C/min N_2 flow rate: 40 L/h Reactor type: electrically heated retort furnace	Mass yield: 73 wt% Energy yield: 80%	Wafiq et al. (2016)
Torrefaction	Temperature: 230–290 °C Run time: 10–30 min Feed: 130 g Heating rate: 20 °C/min N_2 flow rate: 100 mL/min Reactor type: horizontal furnace and tubular quartz reactor	Mass yield: 65.3–92.3 wt% Energy yield: 76.7–96%	Xue et al. (2014b)
Fermentation	Pretreatment method: dilute acid Temperature: 24 °C Fermentation microorganism: *Candida shehatae*	Ethanol: 64–66%	Guo et al. (2008)
Fermentation	Pretreatment method: 0.73 wt% H_2SO_4 blended with trifluoroacetic acid and maleic acid Temperature: 30 °C Fermenting microorganism: *Saccharomyces cerevisiae*	Ethanol: 27–54%	Guo et al. (2012)
Fermentation	Pretreatment method: NaOH Fermenting microorganism: *Saccharomyces cerevisiae* Temperature: 32 °C Time: 48 h	Ethanol: 84.7%	Han et al. (2011)

dilute sulfuric acid or alkaline pretreatment methods require solvent recovery and wastewater disposal, which are often difficult, expensive and energy intensive.

Aqueous ammonia is also a good candidate for pretreatment because it shows higher lignin removal with a 5% mixture of hydrogen peroxide. Ammonia is a suitable pretreatment agent because it is volatile, easily regenerated and weakly reactive with carbohydrates (Yu et al. 2013). As *Miscanthus* has higher cellulose content than other invasive crops, extracting it in liquid solution is beneficial for bioconversion. When untreated *Miscanthus* was hydrolysed, it gave less than 5–10% glucan and xylan conversion (Murnen et al. 2008). On the other hand, ammonia fibre explosion (AFEX) pretreatment significantly increases the enzymatic hydrolysis of *Miscanthus* and enhances the conversion between 30% and 90%, depending on the pretreatment parameters used.

9.4.2 Bioconversion to Bioethanol

The biological production of bioethanol from lignocellulosic biomass is a multistep process, which consists of physicochemical pretreatment, enzymatic saccharification and microbial fermentation. In the first step, depolymerisation of lignin is essential to break the cellulose-hemicellulose-lignin complex in the biomass. In the second step, degradation of structural polysaccharides into fermentable sugars is done via physicochemical and enzymatic pretreatments. In the second step, monomeric sugars are fermented to bioethanol using suitable bacterial or fungal species.

As mentioned earlier, the pretreatment of lignocellulosic biomass followed by enzymatic saccharification can degrade the complex polysaccharides, such as cellulose and hemicellulose, into simple sugars, such as glucose and xylose, which can be converted into bioethanol through fermentation. As mentioned earlier, the pretreatment of holocellulose (cellulose and hemicellulose) and lignin sometimes releases certain intermediate degradation compounds, such as furfural, HMF and phenolics, at moderate to high levels, which inhibit fermenting microorganisms (Hodgson et al. 2010). The neutralisation of inhibitors in the fermentation medium (containing biomass hydrolysate and monomeric sugars) or the separation of inhibitory compounds through adsorption is essential for microbial fermentation (Nanda et al. 2017b, Sarangi and Nanda 2019).

Xylose-containing liquor can be fermented using the yeast *Candida shehatae* because the common baker's yeast, i.e. *Saccharomyces cerevisiae*, lacks the natural ability to ferment C_5 sugars. Nevertheless, glucose, a widespread C_6 sugar present in *Miscanthus*, can efficiently be fermented to bioethanol using *S. cerevisiae*. By using *C. shehatae*, a 64–66% yield of ethanol was obtained from the dilute acid-pretreated *Miscanthus* (Guo et al. 2008). However, the lower yield was due to the presence of inhibitors in the liquor. Similar yield was reported when the fermentation was carried out by *S. cerevisiae* when sulfuric acid and trifluoroacetic acid were used for the pretreatment of biomass (Lee and Kuan 2015). On a theoretical basis, 70% yield can be obtained within 48 h of fermentation (Brosse et al. 2009).

On an economical basis, the operating cost of ethanol fermentation from lignocellulosic biomass (second-generation feedstocks) is still higher than that of starch or sugar-based feedstocks (first-generation feedstocks) (Nanda et al. 2015a). The main bottleneck is to have an efficient technology for biomass pretreatment to maximise fermentable sugar yields, as well as simultaneous fermentation of xylose and glucose by a suitable yeast species (Lee and Kuan 2015). Although the development of genetically engineered fungi is found to resolve the simultaneous fermentation of glucose and xylose at the laboratory scale, their stability, vitality and efficiency are still questionable when applied at the industrial level (Chandel and Singh 2011).

9.5 Thermochemical Conversion of *Miscanthus*

9.5.1 Pyrolysis of *Miscanthus*

Pyrolysis is a process involving thermal decomposition of biomass and organic wastes in the absence of oxygen to produce condensable vapors (bio-oil), biochar and

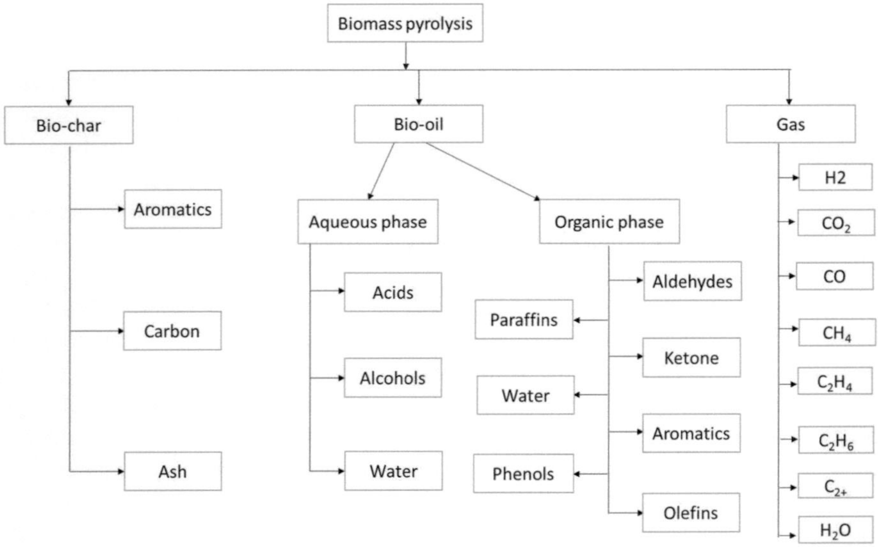

Fig. 9.3 Distribution of components in the pyrolysis of biomass

non-condensable gases, as shown in Fig. 9.3 (Bridgwater 2012). The condensable vapors released from biomass pyrolysis can be quenched at different temperatures into several bio-oil fractions, whereas the solid residues are carbonised to form char (Azargohar et al. 2013). The gas products mostly arise from the carboxylic groups in the unbranched structure of saccharides in biomass. Usually the thermal cracking of hemicellulose releases CO_2 and water vapor. On the other hand, during pyrolysis, cellulose depolymerises, resulting in the cleavage of O–H and C–O groups, thus releasing CO and water vapors (Azargohar et al. 2014).

The gaseous products from lignin include CH_4 and H_2, which mostly arise from the branched polymer of aromatic rings and methoxy groups (–O–CH_3) (Osman et al. 2017). During pyrolysis, cellulose decomposes in the temperature range of 300–350 °C, while hemicellulose degrades at 250–280 °C and lignin cracks at 200–500 °C (Jeguirim and Trouve 2009, Correa et al. 2010). Therefore, at relatively low pyrolysis temperatures, a significant amount of lignin remains non-carbonised, which results in less biochar yield (Oginni et al. 2017).

Depending on the heating rate and vapor residence time, pyrolysis can be classified into slow, intermediate or fast pyrolysis. Moreover, the pyrolysis process can be operated in batch or continuous modes. In the continuous process, the feeding of biomass and the removal of biochar work continuously compared to the batch process. Batch process is mainly considered optimal for biochar production, requires less nitrogen and has abilities to accommodate longer residence times. In contrast, continuous processes require more nitrogen to flush the vapors to the condensers by rapid quenching, resulting in bio-oil as the main product. Moreover, compared to batch pyrolysis processes, which are relatively easier to operate, continuous processes require maintenance as they are prone to plugging by tar in the case of improper insulation of the reactor and tubings.

Depending on the heating rate and vapor residence time, pyrolysis can be classified into slow, fast and intermediate pyrolysis. Fast pyrolysis of *Miscanthus* was carried out at 350–500 °C with different residence times of 1.29–3.87 s (Kim et al. 2014). With the increase in pyrolysis temperature, it was noted that the yield of bio-oil decreased (from 57.2 to 47.7 wt%), whereas this yield of gases increased (from 20.9 to 35.5 wt%). The energy conversion efficiency is also found to be lower in the case of *Miscanthus*-derived bio-oil compared to softwood-derived bio-oil due to the presence of more inorganic components in *Miscanthus* (Kim et al. 2014).

Greenhalf et al. (2013) performed a pyrolysis of *M. giganteus* at 490 °C and obtained the yield of bio-oil and gases in the range of 41–51% and 22–35%, respectively. An increase in vapor residence time showed a decrease in bio-oil yield and an increase in gas yield. Oginni et al. (2017) compared the product yields from the pyrolysis of *Miscanthus* and switchgrass grown on reclaimed mine lands at 500 °C in a fixed bed batch reactor with a heating rate 70 °C/min for 30 min. *Miscanthus* resulted in greater bio-oil yields of 51 wt% compared to switchgrass.

Pham et al. (2018) conducted an oxidative pyrolysis of *Miscanthus* pellets in a fixed bed reactor at 650 °C. In this type of pyrolysis, partial oxidation of biomass occurs auto-thermally and with the aid of low air mass flux. The biochar produced contained 91.2 wt% carbon and 6.6 wt% oxygen. This also attributed to the catalytic effect of ash with the formation of secondary biochar with the condensation and polymerisation of primary tar and contact between volatile matter and biochar particles. Du et al. (2014) performed the pyrolysis of *Miscanthus* in a spouted bed reactor at variable temperatures of 400–600 °C for a reaction time of 10 min. It was reported that as the temperature increased from 400 to 600 °C, the gas yield significantly amplified from 16 to 25 wt%, while the biochar and bio-oil yields decreased from 43 to 35 wt% and from 43 to 33 wt%, respectively.

9.5.1.1 Bio-oil

Bio-oil, a major product of biomass pyrolysis, is a mixture of oxygenated and aromatic compounds. When compared to petroleum-derived oil, bio-oil contains oxygenated compounds as opposed to hydrocarbons. Pyrolysis liquid is composed of both organic and aqueous fractions. These fractions are present due to the condensation process during biomass pyrolysis. While the aqueous fraction contains an acidic phase, the organic fraction consists of an oily phase (Nanda et al. 2016c). Bio-oil cannot be mixed with hydrocarbon liquids because it is composed of oxygenated compounds, which offer potential challenges for direct application.

Water content of bio-oil usually depends on the moisture content of the biomass feedstock. The acidic pH of bio-oil, along with its higher water content and oxygenation, causes many issues relating to storage, polymerisation and low calorific value. The acidic pH also raises concerns for reactor or storage container corrosion. The high concentrations of oxygen, and possible presence of nitrogen and sulphur of certain bio-oils, require upgrading, such as hydrodeoxygenation, hydrodenitrogenation and hydrosulphurisation, to enhance fuel properties and direct use (Zacher et al.

2014). Bio-oil consists of more than 300 different compounds grouped under the classes of aldehydes, ketones, esters, ethers, alcohols, carboxylic acid, nitriles, carbohydrates and hydrocarbons (Czernik and Bridgwater 2004). It should be noted that the bio-oils obtained at longer residence times and higher temperature have higher pH because of lower organic acids. The thermal cracking of organic acids at higher temperatures and longer residence times could further decompose into non-condensable gases (Nanda et al. 2014d).

Pyrolysis temperature is one of the most important variables to determine the yield of bio-oil, biochar and gases. In an experiment on the pyrolysis of *Miscanthus*, it was observed that by increasing the temperature from 350 °C to 550 °C, the biochar yield decreased from 32.4 to 12.1 wt% (Heo et al. 2010). However, maximum bio-oil yield of 69.2 wt% was observed at a temperature of 450 °C, while the yield decreased to 25 wt% at 550 °C. This is because of the increased degree of primary tar produced in the pyrolysis vapors that crack to gases. A rapid increase in aromatic components in the bio-oil occurred at 550 °C (Heo et al. 2010). In contrast, maximum yield of bio-oil of 53.9% in the case of hardwood was recorded at 500 °C, which can be due to the higher content of lignin compared to *Miscanthus* (Mohan et al. 2006). Higher temperatures lead to the breakdown of lignin, which slowly decomposes and results in biochar as the main product.

Bio-oil derived from the pyrolysis of *Miscanthus* at 1.9 s of residence time at 500 °C (31.5 wt%) was higher than the bio-oil derived at 350 °C (25.3 wt%) (Kim et al. 2014). The water content in bio-oils influences its viscosity and flowability (Oginni et al. 2017). The viscosity of *Miscanthus*-derived bio-oil produced at 350 °C (16.5 cSt) was relatively greater than 500 °C (13.9 cSt) (Kim et al. 2014). The HHV of bio-oil generated from *Miscanthus* at 500 °C (17.7 MJ/kg) was greater than that at 350 °C (16.6 MJ/kg). Vapor residence time also has a significant impact on the quality of bio-oil. The water content of bio-oil produced from the pyrolysis of *Miscanthus* at 400 °C at 3.8 s (56.9 wt%) was much higher than that produced at 1.2 s (21.1 wt%) (Kim et al. 2014). However, the viscosity of bio-oil produced at 3.8 s (12.1 cSt) was lower than the bio-oil derived at 1.2 s (16.1 cSt).

Catalytic pyrolysis of *M. giganteus* with two different heating of 10 °C/min and 50 °C/min was studied at 550 °C (Yorgun and Şimşek 2008). In the first case (i.e. 10 °C/min), pyrolysis conversion increased from 59.7 to 79.9 wt%, but the solid yield decreased from 40.2 to 23 wt% with the catalyst loading (activated alumina) increasing from 10 to 100 wt% of the feedstock. In the second case (i.e. 50 °C/min), as the catalyst loading increased in the same range, the pyrolysis conversion and solid, gaseous and liquid product yields were in the range of 65.2–79.4 wt%, 34.8–20.6 wt%, 16–31.7 wt% and 47.2–51 wt%, respectively. As the catalyst loading increased, the increase in the active sites also occurred, which enhanced the rate of depolymerisation and decarbonylation. This is the reason for an increase in gaseous products, a decrease in solid product, and an increase in pyrolysis conversion. The maximum bio-oil yield was observed at 51 wt% with the heating rate of 50 °C/min at 60 wt% catalyst loading (Yorgun and Şimşek 2008). In addition, the bio-oil produced in the presence of catalysts had higher degree of aromaticity than those produced during non-catalytic pyrolysis. Bio-oil produced through non-catalytic pyrolysis contained 63.2 wt% carbon, 8.1 wt% hydrogen and 27.1 wt% oxygen. In contrast, bio-oil produced through catalytic pyrolysis contained lower carbon (51.1 wt%) and hydrogen (6.4 wt%) but higher oxygen (48.4 wt%) levels.

In an experiment on the pressurised pyrolysis of *Miscanthus* using a fixed bed reactor at 550 °C for 25 min with the heating rate of 13 °C/min with 50 cm^3/min flow of N_2, the yield of the bio-oil remained constant at approximately 35 wt% but increased in the case of the pressure above 16 bars. High yield of tar was observed in the pyrolysis experiments performed at atmospheric pressures. With the increase in the pressure during the reaction, an elevation in the carbon content was noted, which also increased the HHV of the liquid product. This can be due to the secondary decomposition reactions during which large amount of oxygen and hydrogen also are removed from the feedstock, thus retaining carbon in the tar or char. This confirms that pressure has an influence on the quality rather than the quantity of the product (Melligan et al. 2011).

9.5.1.2 Biochar

Biochar is the solid carbonaceous product of the pyrolysis, gasification and carbonization of biomass, which attracts widespread attention for its potential environmental and industrial applications (Mohanty et al. 2013, Kang et al. 2019). Biochar is a carbon-rich product obtained from the thermal decomposition of biomass in oxygen-deprived conditions. Biochar can sequester carbon in soil and helps to decrease the amount of net CO_2 emissions into the atmosphere. It can be used for agricultural applications, water purification, catalyst support, electronics and biomedicine (Budaia et al. 2017, Nanda et al. 2016a).

There are some evidence that indicates that biochar not only increases the stability of carbon stocks in the soil but also enhances the nutrient availability more than inorganic fertilisers do. Biochar can enhance soil quality than any other organic soil amendment. Biochar acts as a soil conditioner and organic fertiliser by providing soil organic carbon and improving microbial carbon metabolism and population dynamics (Kwapinski et al. 2010). The microbial response after the addition of biochar to the soil is mainly due to the available carbon and inorganic elements present in the biochar, which alter the soil pH in the relatively long term.

It is also concluded that the content of Gram-positive bacteria, Gram-negative bacteria, actinobacteria and fungi biomarkers has been increased in the biochar-modified soil (Lehmann et al. 2011). The large ratio of fungi/bacteria in the soil amended with biochar concluded that fungi colonisation is very important to break the structure of polyaromatics in the contaminated soils so that they will be available for other microbial groups for decomposition (Nanda et al. 2018). Moreover, it has been noted that regardless of the temperature of the biochar production, biochar application to highly acidic soils caused an increase in all phospholipid fatty acids by microorganisms. Biochar produced at lower temperature are more utilizable by microorganisms (Kwapinski et al. 2010).

The type of feedstock influences the composition and surface area of the biochar. Biochar from crop residues and woody biochar has larger surface area when compared to the ones produced from other sources. Biochar generates oxygen-containing compounds when it undergoes oxidation in the soil, a few examples of which include carboxylic and lactonic acids and phenolic groups. The oxidation in biochar-amended soils occurs through microbial action, organic matter and solutes (Kim et al. 2018). The content of carbon availability instead of pH change is the main point to consider for determining the microbial utilisation of biochar in acidic soils (Luo et al. 2017). In addition to sequestering soil carbon, biochar also provides several other advantages, such as increasing soil fertility and crop productivity. The contribution of biochar in improving crop productivity is achieved through the retention of plant-available nutrients in the rhizosphere, which results to increase in soil pH, cation exchange capacity, water-holding capacity, decrease in greenhouse gas emissions and immobilisation of toxic compounds including heavy metals (Gronwald et al. 2016).

Greater yields of biochar are produced from pyrolysis at moderate temperatures and longer vapor residence time, especially the conditions optimal for slow pyrolysis. Temperature and heating rate are the most considerable factors in pyrolysis, gasification and carbonization, which can alter the yield and properties of the resulting biochar (Azargohar et al. 2019). Biochar from *Miscanthus* is found to have relatively large surface areas (50.9–51.1 m^2/g) when produced at higher temperatures (e.g. 600 °C) and longer residence times (e.g. 60 min) (Kwapinski et al. 2010). These properties are relevant for using biochar as the soil conditioner. The aromaticity of the biochar also determines its thermal stability in the soil at extreme environments (Budaia et al. 2017). The composition of feedstocks (i.e. contents of cellulose, hemicellulose, lignin, minerals and ash) also determines the chemical composition of the resulting biochar. The organic and inorganic components in the original precursor (biomass) act as catalysts during the pyrolysis and carbonization processes to improve the quality of the biochar (Hodgson et al. 2016).

Dependent on pyrolysis temperature, the properties and yield of the biochar from *Miscanthus* varied largely. The yield of biochar from *Miscanthus* reduced from 25.9–26.2 wt% at 500 °C (for 10 min of reaction time) to 19.8–20.2 wt% at 600 °C (at 60 min of reaction time) (Kwapinski et al. 2010). The surface area of *Miscanthus*-derived biochars also showed a rising trend with the increase in the temperature from 500 °C (1.65–1.95 m^2/g) to 600 °C (50.9–51.1 m^2/g). The HHV data showed little difference at variable temperatures of 400 °C (29.4–30.3 MJ/kg), 500 °C (29.9–30.7 MJ/kg) and 600 °C (31.5–32.5 MJ/kg). As temperature plays a significant role for the biochar production and its properties, biochar produced at high temperature has high pH and more compact structure of polyaromatic compounds than biochar produced at lower temperature. It is suggested that the large surface area of biochar is usable for the adsorption of pollutants from wastewater and retaining soluble carbon and nutrients, which is beneficial for soil applications (Luo et al. 2017).

In an assessment, *Miscanthus* cultivated on the contaminated land was found to accumulate more metals in the roots and rhizomes and less in shoots and stems compared to *Miscanthus* cultivated from uncontaminated land (Janus et al. 2017). Biochar produced from *Miscanthus* cultivated on the contaminated land were efficient for the removal of Cd, Pb and Zn from the aqueous solution. Moreover, higher efficiency in the removal of impurities was found in case of biochar produced at higher pyrolysis temperatures. Greater desorption and lower sorption abilities of biochar is suitable for the treatment of wastewater (Janus et al. 2017).

In a study by Yang et al. (2017), biochar was produced from different feedstock (e.g. Masson pine wood, Chinese fir wood, Chinese fir bark, bamboo leaves, bamboo sawdust, *Miscanthus*, pecan shells and rice straw) through slow pyrolysis at 350 °C and 500 °C. Slow pyrolysis at 350 °C showed the biochar yield trend as bamboo sawdust > pecan shells > *Miscanthus* > Masson pine wood > Chinese fir bark > rice straw > Chinese fir wood > bamboo leaves. However, at 500 °C, the yields of all biochars decreased. The yields of biochar from the slow pyrolysis of all these feedstocks differed because of the variable composition of cellulose, hemicellulose and lignin, which have different thermal degradation kinetics. It has been proven in this study that the yield and thermal properties of biochar were vastly affected by the composition of different feedstocks. The HHV of biochars also showed an increasing trend at 350 °C but decreased at 500 °C (Yang et al. 2017).

As opposed to the benefits of biochars, there is a negative influence due to the presence of polycyclic aromatic hydrocarbons (PAH), naturally generated during the pyrolysis process. A few studies have shown that the total bioavailability of the PAHs is low in the biochar. Hence, biochar can play the role of a carbon sink rather than of an organic pollutant (Hale et al. 2012, Mayer et al. 2016). The factors that affect the concentration of PAHs in biochar are temperature, feedstock composition and carrier gas. The levels of PAHs in the biochar usually increase with the rise in pyrolysis temperature. Aromatic hydrocarbons in the feedstocks depend on the process parameters, which determine their fate in the biochar, bio-oil or gases. High carrier gas flows result in decreasing the PAH concentration (Madej et al. 2016).

A comparison between pyrochar and hydrochar obtained from slow pyrolysis and hydrothermal carbonization was made. For pyrochar from *M. giganteus*, a Pyreg reactor was used at 750 °C for 0.75 h, and for the hydrochar mixture, 1 kg of dry *Miscanthus* with 10 kg of water in tubular reactor was used at 200 °C and 2 MPa for 11 h (Gronwald et al. 2016). Hydrochar is produced from hydrothermal carbonization, which is usually performed at low temperature (180–300 °C) in the presence of water under high pressure (2–2.5 MPa) for several hours. The hydrochar was found to have low surface area, low degree of carbonization and less aromatic carbon compared with pyrochar. Moreover, hydrochar had a high ratio of H/C (1.01) than O/C (0.31) from pyrochar (Gronwald et al. 2016). Compared to hydrochar, the application of pyrochar in the soil can retain the carbon for a longer period. On the other hand, hydrochar applied directly to the soil showed a slow release of fertiliser.

Like nitrogen and potassium, phosphorus is one of the essential elements for crop growth and yield. The impact of phosphorous can be seen in more tropical weathered soils where bioavailability of soil happens under natural conditions. The weathering of rocks releases essential elements for the growth of the crop at a slower rate (Trazzi et al. 2016). The surface area of biochar, volatile matter content and surface organic functional groups influence the phosphorus bioavailability in soils amended with biochar. It has been reported that biochar produced from wheat straw can affect phosphorous concentration, depending on its quantity of amendment in the soil. It was noted that fast pyrolysis biochar decreased phosphorous fixation capacity in the soil (Trazzi et al. 2016). Increased temperature and residence time also increased the fixed carbon and surface area of *Miscanthus* biochar, which resulted in higher phosphorus adsorption capacity compared to sugarcane bagasse biochar.

9.5.2 Torrefaction of *Miscanthus*

Torrefaction was the most promising pretreatment process for biomass before pyrolysis. The physicochemical properties of torrefied biomass are like those of coal. Torrefaction is a process of moisture removal at low temperature, and hemicellulose decomposition is another key aspect of this process. By losing CO_2, moisture and other oxygen-containing compounds, the torrefied product demonstrates high energy density than the raw material. Furthermore, torrefaction helps to improve the grinding property of biomass, which helps to reduce electric power consumption. Torrefaction decreases the content of oxygen and increases carbon content due to loss of moisture, CO_2 and CO (Wafiq et al. 2016).

The impact of pressure on the yield of char from the torrefaction of *Miscanthus* was studied at 250 °C with a heating rate of 10 °C/min and N_2 flow rate of 40 L/h for 30 min of reaction time. The results indicated that at 1 bar, the char yield increased from 22 to 29 wt% until 10 bar, beyond which no further increase was observed with rising pressure. The biochar with torrefied *Miscanthus* showed a similar trend of yield increase from 32 to 42 wt% at 1–10 bar and no further increase at higher pressures (Wafiq et al. 2016). On the other hand, liquid products from pyrolysis decreased as the pressure increased for torrefied *Miscanthus* due to secondary pyrolysis reactions. In terms of the porosity of biochar, the porosity of raw *Miscanthus* decreased until 15 bars and increased at 30 bars. However, in the case of torrefied *Miscanthus*, the porosity decreased as the pressure increased.

Raw *Miscanthus* also has some disadvantages, such as low bulk density, low energy density and non-uniform physical and chemical properties, which could lead to storage problem, lower thermal conversion and utilisation. Torrefied *Miscanthus* also has high energy density and bulk density, making it easier for transportation and storage. *Miscanthus* torrefied at 270 °C for a residence time of 30 min achieved good properties like low moisture and hemicellulose contents, lower O/C ratio, as well as porous structure, larger surface area and high alkali metal content in the solid products, which are optimal for pyrolysis or gasification (Xue et al. 2014b). Furthermore, raw and torrefied *Miscanthus* were gasified at 850 °C under N_2 and CO_2 atmosphere. During the torrefaction, the overall O/C ratio decreased while the pore volume and surface area increased, which improved the gasification properties of torrefied *Miscanthus* compared to the raw *Miscanthus*. The penetration of CO_2 inside the porous particles improved, which positively affected gasification reactivity. Regardless, the composition of gas obtained from gasification depends upon the nature and composition of the feedstock.

9.5.3 Liquefaction of *Miscanthus*

The thermochemical technologies for biomass conversion into liquid fuels also include direct liquefaction and hydrothermal liquefaction. In hydrothermal liquefaction and hydrous pyrolysis, the use of water and catalyst is required for the conversion of biomass. According to Balat (2008), there are some reactions that occur during this process, such as the following:

1. Decomposition and reduction of cellulose, hemicellulose, lignin and lipids
2. Hydrolysis of cellulose and hemicellulose to the simple sugars
3. Hydrogenolysis in the presence of hydrogen
4. Reduction of amino acids
5. Reformation reactions
6. Degradation of C–O and C–C bonds
7. Hydrogenation of functional groups

Most typical temperature for liquefaction is between 300 °C and 500 °C. Over the years, many improvements have been made to overcome certain technical problems associated with liquefaction. Another iteration of liquefaction, i.e. hydrothermal liquefaction, involves alcohol and water as the reaction medium. The advantage of such system is that the solvents can be evaporated, recycled and reused (Cheng et al. 2010). It is found that replacing 50% of water with alcohol during liquefaction increased bio-oil yield, whereas replacing water more than 50% can adversely affect the bio-oil yields (Cheng et al. 2010). The drying of biomass is necessary for pyrolysis but not for hydrothermal liquefaction because of water as the aqueous medium. Catalyst requirement is a very essential part of liquefaction. Alkali catalysts, such as Na_2CO_3 and K_2CO_3, aid liquefaction reactions (Nanda et al. 2014c, 2016c).

Hydrothermal liquefaction involves the application of subcritical or supercritical water as the reaction medium (Kamio et al. 2006). The critical temperature (T_c) and critical pressure (P_c) of water determines the subcritical phase ($T_c < 375$ °C and $P_c < 22.1$ MPa) and supercritical phase ($T_c > 375$ °C and $P_c > 22.1$ MPa) of water (Nanda et al. 2019, Rana et al. 2019). Subcritical water acts as a green solvent that dissolves highly complex organic residues to simple hydrocarbons and permanent gases through ionic and free radical mechanisms (Reddy et al. 2014, 2016). Supercritical water has found applications in hydrothermal gasification, liquefaction and oxidation of biomass and organic refuse (Reddy et al. 2015, 2017).

During the liquefaction of *Miscanthus*, it has been noted that by increasing the temperature from 220 to 280 °C, the quantity of residues decreased while the liquid yields increased. According to the value of oil yield, residue content and heating value, 15 min liquefaction time was efficient for *Miscanthus* liquefaction without any catalyst. Reaction temperature (280 °C) and water/ethanol ratio (50%) were the most effective parameters for this process. The optimal results obtained in the case of *Miscanthus* liquefaction was 52% yield of bio-oil having a heating value of 25 MJ/kg when the biomass-to-solvent ratio used was 1:8 with $ZnCl_2$ as the catalyst (Hafez and Hassan 2015).

9.5.4 Gasification of *Miscanthus*

Gasification operates in the presence of a limited supply of oxygen to produce syngas, which mainly contains H_2 and CO, along with other products, such as CO_2, CH_4 and C_{2+} (Okolie et al. 2019a, b). During gasification, carbon conversion is a key feature that is used to determine gasification efficiency (Rana et al. 2018). Biomass gasification is emerging as one of the cleanest technologies to produce biofuels and decrease the dependence on fossil fuels. Because of the high reactivity of biomass char, gasification is gaining widespread interest for energy production. The reactivity of biochar depends upon three basic properties such as chemical structure, porosity and inorganic constituents. *Miscanthus* has been proved as an appropriate biomass for thermochemical conversion because of its higher volatility and low ash content (Karampinis et al. 2012).

The gasification of *Miscanthus* char was investigated above 800 °C specifically in the presence of steam, and the complete burnout of *Miscanthus* was detected at 1000 °C in the blended (air/oxygen) ambience, as well as 1050 °C for the steam medium (Jayaraman and Gökalp 2015). This shows that the amount of oxygen affects reaction rate and temperature. About H_2 gas evolution during the whole process from pyrolysis to gasification of *Miscanthus*, it was observed that it started at 350 °C with the highest concentration at 950 °C and then decreased similar to CO content. It was also observed that the H_2 yield decreased when the gasification medium changed from pure steam to steam enriched with oxygen and air (Jayaraman and Gökalp 2015).

The bubbling fluidised bed is the simplest and most cost-effective method for biomass gasification. Such reactors are flexible to gasify a wide variety of biomass because of high heat transfer and uniform temperature distribution. However, the disadvantage of such reactors is the risk of bed agglomeration because of the presence of sintering ash and high alkali metals in the biomass, which might also affect the reactor material (corrosion) at high temperatures and catalyse or retard the reaction. During *Miscanthus* gasification, agglomeration was observed at low temperature due to the presence of high silica, potassium and sodium content (Xue et al. 2014a). It was reported that in the case of gasification of *M. giganteus*, the quality of product gas deteriorated with the rise in temperature from 645 °C to 726 °C using optimal equivalence ratio of 0.26 and air flow rate of 53 Ndm^3/min. Among all the gases, CO yield decreased from 39.5 to 33.4 vol%, but the concentration of CO_2 increased from 33.3 to 36.8 vol%. The yields of H_2 and hydrocarbons did not show any deviation at higher temperature (700–726 °C) in comparison with 645 °C (Xue et al. 2014a).

9.6 Conclusions

Miscanthus is a lignocellulosic energy crop that can be used to produce biofuels through thermochemical and biochemical conversion technologies. Biochemical conversion technologies have shown promising results in producing ethanol from *Miscanthus* through fermentation. The bioconversion of lignocellulosic biomass to ethanol operates in three steps, which involves pretreatment, enzymatic saccharification and fermentation. On the other hand, thermochemical conversion technologies, such as pyrolysis and liquefaction of *Miscanthus*, are useful for producing energy dense bio-oil, whereas gasification of *Miscanthus* can produce syngas. The syngas can be either used directly for combustion or catalytically converted to hydrocarbon fuels and chemicals through the Fischer-Tropsch process. *Miscanthus*-derived bio-oil produced from pyrolysis and liquefaction can be upgraded to synthetic transportation fuels.

Biochar is a solid residue of pyrolysis, gasification and torrefaction. Biochar generated from *Miscanthus* has been found to have promising properties for application in the environment and industry. The physical and chemical properties of biochar are greatly dependent upon a variety of process parameters, such as temperature, heating rate, residence time, pressure, biomass composition and biomass particle size. Temperature has a direct impact on the surface properties of biochar, like surface area, pH and cation exchange capacity. *Miscanthus* has shown promising results among other energy crops for thermochemical and biochemical conversion, as well as for the generation of value-added products. There is also an opportunity to design technologies, which will be efficient at a commercial level for getting value out of this energy crop.

References

Arnoult S, Hulmel MB (2015) A review on *Miscanthus* biomass production and composition for bioenergy use: genotypic and environmental variability and implications for breeding. Bioenergy Res 8:502–526

Azargohar R, Nanda S, Rao BVSK, Dalai AK (2013) Slow pyrolysis of deoiled canola meal: product yields and characterization. Energy Fuel 27:5268–5279

Azargohar R, Nanda S, Kozinski JA, Dalai AK, Sutarto R (2014) Effects of temperature on the physicochemical characteristics of fast pyrolysis bio-chars derived from Canadian waste biomass. Fuel 125:90–100

Azargohar R, Nanda S, Dalai AK, Kozinski JA (2019) Physico-chemistry of biochars produced through steam gasification and hydro-thermal gasification of canola hull and canola meal pellets. Biomass Bioenergy 120:458–470

Balat M (2008) Mechanisms of thermochemical biomass conversion processes. Energ Sourc A 30:620–635

Bousiosa S, Worrell E (2017) Towards a multiple input-multiple output paper mill: opportunities for alternative raw materials and sidestream valorisation in the paper and board. Resour Conserv Recycl 125:218–232

Bridgwater AV (2012) Review of fast pyrolysis of biomass and product upgrading. Biomass Bioenergy 38:68–94

Brosse N, Sannigrah P, Ragauskas A (2009) Pretreatment of *Miscanthus× giganteus* using the ethanol organosolv process for ethanol production. Ind Eng Chem Res 48:8328–8334
Brosse N, Dufour A, Meng X, Sun Q, Ragauskas A (2012) *Miscanthus*: a fast-growing crop for biofuels and chemicals production. Biofuels Bioprod Biorefin 6:580–598
Budaeva VV, Makarova EI, Gismatulina YA (2015) Integrated flowsheet for conversion of non-woody biomass into polyfunctional materials. Key Eng Mater 670:202–206
Budaia A, Calucci L, Rasse DP, Strand LT, Pengerud A, Wiedemeier D, Abiven S, Forte C (2017) Effects of pyrolysis conditions on Miscanthus and corncob chars characterization by IR, solid state NMR and BPCA analysis. J Anal Appl Pyrolysis 128:335–345
Cappelletto P, Mongardini F, Barberi B, Sannibale M, Brizzi M, Pignatelli V (2000) Papermaking pulps from the fibrous fraction of *Miscanthus x giganteus*. Ind Crop Prod 11:205–210
Chandel AK, Singh OV (2011) Weedy lignocellulosic feedstock and microbial metabolic engineering: advancing the generation of 'biofuel'. Appl Microbiol Biotechnol 89:1289–1303
Cheng S, D'cruz I, Wang M, Leitch M, Xu CC (2010) Highly efficient liquefaction of woody biomass in hot-compressed alcohol-water co-solvents. Energy Fuel 24:4659–4667
Clark LV, Brummer JE, Głowacka K, Hall MC, Heo K, Peng J, Yamada T, Yoo JH, Yu CY, Zhao H, Long SP, Sacks EJ (2014) A footprint of past climate change on the diversity and population structure of *Miscanthus sinensis*. Ann Bot 114:97–107
Correa AC, de Morais Teixeira E, Pessan LA, Mattoso LHC (2010) Cellulose nanofibers from curaua fibers. Cellulose 17:1183–1192
Czernik S, Bridgwater AV (2004) Overview of applications of biomass fast pyrolysis oil. Energy Fuel 18:590–598
Dohleman FG, Long SP (2009) More productive than maize in the Midwest: how does *Miscanthus* do it? Plant Physiol 150:2104–2115
Du S, Sun Y, Gamliel DP, Valla JA, Bollas GM (2014) Catalytic pyrolysis of *Miscanthus × giganteus* in a spouted bed reactor. Bioresour Technol 169:188–197
Finch KBH, Richards RM, Richel A, Medvedovici AV, Gheorghe NG, Verziu M, Coman SM, Parvulescu VI (2012) Catalytic hydro processing of lignin under thermal and ultrasound conditions. Catal Today 196:3–10
Fontoura CF, Brandão LE, Gomes LL (2015) Elephant grass biorefineries: towards a cleaner Brazilian energy matrix? J Clean Prod 96:85–93
Fougere D, Nanda S, Clarke K, Kozinski JA, Li K (2016) Effect of acidic pretreatment on the chemistry and distribution of lignin in aspen wood and wheat straw substrates. Biomass Bioenergy 91:56–68
Ge X, Xu F, Vasco-Correa J, Li Y (2016) Giant reed: a competitive energy crop in comparison with miscanthus. Renew Sust Energ Rev 54:350–362
Greenhalf C, Nowakowski D, Yates N, Shield I, Bridgwater A (2013) The influence of harvest and storage on the properties of and fast pyrolysis products from *Miscanthus x giganteus*. Biomass Bioenergy 56:247–259
Gronwald M, Vos C, Helfrich M, Don A (2016) Stability of pyrochar and hydrochar in agricultural soil—a new field incubation method. Geoderma 284:85–92
Guo GL, Chen WH, Chen WH, Men LC, Hwang WS (2008) Characterization of dilute acid pretreatment of silver grass for ethanol production. Bioresour Technol 99:6046–6053
Guo B, Zhang Y, Ha SJ, Jin YS, Morgenroth E (2012) Combined biomimetic and inorganic acids hydrolysis of hemicellulose in *Miscanthus* for bioethanol production. Bioresour Technol 110:278–287
Hafez I, Hassan EB (2015) Rapid liquefaction of giant miscanthus feedstock in ethanol–water system for production of biofuels. Energ Convers Manage 91:219–224
Hale SE, Lehmann J, Rutherford D, Zimmerman AR, Bachmann RT, Shitumbanuma V, O'Toole A, Sundqvist KL, Arp HPH, Cornelissen G (2012) Quantifying the total and bioavailable polycyclic aromatic hydrocarbons and dioxins in biochars. Environ Sci Technol 46:2830–2838
Han M, Kim Y, Koo BC, Choi GW (2011) Bioethanol production by *Miscanthus* as a lignocellulosic biomass: focus on high efficiency conversion to glucose and ethanol. Bioresources 6:1939–1953

Heo HS, Park HJ, Yim JH, Sohn JM, Park J, Kim SS, Ryu C, Jeon J, Park YK (2010) Influence of operation variables on fast pyrolysis of *Miscanthus sinensis var. purpurascens*. Bioresour Technol 101:3672–3677

Hodgson EM, Lister SJ, Bridgwater AV, Clifton-Brown J, Donnison IS (2010) Genotypic and environmentally derived variation in the cell wall composition of *Miscanthus* in relation to its use as a biomass feedstock. Biomass Bioenergy 34:652–660

Hodgson EM, Nowakowski DJ, Shield I, Riche A, Bridgwater AV, Clifton-Brown JC, Donnison IS (2011) Variation in *Miscanthus* chemical composition and implications for conversion by pyrolysis and thermo-chemical bio-refining for fuels and chemicals. Bioresour Technol 102:3411–3418

Hodgson E, James AL, Ravella SR, Jones ST, Perkins W, Gallagher J (2016) Optimisation of slow-pyrolysis process conditions to maximise char yield and heavy metal adsorption of biochar produced from different feedstocks. Bioresour Technol 214:574–581

Janus A, Pelfrêne A, Sahmer K, Heymans S, Deboffe C, Douay F, Waterlot C (2017) Value of biochars from *Miscanthus x giganteus* cultivated on contaminated soils to decrease the availability of metals in multicontinental aqueous solutions. Environ Sci Pollut Res 24:18204–18217

Jayaraman K, Gökalp I (2015) Pyrolysis, combustion and gasification characteristics of *Miscanthus* and sewage sludge. Energ Convers Manage 89:83–91

Jeguirim M, Trouve G (2009) Pyrolysis characteristics and kinetics of *Arundo donax* using thermogravimetric analysis. Bioresour Technol 100:4026–4031

Jenkins BM, Baxter LL, Miles TR Jr, Miles TR (1998) Combustion properties of biomass fuel processing technology. Fuel Process Technol 54:17–46

Johnson M, Tucker N, Barnes S, Kirwan K (2005) Improvement of the impact performance of a starch-based biopolymer via the incorporation of *Miscanthus giganteus* fibres. Ind Crop Prod 22:175–186

Kamio E, Takahashi S, Noda H, Fukuhara C, Okamura T (2006) Liquefaction of cellulose in hot compressed water under variable temperatures. Ind Eng Chem Res 45:4944–4953

Kang K, Nanda S, Sun G, Qiu L, Gu Y, Zhang T, Zhu M, Sun R (2019) Microwave-assisted hydrothermal carbonization of corn stalk for solid biofuel production: optimization of process parameters and characterization of hydrochar. Energy 186:115795

Karampinis E, Vamvuka D, Sfakiotakis S, Grammelis P, Itskos G, Kakaras E (2012) Comparative study of combustion properties of five energy crops and Greek lignite. Energy Fuel 26:869–878

Kim JY, Oh S, Hwang H, Moon YH, Choi JW (2014) Assessment of miscanthus biomass (*Miscanthus sacchariflorus*) for conversion and utilization of bio-oil by fluidized bed type fast pyrolysis. Energy 76:284–291

Kim H, Kim J, Kim M, Hyun S, Moon DH (2018) Sorption of sulfathiazole in the soil treated with giant *Miscanthus*-derived biochar: effect of biochar pyrolysis temperature, soil pH, and aging period. Environ Sci Pollut Res 25:25681–25689

Kwapinski W, Byrne CMP, Kryachko E, Wolfram P, Adley C, Leahy JJ, Novotny EH, Hayes MHB (2010) Biochar from biomass and waste. Waste Biomass Valor 1:177–189

Lee WC, Kuan WC (2015) *Miscanthus* as cellulosic biomass for bioethanol production. Biotechnol J 10:840–854

Lehmann J, Rillig MC, Thies J, Masiello CA, Hockaday WC, Crowley D (2011) Biochar effects on soil biota—a review. Soil Biol Biochem 43:1812–1836

Lewandowski I, Scurlock JMO, Lindvall E, Christou M (2003) The development and current status of perennial rhizomatous grasses as energy crops in the US and Europe. Biomass Bioenergy 25:335–361

Li HQ, Li CL, Sang T, Xu J (2013) Pre-treatment on *Miscanthus lutarioriparious* by liquid hot water for efficient ethanol production. Biotechnol Biofuels 6:76

Low T, Booth C, Sheppard A (2011) Weedy biofuels: what can be done? Curr Opin Environ Sustain 3:55–59

Luo H, Klein IM, Jiang Y, Zhu H, Liu B, Kenttämaa HI, Abu-Omar MM (2016) Total utilization of *miscanthus* biomass, lignin and carbohydrates. ACS Sustain Chem Eng 4:2316–2322

Luo Y, Dungait JA, Zhao X, Brookes PC, Durenkamp M, Li G, Lin Q (2017) Pyrolysis temperature during biochar production alters its subsequent utilisation by microorganisms in an acid arable soil. Land Degrad Dev 29:2183–2188

Madej J, Hilber I, Bucheli TD, Oleszczuk P (2016) Biochars with low polycyclic aromatic hydrocarbon concentrations achievable by pyrolysis under high carrier gas flows irrespective of oxygen content or feedstock. J Anal Appl Pyrolysis 122:365–369

Mayer P, Hilber I, Gouliarmou V, Hale SE, Cornelissen G, Bucheli TD (2016) How to determine the environmental exposure of PAHs originating from biochar. Environ Sci Technol 50:1941–1948

Melligan F, Auccaise R, Novotny EH, Leahy JJ, Hayes MHB, Kwapinski W (2011) Pressurised pyrolysis of *Miscanthus* using a fixed bed reactor. Bioresour Technol 102:3466–3470

Menon V, Rao M (2012) Trends in bioconversion of lignocellulose: biofuels platform chemicals & biorefinery concept. Prog Energy Combust Sci 38:522–550

Mohan D, Pittman CU, Steele PH (2006) Pyrolysis of wood/biomass for bio-oil: a critical review. Energy Fuel 20:848–889

Mohanty P, Nanda S, Pant KK, Naik S, Kozinski JA, Dalai AK (2013) Evaluation of the physiochemical development of biochars obtained from pyrolysis of wheat straw, timothy grass and pinewood: effects of heating rate. J Anal Appl Pyrolysis 104:485–493

Murnen HK, Balan V, Chundawat SPS, Bals B, da Costa Sousa L, Dale BE (2008) Optimization of ammonia fiber expansion (AFEX) pre-treatment and enzymatic hydrolysis of *Miscanthus x giganteus* to fermentable sugars. Biotechnol Prog 23:846–850

Nanda S, Mohanty P, Pant KK, Naik S, Kozinski JA, Dalai AK (2013) Characterization of north American lignocellulosic biomass and biochars in terms of their candidacy for alternate renewable fuels. Bioenergy Res 6:663–677

Nanda S, Azargohar R, Kozinski JA, Dalai AK (2014a) Characteristic studies on the pyrolysis products from hydrolyzed Canadian lignocellulosic feedstocks. Bioenergy Res 7:174–191

Nanda S, Dalai AK, Kozinski JA (2014b) Butanol and ethanol production from lignocellulosic feedstock: biomass pretreatment and bioconversion. Energ Sci Eng 2:138–148

Nanda S, Mohammad J, Reddy SN, Kozinski JA, Dalai AK (2014c) Pathways of lignocellulosic biomass conversion to renewable fuels. Biomass Conv Bioref 4:157–191

Nanda S, Mohanty P, Kozinski JA, Dalai AK (2014d) Physico-chemical properties of bio-oils from pyrolysis of lignocellulosic biomass with high and slow heating rate. Energy Environ Res 4:21–32

Nanda S, Azargohar R, Dalai AK, Kozinski JA (2015a) An assessment on the sustainability of lignocellulosic biomass for biorefining. Renew Sust Energ Rev 50:925–941

Nanda S, Maley J, Kozinski JA, Dalai AK (2015b) Physico-chemical evolution in lignocellulosic feedstocks during hydrothermal pretreatment and delignification. J Biobased Mater Bioenerg 9:295–308

Nanda S, Dalai AK, Berruti F, Kozinski JA (2016a) Biochar as an exceptional bioresource for energy, agronomy, carbon sequestration, activated carbon and specialty materials. Waste Biomass Valor 7:201–235

Nanda S, Dalai AK, Kozinski JA (2016b) Supercritical water gasification of timothy grass as an energy crop in the presence of alkali carbonate and hydroxide catalysts. Biomass Bioenergy 95:378–387

Nanda S, Kozinski JA, Dalai AK (2016c) Lignocellulosic biomass: a review of conversion technologies and fuel products. Curr Biochem Eng 3:24–36

Nanda S, Reddy SN, Mitra SK, Kozinski JA (2016d) The progressive routes for carbon capture and sequestration. Energ Sci Eng 4:99–122

Nanda S, Dalai AK, Kozinski JA (2017a) Butanol from renewable biomass: highlights on downstream processing and recovery techniques. In: Mondal P, Dalai AK (eds) Sustainable utilization of natural resources. CRC Press, Florida, USA, pp 187–211

Nanda S, Golemi-Kotra D, McDermott JC, Dalai AK, Gökalp I, Kozinski JA (2017b) Fermentative production of butanol: perspectives on synthetic biology. New Biotechnol 37:210–221

Nanda S, Gong M, Hunter HN, Dalai AK, Gökalp I, Kozinski JA (2017c) An assessment of pinecone gasification in subcritical, near-critical and supercritical water. Fuel Process Technol 168:84–96

Nanda S, Rana R, Zheng Y, Kozinski JA, Dalai AK (2017d) Insights on pathways for hydrogen generation from ethanol. Sustain Energ Fuel 1:1232–1245

Nanda S, Dalai AK, Pant KK, Gökalp I, Kozinski JA (2018) An appraisal on biochar functionality and utility in agronomy. In: Konur O (ed) Bioenergy and biofuels. CRC Press, Florida, USA, pp 389–409

Nanda S, Rana R, Hunter HN, Fang Z, Dalai AK, Kozinski JA (2019) Hydrothermal catalytic processing of waste cooking oil for hydrogen-rich syngas production. Chem Eng Sci 195:935–945

Niu Y, Tan H, Hui S (2016) Ash-related issues during biomass combustion: alkali-induced slagging, silicate melt-induced slagging (ash fusion), agglomeration, corrosion, ash utilization, and related countermeasures. Prog Energy Combust Sci 52:1–61

Nsanganwimana F, Pourrut B, Mench M, Douay F (2014) Suitability of *Miscanthus* species for managing inorganic and organic contaminated land and restoring ecosystem services. A review. J Environ Manag 143:123–134

Oginni O, Singh K, Zondlo JW (2017) Pyrolysis of dedicated bioenergy crops grown on reclaimed mine land in West Virginia. J Anal Appl Pyrolysis 123:319–329

Okolie JA, Nanda S, Dalai AK, Kozinski JA (2019a) Optimization and modeling of process parameters during hydrothermal gasification of biomass model compounds to generate hydrogen-rich gas products. Int J Hydrogen Energ. https://doi.org/10.1016/j.ijhydene.2019.05.132

Okolie JA, Rana R, Nanda S, Dalai AK, Kozinski JA (2019b) Supercritical water gasification of biomass: a state-of-the-art review of process parameters, reaction mechanisms and catalysis. Sustain Energ Fuel 3:578–598

Osman AI, Abdelkader A, Johnston CR, Morgan K, Rooney DW (2017) Thermal investigation and kinetic modeling of lignocellulosic biomass combustion for energy production and other applications. Ind Eng Chem Res 56:12119–12130

Pang J, Zheng M, Wang A, Sun R, Wang H, Jiang Y, Zhang T (2014) Catalytic conversion of concentrated *Miscanthus* in water for ethylene glycol production. AICHE J 60:2254–2262

Pham XP, Piriou C, Salvador S, Valette J, Van de Steene L (2018) Oxidative pyrolysis of pine wood, wheat straw and *Miscanthus* pellets in a fixed bed. Fuel Process Technol 178:226–235

Rana R, Nanda S, Kozinski JA, Dalai AK (2018) Investigating the applicability of Athabasca bitumen as a feedstock for hydrogen production through catalytic supercritical water gasification. J Environ Chem Eng 6:182–189

Rana R, Nanda S, Maclennan A, Hu Y, Kozinski JA, Dalai AK (2019) Comparative evaluation for catalytic gasification of petroleum coke and asphaltene in subcritical and supercritical water. J Energ Chem 31:107–118

Reddy SN, Nanda S, Dalai AK, Kozinski JA (2014) Supercritical water gasification of biomass for hydrogen production. Int J Hydrogen Energ 39:6912–6926

Reddy SN, Nanda S, Hegde UG, Hicks MC, Kozinski JA (2015) Ignition of hydrothermal flames. RSC Adv 5:36404–36422

Reddy SN, Nanda S, Kozinski JA (2016) Supercritical water gasification of glycerol and methanol mixtures as model waste residues from biodiesel refinery. Chem Eng Res Des 113:17–27

Reddy SN, Nanda S, Hegde UG, Hicks MC, Kozinski JA (2017) Ignition of n-propanol–air hydrothermal flames during supercritical water oxidation. Proc Combust Inst 36:2503–2511

Robbins MP, Evans G, Valentine J, Donnison IS, Allison GG (2012) New opportunities for the exploitation of energy crops by thermochemical conversion in northern Europe and the UK. Prog Energy Combust Sci 38:138–155

Sarangi PK, Nanda S (2018) Recent developments and challenges of acetone-butanol-ethanol fermentation. In: Sarangi PK, Nanda S, Mohanty P (eds) Recent advancements in biofuels and bioenergy utilization. Springer Nature, Singapore, pp 111–123

Sarangi PK, Nanda S (2019) Recent advances in consolidated bioprocessing for microbe-assisted biofuel production. In: Nanda S, Sarangi PK, Vo DVN (eds) Fuel processing and energy utilization. CRC Press, Florida, USA, pp 141–157

Shooshtarian S, Anderson JA, Armstrong GW, Luckert MKM (2018) Growing hybrid poplar in western Canada for use as a biofuel feedstock: a financial analysis of coppice and single-stem management. Biomass Bioenergy 113:45–54

Su Y, Song K, Zhang P, Su Y, Cheng J, Chen X (2017) Progress of microalgae biofuel's commercialization. Renew Sust Energ Rev 74:402–411

Trazzi PA, Leahy JJ, Hayes MHB, Kwapinski W (2016) Adsorption and desorption of phosphate on biochars. J Environ Chem Eng 4:37–46

van der Weijde T, Kiesel A, Iqbal Y, Muylle H, Dolstra O, Visser RGF, Lewandowski I, Trindade LM (2016) Evaluation of *Miscanthus sinensis* biomass quality as feedstock for conversion into different bioenergy products. GCB Bioenergy 9:176–190

Villaverde JJ, Ligero P, de Vega A (2010) *Miscanthus x giganteus* as a source of bio based products through organosolv fractionation: a mini review. Open Agric J 4:102–110

Wafiq A, Reichel D, Hanafy M (2016) Pressure influence on pyrolysis product properties of raw and torrefied *Miscanthus*: role of particle structure. Fuel 179:156–167

Xue G, Kwapinska M, Horvat A, Li Z, Dooley S, Kwapinski W, Leahy JJ (2014a) Gasification of *Miscanthus x giganteus* in an air-blown bubbling fluidized bed: a preliminary study of performance and agglomeration. Energy Fuel 28:1121–1131

Xue G, Kwapinska M, Kwapinski W, Czajka KM, Kennedy J, Leahy JJ (2014b) Impact of torrefaction on properties of *Miscanthus giganteus* relevant. Fuel 121:189–197

Yang X, Wang H, Strong PJ, Xu S, Liu S, Lu K, Sheng K, Guo J, Che L, He L, Ok YS, Yuan G, Shen Y, Chen X (2017) Thermal properties of biochars derived from waste biomass generated by agricultural and forestry sectors. Energies 10:469

Yi YB, Lee JW, Chung CH (2015) Conversion of plant materials into hydroxymethylfurfural using ionic liquids. Environ Chem Lett 13:173–190

Yorgun S, Şimşek YE (2008) Catalytic pyrolysis of *Miscanthus* × *giganteus* over activated alumina. Bioresour Technol 99:8095–8100

Yu G, Afzal W, Yang F, Padmanabhan S, Liu Z, Xie H, Shafy MA, Bell AT, Prausnitz JM (2013) Pre-treatment of *Miscanthus*×*giganteus* using aqueous ammonia with hydrogen peroxide to increase enzymatic hydrolysis to sugars. J Chem Technol Biotechnol 89:698–706

Yu TE, English BC, He L, Larson JA, Calcagno J, Fu JS, Wilson B (2016) Analyzing economic and environmental performance of switchgrass biofuel supply chains. Bioenergy Res 9:566–577

Zacher AH, Olarte MV, Santosa DM, Elliott DC, Jones SB (2014) A review and perspective of recent bio-oil hydrotreating research. Green Chem 16:491–515

Process Improvements and Techno-Economic Feasibility of Hydrothermal Liquefaction and Pyrolysis of Biomass for Biocrude Oil Production

10

Pravin G. Suryawanshi, Sutapa Das, Venu Babu Borugadda, Vaibhav V. Goud, and Ajay K. Dalai

Abstract

Biocrude oil production from biomass has gained huge attention globally to complement the conventional fuels and reduce the environmental impact caused by fossil fuels. To produce renewable energy from biomass, several technologies have emerged, such as physical (e.g., drying, pressing, crushing, and pelletization), biochemical (e.g., fermentation and anaerobic digestion), and thermochemical (e.g., pyrolysis, gasification, liquefaction, and combustion) pathways. Among all, thermochemical technologies have gained much attention due to their high-energy content products and process efficiency. The biocrude oil produced from pyrolysis and hydrothermal liquefaction has similar chemical properties to conventional liquid fuels. Therefore, this chapter discusses the current status, challenges, opportunities, recent process developments, and techno-economic feasibility of hydrothermal liquefaction and pyrolysis of biomass for biocrude oil production.

Keywords

Biocrude · Hydrothermal liquefaction · Pyrolysis · Lignocellulosic biomass · Techno-economic assessment

P. G. Suryawanshi · S. Das · V. V. Goud
Department of Chemical Engineering, Indian Institute of Technology Guwahati, Guwahati, Assam, India

V. B. Borugadda (✉) · A. K. Dalai
Department of Chemical and Biological Engineering, University of Saskatchewan, Saskatoon, Saskatchewan, Canada
e-mail: vbb123@usask.ca

S. Nanda et al. (eds.), *Biorefinery of Alternative Resources: Targeting Green Fuels and Platform Chemicals*, https://doi.org/10.1007/978-981-15-1804-1_10

10.1 Introduction

Urbanization, population growth, and technology development are causing ever-increasing energy demands. However, conventional fuels are finite, requiring potential replacement in the future (Chen et al. 2015). Thus, researchers all around the world are working toward developing sustainable technologies to derive energy from renewable sources, to minimize energy dependence on depleting fuels. Currently, 10% of the global energy supply comes from biomass-derived energy sources, and biomass has much higher potential to complement fossil fuels if completely utilized (Khan et al. 2009). According to the Union for the Promotion of Oil and Protein Plants (UFOP) global supply report, if all the available biomass can be completely converted to energy, it would yield as high as 1,081,011 tons of oil equivalent (toe). This figure is huge and is nearly ten times the world's present energy needs (Hussain et al. 2017). The main driving force behind the utilization of biomass is its abundant reserves, renewability, and most importantly CO_2 neutrality (Kumar et al. 2010; Nanda et al. 2015a).

Until date, the U.S. and Brazil have commercialized biocrude oil and its derivatives. Canada has also about 190 establishments that are solely involved in producing bio-oil and other value-added products obtained from biomass. Recently, Germany has also invested in this area with an emphasis on biodiesel and biogas. In 2016, globally, 34 million tons of biodiesel was produced, of which 37% (12,610 million tons) was solely produced by EU-28 member countries (Babu and Subramanian 2013). Apart from the U.S. and the European Union, many other countries have begun to contribute toward producing biocrude oil and its derivatives

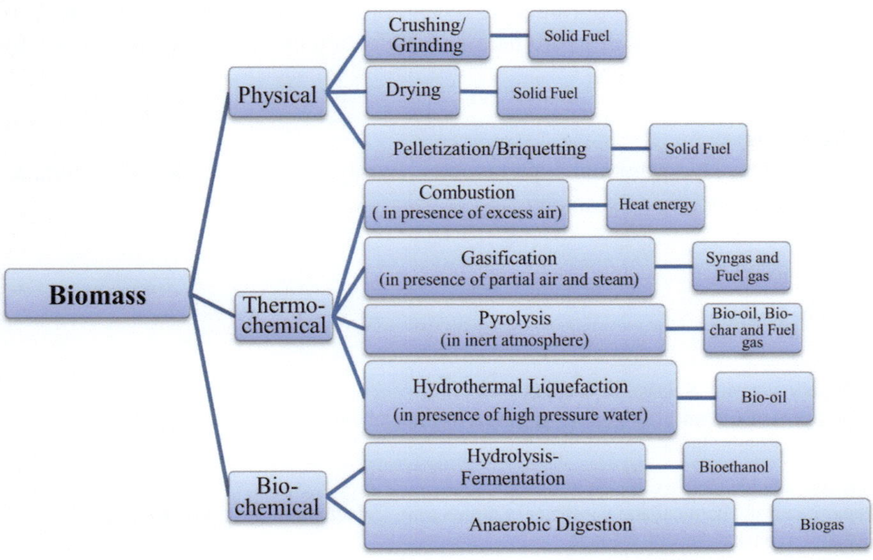

Fig. 10.1 Pathways for the conversion of biomass into bioenergy. (Reproduced with permission from Sharma et al. 2015)

from biomass (Brownsort 2009). The main advantage of using a green raw material is that it has less environmental impact as it produces much less CO_2, SO_x, and NO_x emissions as compared to petroleum oil or coal when gasified, pyrolyzed, or hydrothermally liquefied (Nanda et al. 2016a).

A variety of biomass available for energy purposes includes oil seeds, starch-based feedstocks, lignocellulosic biomass (agricultural crop residues and woody biomass), aquatic biomass, municipal solid waste, and industrial wastes (Nanda et al. 2018a). Lignocellulosic biomass (including herbaceous energy crops, agricultural crop residues, forestry residues, and wood-processing wastes) is one of the most abundant feedstock on earth, which is highly promoted for the sustainable production of fuels and chemicals. At present, there are several biochemical and thermochemical technologies (Fig. 10.1), which can be used to produce biofuels from biomass, such as fermentation, anaerobic digestion, gasification, pyrolysis, and hydrothermal liquefaction. Among these techniques, pyrolysis and hydrothermal liquefaction are found to be the most promising for biocrude oil production and upgradation into liquid transportation fuels (biogasoline, biodiesel, and biojet fuel). Biocrude oil or biodiesel can also be blended with conventional diesel for use in standard diesel engines. The major advantage is that biodiesel has a relatively lower environmental impact compared to petroleum diesel and it is biodegradable without compromising its combustion properties (Babu and Subramanian 2013). It can also be used as a low-carbon alternative to heating oil.

The scope of the chapter is to demonstrate the current state-of-the-art research undertaken in pyrolysis and hydrothermal liquefaction of biomass for the production of liquid transportation fuels and value-added chemicals. Furthermore, the current technical challenges in this field are reviewed for process development. New strategies and their techno-economic feasibility toward commercialization of biofuel products are discussed based on the recent developments in the field.

10.2 Biomass Pyrolysis

Pyrolysis is a thermal decomposition process of dry biomass at elevated temperatures under inert atmosphere. Based on operating process conditions, the pyrolysis process can divided into fast, slow, flash, and hydropyrolysis techniques. Table 10.1

Table 10.1 Process parameters of different pyrolysis processes and their major products

Pyrolysis	Temperature (°C)	Heating rate (°C/s)	Residence time	Major products
Slow pyrolysis	400–600	0.1–0.3	30–60 min	Biochar, biocrude oil, and gas
Fast pyrolysis	700–1200	10–100	<10 s	Biocrude oil
Flash pyrolysis	800–1150	>1000	1 s	Biocrude oil
Hydropyrolysis	400–700	0.2–10	15–30 min	Biocrude oil

References: Manyà (2012), Brownsort (2009)

shows the detailed process conditions and main products in each pyrolysis process. During pyrolysis, the yield of the products depends not only on the type of pyrolysis but also on the nature of biomass feedstock and operating conditions, i.e., temperature, heating rate, pressure, and vapor residence time (Azargohar et al. 2013; Nanda et al. 2014a).

Lignocellulosic biomass is a promising feedstock for biorefineries, where the spectrum of products include biofuels, polysaccharides, fatty acids, proteins, pigments, aromatics, biopolymers, and biomaterials (Fig. 10.2). Lignocellulosic biomass typically consists of cellulose (25–50 wt%), hemicellulose (15–40 wt%), lignin (10–40 wt%), and extractives (up to 15 wt%) with trace amounts of inorganic and mineral matter (Guo et al. 2010). Table 10.2 shows the distribution of cellulose, hemicellulose, and lignin of some chief biomass.

Cellulose is a linear polymer composed of glucose monomers connected by ether bonds, hydrogen bonds, glycosidic linkages, and van der Waals forces (Nanda et al. 2015b). Hemicellulose is a heteropolymer formed from predominantly xylose and mannose monomers. On the other hand, lignin has a complex structure of aromatic polymer formed from phenyl propanol monomers (Fougere et al. 2016; Cao et al. 2018). Cellulose and hemicellulose are utilized effectively for the production of fuels, sugars, and paper. Efficient utilization of lignin is challenging due to its high molecular weight, complex three-dimensional polymeric structure formed from condensation and ether linkages, and heterogeneous nature (Mahmood et al. 2016). Therefore, lignin remains as an underutilized biopolymer (estimated 1.5–1.8 billion tons per year) obtained from many industrial activities (Cao et al. 2017, 2018). Many platform chemicals, such as furfural, guaiacol, and catechol, have been identified from lignocellulosic materials, opening many opportunities to explore these key substances (Kruse et al. 2013; Schuler et al. 2019). Despite extensive research,

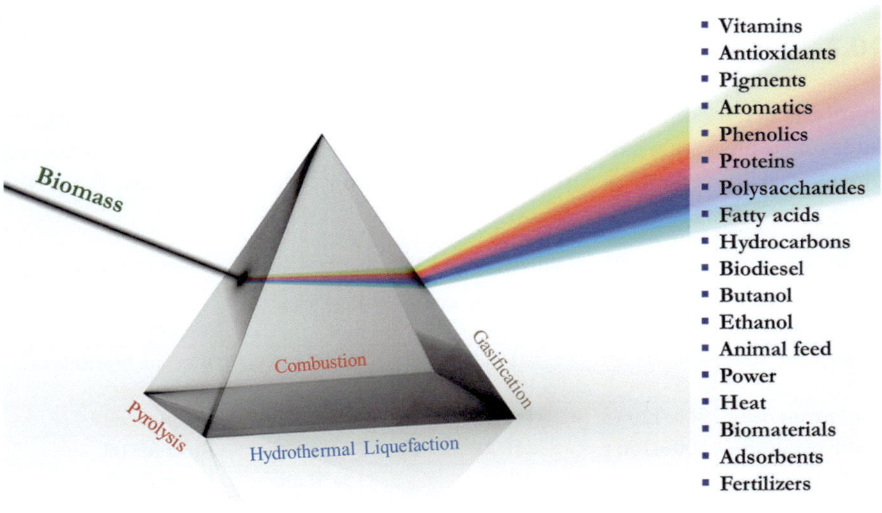

Fig. 10.2 Spectrum of valuable products from the thermochemical processing of biomass

Table 10.2 Distribution of the components of biomass (Reproduced with permission from Wang et al. 2017)

Biomass	Cellulose (wt%)	Hemicellulose (wt%)	Lignin (wt%)	Extractives (wt%)	Biochar (wt %)
Rice straw	37	16	13	13	20
Corn straw	42	23	17	10	7
Wheat straw	37	18	20	4	4
Pine	49	20	25	5	0.3
Bamboo	40	20	20	7	1

the depolymerization of lignin and effective recovery of aromatic monomers from lignin still have not been possible to reach a pilot scale.

Biocrude oil is mainly obtained by thermal degradation of cellulose and hemicellulose, while lignin contributes to biochar production. The extractives present are mainly waxes, fatty acids, and sterols, which can be extracted using various solvents like methanol, ethanol, hexane, water, and supercritical fluids (Nanda et al. 2013). The mass ratio of the organic to inorganic content varies with the biomass, and it greatly affects the pyrolysis process conditions. The pyrolysis of each component has its respective pathway, which affects its product distribution (Fahmi et al. 2007).

Vast research has been carried out on slow, fast, and flash pyrolysis of different biomass. Currently, there is another budding type of pyrolysis, called hydropyrolysis, in which nitrogen is replaced by hydrogen during pyrolysis. In this case, the reaction takes place under hydrogen pressure, which hinders the formation of free radicals; i.e., the formation of unsaturated hydrocarbons is reduced, thereby upgrading the biocrude oil. A study reported by Balagurumurthy et al. (2015) with rice straw as the biomass under both nitrogen and hydrogen environment revealed that biocrude oil obtained under hydrogen pressure was more selective toward the phenolic compounds, but the higher biocrude oil yield is obtained under the conventional nitrogen environment. A study on the pyrolysis of wheat straw and corn stalk suggested that extractives increase the yield of biocrude oil and in turn suppress the yield of biochar and gas (Wang et al. 2017). Another study on pyrolysis of Mongolian pine, confirmed that the lowering of activation energy could increase the acidic products in biocrude oil (Guo et al. 2010). The trace amounts of minerals also affect the pyrolysis process as they act as catalysts to enhance the process.

10.2.1 Biocrude Oil Production Through Pyrolysis

Biocrude oil is produced by fast pyrolysis technique, which involves the rapid breakdown and fragmentation of biomass composed of cellulose, hemicellulose, and lignin. Sharma et al. (2015) reported the detailed mechanism of cellulose,

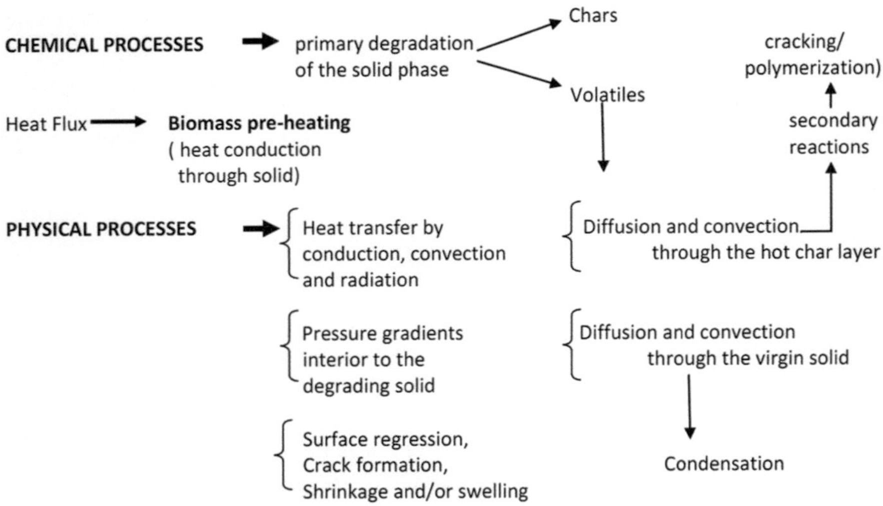

Fig. 10.3 Schematic representation of the physicochemical processes in lignocellulosic biomass during pyrolysis. (Reproduced with permission from Sharma et al. 2015)

hemicellulose, and lignin decomposition during pyrolysis. Figure 10.3 shows the physicochemical changes in biomass during pyrolysis. The obtained products from pyrolysis can be subsequently separated from a gas stream (by-product) by condensation and immediate cooling (Czernik and Bridgwater 2004; Mohan et al. 2006). Since the feedstock composition plays a major role in determining the quality of biocrude oil, the properties are much closely associated with its source. The presence of oxygen and water in biocrude oil is one of the major disadvantages for its immediate application.

Oxygen combines with different carbon components to form a variety of functional groups such as ketones, aldehydes, phenols, sugars, and other aromatic compounds in bio-oil (Bridgwater 2012). Severe drawbacks arise with the high oxygen content in biocrude oil, which not only causes compatibility issues with the fossil fuels but also leads to a massive reduction in energy density (almost to an extent of 50 wt%). To prevent the formation of higher amounts of oxygen content, process parameters such as heating rate and vapor residence time can be altered (Wang et al. 2017). Biocrude oil can be considered as a potential source to complement conventional liquid transportation fuels, which have the capability to match the economical aspect of petroleum derivatives (Bridgwater et al. 1999; Isahak et al. 2012). Table 10.3 represents the comparative differences between biocrude oil and conventional crude oil.

10.2.2 Biochar Production Through Pyrolysis

Biochar is a carbonaceous residue of the biomass decomposition process whose constituents vary considerably depending on the feedstock composition and process parameters employed during pyrolysis, gasification, and carbonization (Mohanty et al. 2013; Azargohar et al. 2014; Nanda et al. 2016a). For instance, the high heating value (HHV) can vary from 20 to 36 MJ/kg, while the carbon content can vary from 53 to 96 wt% (Brownsort 2009). The process conditions primarily determine the different physical, chemical, and mechanical attributes of biochar. Slow pyrolysis not only changes the chemical composition of biochar by varying the C/H and O/H ratios but also significantly contribute toward its physical characteristics by making it highly porous (Azargohar et al. 2019). Volatile removal at these high temperatures equip them with honeycomb-like structures with a high surface area (100–500 m^2/g), which is ideal for pollutant adsorption and filtration applications (Ahmad et al. 2014).

Biochar produced from conventional carbonization or slow pyrolysis techniques has high carbon content (low bio-oil content), which can be utilized mainly for soil supplements in enhancing soil fertility and carbon sequestration (Lee et al. 2013). Another facet of biochar is its high HHV, which endows it with the potential of being a high-energy supplement and possibly replacing coal in certain applications (Worasuwannarak et al. 2007). The biochar yield greatly depends on the pyrolysis temperature and the nature of the biomass, which is shown in Table 10.4.

Table 10.3 Comparison of chemical composition and properties of biocrude oil and conventional crude oil (Reproduced with permission from Mohan et al. 2006)

Properties	Biocrude oil	Crude oil
Water content (wt%)	15–30	0.1
pH	2–3	Neutral
Carbon (wt%)	54–58	85
Hydrogen (wt%)	5.5–7	11
Oxygen (wt%)	35–40	1
Nitrogen (wt%)	0–0.2	0.3
Sulfur (wt%)	Traces	0.5–3
Ash (wt%)	0–0.2	0.1
Higher heating value (MJ/kg)	16–19	40
Specific gravity	1.2	0.94
Viscosity at 50 °C (cP)	40–100	180
Pour point (°C)	−33	−18

Table 10.4 Effects of temperature on the biochar yield

Biomass	Temperature (°C)	Biochar yield (wt%)	References
Rice husk	400–600	25.5–33	Williams and Nugranad (2000)
Corncob	450–1250	5.7–30.6	Demirbas (2004)
Olive husk	450–1250	19.4–44.5	Demirbas (2004)
Pine cone	300–450	26–58	Shabangu et al. (2014)
Sewage sludge	350–950	39–52	Sánchez et al. (2009)

10.2.3 Gaseous Products from Biomass Pyrolysis

The pyrolysis of biomass also results in the formation of certain gaseous products, such as hydrogen (H_2), ammonia (NH_3), methane (CH_4), propane (C_3H_7), ethylene (C_2H_2), carbon monoxide (CO), and carbon dioxide (CO_2). Hydrogen is generally released upon the decomposition and reforming of the aliphatic and aromatic groups. Higher moisture content in the biomass can contribute toward higher H_2 content compared to dried feedstocks (Guoxin et al. 2009; Uddin et al. 2014; Nanda et al. 2017a). CH_4 and CO_2 generally originate after the breakage of methylene and methoxy, bonds while CO and CO_2 are produced from the decomposition of carboxyl and carbonyl groups (Strezov et al. 2008).

A variation in process parameters such as initial moisture content, feedstock particle size, and reaction temperature can significantly affect the gas composition. High moisture-containing feedstocks generally release higher H_2 content in gases. In addition, such feedstocks usually result in a higher extraction of water-soluble components in gases, thereby resulting in reduced gas yields (Dasappa et al. 2004). Furthermore, higher reaction temperatures lead to higher rates of thermal degradation, causing increased volatile products. The gaseous products undergo various reactions such as decarbonylation and dehydrogenation at these conditions (He et al. 2010).

Feedstock size also plays a vital role in determining the overall gas composition as smaller size particles favor the cracking of hydrocarbons, thereby producing more H_2 due to enhanced heating rates. Larger size particles led to lower heating rates, which result in increased biochar yield and decreased gas production because of the low residence time of volatiles in the reactor (Hu and Gholizadeh 2019). Gaseous products from pyrolysis find applications in combustion engines for transportation purposes and in the production of liquid biofuel production through the Fischer-Tropsch (FT) process (Kan et al. 2016).

10.2.4 Challenges in Biomass Pyrolysis

Though pyrolysis processes can be applied to a range of feedstocks, the products obtained are dependent on the composition of the feedstock and are inconsistent in nature. The quality of the feedstock can also be varied by the harvesting culture, collection techniques, and type of pretreatment (mechanical, chemical, or biological). Therefore, the establishment of some standard protocol in this direction can help eradicate the issue of inconsistency. The scale-up to pilot plants from laboratory scale research becomes difficult. Another aspect of biomass, which demands intensive research, is ash content characterization. Elements such as Ca and K, present in ashes, have been reported to facilitate biomass decomposition while assisting in char formation.

Gu et al. (2013) reported that elements Na, K, and Va present in the ash were found to be responsible for the high-temperature corrosion of the reactor. On the other hand, Ca is responsible for hard deposit formation. Such studies pointed toward the harmful effects of ash content, especially when its content is more than

0.1 wt% in biocrude oil. Particular attention has to be diverted toward the ash content in the entrained char as it has been known to catalyze the polymerization reactions in biocrude oil, thereby enhancing its viscosity. In practical applications, negative effects of ash content in the biocrude oil can be directly related to the occurrence of erosion, corrosion, and knocking problems in engines and valves.

The heterogeneous nature of the decomposed biomass products makes the pyrolysis process harder to control and the different constituents of biomass breakdown under reaction conditions. These distinct reaction rates further complicate the pyrolysis process and the selection of reactor design, which determines their thermal processing conditions. Unwanted moisture in the biocrude oil is usually very detrimental to biocrude oil properties as it lowers the heating value and flame temperature. This generally leads to delayed ignition times and subsequently affects the combustion rates in a negative way. Moisture content not only creates filtration problems but also causes heterogeneity in biocrude oil, which can lead to layering or partial separation of different phases. Due to the presence of reactive oxygenated functional groups in biocrude oil, its thermal stability is adversely affected, and this creates problems in its storage.

Another factor, which contributes to slow combustion rates and the sudden appearance of "sparklers," is the existence of suspended char. The abovementioned problems mostly lead to large potential deposits or high levels of CO emissions. Char is also well known for its high catalytic activity and usually assists in the potential cracking of bio-oil components while reducing the condensables' yield up to 20%.

10.2.5 Recent Technological Developments in Biomass Pyrolysis

Although the pyrolysis technique has been extensively studied, it is still behind large-scale commercialization owing to several bottlenecks. Techniques for pyrolysis still demands strict control over the process parameters for the production of conventional products (biochar and biocrude oil). Moreover, the actual relationship between biomass products and pyrolysis conditions remains to be unraveled for achieving a better understanding of the process.

The lack of comprehensive molecular simulations creates difficulty in understanding the products' distribution and the effect of the process parameters on the kinetic measurements (Editors et al. 2016). These are some of the different problems that not only plague the actual scaling up of the process but also harms the overall outlook of this technology. In order to address and tackle these issues, a comprehensive understanding of pyrolysis at different lengths and time scales is required. On a macro-scale, implementing the correct conditions for maximal production demands a better suited reactor design. Moreover, to make the technology commercially competitive, enhancement in the pyrolytic products is required. The by-products themselves need to be upgraded to not only enhance the overall value of the system but also devise ways to construct a circular economy (Jiang et al. 2017). A few of the measures intended to alleviate the problems associated have been discussed in the following section.

10.2.5.1 Reactor Design

To make the pyrolysis process more efficient, heat transfer characteristics of the biomass have to be improved as it possesses poor thermal conductivity. Initial size reduction or comminution of the biomass can alleviate this shortcoming. A few reactors, which have been employed for carrying out the pyrolysis of biomass, include tube furnace and fluidized bed reactor. Although these are the conventional methods utilized, they too suffer from many drawbacks, such as long operational hours, resulting in low-quality products due to the appearance of secondary reactions.

In addition, a large amount of heat is required in pyrolysis to carry out the reactions (Palgan and McCormick 2016). Microwave heating is an attractive alternative to reducing heat losses. Microwave heating delivers enhanced, faster, and uniform heating rates. This technique also improves the quality of the downstream products by keeping the secondary reactions under control (Tsagkari et al. 2016).

10.2.5.2 Biocrude Oil Upgrading Techniques

The techno-economical facet of the pyrolysis process provides the main impetus behind the continual search for an improved process. Over the years, many alternate techniques for upgrading the characteristics of biocrude oil have been proposed, such as catalytic cracking, catalytic pyrolysis, molecular distillation, and esterification, to name a few. In particular, biocrude oil upgradation by employing catalytic cracking involves several steps such as cracking, decarboxylation, hydrodeoxygenation, etc. One of the favored techniques utilized is esterification, which consists of the addition of certain polar solvents (ethanol and furfural) into biocrude oil to homogenize the mixture. This eventually decreases the biocrude oil's density, viscosity, and corrosiveness, thus improving its stability characteristics. Moreover, the addition of solvents increased the organic content of the upgraded bio-oil (Peng et al. 2009). Simpler alternatives for improving biocrude oil properties include the addition of diesel fuels and forming an emulsion. This emulsification process not only is cost-effective but also provides for a short-term route to the deployment of biocrude oil in the diesel engines. Additionally, this process results in the improvement of its properties such as water content, flow properties, and viscosity (Cao et al. 2018).

The drive for discovering better alternatives in bio-oil upgradation has led researchers to employ supercritical fluids (Reddy et al. 2018). Supercritical fluid is a state of fluid that exists beyond its critical temperature and pressure. At this condition, the fluid behaves in an interesting way, wherein it can easily pass through surrounding materials and even diffuse through solids. In particular, the fluids possess the ability to dissolve and liquefy the materials which are not soluble either in a solid or liquid state, thereby enhancing their liquefaction reactions (Zhang et al. 2016). The process involving the utilization of supercritical fluids demands lower operating temperatures and is relatively ecofriendly, but it is more expensive than conventional processes (Nanda et al. 2018b). In order to overcome this shortcoming and make this process economically viable on a large scale, the utilization of certain low-cost chemicals such as methanol and glycerol has gained popularity (Sandun et al. 2006; Reddy et al. 2016).

10.2.5.3 Extraction of Value-Added Chemicals

There are various upgradation techniques available for biocrude oil to enhance its calorific value. However, these processes have become difficult to use due to the high tendencies of biocrude oil to get polymerized to form coke at elevated temperatures. Therefore, a new gateway to extract other value-added products from biocrude oil have come into existence. Sugar or its derivatives present in biocrude oil can be separated from aromatics, as they have distinct polarities, by using water/chloroform extraction. It is a simple extraction procedure where the biocrude oil is mixed with water to form two separate phases. The sugar derivatives are present in the lighter phase, and the aromatics and other compounds go into the heavier phase. Furthermore, by performing acid catalysis of the lighter phase, various chemicals like levulinic acid, furfural, acetic acid, 5-hydroxyfurfural, etc. can be obtained. The heavier phase is further hydrotreated to obtain transportation fuels (Hu et al. 2012).

In a study on the pyrolysis of olive mill wastes, the biocrude oil produced have two distinct phases, i.e., an aqueous phase and an organic phase or the tarry phase. The aqueous phase consisted of monosaccharides, phenolic derivatives, and acetic acid, while the tarry phase consisted of methyl esters. Two liquid–liquid extraction techniques were used mainly, first one at original pH and other with acid-base extraction at pH 12. At the original pH, the nonaqueous phase was dissolved in methanol first before extraction. The results showed that the best method for extraction was the acid-base procedure. Phenolic derivatives were found to be present in both the aqueous phase (hexane) and nonaqueous phase (ethyl acetate phase). The nonaqueous phase (hexane phase) consisted of methyl esters, which could be further used for the production of biodiesel (del Pozo et al. 2018).

10.2.5.4 Techno-Economic Feasibility of Biomass Pyrolysis

To demand widespread usage and compete with the existing conventional technologies, biomass degradation via pyrolysis needs to be revamped in different sectors. The need for developing a deeper understanding of the process can not only help improve the overall processing but also play a vital role in enhancing the economic outlook in terms of the development of different pyrolytic products. The commercial aspect of the pyrolysis process can be quantified when it is employed at an industrial-scale production capacity. To effectively measure its economic potential, the total annual biomass production capacity has to be taken into account. Furthermore, the energy density of conventional fossil fuels is considerably higher than that of biomass. For biofuels to achieve grid parity with fossil fuels, the costs associated with feedstock supply, processing, and postprocessing becomes significant. Its direct implication can be observed on the scale of operations where biomass is being utilized. The average production in a typical oil refinery in the U.S. is 18,000 metric tons per day (MTPD), which is considerably more than the average biomass production plant (1800 dry metric tons per day, DMT) (Hu et al. 2012; Zhang et al. 2016; del Pozo et al. 2018).

There are many factors that can contribute toward the determination of the economic feasibility of pyrolysis, such as reactor configuration, catalyst selection, applied process parameters, rate of pyrolysis, feedstock source, and its associated logistics. Despite several challenges, pyrolysis retains the position of the most promising candidate for an economical thermochemical route of biomass conversion. With certain measures like effective upgradation of product yields, the integration of value-added by-products (biochemicals) could increase the overall economic potential of the refinery. Moreover, utilization of the existing infrastructure of petrochemical refineries for biorefining with minimal reconstruction could result in a positive ecological footprint. Thus, the implementation of the techno-economic analysis method for detailed cost analysis over the operational lifetime can go a long way in successfully implementing the pyrolysis technology.

10.3 Hydrothermal Liquefaction

Hydrothermal liquefaction (HTL) is an efficient thermochemical process in which hydrolysis and pyrolysis are carried out simultaneously (also termed as hydrous pyrolysis). Under these conditions, macromolecules from biomass are converted to produce low molecular weight hydrocarbons. HTL is a versatile process in terms of feedstock quality, which converts high moisture containing (wet biomass) feedstock into biocrude oil, biochar, gases, and aqueous products. This process is advantageous for the utilization of environmentally benign solvents with easily adjustable properties and also contributes to decreasing the greenhouse gas emissions (Arturi et al. 2019).

Thermochemical processes such as pyrolysis, gasification, and combustion use dry feedstock, requiring time, labor, and energy for drying the feedstocks for the removal of moisture and volatile components. However, HTL can process the wet feedstocks where water serves as a reaction medium and catalyst. Therefore, HTL is considered as an eco-friendly, economical, and promising process compared to other thermochemical processes. This chapter emphasizes the improvement of biocrude oil yield and quality by reducing carbon distribution in the aqueous, gaseous, and solid residue phases.

The yield of different products varies substantially with feedstock type, reaction conditions, and reactor configurations (Dimitriadis and Bezergianni 2017). The most sensitive operating parameters to be maintained carefully during hydrothermal liquefaction of biomass are the type of biomass, reaction time (5–120 min), biomass loading (2–30 wt%), reaction temperature (250–370 °C), pressure (5–25 MPa), heating rate, cosolvents, and catalysts (Xue et al. 2016). The effects of these parameters on product formation have been discussed in the literature (Xue et al. 2016; Sandquist et al. 2019). The hydrothermal processing of biomass feedstock for the bulk production of biocrude has been demonstrated in the small batch reactor, as well as in the continuous-flow process, whereas continuous-flow reactors are advantageous for upscaling.

In HTL, the depolymerization of biomass feedstocks occurs in fuels and valuable chemicals in the near-critical or supercritical water (374 °C and 22.1 MPa) medium. Under high pressure and temperature near subcritical conditions, water dissociates to form ionic products, which may catalyze hydrolysis or depolymerization reactions, whereas at supercritical conditions, the free-radical reactions promote degradation (Nanda et al. 2015c, 2016b, 2017b; Song et al. 2017; Cao et al. 2017). By increasing the temperature and pressure of water from normal temperature and pressure conditions to supercritical conditions, the hydrogen bonding is weakened. Therefore, the dielectric constant (relative permittivity) of water decreases from 80.4 F/m to 14.07 F/m. This results in the decreased polarity of water whose value is comparable to organic solvents such as alcohols (Kumar et al. 2010; Pearce et al. 2018; Okolie et al. 2019b). Thus, the hydrophobic organic molecules (called biocrude oil) are solubilized in supercritical water. Subcritical water and supercritical water have been employed for a variety of hydrothermal reactions, such as gasification (Reddy et al. 2016; Rana et al. 2018, 2019; Okolie et al. 2019a), liquefaction (Alper et al. 2019), oxidation (Reddy et al. 2017, 2019), incineration (Reddy et al. 2015), and carbonization (Kang et al. 2019).

In HTL process, the major constituents of lignocellulosic biomass, i.e., cellulose, hemicellulose, and lignin, are converted into liquid biocrude oil, along with aqueous, gaseous, and solid products through a series of mechanisms including hydrolysis, C–C bond cleavage, alkylation, isomerization, deoxygenation, reforming, depolymerization, and repolymerization reactions (Dimitriadis and Bezergianni 2017; Miliotti et al. 2019). The main oily product of this process termed as biocrude oil is mainly composed of catechol, phenols, and methoxyphenols, which can be further upgraded for the production of synthetic transportation fuels and valuable chemicals (Miliotti et al. 2019). Lignin is the most stable compound in lignocellulosic biomass and is difficult to liquefy through HTL, whereas cellulose and hemicellulose have a simpler structure and are less branched than lignin with high reactivity during HTL. HTL is used as an intermediate reaction to produce bio-oil, phenolic compounds, organic acids, furfurals, and cyclopentenes from these biomass constituents.

10.3.1 Biocrude Oil Production Through Hydrothermal Liquefaction

The biocrude oil obtained from HTL is a dark brown and viscous liquid constituting 18–67% of the total weight of the feedstock. The quality and yield of biocrude oil vary with the type of biomass, operating conditions, and type of catalyst or cosolvent used. The HTL biocrude oil contains a large fraction of phenolic compounds where the fraction of polar compounds such as acids and sugars are less. Typically, the energy content is about 30–36 MJ/kg, and the elemental composition is as follows: 64–73 wt% carbon, 8–10 wt% hydrogen, 10–25 wt% oxygen, and 3–5 wt% nitrogen (Jiang et al. 2018). Biocrude oil can be further upgraded to produce commercial grade liquid transportation fuels using catalytic processes such as catalytic cracking, hydrodeoxygenation, desulfurization, and denitrogenation.

After the liquefaction process, product separation is a crucial stage for the recovery of biocrude oil from the aqueous by-products and solid residues, which can be accomplished by using any of the existing technologies such as solvent extraction, centrifugation, filtration, sedimentation, etc. (Funkenbusch et al. 2019; Miliotti et al. 2019).

Where gravimetric separation of biocrude is not possible, a solvent extraction process is efficiently used. Furthermore, biocrude oil upgradation is required to improve the quality and flow properties similar to the properties of commercially used petroleum hydrocarbon fuels.

Hydrotreating (HDT) is a process for the upgradation of biocrude oil by hydrodeoxygenation, which is carried out by treating biocrude oil with hydrogen gas at a temperature range of 330–390 °C and 1.5–3 MPa pressure (Patil and Vaidya 2017, 2018). The dewatering process is necessary prior to the hydrotreating for better purity and less hydrogen consumption (Elliott et al. 2015; Funkenbusch et al. 2019). Aqueous phase by-products can be recycled or mixed with other solvents to extract the value-added chemicals or used effectively for the cultivation of biomass. The aqueous phase obtained from HTL is considered as a useful by-product, which constitutes 20–50% of the total weight of feedstock, depending on the operating conditions and type of biomass. Major chemicals extracted in the aqueous phase contain organic acids, alcohols, ketones, and phenolic compounds.

10.3.2 Gaseous Products from Hydrothermal Liquefaction

The gaseous by-products constitute 5–10% of the total weight of feedstock. CO_2 is the main gaseous product of HTL, followed by H_2, CO, and CH_4 in small fractions (Magdeldin et al. 2018). Similar to hydrothermal liquefaction, hydrothermal gasification is another hydrothermal process, which is aimed to produce gases by treating biomass in subcritical or supercritical water medium under high pressure and high temperature (Gong et al. 2017a, b; Nanda et al. 2018c). Hydrothermal gasification has been proven advantageous for achieving higher thermal efficiency from wet biomass, producing hydrogen-rich gas with low CO, and reducing char formation and negligible levels of heteroatoms like sulfur, nitrogen, and halogens, which are solubilized in aqueous by-products (Reddy et al. 2014).

10.3.3 Solid Residue from Hydrothermal Liquefaction

Biochar is a solid residue from the hydrothermal and thermochemical processing of biomass, which contains high fractions of aromatic carbon (Nanda et al. 2014b). Generally, char from hydrothermal liquefaction is used as a potential source of nutrient for soil amendment. Similar to hydrothermal liquefaction, hydrothermal torrefaction, carbonization, and gasification are other processes in which solid product is targeted for value-added applications. The hydrochar produced from hydrothermal processing has different physicochemical properties than pyrolysis-derived biochar (Azargohar et al. 2019). Furthermore, both hydrochar and biochar have been used for soil amendment, soil fertility, water retention, capacitor application, low-cost adsorbent, heavy metal removal, phosphate removal, and medical applications (Ruan et al. 2014; Nanda et al. 2016c; Fang et al. 2018).

10.3.4 Catalytic Hydrothermal Liquefaction

Catalytic HTL encounters one of the major challenges related to the deactivation and stability of heterogeneous catalysts under high temperature and pressure conditions. Most oxide catalyst support materials undergo a phase transformation and dissolution in the presence of water mainly due to the hydrolysis of metal oxides with adjacent hydrophilic sites, e.g., Si–O–Si and Al–O–Al bonds, thus leading to the loss of pores and surface area (Murphy and Xu 2018). In a study by Sudarsanam et al. (2019), HTL at 200 °C for 10 h using a mesoporous silica catalyst resulted in about 90% loss in the surface area due to the hydrolysis of the silica. The factors affecting catalyst support stability under hydrothermal conditions are the severity of hydrothermal process conditions, pH of the solution, and hydroxyl ions (Xiong et al. 2014; Rinaldi et al. 2016).

The recovery of homogeneous catalysts from the aqueous phase and the degradation of active sites of heterogeneous catalysts raise concerns of high cost and environmental problems. The choice between homogeneous and heterogeneous catalysts is crucial because both types of catalysts have certain advantages and disadvantages. Heterogeneous catalysts (like Fe, Co, and Ni) increase biocrude oil yield and decrease its O, N, and S contents. On the other hand, active sites of homogeneous catalysts are uniformly distributed and remain available until the completion of the process (Chen et al. 2019).

Understanding of the chemical interactions between water, cosolvents, and catalysts is challenging to improve product yields, the stability of catalysts, the efficiency of separation, and the possibility of process integration. The development of a homogeneous catalyst for denitrogenation of biocrude oil is still a research gap (Chen et al. 2019). There is an opportunity for developing multifunctional catalysts targeting green engineering principles to have higher efficiency, long shelf life, reduced reaction time, as well as lower temperature and pressure requirements, which could overall reduce energy and time consumption. More research is expected to be undertaken to investigate the activity and selectivity of the catalyst to mitigate the challenges due to several technical and economic barriers in the commercial production of biofuels and biochemicals (Murphy and Xu 2018).

10.3.5 Challenges and Opportunities of Hydrothermal Liquefaction

Various methodologies have been developed for the conversion of biomass to biofuels and value-added chemicals, but the cost of drying, biomass pretreatment, bio-oil extraction, and upgradation is still high. Therefore, some thermochemical methods aim for the direct conversion of wet biomass to biocrude oil, along with in situ upgrading to biofuels and biochemicals. HTL is considered as a promising technology to advance the applications of biofuels and value-added chemical intermediates. Besides many possibilities, there are some challenges that need to be addressed for the commercial uses of HTL.

10.3.5.1 Biocrude Oil Production

In the last decade, more research has been undertaken to increase biocrude oil yield by reducing the organic content dissolved in the aqueous phase (Chen et al. 2019). Higher yields of biocrude oil can be achieved with HTL when compared to other thermochemical processes, but it cannot be used directly as a transportation fuel. In order to replace the petroleum fuels, advanced biofuels are expected to have superior fuel properties facilitating effective storage, distribution, and combustion (Román-Leshkov et al. 2007). Further improvements are required in the properties of biofuels, such as purity, heating value, and flow properties. Biocrude oil has high nitrogen and oxygen heteroatom contents. Furthermore, the investigation is expected to understand denitrogenation and deoxygenation mechanisms for designing an efficient catalytic process. Therefore, to improve the quality of biocrude oil, further refining and upgrading need to be investigated (Cao et al. 2017). In addition, breakthrough researches are required to develop new methods and technologies to understand the mechanism among the heterogeneous chemical compounds present in the biocrude oil refining and upgrading (Castello et al. 2019).

10.3.5.2 Value-Added Chemicals Production

Maximum utilization of extracted products by comprehensively targeting each functional group to produce value-added products can improve their economic benefits. As shown in Fig. 10.2, the broad spectrum of chemical products possibly produced from lignocellulosic biomass include phenolic compounds, antioxidants, pigments, aromatics, fatty acids, etc. Many reactions (e.g., hydrolysis, alkylation, degradation, and repolymerization) take place during the HTL of biomass through ionic and free radical intermediates, depending on the operating conditions. Although the desired product can be formed in the HTL process, it is difficult to control its further loss due to cleavage or repolymerization reactions (Schuler et al. 2019).

On the other hand, the difficulty to break C–C type bonds in lignin is a challenge to recover valuable functional compounds from lignin, which leads to the formation of oligomers and tar-like molecules (Kruse et al. 2013; Yong and Yukihiko 2013; Schuler et al. 2019). To address these challenges, a systematic optimization of operating conditions for a higher yield of the selective products is desired. Limited research in identifying more numbers of key substances and investigating the reaction pathways to understand their formation gives an opportunity to address these questions.

10.3.6 Recent Technological Advancements in Hydrothermal Liquefaction

10.3.6.1 Catalytic and Cosolvent-Assisted Hydrothermal Liquefaction

Various homogeneous and heterogeneous catalysts have been studied for the HTL of biomass in the last few decades. Homogeneous catalysts are known to increase biocrude oil yield, suppress char formation, and promote water-gas shift reaction

(Toor et al. 2011). Heterogeneous catalysts are used to improve the quality of biocrude oil by deoxygenation, decarboxylation, denitrogenation, and desulfurization mechanisms. Several types of catalysts have been studied to improve the productivity and quality of biocrude oil. Homogeneous catalysts are water soluble at room temperature and include alkali salts (Na_2CO_3 and KOH) and organic acids (CH_3COOH and HCOOH). Heterogeneous catalysts are water insoluble at room temperature and include metal catalysts (Pd, Pt, Ni, Ru, etc. supported on activated carbon, Al_2O_3, and zeolite), transition metal oxides (NiO), insoluble salts or earth metals ($Ca_3(PO4)_2$), and molecular sieves (zeolite and HZSM-5).

Yue et al. (2018) investigated the two-stage HTL of sweet sorghum bagasse to produce biocrude oil followed by its upgradation via catalytic hydrodeoxygenation. First-stage HTL was carried out at 200 °C, and the lignin-rich solid fraction from the first stage was applied to second-stage HTL at 350 °C in the presence of a K_2CO_3 catalyst to obtain biocrude oil. Biocrude oil from the second-stage HTL process was upgraded using 5% Ru/C catalyzed hydrodeoxygenation at 350 °C to form upgraded drop-in biofuel with qualities comparative to the petroleum diesel. Table 10.5 presents the comparative properties of fuels from different stages and sources. The different properties of biocrude oil are taken as average from a variety of biomass feedstocks.

The yield of HTL biocrude oil is higher when compared to the organic solvolytic liquefaction process because water allows high reaction temperature for the liquefaction of biomass (Sandquist et al. 2019). However, cosolvents such as tetralin, acetone, ethanol, and methanol are used to improve HTL biocrude oil properties by the solubilization and stabilization of products formed during HTL and by scavenging the reactive intermediates. Cosolvents used to alter the physical properties of a reaction medium (such as viscosity, ion solvation, and compound solubilization) and the thermodynamic properties (such as activation energy). Proton donor solvents can promote the reduction of reactive intermediates (Arturi et al. 2019).

Table 10.5 Properties of fuels obtained from the HTL of biomass and petroleum source

Properties	Bio-oil (from HTL)	Biodiesel (from second-stage HTL-HDO)	Heavy fuel oil	Diesel
Yield (%)	18–67	41–43	–	–
Carbon (wt%)	63.55–73	85.9–90.5	85	86
Hydrogen (wt%)	7.66–9.76	8.91–9.31	11	13
Oxygen (wt%)	10.5–25.1	0.24–3.88	0–1	0
Nitrogen (wt%)	3.71–4.47	0.58–0.64	0.3	–
Water content (wt%)	1.0–8.7	–	0.1	0.1
Ash (wt%)	0.2–0.8	–	0.1	–
HHV (MJ/kg)	21.15–36.1	42.2–43.3	40	43
Viscosity at 50 °C (cP)	843	23.7–2.5	180	2.5

Note: *HTL* hydrothermal liquefaction, *HDO* hydrodeoxygenation, *HHV* high heating value.
Reference: Cao et al. (2017), Shakya et al. (2018), Yue et al. (2018)

10.3.6.2 Development of Hybrid Catalysts for Better Selectivity and Stability

The stability of heterogeneous catalysts is one of the major issues for the HTL conversion of biomass to biofuels and biochemicals. Recently, new strategies are being developed to improve the stability of catalyst supports under hydrothermal conditions. In the last few decades, significant research and development have been done in the field of catalytic thermochemical upgradation of biomass to improve the yield and selectivity of desired products. However, there is scope for experimental investigations and computational modeling of liquid phase intermolecular forces and catalyst-solvent interactions to accurately predict the overall operating conditions in HTL.

Hybrid catalysts are hybrid materials, composites, or organocatalysts generally synthesized by the immobilization of catalytically active phases (homogeneous or heterogeneous) on mesoporous and microporous supports to enhance selectivity and stability with an ease of separation from the product stream. Catalytic synthesis techniques such as coating, overlayer deposition, and impregnation of heteroatoms have been found effective to mitigate the hydrothermal instability, leaching, and sintering of metal oxide supports or catalytically active sites. Introducing heteroatoms such as La^{3+}, Ga^{3+}, Sm, Ce, and Ti^{4+} has significantly improved the hydrothermal stability of oxide supports (Xiong et al. 2014).

10.3.6.3 Value-Added Chemical Production

The utilization of lignin for value-added chemical production and sustainable resource management becomes possible through technological advancement. Table 10.6 summarizes a few studies that have shown the production of biochemicals through the HTL of biomass. Cardoso et al. (2018) explored the catalytic HTL for the upgradation of Kraft lignin to monoaromatic compounds, particularly guaiacol and creosol. Ni-CeO_2 catalyst supported on carbon nanofiber (Ni-CeO_2/CNF) showed improved yield over 55% and better stability at 400 °C, 23 MPa, and 15 min. The presence of carbon nanofiber (CNF) improved the selectivity of methoxyphenols, whereas CeO_2 improved the selectivity of guaiacol. Creosol was the main product from HTL with a Ni/CNF catalyst.

Trajano et al. (2013) carried out the HTL of *Populus trichocarpa* × *P. deltoides* wood samples to study the depolymerization of wood lignin. About 75% lignin was recovered in the form of phenolic monomers at 180 °C and 1.1 MPa in 192 min with a 5% biomass loading. Under these HTL conditions, it was observed that the product molecules were extracted by depolymerization and redeposited by the condensation mechanism. Among the identified 18 different phenolic monomers, *p*-hydroxybenzoic acid yield was 83% of the total phenolic compounds.

Bbosa et al. (2018) demonstrated the production of acetic acid, catechol, cresols, and phenol in an integrated biorefinery. Román-Leshkov et al. (2007) used a microwave-assisted catalytic HTL to improve the yield and total phenolic content in biocrude oil products from pine-spruce biomass. Kim et al. (2014) studied the antioxidant activity of hydrothermal extracts of watermelon (flesh, white rind, and green rind). Total phenolic content of 7626.5 μg GAE/g, lycopene content of 60 μg/g, and β-carotene content of 4.8 μg/g were extracted under hydrothermal

Table 10.6 Value-added chemicals produced from the hydrothermal liquefaction of biomass

Biochemicals	Feedstock	Experimental conditions	Product yield	References
Acetic acid, catechol, cresols, and phenol	Corn stover lignin	350 °C, 20 MPa	39% (acetic acid), 28% (catechol) 12% (cresols) and 8% (phenol)	Bbosa et al. (2018)
Antioxidants (total phenol content)	Watermelon	300 °C, 30 min, 5 g/5 mL	7626.5 µg GAE/g	Kim et al. (2014)
β-carotene	Watermelon	300 °C, 30 min, 5 g/5 mL	4.8 µg/g	Kim et al. (2014)
Catechol	Beech wood bark lignin	300 °C, 1 wt% KOH, 180 min	30 mg/g	Schuler et al. (2019)
Creosol	Kraft lignin	400 °C, 23 MPa, water, Ni/CNF catalyst, 15 min	30–40%	Cardoso et al. (2018)
Cyclopentenes, furfurals, levulinic acid	Lactose, maltose	350 °C, 2 MPa, 10 wt% biomass, 20 min	6–10%	Fan et al. (2018)
Fluorescent quantum dots	Cocoon silk	200 °C, 72 h, 5% Na_2CO_3, 1 g/10 mL	–	Ruan et al. (2014)
Guaiacol	Beech wood bark lignin	300 °C, 1 wt% KOH, 30 min	28 mg/g	Schuler et al. (2019)
Guaiacol	Kraft lignin	400 °C, 23 MPa, water, Ni-CeO_2/CNF catalyst, 15 min	55%	Cardoso et al. (2018)
Lycopene	Watermelon	300 °C, 30 min, 5 g/5 mL	60 µg/g	Kim et al. (2014)
Phenolic monomers	*Populus trichocarpa* × *P. deltoides* wood	180 °C, 1.1 MPa, 192 min, 5% mass loading, 8 mL	75%	Trajano et al. (2013)
p-hydroxybenzoic acid	*Populus trichocarpa* × *P. deltoides* wood	180 °C, 1.1 MPa, 192 min, 5% mass loading, 8 mL	62%	Trajano et al. (2013)
Piperidines and quinolines	Lysine	350 °C, 2 MPa, 10 wt% biomass, 20 min	5–17%	Fan et al. (2018)
Pyrazine and its derivatives	Lactose+lysine/maltose+lysine	350 °C, 2 MPa, 10 wt% biomass, 20 min	10–39%	Fan et al. (2018)

(continued)

Table 10.6 (continued)

Biochemicals	Feedstock	Experimental conditions	Product yield	References
Syringol	Beech wood bark lignin	300 °C, 1 wt% KOH, 120 min	3.27 mg/g	Schuler et al. (2019)
Total phenol content	Pine and spruce pallets	250 °C, 8 MPa, 10 g/500 mL biomass, 115 min, Ni-Co/Al-Mg catalyst, 2.45 GHz microwave	14%	Remón et al. (2019)

Note: *CNF* carbon nanofiber, *GAE* gallic acid equivalents

conditions of 300 °C, 30 min, and 5 g/5 mL biomass loading. In addition, phenolic compounds such as catechol and its derivatives were identified in the HTL extract of watermelon. These findings reveal the possible applications of HTL for nutraceutical compound extraction.

10.3.6.4 Techno-Economic Feasibility for Hydrothermal Liquefaction

The techno-economic analysis study by Nie and Bi (2018) shows the minimum selling price of HTL biofuel to be about 60–80% higher than that of petroleum fuel. A paradigm shift is needed in the research and development of alternative fuels. A reduction in the overall operating cost for the conversion of biomass to commercial grade biofuel is a key for the sustainability of any biorefinery process. Utmost care is required during the process and equipment design to minimize production cost. The water-insoluble tar-like solid residues formed during the reaction creates the obstruction and risk for the functionality of the equipment. The deposits formed as undesired by-products not only create concern for cleaning but also become the bottleneck for product recovery.

Accurate analysis of the effects of temperature and time is difficult due to the thermal transience during the heating and cooling cycles of the HTL reactor. The equipment design needs highly advanced instrumentation and safety measures to handle high temperature, high pressure, increased corrosion possibilities, and fouling resistance to avoid unexpected accidental scenarios. The selection of operating parameters for process optimization is based on technical perception and knowledge, which may involve a certain degree of biases, thus creating hurdles in the optimum reaction conditions. The requirement of reducing gas utility can increase the operating cost, thus investigating efficient and cost-effective hydrogen donors to mitigate the operating cost (Cao et al. 2017).

The kinetics and mechanistic investigations of complex reactions to develop accurate mathematical models for process optimization can promote the industrial applications of HTL technology. Trained labor versus automation choices can improve production economics. In addition, it is worth mentioning that the hidden cost of the adverse impact on the environment due to fossil fuels is much higher. Thus, government policies (such as carbon tax and biofuel subsidies) for damage control due to climate change can improve the profitability of biofuel production process. However, the opportunity of HTL as a cleaner processing, sustainable development, and feedstock flexibility keeps the gateways open for its large-scale applications.

10.3.6.5 Integrated Biorefinery Approaches for Sustainable Development

The production of any single product from biomass is neither economically feasible nor environmentally viable. Therefore, integrated production of biofuels, biochemicals, and biomaterials for the maximal utilization of biomass is an excellent way to achieve the sustainability and commerciality of the biorefining process (Yadav et al. 2019; Sarangi and Nanda 2019). Bbosa et al. (2018) investigated the integration of HTL in the ethanol production plant. Lignin waste from the ethanol process efficiently produced biochemicals within the ethanol biorefinery using HTL. Ong et al. (2018) evaluated the integration of the pine biomass HTL process to utilize black liquor from Kraft pulp mill to coproduce biocrude oil and biopower. Black liquor from Kraft pulp mill utilized for the co-HTL of pine biomass and part of the heavy biocrude oil from HTL was utilized as feed to the boiler of Kraft pulp mill.

Heat integration reduced the steam demand for the boiler by 3%. Valuable products like BTEX compounds (benzene, toluene, ethylbenzene, and xylenes) can be coproduced using HTL in an integrated Kraft pulp mill (Funkenbusch et al. 2019). Yu et al. (2011) introduced an environment-enhancing-energy treatment process that integrates HTL gaseous by-products (as a CO_2 source) and HTL aqueous by-products (as a nutrient source) for the cultivation of algae for sustainable wastewater treatment and the sequestration of CO_2. Novel approaches can be developed based on interdisciplinary knowledge and applications. HTL process allows the integration of fresh feedstocks, by-products, and residues from other processes for maximum biomass conversion to biofuels and biochemicals. This approach can turn into a sustainable, techno-economically feasible, and cleaner technology in the future.

10.3.6.6 Process and Equipment Design for Hydrothermal Liquefaction

New innovative approaches can significantly improve the profitability and sustainability of the HTL process. Remón et al. (2019) introduced an advanced process intensification approach of microwave-assisted catalytic HTL for the liquefaction of pine-spruce biomass to improve the yield and quality of biocrude oil. Quantitative mathematical models can be developed to optimize HTL process variables from the knowledge of feedstock composition. Yang et al. (2019) developed a quantitative model to predict HTL product yields by understanding the chemical interactions between biomolecules in the biomass. Development of batch to a continuous process for the HTL of biomass has been studied in the recent years, which is found to be advantageous for eliminating the limitations of thermal transience, reactant contact patterns, production capacity, and capital investment cost (Elliott et al. 2015; Castello et al. 2018).

The National Advanced Biofuels Consortium (NABC) in the U.S. developed a continuous flow process for biocrude oil production from loblolly pine and corn stover (Elliott et al. 2015). Comparing the performance of 1 L continuous stirred tank reactor and plug flow reactor, it was recommended to use scalable plug flow reactor for high biocrude oil yield, high carbon efficiency (>50%), and low capital cost. More research is anticipated in the continuous HTL process to optimize phase separation, product yield, space velocity, solid content of slurry, heat recovery, and operating cost. The sensitivity analysis of techno-economic assessment specifies that current technological advancement is the most important factor to make the HTL process competitive and profitable. Therefore, current strategies to increase the productivity of biocrude oil and biofuels to improve the quality of value-added products and the integrated production of valuable products and biofuels are inevitable for technology advancement.

10.4 Conclusions

A wide variety of biomass feedstocks have been processed through pyrolysis and hydrothermal liquefaction over the last few decades. Several experimental and statistical studies have been performed to optimize the effect of operating variables, catalysts, and cosolvents to improve biocrude oil yield and quality. The challenges and opportunities in the pyrolysis and hydrothermal liquefaction of biomass have been addressed. The challenges vary depending on the biomass composition and targeted products in HTL and the pyrolysis process. In the progress to make biofuels produced from HTL and pyrolysis competitive with fossil fuels, the aspect of technological advancement is most sensitive.

Current technological advancements are reviewed, and new strategies are discussed to overcome the existing challenges. From the review of literature, it is evident to suggest that HTL technique can be utilized to improve the yield of biocrude oil from high-moisture containing biomass. Further investigations on the selection of a suitable catalyst and cosolvent are required to ascertain the increase in biocrude oil yield and quality and to reduce the overall cost of biocrude oil production. The developments in hybrid catalytic systems and recent progress are certainly leading to overcoming some universal challenges, which can result in HTL and pyrolysis as commercial biorefining processes. Innovative approaches such as process intensification, mathematical modeling for process simulation, and equipment design for energy efficiency can significantly improve the profitability and sustainability of HTL and the pyrolysis process. An integrated biorefinery approach based on the coproduction of biofuel and valuable platform chemicals from biomass has promising potential to access affordable energy supply with positive effects on the environment and economy.

References

Ahmad M, Rajapaksha AU, Lim JE, Zhang M, Bolan N, Mohan D, Vithanage M, Lee SS, Ok YS (2014) Biochar as a sorbent for contaminant management in soil and water: a review. Chemosphere 99:19–33

Alper K, Tekin K, Karagöz S (2019) Hydrothermal liquefaction of lignocellulosic biomass using potassium fluoride-doped alumina. Energy Fuel 33:3248–3256

Arturi KR, Kucheryavskiy S, Nielsen RP, Maschietti M, Vogel F, Bjelić S, Søgaard EG (2019) Molecular footprint of co-solvents in hydrothermal liquefaction (HTL) of *Fallopia japonica*. J Supercrit Fluids 143:211–222

Azargohar R, Nanda S, Rao BVSK, Dalai AK (2013) Slow pyrolysis of deoiled canola meal: product yields and characterization. Energy Fuel 27:5268–5279

Azargohar R, Nanda S, Kozinski JA, Dalai AK, Sutarto R (2014) Effects of temperature on the physicochemical characteristics of fast pyrolysis bio-chars derived from Canadian waste biomass. Fuel 125:90–100

Azargohar R, Nanda S, Dalai AK, Kozinski JA (2019) Physico-chemistry of biochars produced through steam gasification and hydro-thermal gasification of canola hull and canola meal pellets. Biomass Bioenergy 120:458–470

Babu MKG, Subramanian KA (2013) Alternative transportation fuels: utilisation in combustion engines. CRC Press, Boca Raton, FL, p 464

Balagurumurthy B, Srivastava V, Vinit KJ, Biswas B, Singh R, Gupta P, Kumar KLNS, Singh R, Bhaskar T (2015) Value addition to rice straw through pyrolysis in hydrogen and nitrogen environments. Bioresour Technol 188:273–279

Bbosa D, Mba-Wright M, Brown RC (2018) More than ethanol: a techno-economic analysis of a corn Stover-ethanol biorefinery integrated with a hydrothermal liquefaction process to convert lignin into biochemicals. Biofuels Bioprod Biorefin 12:497–509

Bridgwater AV (2012) Review of fast pyrolysis of biomass and product upgrading. Biomass Bioenergy 38:68–94

Bridgwater AV, Meier D, Radlein D (1999) An overview of fast pyrolysis of biomass. Org Geochem 30:1479–1493

Brownsort P (2009) Biomass pyrolysis processes: performance parameters and their influence on biochar system benefits, M.Sc. Thesis. University of Edinburgh, Edinburgh

Cao L, Zhang C, Chen H, Tsang DCW, Luo G, Zhang S, Chen J (2017) Hydrothermal liquefaction of agricultural and forestry wastes: state-of-the-art review and future prospects. Bioresour Technol 245:1184–1193

Cao Z, Dierks M, Clough MT, Daltro de Castro IB, Rinaldi R (2018) A convergent approach for a deep converting lignin-first biorefinery rendering high-energy-density drop-in fuels. Joule 2:1118–1133

Cardoso A, Ramirez Reina T, Suelves I, Pinilla JL, Millan M, Hellgardt K (2018) Effect of carbon-based materials and CeO_2 on Ni catalysts for Kraft lignin liquefaction in supercritical water. Green Chem 20:4308–4318

Castello D, Pedersen TH, Rosendahl LA (2018) Continuous hydrothermal liquefaction of biomass: a critical review. Energies 11:3165

Castello D, Haider MS, Rosendahl LA (2019) Catalytic upgrading of hydrothermal liquefaction biocrudes: different challenges for different feedstocks. Renew Energy 141:420–430

Chen D, Liu D, Zhang H, Chen Y, Li Q (2015) Bamboo pyrolysis using TG–FTIR and a lab-scale reactor: analysis of pyrolysis behavior, product properties, and carbon and energy yields. Fuel 148:79–86

Chen Y, Cao X, Zhu S, Tian F, Xu Y, Zhu C, Dong L (2019) Synergistic hydrothermal liquefaction of wheat stalk with homogeneous and heterogeneous catalyst at low temperature. Bioresour Technol 278:92–98

Czernik S, Bridgwater AV (2004) Overview of applications of biomass fast pyrolysis oil. Energy Fuel 18:590–598

Dasappa S, Paul PJ, Mukunda HS, Rajan NKS, Sridhar G, Sridhar HV (2004) Biomass gasification technology—a route to meet energy needs. Curr Sci 87:908–916
del Pozo C, Bartrolí J, Puy N, Fàbregas E (2018) Separation of value-added chemical groups from bio-oil of olive mill waste. Ind Crop Prod 125:160–167
Demirbas A (2004) Effects of temperature and particle size on bio-char yield from pyrolysis of agricultural residues. J Anal Appl Pyrolysis 72:243–248
Dimitriadis A, Bezergianni S (2017) Hydrothermal liquefaction of various biomass and waste feedstocks for biocrude production: a state of the art review. Renew Sust Energ Rev 68:113–125
Editors G, Sabev Varbanov P, Liew P-Y, Yong J-Y, Jaromír Klemeš J, Loong Lam H, Goryunov AG, Goryunova NN, Ogunlana AO, Manenti F (2016) Production of energy from biomass: near or distant future prospects? Chem Eng Trans 52:1220–1225
Elliott DC, Biller P, Ross AB, Schmidt AJ, Jones SB (2015) Hydrothermal liquefaction of biomass: developments from batch to continuous process. Bioresour Technol 178:147–156
Fahmi R, Bridgwater AV, Darvell LI, Jones JM, Yates N, Thain S, Donnison IS (2007) The effect of alkali metals on combustion and pyrolysis of Lolium and Festuca grasses, switchgrass and willow. Fuel 86:1560–1569
Fan Y, Hornung U, Dahmen N, Kruse A (2018) Hydrothermal liquefaction of protein-containing biomass: study of model compounds for Maillard reactions. Biomass Conv Bioref 8:909–923
Fang J, Zhan L, Ok YS, Gao B (2018) Minireview of potential applications of hydrochar derived from hydrothermal carbonization of biomass. J Ind Eng Chem 57:15–21
Fougere D, Nanda S, Clarke K, Kozinski JA, Li K (2016) Effect of acidic pretreatment on the chemistry and distribution of lignin in aspen wood and wheat straw substrates. Biomass Bioenergy 91:56–68
Funkenbusch LT, Mullins ME, Vamling L, Belkhieri T, Srettiwat N, Winjobi O, Shonnard DR, Rogers TN (2019) Technoeconomic assessment of hydrothermal liquefaction oil from lignin with catalytic upgrading for renewable fuel and chemical production. WIREs Energ Environ 8:e319
Gong M, Nanda S, Hunter HN, Zhu W, Dalai AK, Kozinski JA (2017a) Lewis acid catalyzed gasification of humic acid in supercritical water. Catal Today 291:13–23
Gong M, Nanda S, Romero MJ, Zhu W, Kozinski JA (2017b) Subcritical and supercritical water gasification of humic acid as a model compound of humic substances in sewage sludge. J Supercrit Fluids 119:130–138
Gu S, Zhou J, Luo Z, Wang Q, Ni M (2013) A detailed study of the effects of pyrolysis temperature and feedstock particle size on the preparation of nanosilica from rice husk. Ind Crop Prod 50:540–549
Guo X, Wang S, Wang K, Liu Q, Luo Z (2010) Influence of extractives on mechanism of biomass pyrolysis. J Fuel Chem Technol 38:42–46
Guoxin H, Hao H, Yanhong L (2009) Hydrogen-rich gas production from pyrolysis of biomass in an autogenerated steam atmosphere. Energy Fuel 23:1748–1753
He M, Xiao B, Liu S, Hu Z, Guo X, Luo S, Yang F (2010) Syngas production from pyrolysis of municipal solid waste (MSW) with dolomite as downstream catalysts. J Anal Appl Pyrolysis 87:181–187
Hu X, Mourant D, Gunawan R, Wu L, Wang Y, Lievens C, Li C-Z (2012) Production of value-added chemicals from bio-oil via acid catalysis coupled with liquid–liquid extraction. RSC Adv 2:9366–9370
Hu X, Gholizadeh M (2019) Biomass pyrolysis: a review of the process development and challenges from initial researches up to the commercialisation stage. J Energy Chem 39:109–143
Hussain A, Arif SM, Aslam M (2017) Emerging renewable and sustainable energy technologies: state of the art. Renew Sust Energ Rev 71:12–28
Isahak WNRW, Hisham MWM, Yarmo MA, Yun Hin T (2012) A review on bio-oil production from biomass by using pyrolysis method. Renew Sust Energ Rev 16:5910–5923
Jiang Y, van der Werf E, van Ierland EC, Keesman KJ (2017) The potential role of waste biomass in the future urban electricity system. Biomass Bioenergy 107:182–190

Jiang W, Kumar A, Adamopoulos S (2018) Liquefaction of lignocellulosic materials and its applications in wood adhesives—a review. Ind Crop Prod 124:325–342
Kan T, Strezov V, Evans TJ (2016) Lignocellulosic biomass pyrolysis: a review of product properties and effects of pyrolysis parameters. Renew Sust Energ Rev 57:1126–1140
Kang K, Nanda S, Sun G, Qiu L, Gu Y, Zhang T, Zhu M, Sun R (2019) Microwave-assisted hydrothermal carbonization of corn stalk for solid biofuel production: optimization of process parameters and characterization of hydrochar. Energy 186:115795
Khan AA, de Jong W, Jansens PJ, Spliethoff H (2009) Biomass combustion in fluidized bed boilers: potential problems and remedies. Fuel Process Technol 90:21–50
Kim S-J, Matsushita Y, Fukushima K, Aoki D, Yagami S, Yuk HG, Lee SC (2014) Antioxidant activity of a hydrothermal extract from watermelons. LWT—Food Sci Technol 59:361–368
Kruse A, Funke A, Titirici M-M (2013) Hydrothermal conversion of biomass to fuels and energetic materials. Curr Opin Chem Biol 17:515–521
Kumar A, Kumar K, Kaushik N, Sharma S, Mishra S (2010) Renewable energy in India: current status and future potentials. Renew Sust Energ Rev 14:2434–2442
Lee Y, Park J, Ryu C, Gang KS, Yang W, Park Y-K, Jung J, Hyun S (2013) Comparison of biochar properties from biomass residues produced by slow pyrolysis at 500°C. Bioresour Technol 148:196–201
Magdeldin M, Kohl T, Järvinen M (2018) Techno-economic assessment of integrated hydrothermal liquefaction and combined heat and power production from lignocellulose residues. J Sustain Dev Energ Water Environ Sys 6:89–113
Mahmood N, Yuan Z, Schmidt J, Xu C (2016) Depolymerization of lignins and their applications for the preparation of polyols and rigid polyurethane foams: a review. Renew Sust Energ Rev 60:317–329
Manyà JJ (2012) Pyrolysis for biochar purposes: a review to establish current knowledge gaps and research needs. Environ Sci Technol 46:7939–7954
Miliotti E, Dell'Orco S, Lotti G, Rizzo AM, Rosi L, Chiaramonti D (2019) Lignocellulosic ethanol biorefinery: valorization of lignin-rich stream through hydrothermal liquefaction. Energies 12:723
Mohan D, Pittman CU, Steele PH (2006) Pyrolysis of wood/biomass for bio-oil: a critical review. Energy Fuel 20:848–889
Mohanty P, Nanda S, Pant KK, Naik S, Kozinski JA, Dalai AK (2013) Evaluation of the physiochemical development of biochars obtained from pyrolysis of wheat straw, timothy grass and pinewood: effects of heating rate. J Anal Appl Pyrolysis 104:485–493
Murphy BM, Xu B (2018) Foundational techniques for catalyst design in the upgrading of biomass-derived multifunctional molecules. Prog Energy Combust Sci 67:1–30
Nanda S, Mohanty P, Pant KK, Naik S, Kozinski JA, Dalai AK (2013) Characterization of North American lignocellulosic biomass and biochars in terms of their candidacy for alternate renewable fuels. Bioenergy Res 6:663–677
Nanda S, Azargohar R, Kozinski JA, Dalai AK (2014a) Characteristic studies on the pyrolysis products from hydrolyzed Canadian lignocellulosic feedstocks. Bioenergy Res 7:174–191
Nanda S, Mohammad J, Reddy SN, Kozinski JA, Dalai AK (2014b) Pathways of lignocellulosic biomass conversion to renewable fuels. Biomass Conv Bioref 4:157–191
Nanda S, Azargohar R, Dalai AK, Kozinski JA (2015a) An assessment on the sustainability of lignocellulosic biomass for biorefining. Renew Sust Energ Rev 50:925–941
Nanda S, Maley J, Kozinski JA, Dalai AK (2015b) Physico-chemical evolution in lignocellulosic feedstocks during hydrothermal pretreatment and delignification. J Biobased Mater Bioenerg 9:295–308
Nanda S, Reddy SN, Hunter HN, Butler IS, Kozinski JA (2015c) Supercritical water gasification of lactose as a model compound for valorization of dairy industry effluents. Ind Eng Chem Res 54:9296–9306
Nanda S, Dalai AK, Berruti F, Kozinski JA (2016a) Biochar as an exceptional bioresource for energy, agronomy, carbon sequestration, activated carbon and specialty materials. Waste Biomass Valor 7:201–235

Nanda S, Isen J, Dalai AK, Kozinski JA (2016b) Gasification of fruit wastes and agro-food residues in supercritical water. Energ Convers Manage 110:296–306
Nanda S, Reddy SN, Mitra SK, Kozinski JA (2016c) The progressive routes for carbon capture and sequestration. Energ Sci Eng 4:99–122
Nanda S, Gong M, Hunter HN, Dalai AK, Gökalp I, Kozinski JA (2017a) An assessment of pinecone gasification in subcritical, near-critical and supercritical water. Fuel Process Technol 168:84–96
Nanda S, Rana R, Zheng Y, Kozinski JA, Dalai AK (2017b) Insights on pathways for hydrogen generation from ethanol. Sustain Energ Fuel 1:1232–1245
Nanda S, Rana R, Sarangi PK, Dalai AK, Kozinski JA (2018a) A broad introduction to first, second and third generation biofuels. In: Sarangi PK, Nanda S, Mohanty P (eds) Recent advancements in biofuels and bioenergy utilization. Springer Nature, Singapore, pp 1–25
Nanda S, Reddy SN, Fang Z, Dalai AK, Kozinski JA (2018b) Hydrothermal events occurring during gasification in supercritical water. In: Hunt AJ, Attard TM (eds) Supercritical and other high-pressure solvent systems: for extraction, reaction and material processing. Royal Society of Chemistry, London, pp 560–587
Nanda S, Reddy SN, Vo DVN, Sahoo BN, Kozinski JA (2018c) Catalytic gasification of wheat straw in hot compressed (subcritical and supercritical) water for hydrogen production. Energ Sci Eng 6:448–459
Nie Y, Bi XT (2018) Techno-economic assessment of transportation biofuels from hydrothermal liquefaction of forest residues in British Columbia. Energy 153:464–475
Okolie JA, Nanda S, Dalai AK, Kozinski JA (2019a) Optimization and modeling of process parameters during hydrothermal gasification of biomass model compounds to generate hydrogen-rich gas products. Int J Hydrog Energy. https://doi.org/10.1016/j.ijhydene.2019.05.132
Okolie JA, Rana R, Nanda S, Dalai AK, Kozinski JA (2019b) Supercritical water gasification of biomass: a state-of-the-art review of process parameters, reaction mechanisms and catalysis. Sustain Energy Fuel 3:578–598
Ong BHY, Walmsley TG, Atkins MJ, Walmsley MRW (2018) Hydrothermal liquefaction of Radiata pine with Kraft black liquor for integrated biofuel production. J Clean Prod 199:737–750
Palgan YV, McCormick K (2016) Biorefineries in Sweden: perspectives on the opportunities, challenges and future. Biofuels Bioprod Biorefin 10:523–533
Patil SJ, Vaidya PD (2017) Catalytic hydrotreatment of jatropha oil over lanthanum hydroxide supported noble metals: effect of promotion with cerium. Chem Select 2:11918–11925
Patil SJ, Vaidya PD (2018) Production of hydrotreated jatropha oil using Co–Mo and Ni–Mo catalysts and its blending with petroleum diesel. Energy Fuel 32:1812–1821
Pearce M, Tonnellier X, Sengar N, Sansom C (2018) Commercial development of bio-combustible fuels from hydrothermal liquefaction of waste using solar collectors. In: AIP Conference Proceedings, vol 2033. AIP Publishing, College Park, p 130011
Peng J, Chen P, Lou H, Zheng X (2009) Catalytic upgrading of bio-oil by H-ZSM-5 in sub- and super-critical ethanol. Bioresour Technol 100:3415–3418
Rana R, Nanda S, Kozinski JA, Dalai AK (2018) Investigating the applicability of Athabasca bitumen as a feedstock for hydrogen production through catalytic supercritical water gasification. J Environ Chem Eng 6:182–189
Rana R, Nanda S, Maclennan A, Hu Y, Kozinski JA, Dalai AK (2019) Comparative evaluation for catalytic gasification of petroleum coke and asphaltene in subcritical and supercritical water. J Energ Chem 31:107–118
Reddy SN, Nanda S, Dalai AK, Kozinski JA (2014) Supercritical water gasification of biomass for hydrogen production. Int J Hydrog Energy 39:6912–6926
Reddy SN, Nanda S, Hegde UG, Hicks MC, Kozinski JA (2015) Ignition of hydrothermal flames. RSC Adv 5:36404–36422
Reddy SN, Nanda S, Kozinski JA (2016) Supercritical water gasification of glycerol and methanol mixtures as model waste residues from biodiesel refinery. Chem Eng Res Des 113:17–27
Reddy SN, Nanda S, Hegde UG, Hicks MC, Kozinski JA (2017) Ignition of n-propanol–air hydrothermal flames during supercritical water oxidation. Proc Combust Inst 36:2503–2511

Reddy SN, Nanda S, Sarangi PK (2018) Applications of supercritical fluids for biodiesel production. In: Sarangi PK, Nanda S, Mohanty P (eds) Recent advancements in biofuels and bioenergy utilization. Springer Nature, Singapore, pp 261–284

Reddy SN, Nanda S, Kumar P, Hicks MC, Hegde UG, Kozinski JA (2019) Impacts of oxidant characteristics on the ignition of n-propanol-air hydrothermal flames in supercritical water. Combust Flame 203:46–55

Remón J, Randall J, Budarin VL, Clark JH (2019) Production of bio-fuels and chemicals by microwave-assisted, catalytic, hydrothermal liquefaction (MAC-HTL) of a mixture of pine and spruce biomass. Green Chem 21:284–299

Rinaldi R, Jastrzebski R, Clough MT, Ralph J, Kennema M, Bruijnincx PCA, Weckhuysen BM (2016) Paving the way for lignin valorisation: recent advances in bioengineering, biorefining and catalysis. Angew Chem Int Ed 55:8164–8215

Román-Leshkov Y, Barrett CJ, Liu ZY, Dumesic JA (2007) Production of dimethylfuran for liquid fuels from biomass-derived carbohydrates. Nature 447:982–985

Ruan S, Wan J, Fu Y, Han K, Li X, Chen J, Zhang Q, Shen S, He Q, Gao H (2014) PEGylated fluorescent carbon nanoparticles for noninvasive heart imaging. Bioconjug Chem 25:1061–1068

Sánchez ME, Menéndez JA, Domínguez A, Pis JJ, Martínez O, Calvo LF, Bernad PL (2009) Effect of pyrolysis temperature on the composition of the oils obtained from sewage sludge. Biomass Bioenergy 33:933–940

Sandquist J, Tschentscher R, del Alamo Serrano G (2019) Hydrothermal liquefaction of organic resources in biotechnology: how does it work and what can be achieved? Appl Microbiol Biotechnol 103:673–684

Sandun F, Sushil A, Chauda C, Murali N (2006) Biorefineries: current status, challenges, and future direction. Energy Fuel 20:1727–1737

Sarangi PK, Nanda S (2019) Recent advances in consolidated bioprocessing for microbe-assisted biofuel production. In: Nanda S, Sarangi PK, Vo DVN (eds) Fuel processing and energy utilization. CRC Press, Boca Raton, FL, pp 141–157

Schuler J, Hornung U, Dahmen N, Sauer J (2019) Lignin from bark as a resource for aromatics production by hydrothermal liquefaction. GCB Bioenergy 11:218–229

Shabangu S, Woolf D, Fisher EM, Angenent LT, Lehmann J (2014) Techno-economic assessment of biomass slow pyrolysis into different biochar and methanol concepts. Fuel 117:742–748

Shakya R, Adhikari S, Mahadevan R, Hassan EB, Dempster TA (2018) Catalytic upgrading of bio-oil produced from hydrothermal liquefaction of *Nannochloropsis* sp. Bioresour Technol 252:28–36

Sharma A, Pareek V, Zhang D (2015) Biomass pyrolysis—a review of modelling, process parameters and catalytic studies. Renew Sust Energ Rev 50:1081–1096

Song W, Wang S, Guo Y, Xu D (2017) Bio-oil production from hydrothermal liquefaction of waste Cyanophyta biomass: influence of process variables and their interactions on the product distributions. Int J Hydrogen Energy 42:20361–20374

Strezov V, Evans TJ, Hayman C (2008) Thermal conversion of elephant grass (*Pennisetum Purpureum Schum*) to bio-gas, bio-oil and charcoal. Bioresour Technol 99:8394–8399

Sudarsanam P, Peeters E, Makshina EV, Parvulescu VI, Sels BF (2019) Advances in porous and nanoscale catalysts for viable biomass conversion. Chem Soc Rev 48:2366–2421

Toor SS, Rosendahl L, Rudolf A (2011) Hydrothermal liquefaction of biomass: a review of subcritical water technologies. Energy 36:2328–2342

Trajano HL, Engle NL, Foston M, Ragauskas AJ, Tschaplinski TJ, Wyman CE (2013) The fate of lignin during hydrothermal pretreatment. Biotechnol Biofuels 6(110):1–16

Tsagkari M, Couturier J-L, Kokossis A, Dubois J-L (2016) Early-stage capital cost estimation of biorefinery processes: a comparative study of heuristic techniques. ChemSusChem 9:2284–2297

Uddin MN, Daud WMAW, Abbas HF (2014) Effects of pyrolysis parameters on hydrogen formations from biomass: a review. RSC Adv 4:10467

Wang S, Dai G, Yang H, Luo Z (2017) Lignocellulosic biomass pyrolysis mechanism: a state-of-the-art review. Prog Energy Combust Sci 62:33–86

Williams PT, Nugranad N (2000) Comparison of products from the pyrolysis and catalytic pyrolysis of rice husks. Energy 25:493–513

Worasuwannarak N, Sonobe T, Tanthapanichakoon W (2007) Pyrolysis behaviors of rice straw, rice husk, and corncob by TG-MS technique. J Anal Appl Pyrolysis 78:265–271

Xiong H, Pham HN, Datye AK (2014) Hydrothermally stable heterogeneous catalysts for conversion of biorenewables. Green Chem 16:4627–4643

Xue Y, Chen H, Zhao W, Yang C, Ma P, Han S (2016) A review on the operating conditions of producing bio-oil from hydrothermal liquefaction of biomass. Int J Energy Res 40:865–877

Yadav P, Reddy SN, Nanda S (2019) Cultivation and conversion of algae for wastewater treatment and biofuel production. In: Nanda S, Sarangi PK, Vo DVN (eds) Fuel processing and energy utilization. CRC Press, Boca Raton, FL, pp 159–175

Yang J, He Q, Corscadden K, Niu H, Lin J, Astatkie T (2019) Advanced models for the prediction of product yield in hydrothermal liquefaction via a mixture design of biomass model components coupled with process variables. Appl Energy 233–234:906–915

Yong TL-K, Yukihiko M (2013) Kinetic analysis of guaiacol conversion in sub- and supercritical water. Ind Eng Chem Res 52:9048–9059

Yu G, Zhang Y, Schideman L, Funk T, Wang Z (2011) Distributions of carbon and nitrogen in the products from hydrothermal liquefaction of low-lipid microalgae. Energy Environ Sci 4:4587–4595

Yue Y, Kastner JR, Mani S (2018) Two-stage hydrothermal liquefaction of sweet sorghum biomass—Part II: Production of upgraded biocrude oil. Energy Fuel 32:7620–7629

Zhang X, Zhang Q, Wang T, Li B, Xu Y, Ma L (2016) Efficient upgrading process for production of low quality fuel from bio-oil. Fuel 179:312–321

Biocrude Oil Production via Hydrothermal Liquefaction of Algae and Upgradation Techniques to Liquid Transportation Fuels

11

Shima Masoumi, Venu Babu Borugadda, and Ajay K. Dalai

Abstract

Hydrothermal liquefaction of algae is regarded as a favorable thermochemical process to produce biocrude oil from biomass with potential to complement conventional crude oil. This chapter discusses the production of biocrude oil via hydrothermal liquefaction of microalgae. Due to the presence of high protein content in algal species, the catalytic removal of heteroatoms is required to make liquid transportation fuels (biodiesel and biogasoline) from algal biocrude oil. Therefore, different upgradation techniques are explored to remove the heteroatoms using various heterogeneous acid catalysts. Special focus is given to the effects of process parameters on hydrothermal liquefaction and upgradation techniques to escalate biocrude oil yield and liquid transportation fuels.

Keywords

Algae · Hydrothermal liquefaction · Liquid transportation fuels · Process parameters · Fuel upgrading

11.1 Introduction

For the last few decades, due to incremental human population, industrialization, and energy consumption, there is a rapid increase in CO_2 emission into the environment (Nanda et al. 2016f). Furthermore, a drastic increase in energy consumption and the lack of sustainable resources have concerned scientists for alternative sources of energy.

S. Masoumi · V. B. Borugadda · A. K. Dalai (✉)
Department of Chemical and Biological Engineering, University of Saskatchewan, Saskatoon, Saskatchewan, Canada
e-mail: ajay.dalai@usask.ca

S. Nanda et al. (eds.), *Biorefinery of Alternative Resources: Targeting Green Fuels and Platform Chemicals*, https://doi.org/10.1007/978-981-15-1804-1_11

Therefore, many researchers have focused on finding an alternative fuel source for commercialization. Waste organic biomass is regarded as an inexhaustible and sustainable future energy source. This includes biomass sources such as forestry wastes, agricultural crop residues, energy crops, microalgae, and animal manure (Nanda et al. 2014b, 2016b, c). Of the candidate biomass feedstocks for biofuel production, much attention is given to microalgae due to faster growth, higher yield, and higher ability to CO_2 sequestration as compared to other biomasses (Duan and Savage 2011).

Algal biofuels, considered as third-generation biofuels, can be obtained through thermochemical conversion processes. Hydrothermal liquefaction (HTL) of algae is considered as a well-known technique to transform algae feedstocks into biocrude oil in water or solvent medium under high pressures and moderate temperatures. Although high biocrude oil yield can be obtained through this process, large amounts of nitrogen, sulfur, and oxygen can still be present in biocrude oil (Guo et al. 2015). This leads to the instability of biocrude oil, which creates many difficulties for its applications. Therefore, subsequent upgrading of biocrude oil and improving its stability will make it more suitable for producing liquid transportation fuels.

Environmentally friendly transportation fuels such as biodiesel obtained from the transesterification of algal biocrude oil is a promising alternative due to its advantages, such as higher flash point and low sulfur content, and it also can be considered as a better lubricant (Toor et al. 2013). Esterification is an important upgrading process for biocrude oil to convert carboxylic acids into desirable esters. Esterification under a supercritical fluid environment with higher temperatures and mass transfer rates to remove organic acids in biocrude oil has attracted much attention. Additionally, the use of catalysts increases the efficiency of the esterification and transesterification of biocrude oil.

Significant research has focused on heterogeneous catalysts due to their separation and reusability over homogeneous catalysts. An option that has great feasibility but has not been fully explored is the preparation of catalysts from sustainable renewable sources (Nanda et al. 2016e). Nanda et al. (2016e) have demonstrated synthesizing nanocatalytic nickel particles in the cell wall of agricultural (wheat straw) and forestry (pinewood) biomass in subcritical and supercritical water for hydrothermal gasification. Moreover, functionalized biochar-based catalysts are considered desirable because of their favorable properties, such as low material cost, high surface area, and thermal stability (Manayil et al. 2016, Nanda et al. 2016a). Thus, the production of biocrude oil from algae via liquefaction and upgradation routes is discussed with a focus on the effect of process parameters on biocrude oil yield and liquid transpiration fuel production technologies.

11.2 First-Generation and Second-Generation Biofuel Refining

Based on the nature of the feedstock used to produce biofuels, they are divided into three generations. First-generation biofuels are obtained from food crops like corn, wheat, and soybean, which can also be consumed as human food. Their use in biorefineries for biofuel production has led to an increase in food prices and created many social, economic, and environmental challenges (Azad et al. 2015, Nanda et al. 2015). In terms of first-generation biofuels, biodiesel and bioethanol are produced from vegetable oils and sugar/starch-based crops (corn, wheat, and potato). Non-edible lignocellulosic biomass, which consists of nonfood crops such as grasses, wood, and agricultural wastes, is considered as second-generation feedstock. Second-generation biofuels are known as advanced biofuels because they are carbon neutral and do not compete with food supply and arable lands. The need for a large area of land with moist soil is one of the disadvantages of first-generation biofuels (Azad et al. 2015). Biofuels derived from marine biomass are considered as third-generation biofuels, and they provide more advantages compared to biofuels generated from previous generations. Microalgae are photosynthetic microorganisms that can also grow in saline environments and transform sunlight and CO_2 to renewable algal biomass and lipids.

The second- and third-generation feedstocks can be converted into biofuels through three main processes: thermochemical process, biological process, and direct combustion (Tsukahara and Sawayama 2005, Nanda et al. 2013, Azargohar et al. 2013). Thermochemical conversion leads to biomethanol, biodiesel, biocrude oil, biosyngas, and biohydrogen production. Gasification, pyrolysis, and liquefaction are the three main routes for biomass thermochemical conversion. Biochemical conversion technologies result in the production of bioethanol, biobutanol, biohydrogen, and biomethane through respective technologies such as ethanol/butanol fermentation, dark/photofermentation, and anaerobic digestion (Nanda et al. 2014a, 2017a, c). In comparison to biochemical technologies, thermochemical processes are preferred due to their ability to convert biomass into transportation fuels with higher heating value (Akia et al. 2014).

Out of all the processes, gasification and pyrolysis require higher temperature and dried biomass as feedstock. During gasification, biomass produces synthesis gas (a mixture of H_2 and CO), which can be converted to liquid fuel over a suitable catalyst via the Fischer-Tropsch synthesis process. Pyrolysis is used to produce bio-oil from dried biomass, whereas hydrothermal liquefaction and gasification can use high-moisture-containing biomass because the medium of reaction is water or solvents in the presence of suitable catalysts. Liquefaction technique is a low-temperature and high-pressure process that can convert the components of biomass into hydrocarbons in water or solvent medium (Dimitriadis and Bezergianni 2017).

Table 11.1 shows the oil content of algal biomass when compared to nonedible feedstocks. Algae can be categorized into microalgae and macroalgae. Compared to microalgae, macroalgae produce superior biomass densities. However, their lipid content is less compared to their high carbohydrate and protein contents. Therefore, it is believed that macroalgae cannot be an economically feasible source of biodiesel production (Van et al. 2014). Biofuel production from algae is promising because of the following advantages:

1. Algae have fast growth rate. It is assessed that compared to crops such as canola (200–450 L/ha), algae could yield 61,000 L/ha.
2. Able to sustain themselves under harsh condition due to their unicellular growth.

3. Have a simple multicellular structure.
4. Short harvesting cycle.
5. Able to capture CO_2.
6. Can be cultivated in nonarable lands.
7. Do not compete for food resources.
8. Can produce a remarkable variety of biofuels such as biodiesel, biogasoline, bioethanol, and biojet fuel.
9. Capable of higher biocrude oil yield.
10. Relatively higher photosynthesis rate compared to terrestrial plants.
11. Algal biofuels are nontoxic and biodegradable, as well as contain less sulfur.

Although the potential for the production of algal biofuels is high, the capital and operating costs are relatively high. It requires further research and innovation to develop sustainable and viable methods of biofuel production on a commercial scale. Currently, a number of companies worldwide are working on the commercial development of algal biofuels, a few of which include ALG Western Oil (South Africa), Algae Link (Netherlands), Affinity Eenergy and Health (Australia), AlgaFuel (Portugal), Algenol (USA), Aurora Algae Inc. (USA), British Petroleum (England), BRTeam (Iran), DENSO Corporation (Japan), Eni (Italy), Greon (Bulgaria), Neste Oil (Finland), OilFox (Argentina), Pond Biofuels (Canada), Total (France), and Varican Aqua Solutions (UK) (Saber et al. 2016).

Typically, algal biomass contains three major compounds in varying proportions: lipids, proteins, and carbohydrates. During photosynthesis, microalgae capture CO_2, resulting in the synthesis of carbohydrates. At this stage, lipid content can be varied based on some stress factors such as nitrogen starvation, which causes the photosynthetic mechanism to switch to accumulating lipids. The lipid content as a precursor for fatty acid depends on the algal strain (Demirbas and Demirbas 2011). The typical oil contents of most widely used microalgae are given in Table 11.2. The

Table 11.1 Oil content of different feedstocks for the production of first-, second-, and third-generation biofuels (Baskar and Aiswarya 2016)

Type of oil	Feedstock	Oil content (wt%)
Edible	Soybean	15–20
	Rapeseed	38–46
	Sunflower	25–35
	Peanut	45–55
	Coconut	63–65
	Palm	30–60
Nonedible	*Jatropha* seed	35–40
	Pongamia pinnata	27–39
	Neem	20–30
	Castor	53
Other sources	Rubber seed	40–50
	Sea mango	54
	Cottonseed	18–25
	Microalgae	30–70

Table 11.2 Oil contents of microalgae species (Shuba and Kifle 2018)

Microalgae	Oil content (wt% on dry basis)
Botryococcus braunii	25–75
Chlorella sp.	28–32
Crypthecodinium cohnii	20
Cylindrotheca sp.	16–37
Dunaliella primolecta	23
Isochrysis sp.	25–33
Monallanthus salina	>20
Nannochloris sp.	20–35
Nannochloropsis sp.	31–68
Neochloris oleoabundans	35–54
Nitzschia sp.	45–47
Phaeodactylum tricornutum	20–30
Schizochytrium sp.	50–77
Tetraselmis suecica	15–23

productivity of algae-derived biofuels is approximately in two orders of magnitude more than that from terrestrial oilseed crops. Algal biodiesel has lower melting point and better cold flow properties owing to the presence of polyunsaturated fatty acids (Demirbaş 2008).

11.3 Hydrothermal Liquefaction of Algae

Although lipid conversion technology via HTL is relatively recognized, it requires algae with a high lipid content to be economically feasible. Thus, rapid reaction and use of feedstocks (considering the high moisture content in algae) with no limitation in terms of lipid content make the HTL process an appropriate method for producing biocrude oil. One of the problems related to the liquefaction of algae is the difficulty to convert high-quality hydrocarbons than lipids in biocrude oil because it contains high concentrations of nitrogen, and consequently the biocrude oil is richer in heteroatoms.

The typical reaction network of algae HTL is presented in Fig. 11.1, which includes the hydrolysis of major components such as lipid, protein, and carbohydrate. The biocrude oil extracted from liquefaction has higher yield and quality, moderate oxygen concentration, and higher heating value (HHV) in the range of 25–35 MJ/kg as compared to those of the traditional pyrolysis biocrude oils (14–20 MJ/kg) (Yang et al. 2016). Furthermore, the advantage of HTL is the formation of distinct oil and water phases, whereas pyrolysis oil contains a substantial quantity of water and oxygenated compounds.

Hydrothermal liquefaction process includes three main stages: depolymerization, decomposition, and recombination. During the depolymerization of biomass, long-chain macromolecules consisting of hydrogen, carbon, and oxygen are

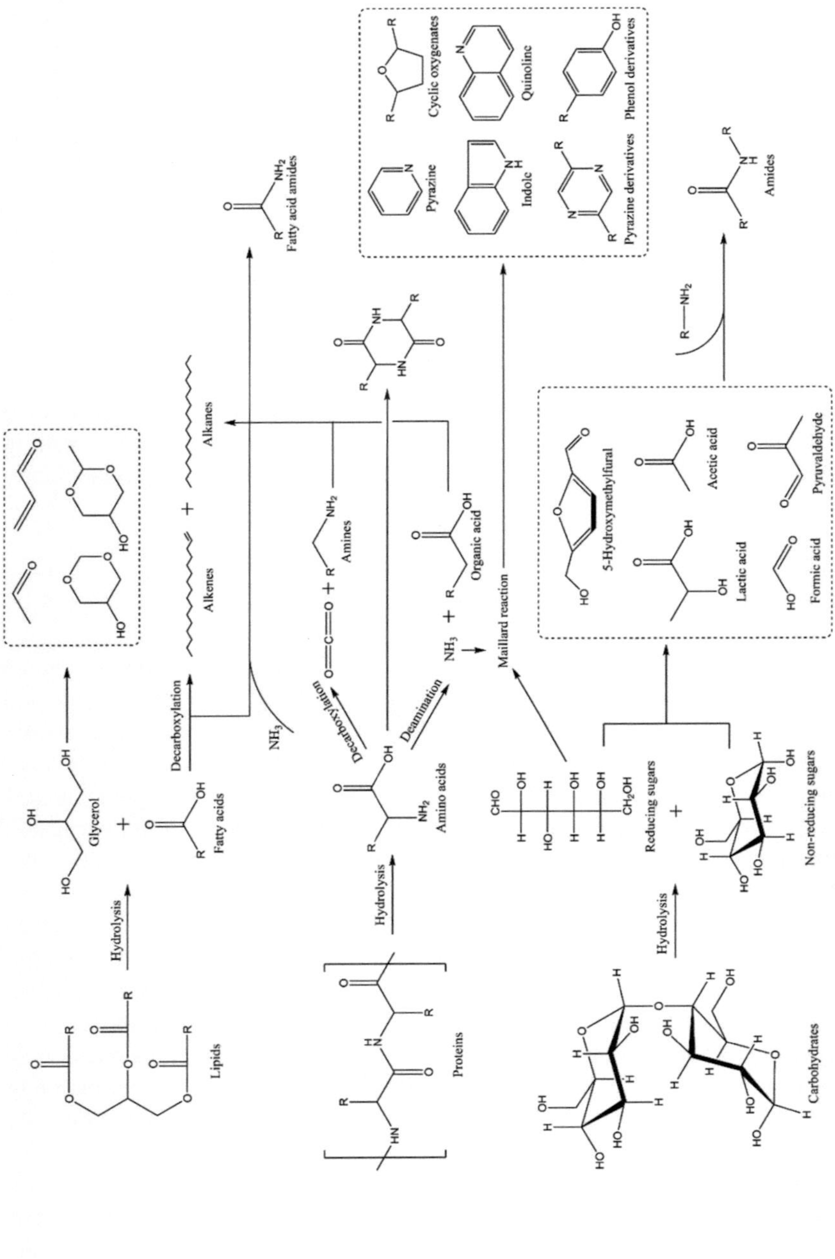

Fig. 11.1 Typical reaction network of algae HTL (Xu et al. 2018)

converted into short-chain molecules under high temperature and pressure conditions. Decomposition of biomass involves dehydration (loss of water molecules), deamination (loss of amino acid content), and decarboxylation (loss of CO_2). The recombination of the fragments forming compounds with high molecular weight occurs when a large number of free radicals are present during the process (Gollakota et al. 2018).

As can be seen in Fig. 11.2, four phases were generated after the HTL process: light gases (principally CO_2), a solid residue (biochar), biocrude oil, and an aqueous phase having a high organic carbon content. The relative reaction rates are strongly dependent on the nature of the feedstocks and processing conditions such as reaction temperature, residence time, and biomass loading affecting the ultimate product distribution and composition (Nanda et al. 2016d). Biochar is a carbon-rich solid product of thermochemical biomass conversion technologies having potential applications in environmental (adsorption of pollutants from wastewater and air, as well as carbon sequestration), agricultural (soil fertility and crop productivity), and material engineering (activated carbon, carbon nanotubes, electrode materials, fuel cells, etc.) (Mohanty et al. 2013, Azargohar et al. 2014, Nanda et al. 2016a).

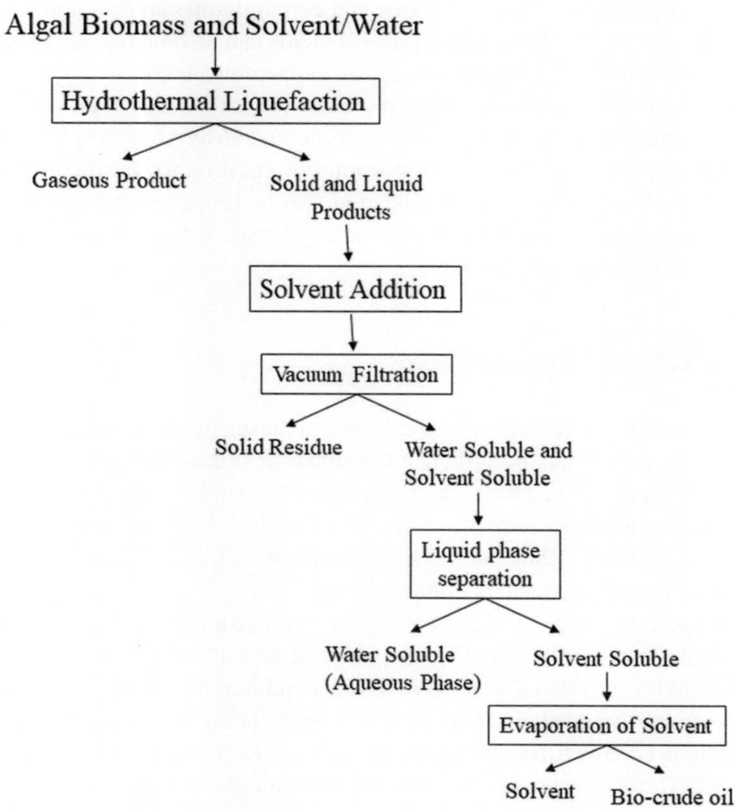

Fig. 11.2 Process flow diagram for the extraction of hydrothermal liquefaction products

11.4 Effects of Hydrothermal Liquefaction Process Parameters on Biocrude Oil Yield

The yield and physicochemical properties of the biocrude oil obtained from the liquefaction of algae are impacted by operating factors such as reaction time, process temperature, solvent type and solvent-to-biomass ratio, algae composition, catalyst type, and loading. This section elaborates the effects of all these process parameters on biocrude oil yield.

11.4.1 Effects of Reaction Temperature

Temperature is considered as an important factor in the safety and economics of industrial operation, suitable range of operating temperature relied on the nature of biomass feedstock, solvent polarity, catalyst loading, and other process factors. The ionic characteristic of water, which changes with temperature, causes different reactions to dominate. At low-temperatures, hydrolysis dominates, thereby biocrude oil yield is reduced. However, biocrude oil yield increases with increasing reaction temperature and then decreases at a certain temperature (Dimitriadis and Bezergianni 2017). The highest biocrude oil yields can be obtained at the temperature range of 250–370 °C. Furthermore, as temperature increases, biocrude with higher quality (higher HHV) is produced, while the carbon and hydrogen contents present in the aqueous phase are reduced. Simultaneously, the nitrogen content in the biocrude oil starts to increase significantly, suggesting higher incorporation of protein-derived molecules (Garcia Alba et al. 2011). It is clear that maximum biocrude yield does not correspond to the best biocrude oil quality, and these two factors must be carefully balanced.

11.4.2 Effects of Reaction Time

Reaction time is a critical factor to be considered during the HTL of algae. To evaluate the process economically, sufficient reaction time is necessary to have maximum biocrude oil through the conversion of algal biomass components. A longer reaction time results in lower biocrude oil yields because of the higher production of gases and aqueous products (Zhou et al. 2010). On the other hand, reduced reaction time leads to lower equipment and operational costs.

Anastasakis and Ross (2011) investigated the optimum reaction time to have higher biocrude oil yield. Their results showed that 15 min at a temperature of 350 °C could be considered as an appropriate condition for the HTL conversion of algae. However, these values are generally reported at the reaction temperature and do not include heating times. Furthermore, the increase in reaction times resulted in increased N/C ratios, whereas the O and H concentrations in the oil dropped. This shows that similar to the reaction temperatures, the holding times need to be carefully adjusted to obtain an optimal balance between biocrude yields and quality.

11.4.3 Effects of Solvent

As can be seen in Fig. 11.3, the efficiency of HTL depends on the critical point of water or other solvents used during the reaction. At near-critical conditions, water, which is a polar solvent, converts into nonpolar solvent due to weak hydrogen bonding (Rana et al. 2018). In this situation, water as a nonpolar solvent is able to dissolve and extract the organic components from the biomass (Peterson et al. 2008, Reddy et al. 2016, 2017, 2019). Furthermore, near critical point, the water dissociation constant (K_w) is higher in several orders of magnitude than at ambient conditions, significantly increasing the ionic products of water, i.e., H^+ and OH^- ions, which help to promote base- and acid-catalyzed reactions, as well as ring-opening reactions (Toor et al. 2013, Gong et al. 2017a, b).

Singh et al. (2015) studied the effects of various solvents such as water and alcohol, including methanol and ethanol, on the product distribution of the HTL process. The results showed that supercritical alcohols used for HTL process are effective to produce liquid hydrocarbons. Furthermore, Biswas et al. (2017) confirmed that biocrude oil yield increases with the use of alcoholic solvents. Zhang et al. (2014) studied liquefaction of algae in ethanol-water and cosolvent system to produce biocrude oil. Their results showed that compared to monosolvent, the mixtures of solvents with different polarities yield higher biocrude oil and less solid residue.

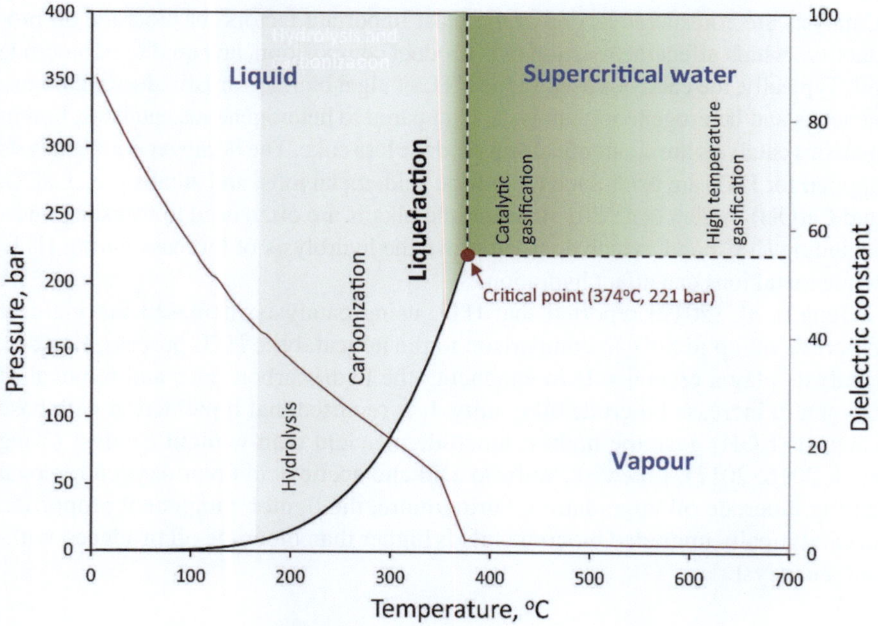

Fig. 11.3 Phase diagram of hydrothermal liquefaction reaction (Tran et al. 2017)

11.4.4 Effects of Algal Composition and Feedstock Loading

Microalgae are predominantly composed of three main biochemical compounds, namely proteins, lipids, and carbohydrates. Their contributions are dependent on the algal species and their growth conditions. Lipid-rich algae are found to have much higher oil yield than protein-rich algae. Some studies have confirmed that lipids are more readily converted into biocrude oil than proteins or carbohydrates (Gollakota et al. 2018). Research on the HTL of algae has been conducted over a wide range of biomass loadings (1–50 wt%) (Li et al. 2010). Peterson et al. (2008) proposed that biomass concentrations should be in the range of 15–20 wt% for higher biocrude oil yield, whereas Barreiro et al. (2013) suggested slightly lower loadings ranging from 5 to 15 wt%, and these studies were mainly focused on *Chlorella*, *Nannochloropsis*, *Dunaliella*, *Spirulina*, and *Phaeodactylum*.

Biller and Ross (2011) converted a number of model compounds, which showed that the highest oil yields were obtained from lipids (55–80%), followed by proteins (11–18%) and carbohydrates (6–15%). They obtained similar biocrude oil quantities from two microalgae, *Chlorella* and *Nannochloropsis*, whereas the obtained yields were different for *Porphyridium* and *Spirulina*. Toor et al. (2013) found that *Nannochloropsis salina* as feedstock seems to be more appropriate than *Spirulina* for biocrude oil production based on the H/C ratio in the biocrude oils.

11.4.5 Effects of Catalysts

Catalysts are considered as one of the most important factors for biocrude oil production, which affect the reaction rate, product composition, and quality of biocrude oil. Typically, the catalysts used in the HTL of algal biomass are divided into homogeneous and heterogeneous catalysts. Compared to heterogeneous catalysts, homogeneous catalysts are economical and produce less coke. The homogeneous catalysts applied for HTL are acids such as sulfuric acid, metal ions, and alkalis (i.e., $CaCO_3$ and $Ca(OH)_2$) (Tian et al. 2014). Acids and alkalis are often used to weaken bonds, including C–C bond, which could improve the hydrolysis of biomass during HTL, while metal ions can affect hydration.

Jena et al. (2012) reported that HTL using catalysts increased the yield of biocrude oil up to 50% in comparison to the noncatalytic HTL process. Besides, catalysts play a crucial role in enhancing the hydrocarbon ratio and removal of oxygen to increase biocrude oil quality. It is reported that liquefaction with base catalyst (KOH) gave the highest biocrude oil yield than without catalyst (Yang et al. 2016, 2017). Likewise, sulfuric acid and acetic acid favor oxygen removal during biocrude oil upgradation. Furthermore, the lighter component proportion in catalytically upgraded biocrude oils is higher than biocrude oil produced without a catalyst.

11.5 Algal Biocrude Oil Upgradation Techniques

Algal biocrude oil produced by HTL process has similar properties to that of crude oil derived from fossils. On the other hand, biocrude oil contains higher oxygen (10–20 wt%) and nitrogen (1–8 wt%). The presence of these heteroatoms causes several undesired properties, as mentioned below, which limit its direct application in compression-ignition (CI) engines (Roussis et al. 2012):

1. High viscosity
2. High corrosiveness because of high amount of fatty acids
3. Thermal and chemical instability
4. Low heating value owing to higher oxygenated compounds

Therefore, the quality of biocrude oil needs to be improved to be used as a synthetic liquid transportation fuel. Due to the identical physicochemical properties of vegetable oils, the technologies used for biodiesel production from plant and vegetable seed oils can be applied to algal biocrude oils. There are varieties of techniques for biocrude oil upgradation, such as solvent addition, emulsification, esterification, transesterification, hydrotreating, hydrodeoxygenation, and catalytic hydrotreating.

11.5.1 Solvent Addition

Solvent addition enhances the quality of the biocrude oil in terms of homogeneity, stability, acidity, and viscosity to be used as liquid transportation fuels (Oasmaa et al. 2004). Further, the addition of polar solvents enhances the heating value of biocrude oils. On the other hand, the addition of polar solvents promotes aging reactions such as polymerization and condensation due to large polar differences among the biocrude oil components and solvents, which results in phase separation and increase in viscosity (Pidtasang et al. 2013). Therefore, to increase stability and decrease viscosity, aging could be slowed through the irreversible reaction of oligomers with low molecular weight reactants or the reversible reaction of acidic oligomers with alcohol, e.g., methanol (Liu et al. 2014). Diebold and Czernik (1997) studied the development of adding solvent to enhance the stability of biocrude oil viscosity during storage. Their results showed that using methanol as the best additive enhances the stability of biocrude oil viscosity during the storage time of 96 h.

Boucher et al. (2000) modified the biocrude oil via solvent addition and found that it meets the ASTM No. 4 specifications of diesel fuel in terms of viscosity. Oasmaa et al. (2004) investigated the upgrading of the biocrude oil quality by methanol addition and found significant reduction in viscosity during the aging of softwood pyrolysis liquids. In addition, the stability of the biocrude oil was enhanced by the addition of more than 10% alcohol; thereby aging reaction was suppressed for an year. Liu et al. (2014) investigated the effects of adding acetone as a solvent on the properties of biocrude oils. The results indicated that with an increase in acetone concentration, the final viscosities decreased. In addition, gas chromatography-mass spectrometry (GC-MS) data indicated that acetone probably suppressed aging reactions of biocrude oil. Therefore, the addition of acetone as solvent revealed significant effects on the properties of biocrude oil.

11.5.2 Emulsification

Emulsification is a feasible method used to promote the application of biocrude oil by blending with other fuels using suitable surfactants (Zhang et al. 2013). It does not require chemical reaction for the application of biocrude oils in CI engines. For having a stable emulsion, the following basic conditions must be provided (Leng et al. 2018):

1. The two liquids must be immiscible.
2. Adequate agitation must be applied to dissolve the liquids.
3. An emulsifying agent may be required.

Although emulsification can increase the ignition characteristics of biocrude oil, it is still difficult to develop other fuel characteristics, such as heating value, to meet desirable requirements. Moreover, it requires a large amount of energy to form an emulsion. Stirring intensity and temperature are found to be the most significant parameters affecting the emulsification process. During emulsification, interfacial tension increases along with rotational speed, which causes to have a better emulsion and adsorption of the emulsifier. It was also reported that mixing multiple surfactants is often more efficient than using a single surfactant due to their mutual interaction, which reduces interfacial tension (Liu et al. 2014).

Al-Sabagh et al. (2011) concluded that compared to individual surfactants (Span 80 or EMAROL 85), blends of the surfactants possess superior thermodynamic properties with less interfacial tension. Jiang and Ellis (2009) found the optimum conditions for a stable emulsion with octanol as an emulsifier was found to be at the surfactant dosage of 4 vol%, 1200 rpm stirring intensity at the emulsifying temperature 30 °C with mixing time of 15 min. Furthermore, the storage and thermal stability of the mixture of biocrude oil and biodiesel were measured up to 180 h. During the storage, water content of the mixtures slightly increased with time, while acid number and viscosity showed an overall decreasing trend.

Wang et al. (2012) studied mechanical and ultrasound emulsification methods to blend biocrude oil and diesel. The emulsions prepared via the ultrasound technique were more stable with lower surface tension and smaller droplet size than mechanical emulsions. Xu et al. (2010) investigated the lubricity of biocrude oil and diesel fuel blend and found that blend showed higher lubricity than conventional diesel fuel. However, compared to conventional diesel fuel, anticorrosion properties of the blend were not acceptable.

11.5.3 Hydrodeoxygenation, Hydrogenation, and Hydrocracking

As discussed earlier, biocrude oil has high oxygen content, which leads to undesirable properties such as chemical instability and low heating value. Hydrotreating is a process used to improve heating value by increasing hydrogen content and reducing O, N, and S through catalytic reaction conditions of pressure up to 20 MPa and temperatures in the range of 300–450 °C. The simplest hydrotreating reaction for biocrude oil is mentioned below:

$$\text{Biocrude oil} + H_2 \rightarrow \text{Upgraded biocrude oil} + H_2O \qquad (11.1)$$

The acceptable amount of oxygen present in hydrocarbon liquid fuels should be less than 1 wt%, but the oxygen content of algal biomass is around 40–60 wt%. Therefore, the main reaction involved in hydrotreating is hydrodeoxygenation due to a significant amount of oxygenated compounds present in biocrude oil. Oxygen can be removed as water, CO_2, and/or CO through a combination of decarbonylation, decarboxylation, and hydrodeoxygenation reactions, shown in Scheme 11.1.

Oxygen removal through CO_2 and CO formation leads to lower carbon yield, so removing oxygen, as water is the preferred route. Hydrogenation can also be used to improve biocrude oil quality. It is believed that as the content of H/C present in liquid fuel increases, the quality of the liquid hydrocarbons also increases. The partial cracking of heavy components is also expected during this process. Therefore, hydrocracking and hydrogenation also occur during hydrotreating (Scheme 11.2). Due to H_2 consumption during hydrocracking and hydrogenation, unsaturated compounds become saturated compounds (Scheme 11.2).

Hydrogenation is carried out at moderate conditions, followed by operation at moderate temperature (300–450 °C) and relatively high pressure (75–300 bar) (Saber et al. 2016). High pressure increases the reaction rate, as well as the solubility of hydrogen in the biocrude oil. It also decreases coking in the reactor. Owing to moderate operating conditions, hydrotreating favors lower coking. Cracking can be

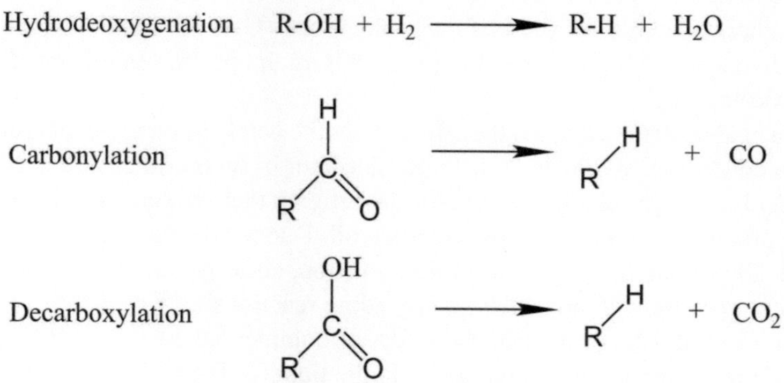

Scheme 11.1 Deoxygenation reactions during hydrotreating technologies, i.e., hydrodeoxygenation, carbonylation, and decarboxylation

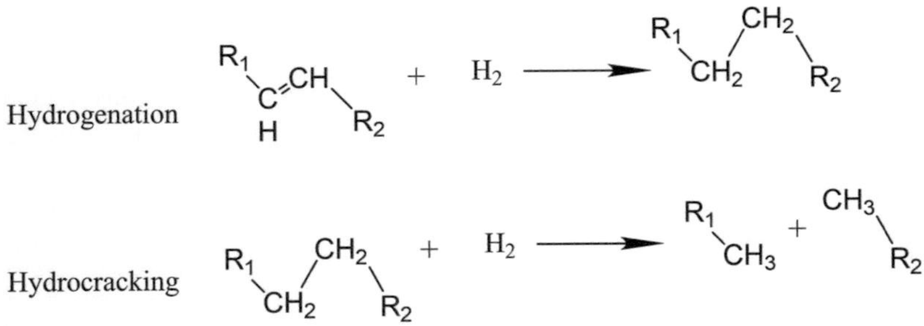

Scheme 11.2 Reactions during hydrogenation, i.e., hydrogenation and hydrocracking

carried out using H-ZSM-5 as catalysts for upgrading biocrude oil. However, hydrotreating and hydrodeoxygenation using zeolite catalysts result in low-grade hydrocarbon fuels; i.e., HHV of these fuels are 25% less than that of biocrude oils produced via catalytic HTL (Mortensen et al. 2011).

11.5.4 Catalytic Hydrotreating

Biocrude oil upgrading through catalytic hydrotreatment promises to produce hydrocarbon-rich fuel. Notable research has focused on the development of catalysts with higher activity and stability for hydrodeoxygenation of biocrude oil, especially at milder reaction conditions (Ramirez et al. 2015). Heterogeneous catalysts have several advantages over traditional homogeneous catalysts such that they can be easily recovered and reused. Different heterogeneous catalysts have been used in biocrude oil upgrading reactions, such as zeolites, noble metals, transition metals, and carbides mostly supported on alumina or activated carbon. CoMo- and NiMo-based catalysts are commercially used for industrial hydrotreating to remove O, N, and S. Due to their instability during the hydrotreating process, some research has focused on noble metal (Pt, Pd, and Ru) catalysts for the hydrotreating of biocrude oil. This helps to convert aromatic compounds into hydrocarbons suitable for diesel fuel applications (Galadima and Muraza 2018). However, due to the higher cost of noble metals, their application is limited to catalytic hydrotreating.

Duan et al. (2016) evaluated the effect of zeolite catalysts on algal biocrude oil at reaction conditions of 400 °C, 6 MPa for 240 min in supercritical water. Nine different zeolites were selected in order to investigate their effects on product yields and the properties of the upgraded biocrude oil. Due to the acidic characteristics of zeolites, all of them improved the denitrogenation, deoxygenation, and desulfurization in comparison with noncatalytic upgrading reactions.

Barreiro et al. (2016) studied the effect of commercial catalysts (Pt/Al_2O_3 and HZSM-5) for biocrude oil upgradation via the liquefaction of *Scenedesmus almeriensis* (freshwater) and *Nannochloropsis gaditana* (marine) algae species. The noncatalyzed reaction of *S. almeriensis* revealed that the highest biocrude oil yield was

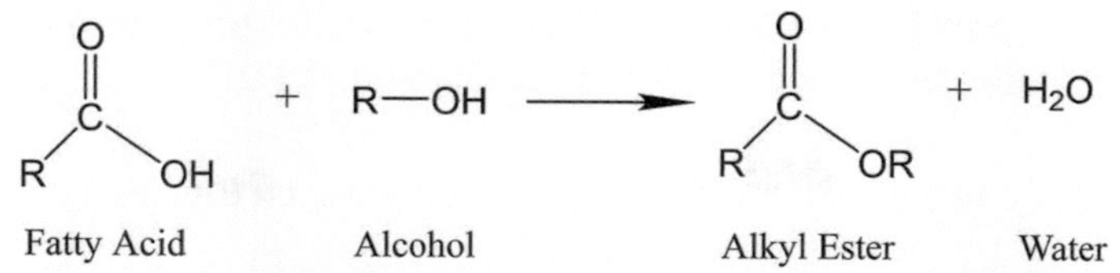

Scheme 11.3 Esterification or transesterification of fatty acid with alcohol

obtained at 4–8 MPa of H_2 pressure, 400 °C of reaction temperature for 4 h in 10 mL micro autoclaves. At these process conditions, catalysts did not show significant activity, and the process was promoted by the temperature rather than the catalysts. The main products obtained during upgradation was found to be 50–70% alkanes, gaseous cracking species, unsaturated fatty acids, and phenols.

Elliott et al. (2013) investigated the catalytic hydroprocessing for algal biocrude oil in a continuous-flow reactor using sulfided CoMo/γ-Al_2O_3 and found that hydrotreating is efficient in removing S and N to undetectable levels. Konwar et al. (2014) studied the hydroprocessing of rapeseed biocrude oil produced via pyrolysis using NiMo/γ-Al_2O_3 catalyst. Their results showed that S, O, and N was reduced to 33.3, 70.8, and 21.1 wt%, respectively. Furthermore, the efficiency of hydroprocessing is enhanced by the removal of lighter hydrocarbons from biocrude oil through fractionation and reducing the biocrude oil LHSV to 0.5 h^{-1}.

Wildschut et al. (2009) achieved 90% hydrodeoxygenation of biocrude oil using Ru supported on carbon, which has a higher conversion in comparison to commercial catalysts (sulfided NiMo/γ-Al_2O_3 and CoMo/γ-Al_2O_3). Tian et al. (2014) examined the catalytic efficiency of Ru, Pd, and Pt on carbon support; Pt/Al_2O_3; Pt/C-sulfide; Rh/Al_2O_3; activated carbon; moly sulfide; moly carbide; Co-Mo/Al_2O_3; and zeolite on the hydrodeoxygenation of biocrude oil. Among all the catalysts studied, Pt supported on carbon catalyst promoted hydrodeoxygenation of biocrude oil.

11.5.5 Esterification and Transesterification

Esterification or transesterification is considered as one of the best routes for increasing the quality of biocrude oil suitable for CI engines. Esterification or transesterification enriches the flow properties (viscosity) and heating value of the biocrude oil by removing glycerol and other carboxylic acid groups. During esterification, fatty acids react with methanol or ethanol in the presence of an acid catalyst to form corresponding methyl or ethyl esters (biodiesel). A wide variety of feedstocks can be used as a source of fatty acids or triglycerides, which include biocrude oil, edible and nonedible oils, animal fats, and lipids from algae. During the transesterification and esterification (Scheme 11.3), different alcohols (methanol, ethanol, propanol, and butanol) were used by varying carbon number to decrease the viscosity and acidity as well as increase the heating value. Methanol is the typical alcohol used during transesterification and esterification due to its low cost. The overall transesterification reaction is shown in Eq. (11.2):

$$\text{Tryglycerides} + 3\,\text{Methanol} \overset{\text{Catalyst}}{\leftrightarrow} \text{Glycerol} + 3\,\text{Methyl esters}\left(\text{Biodiesel}\right) \tag{11.2}$$

During transesterification, excess alcohol has been used to promote alkyl ester production (Eq. 11.2), and the unreacted alcohol can be recovered by Rotavapor and reused. The yield and quality of biodiesel depend on process parameters such as reaction time, temperature, catalyst loading, the type of alcohol, and the nature of feedstocks (Tran et al. 2017). Transesterification process is a catalytic process consisting of the separation of by-products from biodiesel, recovery of glycerol, catalysts, and purification of biodiesel. Conventional transesterification demands higher energy, as well as excessive stirring speed, to mix the feedstocks and alcohols due to their immiscibility. Supercritical fluids promote the reaction due to the unique transportation properties such as gas-like diffusivity and liquid-like density results in dissolving complex feedstocks (Reddy et al. 2014a, b, 2015; Verma et al. 2016; Nanda et al. 2017b; Okolie et al. 2019a, b).

Transesterification can also be performed in supercritical condition to utilize organic acids present in biocrude oil with catalytic and noncatalytic reactions owing to their distinctive properties of dissolving power, faster heat, and mass transfer rates (Demirbaş 2008). The solubility of triglycerides or fatty acids exceptionally increases under supercritical conditions to accelerate the process to a more homogeneous phase (Tran et al. 2017, Nanda et al. 2019). Furthermore, this process requires low energy due to simplified separation and purification steps. It has demonstrated great potential for producing biodiesel with high calorific values and lower viscosity. The conversion of fatty acids and triglycerides into biodiesel is subjected to supercritical process conditions such as reaction temperature and pressure, the nature of alcohol and supercritical solvent, the purity of reactants, the molar ratio of alcohol to oil, and the nature of the feedstocks (Reddy et al. 2018).

The utilization of homogeneous catalysts requires catalyst removal by water washing of biodiesel, which generates a large quantity of wastewater. Therefore, the cost of the production process can be minimized by the utilization of heterogeneous catalysts because they can be easily separated, regenerated, and recycled. Thus, most of the studies concentrated on the improvement of the heterogeneous acidic

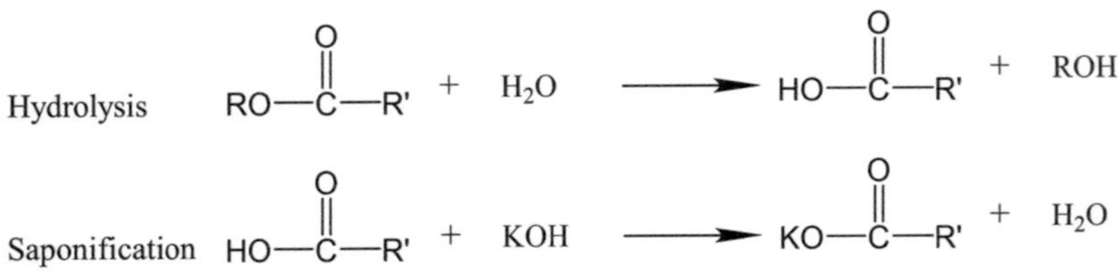

R = Alkyl group of the alcohol
R' = Carbon chain of the fatty acid

Scheme 11.4 Hydrolysis and saponification reactions of fatty acid alkyl esters catalysts for biocrude oil esterification (Hu et al. 2012). Although alkali-catalyzed transesterification of triglyceride is fast, free fatty acids may react with the base catalyst to form soap and water, which leads to the loss of catalysts during the reaction, as shown in Scheme 11.4 (Fukuda et al. 2001). Figure 11.4 shows the schematic representation of biodiesel production via transesterification using heterogeneous catalysts.

Many heterogeneous acidic catalysts such as carbon-based acids and bases, CaO, TiO_2/ZrO_2, MgO, and Amberlyst-15 have been used for biodiesel production with acidic sites on the catalyst surface to adsorb oxygenated compounds. The study by Guo et al. (2015) revealed that 1.56 wt% of catalyst loading, 3.2:10 alcohol-to-biocrude oil molar ratio, 125 min of reaction time at 50 °C of reaction temperature yielded 88.9% of biodiesel. The pros and cons of the transesterification catalysts are tabulated in Table 11.3.

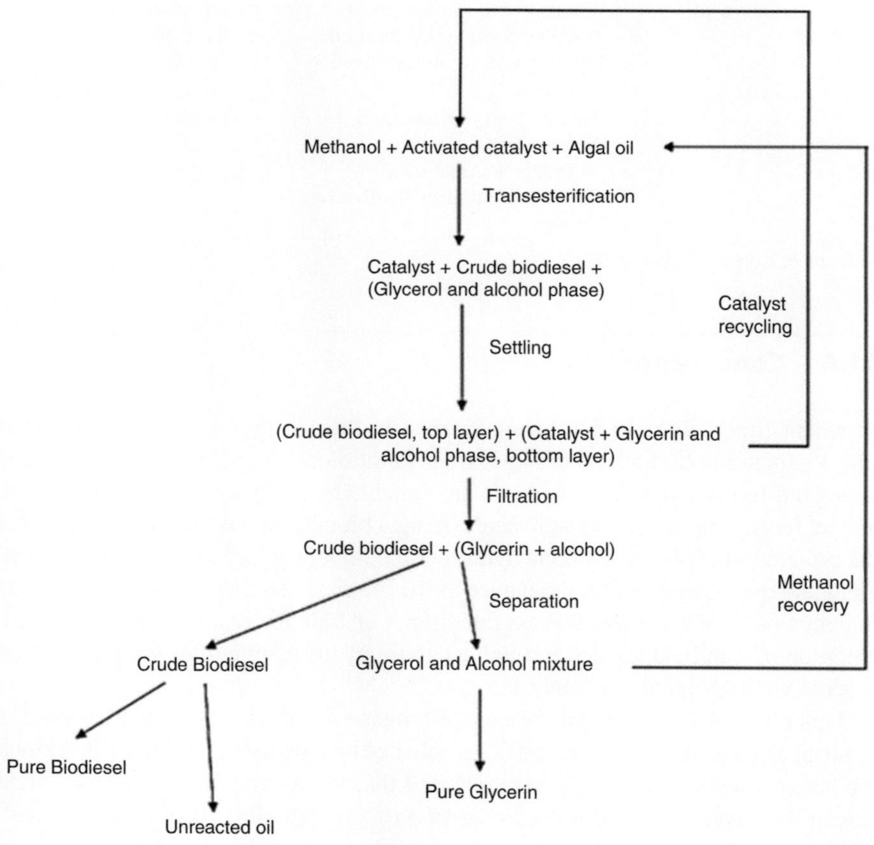

Fig. 11.4 Schematic representation of biodiesel production via the transesterification process using heterogeneous catalysts

Table 11.3 Overview of biodiesel production by transesterification using different catalysts

Type of catalyst	Advantage	Disadvantage
Homogeneous alkali (NaOH and KOH)	• Reacts fast • Inexpensive	• Difficult to reuse • Sensitive to free fatty acids (causes soap formation)
Traditional mineral acids (H_2SO_4 and HCl)	• Insensitive to free fatty acids	• Has slow reaction rate • Corrosive • Difficult to reuse and recover
Heterogeneous alkali catalysts	• Has faster reaction rate compared to acid-catalyzed transesterification	• With expensive synthesis route • Sensitive to water and free fatty acids • Can cause soap formation
Heterogeneous acid catalysts	• Insensitive to free fatty acids • Simultaneously catalyzes both esterification and transesterification	• With expensive synthesis route • Has slow reaction rate
Carbon-based catalysts	• Reusable • Has simple synthesis route • Inexpensive • With high thermal stability • Capable of uniform distribution of active sites	• Has slow reaction rate • Leaches away

Reference: Konwar et al. (2014)

11.6 Conclusions

In recent times, there is a growing interest in the valorization of algal biomass (third-generation biofuels) into liquid transportation fuels owing to higher calorific value and the fact that hydrocarbons are suitable to complement gasoline, diesel, and jet fuels. This chapter established that algal biomass is a potential feedstock for the production of biocrude oil via hydrothermal liquefaction. However, biocrude oil yield and quality depend on the nature of the biomass feedstocks, the nature of the liquefaction catalysts, and process conditions of liquefaction. On the other hand, biocrude oil can be upgraded through catalytic hydrotreating after the liquefaction process via heterogeneous catalysts.

This chapter discussed the various heterogeneous acidic catalysts used for the removal of heteroatoms (N, S, and O) to enhance the quality of biocrude oil. Among the heterogeneous catalysts, it was reported that the maximum biocrude oil yield was in the range of 20–60 wt%. Furthermore, the upgrading of biocrude oil with zeolites proves to be a great means for producing synthetic liquid transportation fuels. Moreover, efficient catalysts and process design need to be developed to enhance biocrude oil yield and the respective hydrocarbon yield for industrial applications. The detailed mechanism that catalyzes hydrocarbon formation in liquefaction and upgradation process, effects of zeolites textural properties for producing specific group of hydrocarbons and their stability in hydrothermal conditions is yet to be understood. Therefore, catalysts, including zeolites, require substantial research and development work for the industrialization and commercialization of liquid transportation fuels derived from algal biomass.

References

Akia M, Yazdani F, Motaee E, Han D, Arandiyan H (2014) A review on conversion of biomass to biofuel by nanocatalysts. Biofuel Res J 1:16–25

Al-Sabagh AM, Emara MM, El-Din MN, Aly WR (2011) Formation of water-in-diesel oil nano-emulsions using high energy method and studying some of their surface active properties. Egypt J Pet 20:17–23

Anastasakis K, Ross AB (2011) Hydrothermal liquefaction of the brown macro-alga *Laminaria saccharina*: effect of reaction conditions on product distribution and composition. Bioresour Technol 102:4876–4883

Azad AK, Rasul MG, Khan MMK, Sharma SC, Hazrat MA (2015) Prospect of biofuels as an alternative transport fuel in Australia. Renew Sust Energ Rev 43:331–351

Azargohar R, Nanda S, Rao BVSK, Dalai AK (2013) Slow pyrolysis of deoiled canola meal: product yields and characterization. Energy Fuel 27:5268–5279

Azargohar R, Nanda S, Kozinski JA, Dalai AK, Sutarto R (2014) Effects of temperature on the physicochemical characteristics of fast pyrolysis bio-chars derived from Canadian waste biomass. Fuel 125:90–100

Barreiro DL, Prins W, Ronsse F, Brilman W (2013) Hydrothermal liquefaction (HTL) of microalgae for biofuel production: state of the art review and future prospects. Biomass Bioenergy 53:113–127

Barreiro DL, Gómez BR, Ronsse F, Hornung U, Kruse A, Prins W (2016) Heterogeneous catalytic upgrading of biocrude oil produced by hydrothermal liquefaction of microalgae: state of the art and own experiments. Fuel Process Technol 148:117–127

Baskar G, Aiswarya R (2016) Trends in catalytic production of biodiesel from various feedstocks. Renew Sust Energ Rev 57:496–504

Biller P, Ross AB (2011) Potential yields and properties of oil from the hydrothermal liquefaction of microalgae with different biochemical content. Bioresour Technol 102:215–225

Biswas B, Kumar AA, Bisht Y, Singh R, Kumar J, Bhaskar T (2017) Effects of temperature and solvent on hydrothermal liquefaction of *Sargassum tenerrimum* algae. Bioresour Technol 242:344–350

Boucher ME, Chaala A, Roy C (2000) Bio-oils obtained by vacuum pyrolysis of softwood bark as a liquid fuel for gas turbines. Part I: properties of bio-oil and its blends with methanol and a pyrolytic aqueous phase. Bioresour Technol 19:337–350

Demirbaş A (2008) Production of biodiesel from algae oils. Energ Sourc A 31:163–168

Demirbas A, Demirbas MF (2011) Importance of algae oil as a source of biodiesel. Energy Convers Manag 52:163–170

Diebold JP, Czernik S (1997) Additives to lower and stabilize the viscosity of pyrolysis oils during storage. Energ Fuels 11:1081–1091

Dimitriadis A, Bezergianni S (2017) Hydrothermal liquefaction of various biomass and waste feedstocks for biocrude production: a state of the art review. Renew Sust Energ Rev 68:113–125

Duan P, Savage PE (2011) Upgrading of crude algal bio-oil in supercritical water. Bioresour Technol 102:1899–1906

Duan P, Xu Y, Wang F, Wang B, Yan W (2016) Catalytic upgrading of pretreated algal bio-oil over zeolite catalysts in supercritical water. Biochem Eng J 116:105–112

Elliott DC, Hart TR, Schmidt AJ, Neuenschwander GG, Rotness LJ, Olarte MV, Holladay JE (2013) Process development for hydrothermal liquefaction of algae feedstocks in a continuous-flow reactor. Algal Res 2:445–454

Fukuda H, Kondo A, Noda H (2001) Biodiesel fuel production by transesterification of oils. J Biosci Bioeng 92:405–416

Galadima A, Muraza O (2018) Hydrothermal liquefaction of algae and bio-oil upgrading into liquid fuels: role of heterogeneous catalysts. Renew Sust Energ Rev 81:1037–1048

Garcia Alba L, Torri C, Samorì C, van der Spek J, Fabbri D, Kersten SR, Brilman DW (2011) Hydrothermal treatment (HTT) of microalgae: evaluation of the process as conversion method in an algae biorefinery concept. Energy Fuel 26:642–657

Gollakota ARK, Kishore N, Gu S (2018) A review on hydrothermal liquefaction of biomass. Renew Sust Energ Rev 81:1378–1392

Gong M, Nanda S, Hunter HN, Zhu W, Dalai AK, Kozinski JA (2017a) Lewis acid catalyzed gasification of humic acid in supercritical water. Catal Today 291:13–23

Gong M, Nanda S, Romero MJ, Zhu W, Kozinski JA (2017b) Subcritical and supercritical water gasification of humic acid as a model compound of humic substances in sewage sludge. J Supercrit Fluids 119:130–138

Guo Q, Wu M, Wang K, Zhang XX (2015) Catalytic hydrodeoxygenation of algae bio-oil over bimetallic Ni–Cu/ZrO_2 catalysts. Ind Eng Chem Res 54:890–899

Hu X, Gunawan R, Mourant D, Lievens C, Li X, Zhang S, Li CZ (2012) Acid-catalysed reactions between methanol and the bio-oil from the fast pyrolysis of mallee bark. Fuel 97:512–522

Jena U, Das KC, Kastner JR (2012) Comparison of the effects of Na_2CO_3, $Ca_3(PO_4)_2$, and NiO catalysts on the thermochemical liquefaction of microalga *Spirulina platensis*. Appl Energy 98:368–375

Jiang X, Ellis N (2009) Upgrading bio-oil through emulsification with biodiesel: mixture production. Energy Fuel 24:1358–1364

Konwar LJ, Boro J, Deka D (2014) Review on latest developments in biodiesel production using carbon-based catalysts. Renew Sust Energ Rev 29:546–564

Leng L, Li H, Yuan X, Zhou W, Huang H (2018) Bio-oil upgrading by emulsification/microemulsification: a review. Energy 161:214–232

Li Y, Wang T, Liang W, Wu C, Ma L, Zhang Q, Jiang T (2010) Ultrasonic preparation of emulsions derived from aqueous bio-oil fraction and 0# diesel and combustion characteristics in diesel generator. Energy Fuel 24:1987–1995

Liu R, Fei W, Shen C (2014) Influence of acetone addition on the physicochemical properties of bio-oils. J Energy Inst 87:127–133

Manayil JC, Inocencio CV, Lee AF, Wilson K (2016) Mesoporous sulfonic acid silicas for pyrolysis bio-oil upgrading via acetic acid esterification. Green Chem 18:1387–1394

Mohanty P, Nanda S, Pant KK, Naik S, Kozinski JA, Dalai AK (2013) Evaluation of the physiochemical development of biochars obtained from pyrolysis of wheat straw, timothy grass and pinewood: effects of heating rate. J Anal Appl Pyrolysis 104:485–493

Mortensen PM, Grunwaldt JD, Jensen PA, Knudsen KG, Jensen AD (2011) A review of catalytic upgrading of bio-oil to engine fuels. Appl Catal A 407:1–19

Nanda S, Mohanty P, Pant KK, Naik S, Kozinski JA, Dalai AK (2013) Characterization of North American lignocellulosic biomass and biochars in terms of their candidacy for alternate renewable fuels. Bioenergy Res 6:663–677

Nanda S, Dalai AK, Kozinski JA (2014a) Butanol and ethanol production from lignocellulosic feedstock: biomass pretreatment and bioconversion. Energ Sci Eng 2:138–148

Nanda S, Mohammad J, Reddy SN, Kozinski JA, Dalai AK (2014b) Pathways of lignocellulosic biomass conversion to renewable fuels. Biomass Conv Biorefin 4:157–191

Nanda S, Azargohar R, Dalai AK, Kozinski JA (2015) An assessment on the sustainability of lignocellulosic biomass for biorefining. Renew Sust Energ Rev 50:925–941

Nanda S, Dalai AK, Berruti F, Kozinski JA (2016a) Biochar as an exceptional bioresource for energy, agronomy, carbon sequestration, activated carbon and specialty materials. Waste Biomass Valor 7:201–235

Nanda S, Dalai AK, Gökalp I, Kozinski JA (2016b) Valorization of horse manure through catalytic supercritical water gasification. Waste Manag 52:147–158

Nanda S, Isen J, Dalai AK, Kozinski JA (2016c) Gasification of fruit wastes and agro-food residues in supercritical water. Energ Convers Manage 110:296–306

Nanda S, Kozinski JA, Dalai AK (2016d) Lignocellulosic biomass: a review of conversion technologies and fuel products. Curr Biochem Eng 3:24–36

Nanda S, Reddy SN, Dalai AK, Kozinski JA (2016e) Subcritical and supercritical water gasification of lignocellulosic biomass impregnated with nickel nanocatalyst for hydrogen production. Int J Hydrogen Energ 41:4907–4921

Nanda S, Reddy SN, Mitra SK, Kozinski JA (2016f) The progressive routes for carbon capture and sequestration. Energ Sci Eng 4:99–122

Nanda S, Golemi-Kotra D, McDermott JC, Dalai AK, Gökalp I, Kozinski JA (2017a) Fermentative production of butanol: perspectives on synthetic biology. New Biotechnol 37:210–221

Nanda S, Gong M, Hunter HN, Dalai AK, Gökalp I, Kozinski JA (2017b) An assessment of pinecone gasification in subcritical, near-critical and supercritical water. Fuel Process Technol 168:84–96

Nanda S, Rana R, Zheng Y, Kozinski JA, Dalai AK (2017c) Insights on pathways for hydrogen generation from ethanol. Sustain Energ Fuel 1:1232–1245

Nanda S, Rana R, Hunter HN, Fang Z, Dalai AK, Kozinski JA (2019) Hydrothermal catalytic processing of waste cooking oil for hydrogen-rich syngas production. Chem Eng Sci 195:935–945

Oasmaa A, Kuoppala E, Selin JF, Gust S, Solantausta Y (2004) Fast pyrolysis of forestry residue and pine. 4. Improvement of the product quality by solvent addition. Energy Fuel 18:1578–1583

Okolie JA, Nanda S, Dalai AK, Kozinski JA (2019a) Optimization and modeling of process parameters during hydrothermal gasification of biomass model compounds to generate hydrogen-rich gas products. Int J Hydrogen Energ. https://doi.org/10.1016/j.ijhydene.2019.05.132

Okolie JA, Rana R, Nanda S, Dalai AK, Kozinski JA (2019b) Supercritical water gasification of biomass: a state-of-the-art review of process parameters, reaction mechanisms and catalysis. Sustain Energ Fuel 3:578–598

Peterson AA, Vogel F, Lachance RP, Fröling M, Antal MJ Jr, Tester JW (2008) Thermochemical biofuel production in hydrothermal media: a review of sub-and supercritical water technologies. Energy Environ Sci 1:32–65

Pidtasang B, Udomsap P, Sukkasi S, Chollacoop N, Pattiya A (2013) Influence of alcohol addition on properties of bio-oil produced from fast pyrolysis of eucalyptus bark in a free-fall reactor. J Ind Eng Chem 19:1851–1857

Ramirez JA, Brown RJ, Rainey TJ (2015) A review of hydrothermal liquefaction bio-crude properties and prospects for upgrading to transportation fuels. Energies 8:6765–6794

Rana R, Nanda S, Kozinski JA, Dalai AK (2018) Investigating the applicability of Athabasca bitumen as a feedstock for hydrogen production through catalytic supercritical water gasification. J Environ Chem Eng 6:182–189

Reddy SN, Ding N, Nanda S, Dalai AK, Kozinski JA (2014a) Supercritical water gasification of biomass in diamond anvil cells and fluidized beds. Biofuels Bioprod Biorefin 8:728–737

Reddy SN, Nanda S, Dalai AK, Kozinski JA (2014b) Supercritical water gasification of biomass for hydrogen production. Int J Hydrogen Energ 39:6912–6926

Reddy SN, Nanda S, Hegde UG, Hicks MC, Kozinski JA (2015) Ignition of hydrothermal flames. RSC Adv 5:36404–36422

Reddy SN, Nanda S, Kozinski JA (2016) Supercritical water gasification of glycerol and methanol mixtures as model waste residues from biodiesel refinery. Chem Eng Res Des 113:17–27

Reddy SN, Nanda S, Hegde UG, Hicks MC, Kozinski JA (2017) Ignition of n-propanol–air hydrothermal flames during supercritical water oxidation. Proc Combust Inst 36:2503–2511

Reddy SN, Nanda S, Sarangi PK (2018) Applications of supercritical fluids for biodiesel production. In: Sarangi PK, Nanda S, Mohanty P (eds) Recent advancements in biofuels and bioenergy utilization. Springer Nature, Singapore, pp 261–284

Reddy SN, Nanda S, Kumar P, Hicks MC, Hegde UG, Kozinski JA (2019) Impacts of oxidant characteristics on the ignition of *n*-propanol-air hydrothermal flames in supercritical water. Combust Flame 203:46–55

Roussis SG, Cranford R, Sytkovetskiy N (2012) Thermal treatment of crude algae oils prepared under hydrothermal extraction conditions. Energy Fuel 26:5294–5299

Saber M, Nakhshiniev B, Yoshikawa K (2016) A review of production and upgrading of algal bio-oil. Renew Sust Energ Rev 58:918–930

Shuba ES, Kifle D (2018) Microalgae to biofuels: 'promising' alternative and renewable energy, review. Renew Sust Energ Rev 81:743–755

Singh R, Bhaskar T, Balagurumurthy B (2015) Effect of solvent on the hydrothermal liquefaction of macro algae *Ulva fasciata*. Process Saf Environ Prot 93:154–160

Tian C, Li B, Liu Z, Zhang Y, Lu H (2014) Hydrothermal liquefaction for algal biorefinery: a critical review. Renew Sust Energ Rev 38:933–950

Toor SS, Reddy H, Deng S, Hoffmann J, Spangsmark D, Madsen LB, Rosendahl LA (2013) Hydrothermal liquefaction of *Spirulina* and *Nannochloropsis salina* under subcritical and supercritical water conditions. Bioresour Technol 131:413–419

Tran DT, Chang JS, Lee DJ (2017) Recent insights into continuous-flow biodiesel production via catalytic and non-catalytic transesterification processes. Appl Energy 185:376–409

Tsukahara K, Sawayama S (2005) Liquid fuel production using microalgae. J Jpn Pet Inst 48:251–259

Van HJW, Huijgen WJJ, López-Contreras AM (2014) Opportunities and challenges for seaweed in the biobased economy. Trends Biotechnol 32:231–233

Verma P, Sharma MP, Dwivedi G (2016) Impact of alcohol on biodiesel production and properties. Renew Sust Energ Rev 56:319–333

Wang XY, Guo ZG, Wang SR (2012) Emulsion fuels production between diesel and bio-oil middle fraction from molecular distillation. Adv Mater Res 534:151–155

Wildschut J, Mahfud FH, Venderbosch RH, Heeres HJ (2009) Hydrotreatment of fast pyrolysis oil using heterogeneous noble-metal catalysts. Ind Eng Chem Res 48:10324–10334

Xu Y, Wang Q, Hu X, Li C, Zhu X (2010) Characterization of the lubricity of bio-oil/diesel fuel blends by high frequency reciprocating test rig. Energy 35:283–287

Xu D, Lin G, Guo S, Wang S, Guo Y, Jing Z (2018) Catalytic hydrothermal liquefaction of algae and upgrading of biocrude: a critical review. Renew Sust Energ Rev 97:103–118

Yang L, Nazari L, Yuan Z, Corscadden K, Xu CC (2016) Hydrothermal liquefaction of spent coffee grounds in water medium for bio-oil production. Biomass Bioenergy 86:191–198

Yang W, Li X, Zhang D, Feng L (2017) Catalytic upgrading of bio-oil in hydrothermal liquefaction of algae major model components over liquid acids. Energ Convers Manage 154:336–343

Zhang L, Liu R, Yin R, Mei Y (2013) Upgrading of bio-oil from biomass fast pyrolysis in China: a review. Renew Sust Energ Rev 24:66–72

Zhang C, Duan P, Xu Y, Wang B, Wang F, Zhang L (2014) Catalytic upgrading of duckweed biocrude in subcritical water. Bioresour Technol 166:37–44

Zhou D, Zhang L, Zhang S, Fu H, Chen J (2010) Hydrothermal liquefaction of macroalgae *Enteromorpha prolifera* to bio-oil. Energy Fuel 24:4054–4061

Co-pyrolysis of Lignocellulosic Biomass and Polymeric Wastes for Liquid Oil Production

12

Wan Adibah Wan Mahari, Shin Ying Foong, and Su Shiung Lam

Abstract

Co-pyrolysis has been researched as a promising technique for thermochemical conversion of biomass with plastics and other organic wastes into potentially useful products due to the ability to recover energy from these materials. This chapter presents a review of the development and application of co-pyrolysis as a next-generation thermochemical conversion technique for energy recovery from biomass waste materials (e.g., food waste, waste cooking oil etc.). This includes a discussion on the characteristics of co-pyrolysis, the key process parameters, the issues, and challenges arising from the application of co-pyrolysis, and the features and combustion performance of the liquid oil product. Co-pyrolysis using microwave technique shows ability to rectify certain limitations shown by conventional pyrolysis with a potential as a viable means to produce next-generation fuels from biomass wastes. This co-pyrolysis approach is capable of increasing the yield and enhancing the quality of the bio-oil product. Thus, it can be a potentially sustainable and feasible approach for energy recovery from biomass and wastes polymers.

Keywords

Pyrolysis · Co-pyrolysis · Microwave · Biomass · Plastics · Bio-oil

W. A. W. Mahari · S. Y. Foong · S. S. Lam (✉)
Pyrolysis Technology Research Group, Institute of Tropical Aquaculture and Fisheries, Universiti Malaysia Terengganu, Kuala Terengganu, Terengganu, Malaysia
e-mail: lam@umt.edu.my

S. Nanda et al. (eds.), *Biorefinery of Alternative Resources: Targeting Green Fuels and Platform Chemicals*, https://doi.org/10.1007/978-981-15-1804-1_12

12.1 Introduction

The global production of biomass waste such as waste cooking oil and household organic waste is estimated around 29 million tons each year (Maddikeri et al. 2012). Waste cooking oil is commonly discharged into plumbing system, but this practice has led to the accumulation of fatberg that can cause sewer blockage. Owing to these problems, valorization of biomass waste materials is proposed to transform these waste materials into second-generation fuel. Valorization of waste materials can be defined as an increase in the value of worthless waste materials by producing value-added products that can be used in many applications (e.g., energy source). In addition, second-generation biofuel is referred to the fuels produced from non-food materials such as lignocellulosic biomass or other organic residues including waste cooking oil by several valorization techniques.

Recently, pyrolysis technique has received more attention in transforming biomass wastes into an energy source with improved fuel properties (Lam et al. 2019, 2016; Lam and Chase 2012). Pyrolysis is a thermochemical conversion method performed in an inert atmosphere (e.g., vacuum atmosphere or nitrogen atmosphere) at high process temperatures (400–900 °C) to generate three types of fuels (e.g., liquid fuel, gaseous fuel, and solid fuel) (Lam et al. 2016; Nam et al. 2017). The distinct advantages shown by pyrolysis may lead to the potential for a greater production of desirable pyrolysis products that can be used in many applications such as diesel engines, boilers, and turbines for the generation of power and electricity, while at the same time serving the purpose of recovery and treatment of these wastes. Currently, there are a few pyrolysis technologies developed to treat and recover the energy content from biomass wastes such as waste cooking oil and food waste.

12.2 Co-pyrolysis of Waste Biomass and Polymeric Wastes

In recent years, co-pyrolysis of waste materials with other lignocellulose or triglyceride materials has attracted attention. Table 12.1 shows numerous studies on co-pyrolysis of biomass, plastic wastes, and other waste materials to produce pyrolysis products that can be used as chemical additives or fuels. Several studies have revealed that co-pyrolysis of biomass wastes is capable of enhancing the yield and properties of liquid oil, which are nearly comparable to commercial transportation-grade fuel.

Co-pyrolysis is an effective upgrading technique that not only increases the yield of liquid oil, but also improves the properties of liquid oil by reducing the concentrations of oxygenated compounds and increasing the levels of aliphatic hydrocarbon components and the overall calorific value of the oil. Co-pyrolysis of biomass and waste materials has been demonstrated to be a reliable technique in improving the stability of liquid oil through the interactions of hydrocarbon radicals that reduce the oxygen content in liquid oil, thereby improving its stability as a high-grade fuel. Many studies have revealed that the key factor that contributes to the accomplishment of this technique is the synergistic effect and the interactive reaction mechanisms during co-pyrolysis of biomass and waste materials.

Table 12.1 Summary of studies on co-pyrolysis of biomass, plastic, and other types of waste materials

Feedstocks			Process	Product yield (wt%)			Elemental composition (wt%)					Calorific value	
Biomass	Plastic	Other wastes	configuration	Oil	Gas	Char	Carbon	Hydrogen	Nitrogen	Sulfur	Oxygen	(MJ/kg)	References
–	High-density polyethylene	Waste cooking oil	Quartz reactor (heated by microwave)	81.0	18.0	1.0	85.0	14.0	1.0	0.0	0.0	46.0	Mahari et al. (2018a)
Potato skin	High-density polyethylene	–	Stainless steel reactor (heated by furnace)	50.9	9.16	21.0	80.5	12.5	0.3	–	6.7	44.0	Önal et al. (2012)
Wood	Polystyrene	–	Fixed bed reactor (heated by furnace)	58.2	21.5	20.2	–	–	–	–	–	–	Ephraim et al. (2018)
Wood	Polyvinyl chloride	–	Fixed bed reactor (heated by furnace)	78.6	8.30	13.1	46.6	5.9	0.7	0.0	29.4	–	Ephraim et al. (2018)
Sugarcane bagasse	Low-density polyethylene	–	Stainless steel semi-batch reactor	52.8	39.0	8.2	75.4	9.3	0.2	0.0	15.1	40.0	Dewangan et al. (2016)
Rice bran	Polypropylene	–	Semi-batch reactor (heated by furnace)	80.5	–	–	83.5	16.1	0.0	0.3	0.0	43.76	Kumari and Singh (2019)

(continued)

Table 12.1 (continued)

Feedstocks				Product yield (wt%)			Elemental composition (wt%)					Calorific value (MJ/kg)	References
Biomass	Plastic	Other wastes	Process configuration	Oil	Gas	Char	Carbon	Hydrogen	Nitrogen	Sulfur	Oxygen		
Biomass	Low-density polyethylene	–	Dropdown tube pyrolyzer	64.1	7.98	9.32	–	–	–	–	–	–	Yang et al. (2016)
Bamboo sawdust	–	Waste tire	Quartz reactor (heated by microwave)	30.2	33.0	37.0	–	–	–	–	–	–	Wang et al. (2017b)
Grape seed	–	Waste tire	Fixed bed reactor (heated by furnace)	40.3	23.5	32.4	65.1	10.3	2.9	0.1	21.5	32.6	Sanahuja-Parejo et al. (2018)
Kitchen waste	–	Microalgae	Fixed bed reactor (heated by furnace)	57.0	–	–	–	–	–	–	–	33.9	Chen et al. (2018)
–	High-density polyethylene	Newspaper waste	Quartz tube reactor (heated by furnace)	68.4	25.0	6.6	84.4	12.9	0.3	–	7.7	34.79	Chen et al. (2016)

It is worth mentioning that there are several factors influencing the synergistic effects during the co-pyrolysis process such as the type of feedstock, feedstock mixture ratio, reactor temperature, heating rates, reactor pressure, and reaction time. In addition, co-pyrolysis of biomass and waste materials has shown an effective alternative for waste management problems in reducing the massive volumes of waste materials. The use of biomass waste and household waste materials for co-pyrolysis can reduce the cost for waste disposal and lessen the amount of space or land needed for landfilling of the wastes. Therefore, co-pyrolysis of biomass and waste materials could offer the advantages in waste management and recovering energy from biomass and waste materials. Nevertheless, it is crucial to understand the reaction mechanism and interactions during the co-pyrolysis process to improve the overall planning, operations, process design, and development of the process.

Chen et al. (2014) investigated co-pyrolysis of corncob and waste cooking oil in a fixed-bed reactor. The influence of different temperatures (500–600 °C) and corn cob/waste cooking oil mass ratios (1:0, 1:01, 1:0.5, 1:1 and 0:1) were investigated. A stainless-steel tube was used as the reactor and temperature controller was used to control the furnace temperature. The findings showed that the temperature of 550 °C with corn cob-to-waste cooking oil mass ratio of 1:1 produced the highest bio-oil yield (68.9 wt%). The bio-oil comprised of fatty acids, aldehydes, alcohols, phenols, ketones, and furans with a high calorific value of 32.8 MJ/kg.

Tang et al. (2018) studied the co-pyrolysis of food waste (soybean protein) and plastic wastes (Polyvinyl chloride, PVC) in a fixed bed reactor. Nitrogen was purged at a flow rate of 100 mL/min to maintain an inert environment during co-pyrolysis. Different ratios of food waste and plastic waste (4:1, 1:1 and 1:4) as well as process temperatures (400–600 °C) were investigated in this study. It was revealed that the addition of the high proportion of PVC in the food waste (1:4 of food waste-to-PVC ratio) had accelerated the occurrence and intensity of co-pyrolysis reactions that led to the reduction of tar and char yields (Tang et al. 2018). Aromatic compounds such as benzenes and polycyclic aromatic hydrocarbons (PAHs) dominated the tar product. This may be attributed to the release of HCl from PVC during co-pyrolysis reactions that promoted the conversion of light tar compounds into heavy compounds (e.g., benzenes and PAH). The addition of PVC into food waste also contributed to the reduction of nitrogen-containing compounds (amine-N, pyridine-N, pyrrole-N, indol-N, and nitrile-N) in the tar product (from 66% to 15%). This was due to the formation of HCl form PVC that acted like an acid catalyst to suppress the formation of nitrogen-containing compounds during co-pyrolysis. Nevertheless, no chlorinated compounds in the tar product were detected. It should be noted that PVC contains halogen (Cl^-). Thus, dehydrochlorination may occur at high process temperatures (Sharuddin et al. 2016). Therefore, gases product may be released and lead to low tar or liquid yields. The gases product may comprise of chlorine that could lead to the corrosiveness of the pyrolysis system and product storage. Kim (2001) who studied pyrolysis of PVC waste also supported this view.

Mahari et al. (2018a) studied the co-pyrolysis of waste cooking oil and plastic waste (high-density polyethylene, HDPE) in a microwave pyrolysis system. Different ratios of waste cooking oil and plastic waste (1:2, 1:1.5, 1:1, 1.5:1, and 2:1) were investigated in this study. It was revealed that a high yield of liquid oil (up to 81 wt%) with desirable fuel properties was obtained. It should be noted that plastic waste has higher carbon (85 wt%) and hydrogen content (15 wt%), but possess lower oxygen content (<1 wt%) compared to waste cooking oil (carbon: 72 wt%, hydrogen: 13 wt%, and oxygen: 15 wt%). Hence, plastic waste could provide more hydrogen to the waste cooking oil to improve the hydrocarbon content (hydrogen and carbon) in the liquid oil.

Plastic waste has nearly no oxygen compared to waste cooking oil, thereby leading to a positive synergistic effect to reduce the oxygenated compounds and improve the fuel properties of co-pyrolysis liquid oil. In particular, this can contribute to the high calorific value of liquid oil (46 MJ/kg) obtained from co-pyrolysis process compared to that obtained from microwave pyrolysis of waste cooking oil alone (40–43 MJ/kg) (Lam et al. 2017) and microwave pyrolysis of plastic waste alone (41 MJ/kg) (Undri et al. 2014). Wang et al. (2017a) studied on the co-pyrolysis of waste vegetable oil and HDPE to produce hydrocarbon fuels. It was reported that the co-pyrolysis process produced 60 wt% of liquid oil at the process temperature of 430 °C (Wang et al. 2017a). The liquid oil consisted of high composition of aliphatic hydrocarbons (e.g., alkanes, cycloalkanes and olefins) and low oxygenated compounds (<5%). The low oxygenated compounds in the liquid oil resulted in the low viscosity of liquid oil (3.31 mm^2/s), which is nearly comparable to the viscosity of diesel (3–8 mm^2/s). The low oxygenated compounds also led to the high calorific value of liquid oil (46.2 MJ/kg), which is higher than biodiesel (38 MJ/kg) and diesel (45 MJ/kg) (Wang et al. 2017a).

Önal et al. (2012) studied the co-pyrolysis of potato skin and HDPE waste in an electric furnace. The potato skin and HDPE were mixed with different ratios (1:2, 1:1 and 2:1). It was reported that higher yield of liquid oil (50.9 wt%) was obtained from the co-pyrolysis of potato skin and HDPE compared to that obtained from the pyrolysis of potato skin alone (~24 wt%). This was attributed to the high hydrogen content of HDPE that enhanced the hydrogenation reaction during the co-pyrolysis process, thereby increasing the yield of liquid oil (Önal et al. 2012).

Yang et al. (2016) investigated the co-pyrolysis of lignocellulosic biomass (e.g., cedar wood and sunflower stalk) and low-density polyethylene (LDPE) for liquid oil production. The pyrolysis of cedar wood and sunflower stalk alone produced 38.8 wt% and 29.9 wt% of liquid oil, respectively, whereas the pyrolysis of LDPE alone produced 83.3 wt% of liquid oil. It was revealed that co-pyrolysis of cedar wood with LDPE and co-pyrolysis of sunflower stalk with LDPE improved the production of co-pyrolysis liquid oil up to 64.1 wt% and 30.4 wt%, respectively. The studies by Önal et al. (2012) and Yang et al. (2016) clearly showed the positive synergistic effects of the co-pyrolysis of biomass materials (potato skin, cedar wood, and sunflower stalk) with plastic material (HDPE and LDPE), which enhanced the yield of co-pyrolysis liquid oil.

Chen et al. (2016) studied the co-pyrolysis of newspaper waste and HDPE for liquid oil production. The newspaper waste-to-HDPE ratios investigated were 1:0, 2:1, 1:1, 1:2, and 0:1. The results showed that the synergistic effect occurring at 1:2 newspaper waste-to-HDPE ratio produced the highest yield of liquid oil (68.4 wt%)

with lower yield of gases (~25 wt%) and solid char (~6.6 wt%). This was attributed to the transfer of hydrogen atoms and generation of free radicals that led to the cross-reactions between newspaper waste and HDPE. These reactions interfered with the degradation of functional groups bound to the cellulose structure of newspaper waste and inhibited the release of gaseous products while favoring the formation of organic compounds in the liquid oil. In addition, the liquid oil obtained from the co-pyrolysis of newspaper waste and HDPE (68.4 wt%) was higher than that obtained from the pyrolysis of newspaper waste alone (40.5 wt%) and HDPE alone (52.8 wt%). The positive synergistic effects of newspaper waste and HDPE were also observed in the fuel properties of co-pyrolysis liquid oil. The viscosity (5.8–14.5 mm^2/s) and total acid number (8.4–15.2 mg KOH/g) of the co-pyrolysis liquid oil were significantly lower than that obtained from the pyrolysis of newspaper waste alone (viscosity: 68.4 mm^2/s and total acid number: 82.3 mg KOH/g). The co-pyrolysis liquid oil also consisted of higher carbon (up to 84.4 wt%) and hydrogen (up to 12.9 wt%) contents, but lower oxygen content (up to 7.69 wt%) compared to that obtained from the pyrolysis of newspaper waste alone (carbon: 37.4 wt%, hydrogen: 7.59 wt%, oxygen: 54.6 wt%). As a result, the co-pyrolysis liquid oil possessed higher calorific value (up to 34.79 MJ/kg) compared to the pyrolysis oil obtained from the pyrolysis of newspaper waste alone (16.98 MJ/kg) (Chen et al. 2016).

Wang et al. (2017b) investigated the co-pyrolysis of bamboo sawdust and waste tires in microwave pyrolysis system for liquid oil production. The pyrolysis of bamboo sawdust alone produced approximately 29 wt% of liquid oil, 42 wt% of gas, and 26 wt% of char. On the other hand, the pyrolysis of waste tire alone produced nearly 36 wt% of liquid oil, 24 wt% of gas, and 43 wt% of char. It was revealed that the addition of waste tire to bamboo sawdust during the microwave co-pyrolysis process increased the yield of co-pyrolysis liquid oil up to 30.2 wt%. This was attributed to the transfer of hydrogen atoms that promoted radical reactions and enhanced the formation of volatiles during the co-pyrolysis reaction, which later condensed into liquid oil product. The composition of the co-pyrolysis oil can be classified into several groups such as aliphatic hydrocarbons, aromatic hydrocarbons, oxygenated compounds, PAHs, and nitrogen-containing compounds. Nevertheless, the composition of oxygenated compounds (>10 wt%) and PAHs (3–58 wt%) in the co-pyrolysis liquid oil was high. Thus, the upgrading of co-pyrolysis liquid oil is required to improve its properties and composition.

Sanahuja-Parejo et al. (2018) reported on the co-pyrolysis of grape seeds and waste tires in a fixed-bed reactor to produce drop-in biofuels. The pyrolysis of grape seeds alone produced 38.8 wt% of liquid oil, whilst the pyrolysis of waste tires produced 43.7 wt% of liquid oil. Interestingly, the addition of waste tires to the grape seeds during co-pyrolysis had improved the yield of co-pyrolysis liquid oil up to 40.3 wt%. Nevertheless, the addition of waste tires to the grape seeds during co-pyrolysis had decreased the hydrocarbon composition (hydrogen and carbon content) and increased the oxygen content in co-pyrolysis liquid oil. As a result, the co-pyrolysis liquid oil possessed lower calorific value (32.6 MJ/kg) compared to the original grape seeds (36.8 MJ/kg) and waste tires (43.3 MJ/kg) (Sanahuja-Parejo et al. 2018). This result indicated that the combination of biomass materials and other hydrogen-rich materials could also lead to some negative synergistic effects during the co-pyrolysis.

12.3 Physicochemical Properties of Co-pyrolysis Bio-oil

The liquid oil obtained from co-pyrolysis of biomass and organic wastes is yellowish to brownish colored oil produced after co-pyrolysis of biomass and waste materials. The physical and chemical properties of liquid oil are dependent on the characteristics of the feedstock used. Table 12.2 summarizes the physical properties of liquid oil obtained from the co-pyrolysis of biomass and waste materials.

The calorific value is determined by the combustion of liquid oil with oxygen that releases a certain amount of energy. The calorific value of the liquid oil can be an indicator of the energy output for combustion in the fuel engine. It was found that the liquid oil obtained from the co-pyrolysis of corn cob (biomass) and waste cooking oil (biomass) possessed lower calorific value (32.8 MJ/kg) (Chen et al. 2014) compared to other liquid products obtained from the co-pyrolysis of biomass and plastic wastes. This was attributed to the high content of acidic compounds (e.g., fatty acids) in the liquid oil, which dominated the composition of liquid oil (57 wt%), thereby deteriorating its fuel properties.

Liquid oil obtained from the co-pyrolysis of biomass (e.g., waste cooking oil and sugarcane bagasse) and polyethylene (e.g., HDPE and LDPE) possessed high calorific values (40–46 MJ/kg). This is attributed to the high content of aliphatic hydrocarbons (39.5–64 wt%), low content of aromatic hydrocarbons (up to 12 wt%), and low content of acidic compounds (<3 wt%) in the liquid products. Co-pyrolysis of biomass (e.g., rice husk and rice bran wax) and polypropylene (PP) also produced a liquid oil with high calorific value (42–43.8 MJ/kg). This was due to the high content of aliphatic hydrocarbons (33–39 wt%), low content of aromatic hydrocarbons (<4 wt%), and low content of acidic compounds (<2 wt%). Co-pyrolysis of biomass (grape seeds) and waste tire also produced liquid oil with high calorific value (40.4 MJ/kg). This was attributed to the high content of aliphatic hydrocarbons (64 wt%), low content of acidic compounds (<3 wt%), and low content of oxygenated compounds (<3.5 wt%).

Viscosity is defined as the liquid's resistance to flow, which is determined by measuring the time taken by a liquid to flow via an injector nozzle of an engine (Mahari et al. 2018a). Table 12.2 reveals that the liquid oils produced from the co-pyrolysis of biomass and waste materials possess low viscosities (3–5 mm^2/s), which are comparable to diesel, biodiesel, and gasoline (2–6 mm^2/s). The total acid number (TAN) is determined to measure the acidity value of pyrolysis oil. The liquid products obtained from the co-pyrolysis of corn cob with waste cooking oil, newspaper waste with HDPE, and grape seed with waste tire showed high TAN, which was more than 24 mg KOH/g (Table 12.2). These results suggest that the co-pyrolysis liquids produced from these feedstocks need to be upgraded to reduce the acid number.

Table 12.2 Comparison of the physicochemical properties of liquid oil obtained from co-pyrolysis process

Properties	Corn cob and waste cooking oil (Chen et al. 2014)	Sugarcane bagasse and low-density polyethylene (Dewangan et al. 2016)	Newspaper waste and high-density polyethylene (Uzoejinwa et al. 2018)	Waste cooking oil and high-density polyethylene (Mahari et al. 2018b)	Rice husk and polypropylene (Suriapparao et al. 2018)	Ground-nut shell and polystyrene (Suriapparao et al. 2018)	Grape seeds and waste tire (Sanahuja-Parejo et al. 2018)	PP waste and rice bran wax (Alagu and Sundaram 2018)	Diesel (Mahari et al. 2018b)
Calorific value (MJ/kg)	32.8	40.0	28.2–34.4	46.1	42.0	39.6	40.4	43.8	45.0
Dynamic viscosity (mm^2/s)	3.1	4.09	10.3–18.8	–	–	–	–	5.03	2.0–4.0
Density (g/cm^3)	1.03	–	1.14–1.2	0.760	0.963	1.161	–	0.80	0.82–0.86
Total acid number (mg KOH/g)	25	–	17.8–24.5	–	1.63	0.63	14.2	–	–
Aliphatic hydrocarbons (wt%)	21.2	39.5	58	64	33.2	1.0	64.6	38.9	–
Aromatics hydrocarbons (wt%)	3.98	12.0	–	–	4.3	48.9	14.3	–	–
Acids (wt%)	57.1	0.87	2	3	–	–	3.0	1.91	–
Amines (wt%)	–	–	–	–	–	–	–	19.8	–
Furans (wt%)	–	–	2	–	0	0.2	–	–	–
Oxygenates (wt%)	4.71	6.44	3	5	2.7	0.9	3.5	12.2	–

(continued)

Table 12.2 (continued)

Properties	Corn cob and waste cooking oil (Chen et al. 2014)	Sugarcane bagasse and low-density polyethylene (Dewangan et al. 2016)	Newspaper waste and high-density polyethylene (Uzoejinwa et al. 2018)	Waste cooking oil and high-density polyethylene (Mahari et al. 2018b)	Rice husk and polypropylene (Suriapparao et al. 2018)	Ground-nut shell and polystyrene (Suriapparao et al. 2018)	Grape seeds and waste tire (Sanahuja-Parejo et al. 2018)	PP waste and rice bran wax (Alagu and Sundaram 2018)	Diesel (Mahari et al. 2018b)
Esters (wt%)	3.51	–	–	–	–	–	5.3	1.84	–
Alcohols (wt%)	2.68	5.87	–	–	–	–	–	15.8	–
Phenols (wt%)	3.60	24.9	1	–	0	0.4	6.2	–	–
Others (wt%)	3.22	10.4	21	28	60	51.4	3.1	9.55	–

Table 12.2 also shows that the chemical composition of co-pyrolysis oil that can be classified into several components such as aliphatic hydrocarbons, aromatic hydrocarbons, acids, amines, furans, oxygenates, esters, alcohol, phenols, and other minor and unidentified compounds. The aliphatic hydrocarbons can be derived from alkanes, alkenes, alkynes, and cycloalkanes. The carbon component of aliphatic hydrocarbons in liquid oil can range from C_5 to C_{30} hydrocarbons, depending up on the thermal cracking of the waste materials. The hydrocarbon compositions of the liquid oil can be classified into three petrochemical fractions according to the carbon fragments. The carbon component ranging from C_4 to C_{12} is referred as gasoline, C_{11}–C_{15} is referred as kerosene, and C_{15}–C_{24} is referred as diesel (Lam et al. 2012a).

The liquid oil is also composed of aromatic hydrocarbons, which can be derived from benzenes and its derivatives (e.g., toluene, xylene, and phenol). Amines can be derived from nitrogen-containing compounds such as amine-N, pyridine-N, pyrrole-N, indol-N, and nitrile-N. The oxygenates can be derived from oxygen-containing compounds such as ketones and aldehydes. The acid compounds in liquid oil can be derived from fatty acid obtained from triglycerides (e.g., waste cooking oil) (Nanda et al. 2019). The presence of both the acids and oxygenate compounds is not desirable in the pyrolysis liquid. The high composition of acids and oxygenate compounds indicates the high content of oxygen in the liquid oil, which makes the liquid oil reactive and not stable, thus may undergo several unfavorable changes in their characteristics during storage and combustion. In addition, the acids and oxygenate compounds can cause corrosion and oxidation during the oil storage, which may lead to the deterioration of fuel quality.

12.4 Engine Performance and Emission Characteristics

Kalargaris et al. (2017) investigated the performance, combustion, and emission characteristics of a diesel engine fueled by pyrolysis oils obtained from pyrolysis of LDPE and ethylene-vinyl acetate plastics. The pyrolysis oil blended with diesel (termed as PPO 25, PPO 50, PPO 75, PPO 90, and PPO 100) was tested on an AKSA four-cylinder diesel engine with a bore of 110 mm and stroke of 120 mm, rated power of 68 kW, rated speed of 1500 rpm, and injection pressure of 240 bar. There were three types of pyrolysis oil tested in this study, namely LDPE700, EVA900, and EVA900 75. The pyrolysis oil obtained from pyrolysis of LDPE at 700 °C is termed as LDPE700, pyrolysis oil obtained from pyrolysis of ethylene-vinyl acetate plastics at 900 °C is termed as EVA900, while 75% of pyrolysis oil obtained from pyrolysis of ethylene-vinyl acetate plastics at 900 °C blended with 25% of diesel fuel is termed as EVA900 75. LDPE700 demonstrated an excellent combustion characteristic, short ignition delay, and the brake thermal efficiency was almost like diesel fuel operation. LDPE700 also released lower amounts of NO_x, CO, and CO_2, but higher amounts of unburnt hydrocarbons compared to diesel. In contrast, EVA900 and EVA900 75 showed longer ignition delay, lower brake thermal efficiency (1.5–2%), higher emission of NO_x and unburnt hydrocarbons compared to diesel fuel. Nevertheless, EVA900 has reduced the emission of CO and

CO_2. The addition of 25% of diesel fuel to EVA900 (EVA900 75) showed no improvement in the engine performance.

In other work, Alagu and Sundaram (2018) investigated the combustion, performance, and emission characteristics of a compression ignition (CI) engine using pyrolysis oil derived from neem seed. The pyrolysis oil was blended with diesel fuel at proportions of 5% (termed as NB5) and 10% (termed as NB10). Kirloskar four-cylinder diesel engine with a bore of 87.5 mm and stroke of 100 mm, rated power of 4.4 kW, rated speed of 1500 rpm, and injection pressure of 200 bar were used in this study. The brake thermal efficiency for NB5 and NB10 (up to 3.7%) at full engine load was higher than diesel fuel. In addition, the use of NB5 and NB10 in the engine has reduced the emissions of hydrocarbons (46.9%), CO (42.2%), CO_2 (29.8%), and NO_x (20.7%) at full engine load condition. This study demonstrated that pyrolysis oil obtained from pyrolysis of neem seed in fixed bed reactor showed a promising potential to be used in fuel engine as a renewable and sustainable green fuel.

In summary, most of the studies showed that the engine performance of pyrolysis oil was less efficient compared to diesel fuel due to low brake thermal efficiency, low cetane number, and high viscosity of pyrolysis oil, and high emissions of HC, CO, CO_2, and NO_x. Nevertheless, a few studies have also revealed the excellent performance of pyrolysis oil in diesel engine. The pyrolysis oil was reported to increase the brake thermal efficiency and reduce the emission of undesirable gases. Therefore, the use of pyrolysis oil as fuel is believed as a promising and sustainable technique to recover the energy from waste materials, while simultaneously reducing the volume of waste disposed of in landfill, which could release more harmful gases (CO_2, CH_4, SO_x, and NO_x) over the long term.

12.5 Conclusions

It is revealed that co-pyrolysis demonstrates a promising fuel processing technology for the recovery of sustainable energy from lignocellulosic biomass and polymeric wastes. Despite the numerous advantages reported by current pyrolysis technologies, the existence of high oxygen content, high water content, high PAH concentration in liquid oil, and poor physical properties (e.g., high viscosity and high acid value) could lead to many undesirable effects on the fuel engine (e.g., instability, corrosiveness, oxidation and poor atomization). Therefore, it would be worth to investigate co-pyrolysis techniques, which is a promising upgrading technique to improve the yield and properties of liquid oil, and discover the feasibility of this approach in the treatment and recovery of biomass materials. Nonetheless, there are several issues such as maintaining the co-pyrolysis condition, the influence of operating parameters on product distribution and production, which necessitate to be addressed to improve this co-pyrolysis approach.

References

Alagu RM, Sundaram EG (2018) Preparation and characterization of pyrolytic oil through pyrolysis of neem seed and study of performance, combustion and emission characteristics in CI engine. J Energy Inst 91:100–109

Chen G, Liu C, Ma W, Zhang X, Li Y, Yan B, Zhou W (2014) Co-pyrolysis of corn cob and waste cooking oil in a fixed bed. Bioresour Technol 166:500–507

Chen W, Shi S, Zhang J, Chen M, Zhou X (2016) Co-pyrolysis of waste newspaper with high-density polyethylene: synergistic effect and oil characterization. Energ Convers Manage 112:41–48

Chen L, Yu Z, Fang S, Dai M, Ma X (2018) Co-pyrolysis kinetics and behaviors of kitchen waste and *Chlorella vulgaris* using thermogravimetric analyzer and fixed bed reactor. Energ Convers Manage 165:45–52

Dewangan A, Pradhan D, Singh RK (2016) Co-pyrolysis of sugarcane bagasse and low-density polyethylene: influence of plastic on pyrolysis product yield. Fuel 185:508–516

Ephraim A, Minh DP, Lebonnois D, Peregrina C, Sharrock P, Nzihou A (2018) Co-pyrolysis of wood and plastics: influence of plastic type and content on product yield, gas composition and quality. Fuel 231:110–117

Kalargaris I, Tian G, Gu S (2017) Experimental evaluation of a diesel engine fuelled by pyrolysis oils produced from low-density polyethylene and ethylene–vinyl acetate plastics. Fuel Process Technol 161:125–131

Kim S (2001) Pyrolysis kinetics of waste PVC pipe. Waste Manag 21:609–616

Kumari AN, Singh RK (2019) Co-pyrolysis of waste polypropylene and rice bran wax– production of biofuel and its characterization. J Energy Inst 92:933–946

Lam SS, Chase HA (2012) A review on waste to energy processes using microwave pyrolysis. Energies 5:4209–4232

Lam SS, Wan Mahari WA, Ok YS, Peng W, Chong CT, Ma NL, Chase HA, Liew Z, Yusup S, Kwon EE, Tsang DCW (2019) Microwave vacuum pyrolysis of waste plastic and used cooking oil for simultaneous waste reduction and sustainable energy conversion: Recovery of cleaner liquid fuel and techno-economic analysis. Renew Sust Energ Rev 115:109359

Lam SS, Russell AD, Lee CL, Chase HA (2012a) Microwave-heated pyrolysis of waste automotive engine oil: influence of operation parameters on the yield, composition, and fuel properties of pyrolysis oil. Fuel 92:327–339

Lam SS, Russell AD, Lee CL, Lam SK, Chase HA (2012b) Production of hydrogen and light hydrocarbons as a potential gaseous fuel from microwave-heated pyrolysis of waste automotive engine oil. Int J Hydrogen Energ 37:5011–5021

Lam SS, Liew RK, Jusoh A, Chong CT, Ani FN, Chase HA (2016) Progress in waste oil to sustainable energy, with emphasis on pyrolysis techniques. Renew Sust Energ Rev 53:741–753

Lam SS, Mahari WAW, Jusoh A, Chong CT, Lee CL, Chase HA (2017) Pyrolysis using microwave absorbents as reaction bed: an improved approach to transform used frying oil into biofuel product with desirable properties. J Clean Prod 147:263–272

Maddikeri GL, Pandit AB, Gogate PR (2012) Intensification approaches for biodiesel synthesis from waste cooking oil: a review. Ind Eng Chem Res 51:14610–14628

Mahari WAW, Chong CT, Cheng CK, Lee CL, Hendrata K, Yek PNY, Ma NL, Lam SS (2018a) Production of value-added liquid fuel via microwave co-pyrolysis of used frying oil and plastic waste. Energy 162:309–317

Mahari WAW, Chong CT, Lam WH, Anuar TNST, Ma NL, Ibrahim MD, Lam SS (2018b) Microwave co-pyrolysis of waste polyolefins and waste cooking oil: influence of N_2 atmosphere versus vacuum environment. Energ Convers Manage 171:1292–1301

Nam WL, Phang XY, Su MH, Liew RK, Ma NL, Rosli M, Lam SS (2017) Production of bio-fertilizer from microwave vacuum pyrolysis of palm kernel shell for cultivation of oyster mushroom (*Pleurotus ostreatus*). Sci Total Environ 624:9–16

Nanda S, Rana R, Hunter HN, Fang Z, Dalai AK, Kozinski JA (2019) Hydrothermal catalytic processing of waste cooking oil for hydrogen-rich syngas production. Chem Eng Sci 195:935–945

Önal E, Uzun BB, Pütün AE (2012) An experimental study on bio-oil production from co-pyrolysis with potato skin and high-density polyethylene (HDPE). Fuel Process Technol 104:365–370

Sanahuja-Parejo O, Veses A, Navarro MV, López JM, Murillo R, Callén MS, García T (2018) Catalytic co-pyrolysis of grape seeds and waste tyres for the production of drop-in biofuels. Energ Convers Manage 171:1202–1212

Sharuddin SDA, Abnisa F, Wan Daud WMA, Aroua MK (2016) A review on pyrolysis of plastic wastes. Energ Convers Manage 115:308–326

Suriapparao DV, Boruah B, Raja D, Vinu R (2018) Microwave assisted co-pyrolysis of biomasses with polypropylene and polystyrene for high quality bio-oil production. Fuel Process Technol 175:64–75

Tang Y, Huang Q, Sun K, Chi Y, Yan J (2018) Co-pyrolysis characteristics and kinetic analysis of organic food waste and plastic. Bioresour Technol 249:16–23

Undri A, Frediani M, Rosi L, Frediani P (2014) Reverse polymerization of waste polystyrene through microwave assisted pyrolysis. J Anal Appl Pyrolysis 105:35–42

Uzoejinwa BB, He X, Wang S, El-Fatah Abomohra A, Hu Y, Wang Q (2018) Co-pyrolysis of biomass and waste plastics as a thermochemical conversion technology for high-grade biofuel production: recent progress and future directions elsewhere worldwide. Energ Convers Manage 163:468–492

Wang Y, Dai L, Fan L, Cao L, Zhou Y, Zhao Y, Liu Y, Ruan R (2017a) Catalytic co-pyrolysis of waste vegetable oil and high density polyethylene for hydrocarbon fuel production. Waste Manag 61:276–282

Wang Y, Dai L, Fan L, Duan D, Liu Y, Ruan R, Yu Z, Liu Y, Jiang L (2017b) Microwave-assisted catalytic fast co-pyrolysis of bamboo sawdust and waste tire for bio-oil production. J Anal Appl Pyrolysis 123:224–228

Yang J, Rizkiana J, Widayatno WB, Karnjanakom S, Kaewpanha M, Hao X, Abudula A, Guan G (2016) Fast co-pyrolysis of low density polyethylene and biomass residue for oil production. Energ Convers Manage 120:422–429

13 Conversion of Waste Biomass to Bio-oils and Upgradation by Hydrothermal Liquefaction, Gasification, and Hydrodeoxygenation

Vikranth Volli, Anjani Ravi Kiran Gollakota, Mihir Kumar Purkait, and Chi-Min Shu

Abstract

Biofuels produced from biomass are clean, renewable, and eco-friendly alternatives to the conventional fossil fuels in the transportation sector. However, the presence of high-water content, low pH, high viscosity, and oxygenates limits the direct use of biofuel in vehicular engines. The in situ and ex situ catalytic as well as noncatalytic hydrothermal upgradation of bio-oil (converting into hydrocarbons or less oxygenated compounds) are very promising. The recent advances in thermochemical conversion processes, improved strategies in feedstock pretreatment, and optimized use of both homogeneous and heterogeneous catalysts have enhanced the fuel properties of biofuels. The available literature was reviewed extensively to perceive the pros and cons in the selection of the suitable upgrading process to produce the bio-oil based on the end use. In this chapter, the technical developments toward improving the bio-oil properties, both in quality and quantity, the influence of process parameters, reactor configurations, and their primal source were discussed in detail. By comparing the various conversion and upgrading technologies, hydrodeoxygenation is considered as the prominent alternative and the latest technology in contrast to gasification and liquefaction. However, the complexities of the hydrodeoxygenation mechanism, optimal processing conditions, and the choice of the catalysts are yet to be understood. Furthermore, the chapter points out the main barriers for the commercialization of bio-oil upgrading technologies for the future.

V. Volli (✉) · A. R. K. Gollakota · C.-M. Shu (✉)
Department of Safety, Health and Environmental Engineering, National Yunlin University of Science and Technology, Douliou, Taiwan, China
e-mail: vikranth@yuntech.edu.tw; shucm@yuntech.edu.tw

M. K. Purkait
Department of Chemical Engineering, Indian Institute of Technology Guwahati, Guwahati, Assam, India

S. Nanda et al. (eds.), *Biorefinery of Alternative Resources: Targeting Green Fuels and Platform Chemicals*, https://doi.org/10.1007/978-981-15-1804-1_13

Keywords
Biomass · Bio-oil · Thermochemical conversion · Pyrolysis · Gasification · Hydrodeoxygenation

13.1 Introduction

The need for alternative energy sources, exhausting conventional fossil fuels, and the colossal increase in population growth over the decades have impelled the focus on developing efficient processes, innovative technologies, and sustainable growth in energy production. The rate of energy consumption substantially increased up to 5755, 1768, 382, 87, 2121, and 463 Mtoe for Asia, Europe, Africa, Australia, North America, and South America, respectively, during the period of 2000–2017. The distribution of total energy supply increased from 97, 152, 86.7, 28.3, and 55 Mtoe in 2000 to 164, 179, 121, 27.7, and 80.8 Mtoe in 2014 with a growth rate of 3.8, 1.1, 2.4, 0.2, and 14.4% for coal, oil, natural gas, nuclear, and renewable energy, respectively (International Energy Outlook 2017). The continental distribution of total energy consumption for the year 2016 with the contribution of renewables is presented in Fig. 13.1.

The increased consumption of fossil fuels with the growing energy needs along with the consequential greenhouse gas emissions (CO, CO_2, NO_x, and SO_2) from industries and automotives has significantly influenced the growth in pollution and global warming. This complex situation has resulted in exploring the renewable

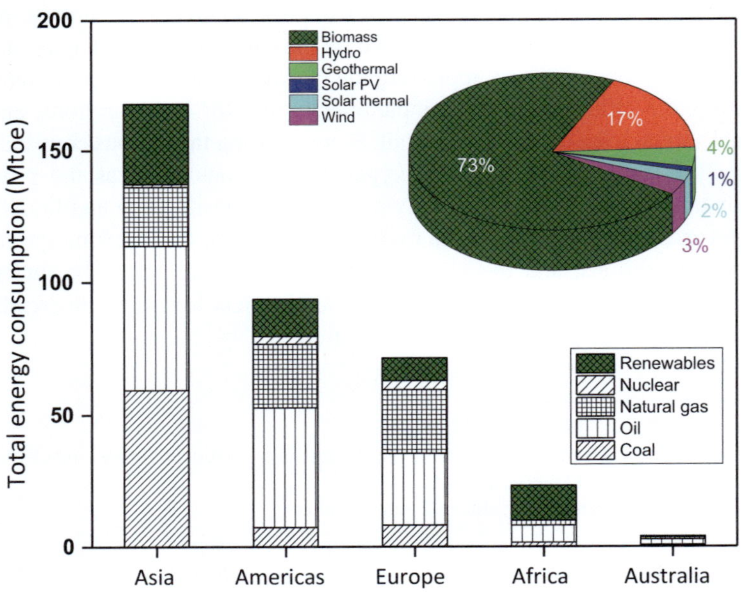

Fig. 13.1 Total energy consumption in different continents and the distribution of renewable energy in 2016

energy sources from biomass, hydro-, geothermal, solar, and wind energy. Among these various alternatives, bioenergy from biomass would replace nearly 27% of the world's transportation fuel and cater 10% of the global energy requirements in the near future (He et al. 2018). Waste organic biomass is a renewable energy source, widely available in nature, and is considered carbon neutral. Biomass is defined as the organic or biodegradable fraction derived from living organisms, residues of biological origin from agricultural, forestry, industrial, and related municipal wastes. Suitable pretreatment of biomass would result in products including transportation fuels, chemicals, biomaterials, and thermal energy.

13.2 Classification of Waste Organic Biomass

The different sources of biomass include wood, agricultural crop residues, food processing wastes, aquatic plants, animal wastes, algae, and their waste by-products. Biomass resources vary with geographical location and are categorized based on their origin, chemical composition, and end use. Biomass can be broadly classified into primary, secondary, and tertiary wastes. The primary wastes are produced directly by the process of photosynthesis (woody, herbaceous, forestry, and agricultural crop residues). The secondary wastes are processed from primary biomass (e.g., sawdust, black liquor, and animal waste). On the other hand, tertiary wastes include by-products and residue streams (e.g., vegetable oils, legumes, and packaging wastes). However, industrial-based classification of raw biomass is as follows (Sánchez et al. 2019):

1. Wood or forest products are one of the largest contributors to lignocellulosic biomass. The wood processing wastes are available in four major forms including firewood, wood logs, wood chips, and pellet.
2. Energy crops are nonfood crops, such as short rotation woody crops, mostly classified into second-generation feedstocks (switchgrass, *Miscanthus*, elephant grass, and hybrid poplar).
3. Agricultural residues are produced after harvesting of agricultural products including sugarcane, cotton, maize, cereals, wheat, palm, peanut, and jute (Ashter 2018).
4. Food and industrial wastes are the by-products generated from municipal waste streams and processing industries (restaurants, paper, pulp, tea and coffee processing, meat production, domestic and household wastes) (Wang et al. 2017).

Except for the first-generation feedstocks (corn, oil palm, soybean, and rapeseed), the individual contribution of aforementioned biomass to bioenergy on a global scale has now been well assessed. The effective conversion of biomass to bioenergy depends on the physicochemical composition of feedstocks, which varies with geographical location and origin. However, the exploration for new and efficient technologies to process biomass for the production of liquid biofuels is still limited and may become economically viable in the near future.

13.3 Characterization of Biomass

13.3.1 Proximate Analysis

The physical characteristics of biomass include moisture content (external and inherent), volatile matter (condensable vapors and gases), ash (inorganic residues), and fixed carbon (nonvolatile fractions). Higher the volatility, greater the bio-crude generation from biomass. American Society for Testing and Materials (ASTM) E1358-97, E872-82, and E1755-01 standards are used to measure the moisture (105 ± 3 °C for 16 h), volatile matter (950 ± 25 °C for 7 min), and ash content (575 ± 25 °C for 3 h) in the biomass while fixed carbon is calculated by the difference on dry weight basis (Cai et al. 2017). The range of moisture, volatile matter, ash, and fixed carbon for woody biomass is typically 1.5–9.5, 65–79, 0.5–1.7, and 13–26 wt%, respectively. Energy crops typically contain 2.5–10 wt% moisture, 70–85 wt% volatile matter, 0.5–1.7 wt% ash, and 12–18 wt% fixed carbon. On the other hand, agricultural residues usually contain 6–10 wt% moisture, 28–85 wt% volatile matter, 2–10 wt% ash, and 4.9–28 wt% fixed carbon. Last but not the least, industrial wastes comprise of 1.5–11 wt% moisture, 29–78 wt% volatile matter, 4.5–39 wt% ash, and 3.5–16 wt% fixed carbon (Cai et al. 2017; Dhyani and Bhaskar 2018).

13.3.2 Ultimate Analysis and Heating Value

The elemental composition of biomass is the measure of carbon (C), hydrogen (H), nitrogen (N), sulfur (S), and oxygen (O) on dry mass basis. Typically, lignocellulosic biomass consists of 40–55 wt% C, 35–40 wt% O, 3–7 wt% H, 0.5–1.5 wt% N, and 0.1–2% S. For example, wood bark contains 53.1 wt% C, 6.1 wt% H, 0.2 wt% N, and 40.6 wt% O; flax straw contains 43.1 wt% C, 6.2 wt% H, 0.7 wt% N, and 49.9 wt% O; rice husk contains 36.9 wt% C, 5 wt% H, 0.4 wt% N, and 37.9 wt% O; switchgrass contains 48.6 wt% C, 5.5 wt% H, 0.5 wt% N, and 39.5 wt% O; and swine manure contains 30.7 wt% C, 4.4 wt% H, 2.5 wt% N, and 61.7 wt% O (Ashter 2018). However, the sulfur content in biomass is lower in beechwood, almond shell, flax straw, bamboo, and orchard grass as 0.7, 0.3, 0.09, 0.26, and 0.03 wt%, respectively. Ash vitrification of inorganic compounds (Na and K) improves the environmental performance.

Apart from the presence of macronutrients (P, K, Ca, and Mg), the trace elements (Na, Cl, Fe, Mn, Cu, Zn, and Si) are also present in lignocellulosic biomass. The presence of respective inorganic content in Albizia, Iroko, and Spruce biomass was Na (22.9, 32.1, and 25.9 ppm), Zn (5.7, 8.3, and 11.4 ppm), Cu (5.1, 0.5, and 0.1 ppm), Fe (331.2, 379.6, and 9 ppm), Al (188.4, 209.1, and 4.6 ppm), and Mg (171.5, 790.5, and 101.3 ppm) (Cai et al. 2017).

Leguminous-based biomass is rich in amine- and carboxyl functional groups that contribute to the enzymatic activity and protein content with a relatively high nitrogen composition (16 wt%) and are responsible for the increase in higher heating value (HHV) when compared to cellulose. Higher heating value is the measure of the energy content of biomass and the total amount of heat energy available including the latent heat of vaporization in the fuel and reaction product. The empirical equations are developed based on the elemental composition of biomass to determine the HHV (Dhyani and Bhaskar 2018). The HHV for biomass is in the range of 12.5–24 MJ/kg. Woody biomass, agricultural residues, energy crops, and industrial wastes have an average HHV of 14.2–17.8, 15.6–21.7, 17.5–24.9, and 6.5–17.1 MJ/kg, respectively (García et al. 2017).

$$\begin{aligned}\mathrm{HHV}\left(\mathrm{MJ}/\mathrm{kg}\right) = {} & 0.3491\left(\mathrm{C\,wt\%}\right) + 1.1783\left(\mathrm{H\,wt\%}\right) - 0.1\left(\mathrm{S\,wt\%}\right) \\ & -0.013\left(\mathrm{O\,wt\%}\right) - 0.0151\left(\mathrm{N\,wt\%}\right) - 0.0211\left(\mathrm{Ash\,wt\%}\right)\end{aligned} \tag{13.1}$$

13.3.3 Chemical Composition

It is known that biomass can be characterized based on the presence of extractives, hemicellulose, cellulose, and lignin contents. The major chemical components of biomass feedstock include lignocellulose, starch, sugars, carbohydrates, lipids, proteins, fatty acids, ash as well as organic and inorganic compounds. Biomass grouped on the basis of the aforementioned are classified as lignocellulose-based (straw, wood, herbaceous, and energy grasses); sugar-based (mono-/disaccharides from beet and sugarcane); starch-based (polysaccharides from wheat, corn, potato, and artichoke); oil-based (rapeseed, sunflower, and microalgae and macroalgae); and protein-based (soybean, legumes, animal and aquatic waste) biomass (Tekin et al. 2014).

Cellulose exists in both crystalline and amorphous forms as linear chain monomolecular polysaccharide glucose units (natural and most abundant) and has a high degree of polymerization (ca. 15,000) (Nanda et al. 2015). Cellulose consists of two anhydroglucose units (rotation of anhydroglucose by 180° to the left and right) linked to a helical axis. Biomass consists of 40–60 dry wt% of high-molecular weight (≥106) of β-glycosidic-linked cellulose chains (Wang et al. 2017). For one single glucosyl unit in cellulose, two intra- and one intermolecular hydrogen bonds with van der Waals forces contribute to the significant thermal stability (crystallinity index of 80%) than the amorphous region (packed cellulose structure).

Hemicellulose is an amorphous short-chain heteropolysaccharide (branched structure) with the degree of polymerization in the range of 50–200. It acts as a connective linkage between cellulose and lignin. The typical composition of hemicellulose is the monosaccharide unit comprising pentose and hexose sugars (L-arabinose and D-xylose, D-galactose, D-glucose, and D-mannose). Hemicellulose in biomass can range up to 20–40 wt% and can be extracted by using hot alkali, ozonolysis, enzymatic hydrolysis, and pressurized hot water. However, complete extraction or separation of hemicellulose cannot be achieved because of its accessible inclination to decomposition (Dhyani and Bhaskar 2018).

Lignin is an amorphous three-dimensional aromatic polymer matrix of *p*-hydroxyphenyl, guaiacyl, and syringyl units as its main composition with values of degree of polymerization between 6 and 41. Ether and ester bonds as well as other C–C bonds are the major linkages in lignin, apart from methoxy, hydroxyl, carboxyl, and carbonyl oxygenated functional groups attached to the propyl side chain resulting in high structural rigidity (Tekin et al. 2014). Softwood and hardwood consist of guaiacyl and syringyl units and are present. The fraction of lignin can be in the range of 10–25 wt% in dry biomass. Kalson method, Kraft process, Bjorkman, organosolv, and sulfate pulping processes are used to separate lignin. Lignin is a biopolymer and highly branched polyphenolic constituent of biomass and is more difficult to be dehydrated than cellulose and hemicelluloses.

Extractives are the nonstructural components of biomass consisting of lipids, waxes, fats, and pigments. They are mainly classified into aliphatic, phenolic, terpenes, and terpenoid compounds. The typical composition of the extractives includes triglycerides, hydroxy fatty acids, resins, and sterols that are responsible for high viscosity in bio-oil, which contribute to a maximum of 2–10 wt% in biomass (Arregi et al. 2018). The presence of extractives promotes the decomposition of cellulose and hemicellulose.

Physically, biomass has an extremely complex structure with an uneven distribution of cellulose, hemicellulose, and lignin. Cellulose microfibrils form the cell wall framework coated with hemicellulose (hydrogen bonds) and the empty spaces filled with lignin linker materials (both covalent and hydrogen bonds). Based on the type of biomass, the contents of extractives, hemicellulose, cellulose, and lignin vary from 2 to 10 wt%, 15 to 30 wt%, 40 to 60 wt%, and 10 to 25 wt%, respectively, on dry weight basis. The presence of small fractions of nonstructural components, carbohydrates, inorganics, trace elements, proteins, and HHV makes them an inexpensive renewable energy resource. The effective utilization of biomass as a feedstock for energy production (bio-oil, fuel gases, and char) could be a value-added alternative. The presence of excess moisture, varying nature of feedstocks, oxygenates, and acidic compounds in bio-oil pose a major challenge in the effective utilization of biomass. Currently, the growing concern on the economical uses of biomass, its seasonal availability, and environmental impacts have attracted tremendous attention in the scientific community.

13.4 Biomass Conversion Processes

Energy from biomass can be derived in various forms. The form of energy determines the conversion process and depends on the availability, type, and quantity of the feedstock. It depends on the desired energy form, end use, project factors, environmental, and economic conditions (Adams et al. 2017). The biomass conversion processes can be broadly classified into three types such as physicochemical, biochemical, and thermochemical processes (Fig. 13.2).

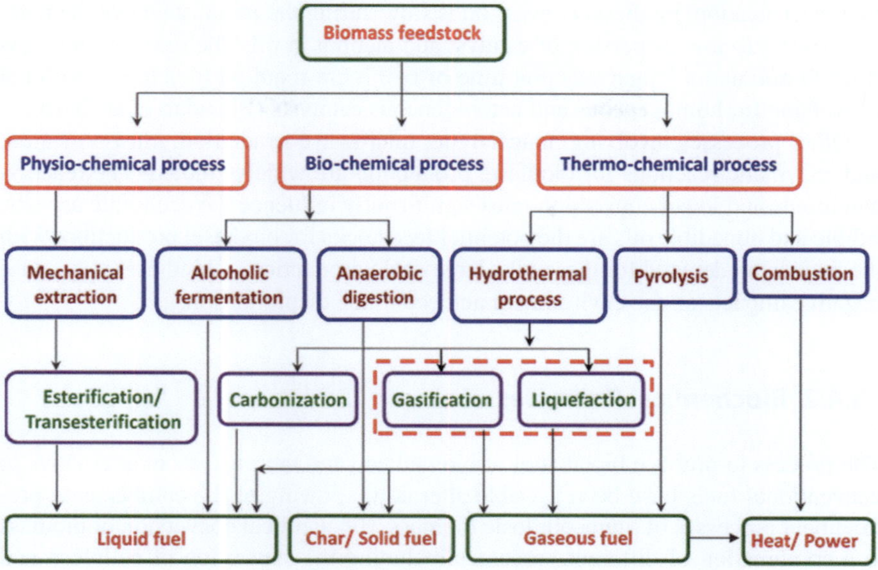

Fig. 13.2 Different biomass conversion processes to value-added products

13.4.1 Physicochemical Process

Physicochemical conversion of biomass includes the extraction of oils from oil-based feedstocks, such as sunflower, palm, hemp, flax, canola, rice bran, soybean, mustard, rapeseed, and waste vegetable oils. The extracted vegetable oils are highly viscous and the presence of free fatty acid content makes them unfit for direct use in diesel engines. Converting these feedstocks to biodiesel using catalyzed reaction (e.g., esterification or transesterification) to corresponding methyl ester has proven to be advantageous because of its reduction in pollution levels. The process of transesterification involves the mixing of alcohol, catalyst, and purification of methyl ester (alcohol removal, glycerine neutralization, and water washing) (Gebremariam and Marchetti 2018).

Transesterification is mostly influenced by the alcohol-to-oil molar ratio, catalyst concentration, time, and temperature of reaction. In stoichiometric conditions, to produce 3 mol of fatty acid esters, 1 mol of triglyceride requires 3 mol of alcohol and depends on the type of catalyst used. The optimal alcohol-to-triglycerides ratio for the formation of biodiesel is 6:1. Moreover, the conversion of triglyceride increases with catalyst concentration. For conventional homogeneous (NaOH, KOH, HCl, and H_2SO_4) and heterogeneous (CaO, Ca/TiO_3, Ca/ZrO_3, and CaO/CeO_2) acid/base catalysts, maximum conversion can be achieved at 1.5 and 10 wt%, respectively (Volli et al. 2019). Higher temperatures also reduce the viscosity of the oil affecting the reaction time. The range of optimal temperatures to achieve maximum conversion is the temperature close to the boiling point of the type of alcohol used (50–60 °C). Furthermore, for a conventional homogeneous catalytic transesterification, biodiesel conversion is slow during the initial stages of the reaction, owing to low dispersion of catalyst and alcohol in oil. The reaction time less than 90 min and a longer reaction time of 6–8 h are required to achieve maximal conversion for homogeneous and heterogeneous catalysts (Bhandari et al. 2015).

Other processes involving noncatalytic, microwave irradiation, ultrasonication, and enzymatic reactions for biodiesel production are widely studied. The reaction conditions and lower conversion rates significantly influence the economic aspects. Edible and nonedible oils are the potential feedstocks for biodiesel production. With the increasing demand for these oils, large-scale production of biodiesel may cause negative impacts on the environment and economic disproportion.

13.4.2 Biochemical Processes

The process to produce bioethanol or biomethane and their use as an alternative to conventional fuels have been a viable alternatives, owing to the cumbersome pretreatment processes of lignocellulosic biomass. Biochemical conversion of biomass is a combination of different processes including the separation of cellulose and hemicellulose, enzymatic hydrolysis, and fermentation of hexose and pentose (Ghosh et al. 2017). However, biochemical processes are classified into two main technologies, namely anaerobic digestion and alcoholic fermentation.

13.4.2.1 Anaerobic Digestion

Anaerobic digestion is considered as the natural process that utilizes bacteria and other microorganisms to convert organic matter to gaseous products (biogas) and nitrogen-rich liquid digestate in the absence of oxygen. It is the most prevalent stabilization procedure to produce biogas, in which the dissolved organic compound at high temperatures generates biogas with increased carbon-to-nitrogen (C/N) ratio. The feedstock for anaerobic digestion includes animal manure, harvest remains, sewer sludge, slaughterhouse wastes, food wastes, pulp and paper wastes, and dairy industrial wastes (Shin et al. 2017).

Digesters with pretreated (mechanical preheating and thermal treatment) biomass are usually used to produce biogas. The degradation pathway is classified into hydrolysis, acidogenesis, aceteogenesis, and methanogenesis, which depends on the type of bacteria used. The feedstock usually consists of 60–80 wt% moisture with the heating value of biogas as 20–40 wt% of the feedstock (Chen and Wang 2017). The biogas thus generated can be directly used for cooking, household heating, and as gas fuel in engines or can be further upgraded to natural gas.

13.4.2.2 Alcoholic Fermentation

The process to produce bioethanol from feedstock with sugars and starch including sugarcane, corn, sweet potatoes, wheat, barley, oat, and rice is alcoholic fermentation. It is a two-step process that involves enzymatic hydrolysis followed by fermentation. The optimum temperatures for fermentation and hydrolysis are 30 and 50 °C, respectively, at pH 4.5–6.0. The most commonly used microorganisms for enzymatic saccharification are *Saccharomyces cerevisiae*, *Trichoderma reesei*, *Candida brassica*, and *Clostridium* (Chen and Wang 2017). Industrial by-products specifically molasses can be converted to bioethanol.

The factors effecting the conversion of bioethanol are biomass pretreatment, temperature, pH, enzymes (hydrolytic, cellulolytic, and ligninolytic), and substrate concentrations. Stoichiometrically, 1 mol of glucose gives 1 mol of adenosine triphosphate (ADP), 2 mol of CO_2, and 1 mol of ethanol. Reactor configurations, namely fluidized-bed bioreactors, immobilized cell and tower reactors, are exceptionally promising for better conversion of bioethanol. The presence of lignin restricts the decomposition of hemicellulose and cellulose to fermentable sugars. Enzymatic hydrolysis at lower temperatures reduces the bioethanol formation, as it is the rate-limiting step during hydrolysis. The selection of appropriate thermal tolerant enzyme for hydrolysis, detailed studies of the optimal composition of feedstock based on geographical location, and their pretreatment could be contributing factors for the improved formation of bioethanol (Shin et al. 2017).

13.4.3 Thermochemical Process

Thermochemical process involves thermal energy to convert biomass chemically into energy and chemical products. They are classified into three major categories combustion, pyrolysis, and hydrothermal process. Biomass also includes feedstocks with high moisture content (anaerobic digestate, manure, food waste, aquatic biomass, and sewage sludge). The conversion of such feedstocks and organic wastes at high temperatures (180–360 °C) and pressures (4–22 MPa) to biofuels is also achieved through hydrothermal technologies (Tekin et al. 2014). Hydrothermal treatment of biomass is classified into three categories, namely carbonization, gasification, and liquefaction.

13.4.3.1 Combustion

The traditional process of burning biomass (exothermic reaction) in the presence of excess oxygen (5–50%) to generate energy in the form of heat is called combustion. The common feedstocks used for combustion are woody biomass and agricultural crop residues. Direct heating of biomass results in evaporative, decomposition, surface, and smoldering combustion. Strong turbulence of air as mixture and longer residence times provide enough oxygen to improve combustion (Ashter 2018). Biomass combustion process generates gaseous products at 800–1000 °C that can directly be used to produce electricity.

Combustion is classified into four stages, such as drying, pyrolysis, volatiles' release, and char combustion, which are influenced by temperature, biomass particle size, feedstock composition, and combustion atmosphere. Suspension and fluidized-bed furnaces are widely used for bulk biomass feedstock and require preheating for efficient energy production. However, high particulates, ash handling, and emissions of CO, CO_2, and NO_x make the combustion process challenging and non-environmental friendly. Selective non-catalytic reduction technique that reduces NO_x to N_2 in the presence of amine-based reagent (urea and ammonia) is used. The flue gas recirculation has been found to be promising (Tekin et al. 2014).

13.4.3.2 Pyrolysis

Pyrolysis is a widely accepted process preceded by combustion and gasification due to its versatility and flexibility in the choice of feedstock (Onay and Kockar 2003). The word "pyro" and "lysis" mean "fire" and "decomposition," respectively, in the absence of oxygen to produce liquid (bio-oil), solid (char), and gaseous (H_2, CO_2, CO, CH_4, and NH_3) products (Dhyani and Bhaskar 2018). The reaction temperature, heating rate, vapor residence time, particle size, biomass composition, and feed rate influence pyrolysis. Lower process temperatures (<300 °C) and longer residence time (10–60 min) favor char production (80 wt%), whereas higher reaction temperatures (750–900 °C) and longer residence time favor the formation of gaseous products (96 wt%). On the other hand, moderate temperatures (400–550 °C) and short residence time favor liquid product (bio-oil) yields.

Pyrolysis is classified into slow or conventional pyrolysis (heating rate: 3–5 °C/min and temperature: 300–400 °C), fast pyrolysis (heating rate: 50–90 °C/min and temperature: 450–600 °C), and flash pyrolysis (heating rate $\geq$ 100 °C/min and temperature $\geq$ 750 °C). Alternate to conventional heating systems, the microwave-based technology has already been successfully utilized in biomass pyrolysis for the production of biochar and bio-oil (Volli and Singh 2012). Different reactor configurations including fluidized-bed, auger, ablative (vortex/cone), bubbling, rotating cone, and circulating-bed reactors are used for pyrolysis.

The fraction of cellulose, hemicellulose, and lignin as well as their interactions are responsible for the pyrolytic behavior and provides the flexibility of using dry and wet biomass. Hemicellulose (D-xylose) upon pyrolysis forms a reaction intermediate xylulose, which undergoes hydrogenation, cyclization, and dehydration to generate furfurals with apparent activation energy of 259.8 kJ/mol. Homolysis and concerted mechanisms are widely studied for the pyrolysis of lignin (Wang et al. 2017). The interactions of hemicellulose, cellulose, lignin and the presence of oxygenated compounds braced the use of catalyst to understand and improve the distribution of pyrolysis products. The major catalytic reaction pathways for biomass include deoxygenation, cracking, aromatization, ketonization, aldol condensation, hydrotreating, and steam reforming.

The use of inorganic salts (K, Ca, Na, and Mg), metal oxides (SiO_2, Al_2O_3, NiO, CuO, MoO_3, CeO_2, ZrO_2, and TiO_2), zeolites (Ni/Co/Fe/Zr/HZSM-5), the in situ catalytic pyrolysis (catalyst mixed with feedstock), and ex situ catalytic pyrolysis (pyrolysis of feedstock and catalytic conversion performed in separate reactors) have been extensively studied (Guedes et al. 2018). A considerable reduction in the formation of coke, phenols, and increase in aromatic hydrocarbons with change in chemical pathway were observed in catalytic pyrolysis. Copyrolysis was also suggested as an alternative to pyrolysis. Copyrolysis of biomass, polymers (polystyrene, high-density polyethylene, and low-density polyethylene), waste plastics, and coal has been found to significantly improve the polyaromatic content in the bio-oil (Abnisa and Wan Daud 2014).

13.4.3.3 Hydrothermal Carbonization

It is an inexpensive technique where biomass is treated in the presence of water at temperatures of 180–300 °C with residence time of 5–240 min to produce hydrochar (up to 90%). Both wet biomass (animal manure, human waste, sewage sludge, municipal solid waste, and algal residues) and dry biomass (agricultural and forest wastes) can be used for carbonization. Carbonization involves three important steps, such as dehydration, polymerization, and carbonization of organic matter, to form spherical (micron-sized with multiple polar functional groups) particles with heating value of 20–29 MJ/kg (He et al. 2018). Moisture content, biomass particle size, heating rate, operating temperature, precursor concentration, and reaction time greatly influence carbonization process. Hydrochars with lower hydrogen/carbon (H/C) and oxygen/carbon (O/C) molar ratios (equivalent to coal and lignite) are obtained with high temperature carbonization (Kang et al. 2019). Hydrothermal carbonization has an ability to produce high-energy content hydrochars at relatively lower residence times.

13.4.3.4 Gasification

Gasification is the most common thermochemical biomass conversion process that generates energy in the form of combustible gases (syngas or producer gas) comprising of CO (25–40 vol%), H_2 (30–50 vol%), CH_4 (6–15 vol%), and CO_2 (8–20 vol%) with heating values and conversion efficiencies of 0.4–10 MJ/m^3 and 50–70%, respectively (Juárez et al. 2012). Gasification involves vaporization, devolatilization, pyrolysis, cracking of tars, reduction, and reaction step at 700–1200 °C using air, oxygen, steam, and other gasifying agents. The factors affecting gasification process are the type of biomass feedstock, moisture content, particle size, temperature, and steam-to-biomass ratio. Different reactor configurations, such as fixed-bed (updraft and downdraft), fluidized-bed, entrained flow, spouted-bed, rotary kiln, plasma, and dual fluidized-bed (bubbling and fluidized-bed) reactors, are widely used.

Attempts have been made to co-gasify coal with cellulosic materials such as rice straw, corn straw, sawdust, pine sawdust, bean stack, peanut stalk, spent mushroom compost, and *Cryptomeria* wood (Widjaya et al. 2018). Natural minerals (dolomite and olivine) and nickel-based catalysts are widely used for their superior activity in tar cracking and reforming. Few studies have reported the use of secondary catalysts (Mg, Ca, K, and La supported on Al_2O_3) and supercritical water gasification (temperature >374 °C; pressure >22.1 MPa) as an alternative to conventional steam gasification (Reddy et al. 2014; Nanda et al. 2017a).

13.4.3.5 Hydrothermal Liquefaction

Hydrothermal liquefaction is the process of converting biomass into liquid oils in pressurized water environment (5–20 MPa) at 300–400 °C for longer residence time (0.2–1 h) with biomass-to-water ratio from 3:1 to 10:1 (Beckman and Elliott 1985). For woody and lignocellulosic biomass, feedstock pretreatment (particle size, contaminant removal, and alkaline treatment) is generally preferred. Biomass decomposes into monomeric units and undergoes dehydration, dehydrogenation, deoxygenation, decarboxylation, and deamination to produce liquid products (35–60 wt%) with heating value in the range of 20–35 MJ/kg. The use of methanol, ethanol, glycerol, acetone, phenol, ethylene glycol, and water at subcritical and supercritical water conditions is also evaluated.

Furthermore, to increase the bio-oil production, homogeneous (NaOH, KOH, $KHCO_3$, K_2CO_3, and Na_2CO_3) and heterogeneous (Ni and Zn/HZSM-5, Mo, Co, Ru, and Fe) solid acid catalysts were also applied (Tekin et al. 2014). However, the complex composition of feedstock, undesirable reaction intermediates, and excess oxygenates restricts the clear understanding of the reaction mechanism of hydrothermal liquefaction. Bio-oils obtained from hydrothermal processing of biomass can be improved and used as a replacement to conventional fuels through certain upgrading technologies.

13.5 Bio-oil Upgrading Processes

The transformation of biomass feedstock to achieve maximum bio-crude conversions and their extensive utilization can be accomplished by the aforementioned technologies. However, the following limitation restricts the direct use of bio-crude. The physical appearance of the bio-oil varies from dark brown through black depending on the presence of micro-carbon with a smoky odor. The chemical composition of bio-oil is intricate due to the presence of more than 400 different oxygenated hydrocarbons (Lopes et al. 2012) such as R–COOH (carboxylic acid), –CHO (aldehydes), C_nH_{2n+1} OH (alcohols), CH_3–C(=O)–CH_3 (ketones), RCOOR′ (esters), $C_4H_{10}O$ (ethers), $C_{12}H_{22}O_{11}$ (sugars), C_6H_5OH (phenols), C_4H_4O (furans), CH_2O (carbohydrates), multifunctional compounds, and large oligomers.

Moisture content, acids, aldehydes, and large molecular oligomers are pivotal factors for the deprived fuel properties of bio-oils. The presence of these undesired properties tends to impart low heating value (<50% compared to conventional fuels), immiscibility with hydrocarbons, thermal and chemical instabilities, higher density (1200 kg/m^3), oxygenates, higher acidic nature (pH: 2–3.7), viscosity (25–1000 cP), and corrosive nature to the bio-crudes (Chaiwat et al. 2013).

Various physical and chemical approaches toward the bio-crude upgrading ought to use as a commercial green fuel replacing the fossil feedstocks. This is achieved via various physical, chemical, and catalytic upgrading processes (Si et al. 2017). Physical upgrading includes filtration, solvent addition, and emulsion, whereas chemical upgrading involves esterification, mild cracking, gasification, catalytic upgrading or hydrodeoxygenation (HDO), zeolite cracking, and steam reforming. Among all, the most effective, advanced, and novel processes include hydrodeoxygenation, hydrothermal liquefaction, and gasification.

13.5.1 Hydrodeoxygenation (HDO)

As per the standards of the American Petroleum Institute (API), the normal oxygen content of bio-oil should not be more than 1.2%. However, bio-crude exceeds these limits, consequently tethering the possibilities of their replacement to fossil fuels. Further, the incessant growth of hydrogen technology, advanced catalysis, and HDO technology has become the trending and top among the others in producing second-generation fuels. Hence, the HDO technology has engrossed the attention of researchers due to its tactical importance. By nature, HDO is an exothermic reaction that involves high temperature and pressure causing carbon deposition on the catalyst surface and affecting the catalytic activity.

Furimsky (1983) presented a comprehensive framework, such as mechanism, catalyst, and the respective kinetics of the HDO process. Later, the research on the HDO received very little attention due to the complexities involved in the mechanism untill the revisiting studies of Furimsky (2000), which redefined the simplification of the process mechanism. Since then, considerable attempts have been made toward the objective of optimizing the process parameters such as temperature, catalysts, pressures, types of reactors, etc. An interesting outcome of these research studies over the decades revealed that the HDO process might not be directly applicable to all the organic and inorganic species (more than 300) which is tedious, so the redefined research was targeted on the specific compounds that hamper the commerciality of bio-crudes, especially the phenolic compounds. Table 13.1 presents the summary of some important findings related to HDO.

Hydrodeoxygenation, as the name indicates, is a combination of hydrogenation and deoxygenation. The combination of these processes tends to cleave the H_2 and O_2 atoms, to stabilize the unsaturated bonds, and reduce the oxygenates of bio-crude. HDO chemical reaction in the presence of H_2 occurs in congestion with hydrodesulfurization (HDS) and hydrodenitrogenation (HDN) during the catalytic hydrotreatment of the liquid fuels. For better understanding, the process of HDO revolves around the critical factors and the optimal conditions that help in the complete elimination of oxygenates in the form of water. Additionally, HDO is similar to the HDS reaction, which involves sulfide catalyst that has good stability and high conversion tendency of bio-crude (Si et al. 2017). The methodology of HDO is extremely productive due to the higher conversion rates of carbon to useful hydrocarbons. The detailed chemistry of HDO is illustrated in Fig. 13.3.

Catalytic upgrading of bio-crude is a complex phenomenon due to the composition and the typical operating process parameters including higher temperatures (300–600 °C) and pressure (10–25 MPa) under the influence of H_2 gas in the presence of catalyst to eliminate the O_2 molecules (Gollakota et al. 2016). The overall reaction of HDO is presented in the following Eq. (13.2):

$$CH_2O(\text{carbohydrate}) + H_2 \xrightarrow{\text{Catalyst}} CH_2 + H_2O \tag{13.2}$$

Table 13.1 Comparison of representative biomass, observations, and optimum parameters for hydrodeoxygenation

Feedstock	Temperature (°C)	Pressure (MPa)	Catalyst	Observation	References
Rice stalk	370	–	NiS/MoS	The effect of zerovalent Al on in situ HDO of bio-oil revealed that this reaction could yield a significant amount of 98% H_2 regeneration that could serve the cumulative HDO batch	Yang et al. (2019)
Pine sawdust	100	3	MoNi/γ--Al_2O_3	The effect of the catalytic promoter (Mo) on the HDO of bio-oil was examined by considering acetic acid as a model compound and converting to methyl acetate	Xu et al. (2010)
Switchgrass	320	14	HZSM-5	The suitability of the feedstock, catalyst, and H_2 consumption rate for the HDO process were critically emphasized. The results indicated that the upgrading is independent of the feedstock and catalyst types	Elkasabi et al. (2014)
Pyrolysis oil	300–350	–	Ru/C	A unique approach of using HDO oil as a solvent to treat the pyrolytic oil showed a substantial improvement in the heating values compared to the conventional solvents	Ahmadi et al. (2017)
Bio-oil	300	3	Noble metal	Interesting facts were revealed using noble metals as catalyst for the HDO process that unaltered the physicochemical composition of the upgraded bio-oil upon extended periods of storage	Oh et al. (2016)

(continued)

Table 13.1 (continued)

Feedstock	Temperature (°C)	Pressure (MPa)	Catalyst	Observation	References
Pinyon-juniper	450	6.2	Ni/RM	The addition of red mud to the catalyst during the HDO process significantly increased the heating value and drastically reduced the water content for commercial scaling	Jahromi and Agblevor (2018)
Wheat straw	400–450	0.101	MoO_3	A comparative study on the HDO of the model compounds and the pyrolysis vapors was performed under the influence of MoO_3 catalyst. The results revealed that higher degree of upgrading is possible for the liquid masses rather than vapors	Zhou et al. (2016)
Anaerobic digestate	229–369	14	Ni-Mo/Al_2O_3	The study elucidates that the upgrading of bio-oil obtained via thermal catalytic reforming yields better HHV when compared to the raw crude from pyrolysis process	Neumann et al. (2016)
Woody biomass	450	0.101	Ni-ZSM-5	A high Si/Al ratio zeolite catalyst was used to upgrade the pyrolysis bio-oil drastically reducing the acidic and furan compounds, thereby increasing the hydrocarbon fractions	Veses et al. (2015)
Cornstalk	280–370	4	Bimetallic ammonium nickel molybdate	The effect of temperatures on the upgrading of bio-oil was evaluated with a decrease in the yield and contrasting increase in the heating value as a function of the temperature	Yang et al. (2016)

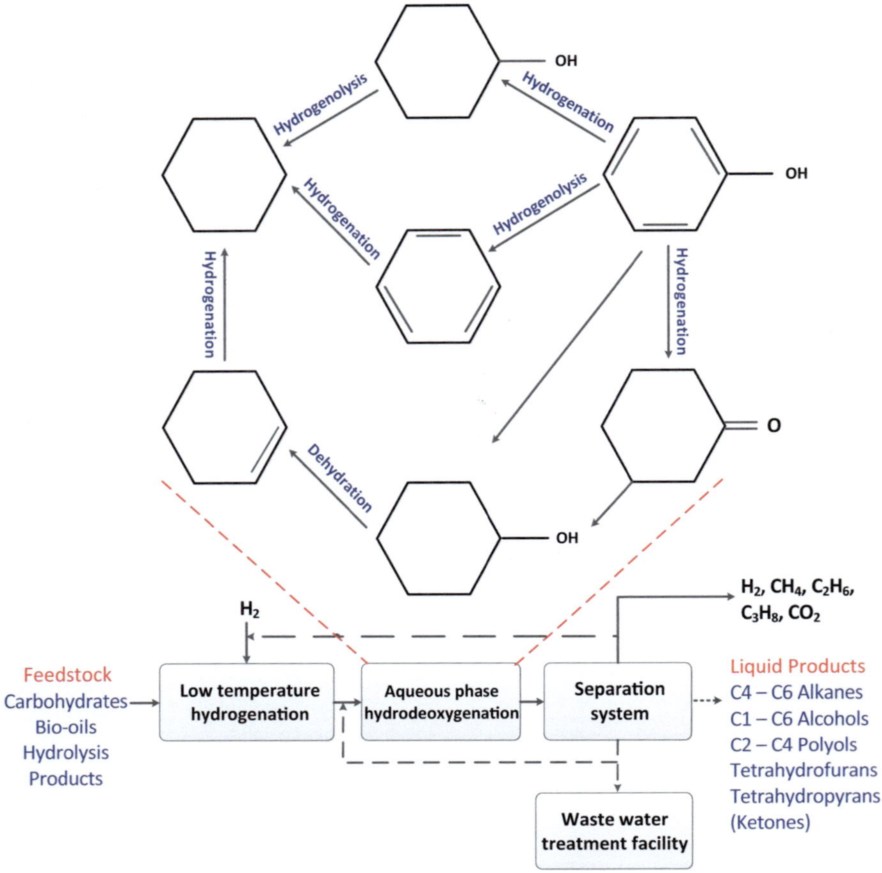

Fig. 13.3 Reaction mechanism of hydrodeoxygenation

Unlike other processes, HDO involves simultaneous secondary reactions targeting to avoid the deleterious contents of the bio-crude (Choudhary and Phillips 2011).

1. Removal/separation of water.
2. Dehydration is a conversion mechanism that involves the loss of water molecule from reacting molecule or ion.
3. Decarboxylation is the process of eliminating carboxyl group that liberates CO_2 from the bio-crude.
4. Hydrogenation is a chemical reaction to treat the unsaturated compounds with H_2.
5. Hydrogenolysis is the process of splitting C–O bonds and releasing O_2 in the form of H_2O.
6. Hydrocracking is a phenomenon of splitting macromolecules into smaller fragments.

The critical parameters that influence the HDO process are the operating temperature, H_2 flow rate, catalysts' active sites, residence time, pressure, and catalyst stability. Among all these factors, catalyst loading plays a crucial role. Lower ratio of catalyst loading favors the dispersion of active components, thereby restricting the formation of active sites. The higher catalytic loading also reduces the dispersion rate, subsequently increasing the crystal size (Gollakota et al. 2016). The second aspect is the choice of the catalyst, i.e., the sensitivity of the catalyst to the operating parameters, which plays a significant role in the catalyst deactivation.

Higher temperatures, pressures, and short residence times favor superior catalytic activity. However, prolonged reaction time results in catalytic poising and deactivation. The widely used catalysts for HDO process are Pt, Ni, and Co with supports of Al_2O_3 due to higher catalytic active and thermal stability. However, in the recent times, the trend of using noble metal catalysts is increasing due to the inert nature of the catalyst, i.e., the regeneration of the catalyst with the same activity near to the original characteristics. On the other hand, the complexity of the secondary reactions and undesired product phase is also a problem.

13.5.2 Hydrothermal Liquefaction

The pioneering research on hydrothermal liquefaction (HTL) dates back to the early nineteenth century. However, the Pittsburgh Research Institute initiated the commercial utilization of HTL process in 1970. Since then, many revolutionary attempts were made to optimize the physicochemical properties of feedstocks, products, treatment procedures, catalysts, and operating parameters (Schaleger et al. 1982; Thigpen 1982; Beckman and Elliott 1985). The main drawback of this mechanism is lenience toward the wet feedstock compared to the dry biomass. Table 13.2 explains the several other research findings related to hydrothermal liquefaction.

Hydrothermal liquefaction is the conversion technique of both wet and dry biomass feedstock to liquid product along with the traces of the gaseous stream. It is known that the composition of wet biomass are lipids, fats, proteins, and amino acids which are nonpolar, resembling aliphatic compounds otherwise known as triglycerides. The typical HTL process operates at the temperatures ranging between 250 and 350 °C (Shakya et al. 2015) and the higher pressures of 10–25 MPa (Chiaramonti et al. 2016) to separate maximum moisture content and reduce the oxygenates. At normal temperature and pressure conditions, fats and strong peptide chains of amino acids are highly insoluble in solvents, thus demanding for higher temperature and pressure scenarios.

At higher temperatures, the lipid content gradually turns into bio-crude. However, there is no change in the glycerol content, except being a water-soluble compound. Furthermore, the peptide chains of protein fraction undergo decarboxylation and deamination reactions forming C_nH_{2n+2} (hydrocarbons), amines, C_2H_4O (aldehyde), and acids. The characteristic features of the HTL process are the higher energy contents for the product phase and possessing enhanced heat recovery options. The detailed mechanism of the HTL process comprising three major steps, namely

Table 13.2 Comparison of process parameters, feedstock, and observations for hydrothermal liquefaction

Feedstock	Temperature (°C)	Pressure (MPa)	Time (min)	Observation	References
Seaweed meal	350	16.5	15	A comparative study on heating rates of bio-oil with and without catalysts was performed. Higher heating rates were achieved without catalyst support	Bach et al. (2014)
Anaerobic sludge Swine manure	350	5–8	60	The effect of the temperature on the bio-oil yield was critically analyzed and it was concluded that the increase in temperature resulted in higher yield of bio-oil and the composition resembled the petroleum crude, except for a slight deviation with the N_2 content	Shakya et al. (2015)
Chaetomorpha, *Cladophora*, *Oedogonium*, *Derbesia*, and *Ulva*	350	14–17	15	The influence of biomass composition on the bio-crude yield for different algal feedstocks was evaluated. The main research observation is that the biomass with lower N_2 content tends to yield higher energy	Neveux et al. (2014)
Scenedesmus almeriensis and *Nannochloropsis gaditana*	350	2	15	The research elucidates the effect of separation methods on bio-oil yields. Two techniques, namely gravity separation and organic solvent separation, were tested. It was observed that the addition of solvents reduced the industrial scalability	López et al. (2015)

depolymerization, decomposition, and recombination, followed by hydrotreating is illustrated in Fig. 13.4.

Depolymerization involves the sequential dissolving of macromolecules to overcome the difficulties to obstinate the physicochemical properties of the biomass. The detailed chemistry of depolymerization involves the splitting of the ether linkages of lignin in the biomass into methoxyl groups. The second step is the splitting

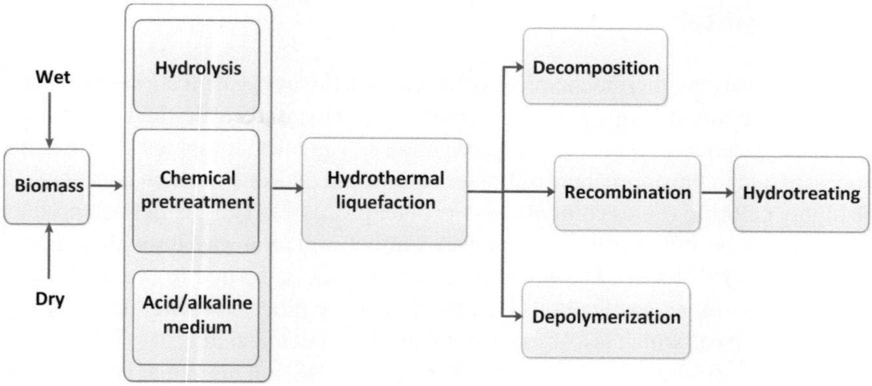

Fig. 13.4 Reaction stages involved in hydrothermal liquefaction

of the side chains into various free radicals and hydrogenated compounds into phenolic and alcohol derivatives. Finally, the ether bonds of the methoxyl groups convert into monophenols (Akash 2015).

Decomposition is the process of breaking large compounds into smaller components via three major steps, namely dehydration, decarboxylation, and deamination. During dehydration and decarboxylation, the oxygenated compounds are separated from biomass in the form of water and CO_2 while the macromolecules are hydrolyzed into polar oligomers and monomers. The moisture content of biomass breaks the hydrogen bonds easily at the prevailing temperatures of HTL process and forms the glucose monomers. The majorly degraded or decomposed products of the biomass are polar organic molecules, furfural, glycoaldehydes, phenols, and organic acids (Gollakota et al. 2018).

Recombination is the third step of the HTL process, which includes polymerization and rearrangement of the monomers into new molecules. The structural arrangement of the free radicals in the biomass organic matrix is solely dependent on the presence or the absence of the H_2 compound. The presence of the H_2 results in the stable structural matrix of constant molecular weight and the unavailability of H_2 leads to an increase in the concentration by the fragments' recombination, thereby forming higher molecular weight species otherwise known as char (Gollakota et al. 2018).

The pivotal factors influencing the HTL process are temperature and residence time, whereas quantity of biomass and solvent have a diminutive contribution. Low reaction temperature causes the incomplete conversion of the biomass macromolecules with lower yields, despite longer residence time. Contrary to this, higher temperatures and reaction times promote effective deoxygenation, thereby increasing heating value and energy recovery of the end product. Nonetheless, higher temperatures and residence time yield significant quantity of the water-soluble fractions, increasing the nitrogen content, which is undesirable. On the other hand, the addition of cosolvents proved to render enhanced yields. Furthermore, the addition of cosolvents drastically lessens the char formation because the solvent does not participate in the chemical reaction but facilitates the extraction of the organic molecules.

13.5.3 Gasification

Among the various thermochemical conversion pathways, gasification is an autothermic procedure in terms of energy balance. The energy recovery and heat capacity of the gaseous products from gasification are greater compared to the other processes due to the optimal utilization of carbon and hydrogen, which contribute to the higher calorific values compared to the other processes. The gasification product (syngas) can be effectively converted to synthetic natural gas through catalytic methanation of CO and CO_2, which is quite complex compared to the other techniques. Furthermore, gasification is the most effective process for producing hydrogen from biomass, which is a key intermediate in the chemical industry. The process of gasification originated before World War II, i.e., 1798, but the actual implementation was around the 1860s, and gradually evolved as the key resource of H_2 and CO (Basu 2010). During the early twentieth century, many rural and urban areas in the USA implemented an integrated gas supply for cooking and lighting. Later, during the second phase of the twentieth century, i.e., in 1973, the severe crisis of oil shortage gave the real start to this gasification technology as an alternate to the conventional fossil fuel resources. At present, the technology is completely matured for generating high-quality syngas from biomass. A detailed summary of literature pertaining to the gasification of various biomass feedstocks is presented in Table 13.3.

Gasification process uses a wide range of feedstocks, namely natural gas, coal, petroleum, pet coke, biomass, and industrial wastes. The process of gasification takes place at higher temperatures (600–1000 °C) under the influence of an oxidizing agent or gasification agent. The typical oxidizing agents for the gasification process are air, steam, N_2, CO_2, O_2, or a combination of all of the above (Juárez et al. 2012). The presence of oxidizing agent, high temperature conditions decompose the heavier polymeric fractions of biomass into lighter hydrocarbons, gases (a mixture of CO, H_2, CH_4, CO_2, and C_{2+}), and solid residues (char, ash, and tar). The overall gasification process is illustrated in Fig. 13.5 and the reaction scheme is shown in the following Eq. (13.3):

$$CH_xO_y\left(\text{biomass}\right)+O_2+H_2O \overset{\text{Catalyst}}{\rightarrow} CH_4+CO+CO_2+H_2+H_2O+C+\text{tar} \quad (13.3)$$

Gasification involves four inherent procedures, such as drying, pyrolysis, combustion, and reduction as presented in Fig. 13.5 (Kumar et al. 2009). Drying process of biomass reduces the moisture content to the range less than 5% at the temperatures of 100–200 °C. During devolatilization or pyrolysis stage, the volatile matter of the biomass is reduced by the thermal decomposition in the absence of oxygen and air. Subsequently, the biomass is turned into solid charcoal, thereby releasing the hydrocarbon gaseous vapors, and upon condensation at low temperatures, it generates liquid tars. Contrary to the previous stage, oxidation is a complex chemical reaction of solid biomass in the presence of oxygen converting carbon into CO_2 and CO while exerting large amount of thermal energy, as shown in the following equations. Several reactions, such as partial oxidation, complete oxidation, steam

Table 13.3 Summary of some notable studies on the gasification of biomass

Feedstock	Temperature (°C)	Pressure (MPa)	Gasifying agent	Observation	References
Rice straw	1000–1200	0.1	Oxygen	The influence of the gasification agent and temperature was evaluated	Gu et al. (2019)
Fruit waste	–	–		A comparative study on the process solutions for producing H_2 was assessed from biomass to achieve the target of 10 MW H_2 energy	Langè and Pellegrini (2013)
Palm shell	750–900	0.1	Steam	The newly developed trimetallic catalyst (nano-NiLaFe/γ-Al_2O_3) was used for upgrading of the biomass steam gasification. A significant removal of over 99% tar was reported with an increase in H_2 volume up to 17%	Li et al. (2009)
Wheat straw	700–750	0.1	Water	The exergy analysis was performed for different gasification configurations projecting the optimized processing conditions for the efficient tar removal and higher gas phase generation	Li et al. (2019)
Kenaf and Sorghum	700–900	0.1	Steam	The research is a sensitivity analysis of lignocellulosic biomass, biomass hydolysate, and coal. The results indicated that hydrolysates were more sensitive to degradation, yield higher fractions of H_2 and gas phase	Seçer et al. (2018)
Sawdust	660	–	Air	The effect of gasification temperature altered the composition of the gases. H_2 promotion ratio was a strong function of condensable product phase	Huang et al. (2012)

(continued)

Table 13.3 (continued)

Feedstock	Temperature (°C)	Pressure (MPa)	Gasifying agent	Observation	References
Municipal solid wastes	–	0.1	Air	A mathematical and numerical approach to study the feasibility and potential of municipal solid waste gasification for energy generation was performed	Thakare and Nandi (2015)
Pinewood char	800–950	0.1	Oxygen	The effect of potassium salts' impregnation on the catalyst substantially improved the syngas composition. The influence of various parameters, such as particle size, temperature, and gasifying agents, was also explained	Jia et al. (2017)
Karanja press seed cake	400–1000	0.1	Oxygen and steam	The research was focused toward the sensitivity analysis of the critical parameters on the gasification process. A numerical approach of gasification process was introduced for future research prospects	Shah et al. (2018)
Sugarcane bagasse	1000	0–1.5	Oxygen	The potential of sugarcane bagasse gasification for higher heating value material was explored. The choice of the gasifier was also detailed along with the optimal operating conditions	Motta et al. (2019)

reforming, Boudouard reaction, water–gas shift reaction, hydrogenation, and methanation, are involved in gasification process. Reduction is the process of stripping oxygen atoms off the combustion products of hydrocarbons to return the molecules to form the gaseous stream. During the reduction reaction, the char and tar produced from the oxidation release gases at 600–950 °C. The typical reactions including the heterogeneous water–gas shift reaction take place to produce CO and H_2.

Partial oxidation:

$$2C + O_2 \rightarrow 2CO \tag{13.4}$$

Complete oxidation:

$$C + O_2 \rightarrow CO_2 \tag{13.5}$$

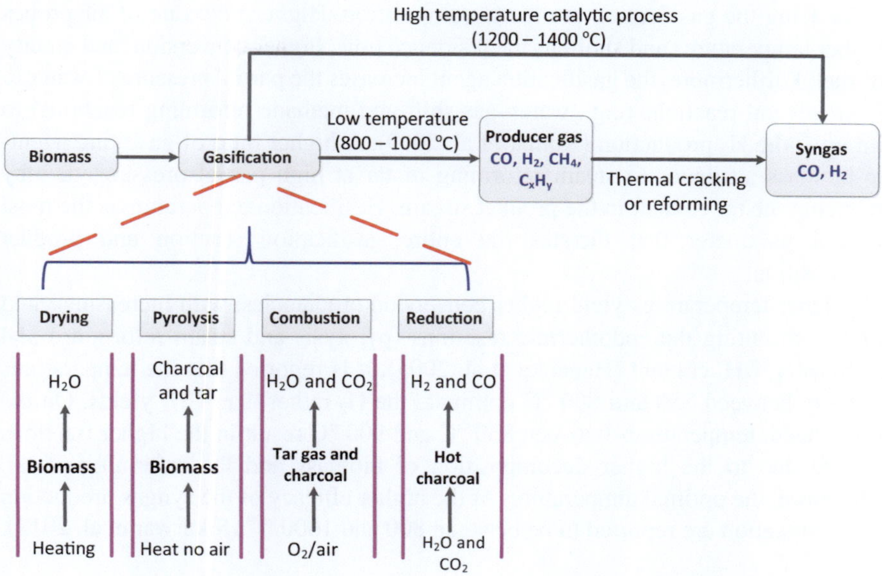

Fig. 13.5 Feedstock, process parameters, and products of biomass gasification process

Steam reforming reaction:

$$CH_4 + H_2O \rightarrow CO + 3H_2 \tag{13.6}$$

Boudouard reaction:

$$C + CO_2 \rightarrow 2CO \tag{13.7}$$

Water–gas reaction:

$$C + H_2O \rightarrow CO + H_2 \tag{13.8}$$

Water–gas shift reaction:

$$CO + H_2O \rightarrow CO_2 + H_2 \tag{13.9}$$

Critical factors that influence the gasification process include the design of the gasifier, gasification temperature, feed rate of biomass, oxidizing agents, type, and load of the catalyst as well as biomass type and properties (Nanda et al. 2017b). Biomass feed rate is a vital parameter and overloading/overfeeding results in chocking of the reactor, thereby reducing the conversion efficiency, and the yields. The choice of biomass is another aspect that needs careful attention due to its complexity in composition.

Various studies have revealed that cellulosic biomass tends to have a higher carbon conversion to CO and CH_4 as well as lower CO_2 and H_2, in comparison to xylan and lignin fractions (Couhert et al. 2009). The equivalence ratio (ER) and the superficial velocity (SV) are also important factors during the combustion reaction influencing the gasification products' composition. Higher flow rate of air probes higher temperatures and shortens the residence time, higher conversion, and quality of fuel. Furthermore, the gasification agent increases the partial pressure of water to favor several reactions (e.g., water–gas shift and methane reforming reactions) to improve the H_2 production (Kumar et al. 2009). A higher ratio of gasifying agent-to-biomass attributes to steam reforming of tar at high partial pressure, thereby reducing the tar content in the product stream. Gasification temperature is the most crucial parameter that dictates the entire gasification reaction and product composition.

Higher temperatures yield higher conversion efficiencies, with increasing yield of H_2 attributing the endothermic reactions (pyrolysis and steam reforming) and decreasing CH_4 content (Hanaoka et al. 2005). It is reported that the temperatures ranging between 750 and 800 °C dominate the H_2 rather than CH_4 yields. On the other hand, temperatures between 850 °C and 900 °C result in the higher fractions of CO due to the higher decomposition of biomass and the reforming of tar. Therefore, the optimal temperatures of the higher efficacy of the syngas production via gasification are reported to be between 800 and 1000 °C (Sikarwar et al. 2017).

13.6 Future Prospects and Recommendations

In summary, there exist substantial merits and demerits for all the aforementioned methodologies holding back the real-time applicability of biofuels. Biomass gasification process still suffers from the maintenance, operating costs, and lack of generic gasifiers for efficient biomass conversion. To surpass these limitations, integrated gasification combined cycle (IGCC) technology, a post treatment of syngas, is used to ensure the versatility of the process for efficient conversion and niche applications. On the other hand, hydrothermal liquefaction holds the key problem of excess usage of solvents and the batch system of processing. To date, a majority of the hydrothermal liquefaction research is carried out on algal biomass in small-scale batch autoclaves. Moreover, it is suggested that product separation followed by the thermochemical conversion can significantly improve the yield of the products, thereby improving the overall economic balance of the process.

Finally, upgrading biofuels via hydrodeoxygenation requires relatively higher hydrogen at higher pressure to convert the oxygenated compounds of bio-oils to oxygen-free products. In the interim, the higher hydrogen pressure leads to the complete hydrogenation of unsaturated hydrocarbon products. Furthermore, at elevated temperatures, and due to the thermodynamic limitations, phenol is the main compound of bio-crude, which is a complex compound from which the separation of oxygen is difficult. Despite these limitations, hydrodeoxygenation holds the key and futuristic approach, as most of the biomass is from dried sources. However, in the case of the wet feedstocks, hydrothermal liquefaction holds the key that can produce green fuels with better properties. Tables 13.4, 13.5, and 13.6 summarize the proximate and ultimate analyses of the various feedstocks and the conversion and upgrading products because of gasification, hydrothermal liquefaction, and hydrodeoxygenation, respectively.

Table 13.4 Comparison of improved physicochemical properties of feedstocks and respective products from gasification

Feedstock	Feedstock properties								Product properties				References
	C	H	N	S	O	A	M	HHV	H_2	CO	CH_4	CO_2	
Rice straw	47.5	5.7	0.2	0.73	40.1	5.6	9.3	19.1	30.1	37.4	7.6	23.4	Gu et al. (2019)
Fruit waste	45.1	6.4	0.2	1.1	47.3	7.5	60.1	–	71.8	1.4	1.5	25.1	Langè and Pellegrini (2013)
Palm shell	53.7	7.2	–	0.5	36.3	2.2	5.7	–	36.5	25.8	10.2	24.5	Li et al. (2009)
Wheat straw	36.5	4.9	0.5	–	40.5	7.7	9.6	–	0.04	8.7	2.8	19.1	Li et al. (2019)
Kenaf	48.6	5.7	0.1	–	42.5	3.1	10.4	–	79.4	1.7	12.1	6.8	Seçer et al. (2018)
Sorghum	39.8	5.2	0.8	–	45.8	8.4	8.1	–	28.3	43.2	9.1	14.5	
Sawdust	47.9	5.9	0.4	–	45.8	0.6	6.3	–	24.2	39.5	6.2	30.1	Huang et al. (2012)
Municipal solid waste	30.3	3.4	1.4	–	35.8	29.1	–	–	7.5	20.2	1.5	7.9	Thakare and Nandi (2015)
Pinewood char	87.2	2.4	1.1	–	10.3	2.2	–	31.2	62.7	14.9	1.2	21.2	Jia et al. (2017)
Karanja press seed cake	47.2	6.5	4.3	–	42.1	4.1	7.2	18.3	37.2	48.7	2.3	4.2	Shah et al. (2018)
Sugarcane bagasse	46.9	5.7	0.2	0.02	44.1	2.9	50.1	18.5	35.9	12.7	5.5	42.6	Motta et al. (2019)

Notes: Carbon (C), hydrogen (H), nitrogen (N), sulfur (S), oxygen (O), ash (A), moisture (M), and higher heating value (HHV)

Table 13.5 Comparison of improved physicochemical properties of algal feedstocks and respective products from hydrothermal liquefaction

Feedstock	Feedstock properties								Product properties								References
	C	H	N	S	O	A	M	HHV	C	H	N	S	O	A	M	HHV	
Seaweed meal	39.4	5.1	2.9	0.6	52.1	16.5	–	14.4	75.5	9.1	3.65	0.6	11.6	–	–	35.97	Bach et al. (2014)
Nannochloropsis	56.8	9.3	10.1	0.3	19.9	3.42	68.8	24.1	79.3	11.9	5.2	0.2	3.1	0.2	7.1	34.8	Shakya et al. (2015)
Pavlova	54.3	8.6	8.67	0.8	24.1	3.4	17.7	22.6	78.3	10.1	4.7	0.2	6.3	0.4	5.6	32.2	
Chaetomorpha	26.5	4.1	3.4	2.1	31.1	36.6	5.1	10.3	70.9	7.7	6.8	0.1	11.4	–	–	32.5	Neveux et al. (2014)
Cladophora	30.9	5.1	5.2	2.3	34.9	25.5	6.7	12.7	71.6	8.1	7.1	0.9	10.6	–	–	33.3	
Oedogonium	36.6	5.7	4.8	0.4	30.9	20.6	6.5	15.8	72.1	8.1	6.3	0.8	10.4	–	–	33.7	
Derbesia	29.2	4.8	4.5	2.8	27.4	34.7	6.4	12.4	73.1	7.5	6.5	0.7	10.6	–	–	33.2	
Ulva	27.7	5.5	3.5	5.1	41.1	30.7	7.2	11.7	72.6	8.2	5.8	0.4	11.1	–	–	33.8	
Scenedesmus	38.1	5.6	5.5	0.5	30.4	20.1	–	16.8	74.9	9.1	5.9	0.7	9.6	–	–	36.2	López et al. (2015)
Nannochloropsis gaditana	47.6	7.5	6.9	0.5	25.1	12.4	–	23.1	76.1	10.3	4.5	0.4	8.8	–	–	38.1	

Notes: Carbon (C), hydrogen (H), nitrogen (N), sulfur (S), oxygen (O), ash (A), moisture (M), and higher heating value (HHV)

Table 13.6 Comparison of improved physicochemical properties of feedstocks and respective products from hydrodeoxygenation

	Feedstock properties								Product properties								
Feedstock	C	H	N	S	O	A	M	HHV	C	H	N	S	O	A	M	HHV	References
Rice stalk	39.4	5.5	1.2	0.1	53.6	15.8	5.9	14.5	80.4	8.7	2.1	1.6	7.1	–	–	35.8	Yang et al. (2019)
Pine sawdust	53.6	6.2	0.02	–	40.1	–	35.5	13.9	53.8	6.9	0.01	–	39.2	–	41.5	14.1	Xu et al. (2010)
Switchgrass	53.8	5.3	0.5	–	33.2	–	7.9	–	72.1	7.8	0.8	–	19.2	–	2.3	–	Elkasabi et al. (2014)
Pyrolysis oil	38.2	7.6	–	–	54.1	–	27.9	–	71.1	8.3	–	–	20.4	–	4.1	–	Ahmadi et al. (2017)
Bio-oil	40.2	7.5	0.2	–	52.2	–	17.7		18.2	55.1	9.1	0.4	–	35.6	2.3	23.4	Oh et al. (2016)
Pinyon-juniper	54.4	6.2	0.1	–	38.6	0.5	6.6	19.3	83.1	16.5	0.3	–	–	–	–	45.8	Jahromi and Agblevor (2018)
Wheat straw	51.3	5.7	0.2	0.03	40.5	2.7	7.1	–	67.8	7.9	2.3	–	22.1	29.3	18.2	–	Zhou et al. (2016)
Anaerobic digestate	34.8	4.3	1.9	0.4	26.9	16.7	15.	–	86.2	13.1	0.1	–	0.80	–	–		Neumann et al. (2016)
Woody biomass	60.1	7.4	0.3	0.04	32.1	–	11.1	25.1	70.1	7.9	0.20	0.03	22.1	–	–	–	Veses et al. (2015)
Cornstalk	41.2	5.4	0.8	0.2	52.4	–	–	14.5	72.1	7.7	0.02	0.01	20.2	–	25.6	32.1	Yang et al. (2016)

Notes: Carbon (C), hydrogen (H), nitrogen (N), sulfur (S), oxygen (O), ash (A), moisture (M), and higher heating value (HHV)

13.7 Conclusions

Anticipated depletion of the fossil fuels and growing energy demand, especially from transportation sectors, have generated consideration for alternate fuels from biomass as a definitive answer to the declining oil production. Additionally, biofuels are the imperative choices to seek the complete decarbonization of the transportation sector. With the devoted interest in this area, various biomass to bio-crude technologies are discussed in this chapter. This led to the development of several upgrading technologies but the targeted perspective of clean energy and substantial replacement to the conventional fossil resources is not yet commercially feasible.

While some intriguing data have been made available by previous biomass upgrading strategies, several gaps still exist in this area. These gaps are divided into two groups (optimization and technology gaps), both contingent upon the feedstock utilized for producing renewable fuels. The optimization gaps of the biomass upgrading technologies can be addressed by means of additional research and through understanding of the inhibition effects, while the technological gap of bio-crude upgrading is exceedingly challenging. From the technical standpoint, there exists a substantial gap in the choice of the catalyst, optimal reactors, and varied operating parameters, which restrict the biofuels to reach the energy nexus. On the other hand, the nontechnical perspectives are to be considered in competencies to reach the desired objective of commercializing the biofuels. Despite uncertainties allied with renewable energy, the certainty factor of the research on renewable fuels will be vibrant and evergreen, as there is a scope to find the better way of coupling the economic and environmental factors and the associated challenges.

Acknowledgements The authors greatly appreciate the PS&DPL group, National Yunlin University of Science and Technology, Taiwan, for their cooperation and support.

References

Abnisa F, Wan Daud WMA (2014) A review on co-pyrolysis of biomass: an optional technique to obtain a high-grade pyrolysis oil. Energy Convers Manag 87:71–85

Adams P, Bridgwater T, Lea-Langton A, Ross A, Watson I (2017) Biomass conversion technologies. In: Thornley P, Adams P (eds) Greenhouse gas balances of bioenergy systems. Academia Press, Elsevier Inc., Cambridge, pp 107–139

Ahmadi S, Reyhanitash E, Yuan Z, Rohini S, Xu C (2017) Upgrading of fast pyrolysis oil via catalytic hydrodeoxygenation: effects of type of solvents. Renew Energy 114:376–382

Akash B (2015) Thermochemical depolymerization of biomass. Procedia Comput Sci 52:827–834

Arregi A, Amutio M, Lopez G, Bilbao J, Olazar M (2018) Evaluation of thermochemical routes for hydrogen production from biomass: a review. Energy Convers Manag 165:696–719

Ashter SA (2018) Biomass conversion approaches. In: Tillman DA, Duong DND, Harding NS (eds) Technology and applications of polymers derived from biomass. Elsevier Inc., Cambridge, pp 75–110

Bach QV, Sillero MV, Tran KQ, Skjermo J (2014) Fast hydrothermal liquefaction of a Norwegian macro-alga: screening tests. Algal Res 6:271–276

Basu P (2010) Gasification theory and modeling of gasifiers. In: Biomass gasification and pyrolysis: practical design and theory. Academia Press, Elsevier Inc., Cambridge, pp 117–166

Beckman D, Elliott DC (1985) Comparisons of the yields and properties of the oil products from direct thermochemical biomass liquefaction processes. Can J Chem Eng 63:99–104

Bhandari R, Volli V, Purkait MK (2015) Preparation and characterization of fly ash based mesoporous catalyst for transesterification of soybean oil. J Environ Chem Eng 3:906–914

Cai J, He Y, Yu X, Banks SW, Yang Y, Zhang X, Yu Y, Liu R, Bridgewater AV (2017) Review of physicochemical properties and analytical characterization of lignocellulosic biomass. Renew Sust Energ Rev 76:309–322

Chaiwat W, Gunawan R, Gholizadeh M, Li X, Lievens C, Hu X, Wang Y, Mourant D, Rossiter A, Bromly J, Li CZ (2013) Upgrading of bio-oil into advanced biofuels and chemicals. Part II. Importance of holdup of heavy species during the hydrotreatment of bio-oil in a continuous packed-bed catalytic reactor. Fuel 112:302–310

Chen H, Wang L (2017) Microbial fermentation strategies for biomass conversion. In: Technologies for biochemical conversion of biomass. Academia Press, Elsevier Inc., Cambridge, pp 165–196

Chiaramonti D, Prussi M, Buffi M, Rizzo AM, Pari L (2016) Review and experimental study on pyrolysis and hydrothermal liquefaction of microalgae for biofuel production. Appl Energy 185:1–10

Choudhary TV, Phillips CB (2011) Renewable fuels via catalytic hydrodeoxygenation. Appl Catal A Gen 397:1–12

Couhert C, Commandre JM, Salvador S (2009) Is it possible to predict gas yields of any biomass after rapid pyrolysis at high temperature from its composition in cellulose, hemicellulose and lignin? Fuel 88:408–417

Dhyani V, Bhaskar T (2018) A comprehensive review on the pyrolysis of lignocellulosic biomass. Renew Energy 129:695–716

Elkasabi Y, Mullen CA, Pighinelli ALMT, Boateng AA (2014) Hydrodeoxygenation of fast-pyrolysis bio-oils from various feedstocks using carbon-supported catalysts. Fuel Process Technol 123:11–18

Furimsky E (1983) The mechanism of catalytic hydrodeoxygenation of furan. Appl Catal 6:159–164

Furimsky E (2000) Catalytic hydrodeoxygenation. Appl Catal A Gen 199:147–190

García R, Pizarro C, Lavín AG, Bueno JL (2017) Biomass sources for thermal conversion. Techno-economical overview. Fuel 195:182–189

Gebremariam SN, Marchetti JM (2018) Economics of biodiesel production: review. Energy Convers Manag 168:74–84

Ghosh S, Chowdhury R, Bhattacharya P (2017) Sustainability of cereal straws for the fermentative production of second generation biofuels: a review of the efficiency and economics of biochemical pretreatment processes. Appl Energy 198:284–298

Gollakota ARK, Reddy M, Subramanyam MD, Kishore N (2016) A review on the upgradation techniques of pyrolysis oil. Renew Sust Energ Rev 58:1543–1568

Gollakota ARK, Kishore N, Gu S (2018) A review on hydrothermal liquefaction of biomass. Renew Sust Energ Rev 81:1378–1392

Gu H, Tang Y, Yao J, Chen F (2019) Study on biomass gasification under various operating conditions. J Energy Inst 92:1329–1336

Guedes RE, Lunaa AS, Torres AR (2018) Operating parameters for bio-oil production in biomass pyrolysis: a review. J Anal Appl Pyrolysis 129:134–149

Hanaoka T, Yoshida T, Fujimoto S et al (2005) Hydrogen production from woody biomass by steam gasification using CO_2 sorbent. Biomass Bioenergy 28:63–68

He C, Tang C, Li C, Yuan J, Tran KQ, Bach QV, Qiu R, Yang Y (2018) Wet torrefaction of biomass for high quality solid fuel production: a review. Renew Sust Energ Rev 91:259–271

Huang BS, Chen HY, Kuo JH, Kamei K, Harada M, Suzuki Y, Hatano H, Yokoyama SY, Minowas T (2012) Catalytic upgrading of syngas from fluidized bed air gasification of sawdust. Bioresour Technol 110:670–675

International Energy Outlook (2017) Energy Information Administration. USA

Jahromi H, Agblevor FA (2018) Hydrodeoxygenation of pinyon-juniper catalytic pyrolysis oil using red mud-supported nickel catalysts. Appl Catal B Environ 236:1–12

Jia S, Ning S, Ying H, Sun Y, Xu W, Yin H (2017) High quality syngas production from catalytic gasification of woodchip char. Energy Convers Manag 151:457–464
Juárez MC, Morales MP, Muñoz P, Mendivil MA (2012) Biomass gasification for electricity generation: review of current technology barriers. Renew Sust Energ Rev 18:174–183
Kang K, Nanda S, Sun G, Qiu L, Gu Y, Zhang T, Zhu M, Sun R (2019) Microwave-assisted hydrothermal carbonization of corn stalk for solid biofuel production: optimization of process parameters and characterization of hydrochar. Energy 186:115795
Kumar A, Jones DD, Hanna MA (2009) Thermochemical biomass gasification: a review of the current status of the technology. Energies 2:556–581
Langè S, Pellegrini LA (2013) Economic analysis of a combined production of hydrogen-energy from empty fruit bunches. Biomass Bioenergy 59:520–531
Li J, Yin Y, Zhang X, Liu J, Tan R (2009) Hydrogen-rich gas production by steam gasification of palm oil wastes over supported tri-metallic catalyst. Int J Hydrog Energy 34:9108–9115
Li Q, Song G, Xiao J, Sun T, Yamg K (2019) Exergy analysis of biomass staged-gasification for hydrogen-rich syngas. Int J Hydrog Energy 44:2569–2579
Lopes JM, Cerqueira HS, Ribeiro MF (2012) Bio-oils upgrading for second generation biofuels. I&EC Res 52:275–287
López BD, Riede S, Hornung U, Kruse A, Prins W (2015) Hydrothermal liquefaction of microalgae: effect on the product yields of the addition of an organic solvent to separate the aqueous phase and the biocrude oil. Algal Res 12:206–212
Motta IL, Miranda NT, Maciel FR, Maciel MRW (2019) Sugarcane bagasse gasification: simulation and analysis of different operating parameters, fluidizing media, and gasifier types. Biomass Bioenergy 122:433–445
Nanda S, Maley J, Kozinski JA, Dalai AK (2015) Physico-chemical evolution in lignocellulosic feedstocks during hydrothermal pretreatment and delignification. J Biobased Mater Bioenerg 9:295–308
Nanda S, Gong M, Hunter HN, Dalai AK, Gökalp I, Kozinski JA (2017a) An assessment of pinecone gasification in subcritical, near-critical and supercritical water. Fuel Process Technol 168:84–96
Nanda S, Rana R, Zheng Y, Kozinski JA, Dalai AK (2017b) Insights on pathways for hydrogen generation from ethanol. Sustain Energ Fuel 1:1232–1245
Neumann J, Jäger N, Apfelbacher A, Saschner R, Binder S, Hornung A (2016) Upgraded biofuel from residue biomass by thermo-catalytic reforming and hydrodeoxygenation. Biomass Bioenergy 89:91–97
Neveux N, Yuen AKL, Jazrawi C, Magnusson M, Haynes BS, Masters AF, Montoya A, Paul MA, Maschmeyer T, de Nys R (2014) Biocrude yield and productivity from the hydrothermal liquefaction of marine and freshwater green macroalgae. Bioresour Technol 155:334–341
Oh S, Choi HS, Kim UJ, Choi IG, Choi JW (2016) Storage performance of bio-oil after hydrodeoxygenative upgrading with noble metal catalysts. Fuel 182:154–160
Onay O, Kockar OM (2003) Slow, fast and flash pyrolysis of rapeseed. Renew Energy 28:2417–2433
Reddy SN, Nanda S, Dalai AK, Kozinski JA (2014) Supercritical water gasification of biomass for hydrogen production. Int J Hydrogen Energ 39:6912–6926
Sánchez J, Curt MD, Robert N, Fernández J (2019) Biomass resources. In: The role of bioenergy in the bioeconomy. Academia Press, Elsevier Inc, Cambridge, pp 25–111
Schaleger LL, Figueroa C, Davis HG (1982) Direct liquefaction of biomass: results from operation of continuous bench scale unit in liquefaction of water slurries of Douglas fir wood. In: Symposium on biotechnology in energy production and conservation, Gatlinburg, TN, pp 1–28
Seçer A, Küçet N, Fakı E, Hasanoğlu A (2018) Comparison of co–gasification efficiencies of coal, lignocellulosic biomass and biomass hydrolysate for high yield hydrogen production. Int J Hydrog Energy 43:21269–21278
Shah K, Dhanavath KN, Bankupalli S, Parthasarathy R (2018) Oxygen–steam gasification of karanja press seed cake: fixed bed experiments, ASPEN plus process model development and benchmarking with saw dust, rice husk and sunflower husk. J Environ Chem Eng 6:3061–3069

Shakya R, Whelen J, Adhikari S, Mahadevan R, Neupane S (2015) Effect of temperature and Na_2CO_3 catalyst on hydrothermal liquefaction of algae. Algal Res 12:80–90
Shin JD, Xu C, Kim SH, Kim H, Mahmood N, Lim M (2017) Biomass conversion of plant residues. In: Grumezescu AM, Holban AM (eds) Food bioconversion. Academia Press, Elsevier Inc., Cambridge, pp 351–383
Si Z, Zhang X, Wang C, Ma L, Dong R (2017) An overview on catalytic hydrodeoxygenation of pyrolysis oil and its model compounds. Catalysts 7:169
Sikarwar VS, Zhao M, Fennell PS, Shah N, Anthony EJ (2017) Progress in biofuel production from gasification. Prog Energy Combust Sci 61:189–248
Tekin K, Karagoz S, Bekta S (2014) A review of hydrothermal biomass processing. Renew Sust Energ Rev 40:673–687
Thakare S, Nandi S (2015) Study on potential of gasification technology for municipal solid waste (MSW) in Pune city. Energy Procedia 90:509–517
Thigpen PL (1982) An investigation of liquefaction of wood at the biomass liquefaction facility, Albany, Oregon, Battelle Pacific Northwest Laboratories. Department of Energy Contract AC01-78ET 23032 Wheelabrator Cleanfuel Corporation, United States
Veses A, Puértolas B, Callén MS, García T (2015) Catalytic upgrading of biomass derived pyrolysis vapors over metal-loaded ZSM-5 zeolites: effect of different metal cations on the bio-oil final properties. Microporous Mesoporous Mater 209:189–196
Volli V, Singh RK (2012) Production of bio-oil from de-oiled cakes by thermal pyrolysis. Fuel 96:579–585
Volli V, Kumar Purkait M, Shu CM (2019) Preparation and characterization of animal bone powder impregnated fly ash catalyst for transesterification. Sci Total Environ 669:314–321
Wang S, Dai G, Yang H, Luo Z (2017) Lignocellulosic biomass pyrolysis mechanism: a state-of-the-art review. Prog Energy Combust Sci 62:33–86
Widjaya ER, Chen G, Bowtell L, Hills C (2018) Gasification of non-woody biomass: a literature review. Renew Sust Energ Rev 89:184–193
Xu Y, Wang T, Ma L, Zhang Q, Liang W (2010) Upgrading of the liquid fuel from fast pyrolysis of biomass over MoNi/γ-Al_2O_3 catalysts. Appl Energy 87:2886–2891
Yang T, Jie Y, Li B, Kai X, Yan Z, Li R (2016) Catalytic hydrodeoxygenation of crude bio-oil over an unsupported bimetallic dispersed catalyst in supercritical ethanol. Fuel Process Technol 148:19–27
Yang T, Shi L, Li R, Li B, Kai X (2019) Hydrodeoxygenation of crude bio-oil in situ in the bio-oil aqueous phase with addition of zero-valent aluminum. Fuel Process Technol 184:65–72
Zhou G, Jensen PA, Le DM, Knidsen NO, Jensen AD (2016) Atmospheric hydrodeoxygenation of biomass fast pyrolysis vapor by MoO_3. ACS Sustain Chem Eng 4:5432–5440

Upgrading of Bio-oil from Biomass Pyrolysis: Current Status and Future Development

14

Quang Thang Trinh, Arghya Banerjee, Khursheed B. Ansari, Duy Quang Dao, Asmaa Drif, Nguyen Thanh Binh, Dang Thanh Tung, Phan Minh Quoc Binh, Prince Nana Amaniampong, Pham Thanh Huyen, and Minh Thang Le

Q. T. Trinh
Cambridge Centre for Advanced Research and Education in Singapore (CARES), Campus for Research Excellence and Technological Enterprise (CREATE), Singapore, Singapore

A. Banerjee
Department of Chemical Engineering, Birla Institute of Technology and Science (BITS) Pilani, Pilani, Rajasthan, India

K. B. Ansari
Department of Chemical Engineering, Aligarh Muslim University, Aligarh, Uttar Pradesh, India

D. Q. Dao
Institute of Research and Development, Duy Tân University, Thanh Khê, Đà Nẵng, Vietnam

A. Drif
INCREASE, Université de Poitiers, Poitiers, France

N. T. Binh
Faculty of Chemistry, Vietnam National University, Hà Nội, Vietnam

D. T. Tung · P. M. Q. Binh
Vietnam Petroleum Institute, Vietnam National Oil & Gas Group, Hà Nội, Vietnam

P. N. Amaniampong
Institut de Chimie des Milieux et Matériaux de Poitiers (IC2MP), Université de Poitiers, Poitiers, France

P. T. Huyen · M. T. Le (✉)
School of Chemical Engineering, Hanoi University of Science and Technology, Hà Nội, Vietnam
e-mail: thang.leminh@hust.edu.vn

S. Nanda et al. (eds.), *Biorefinery of Alternative Resources: Targeting Green Fuels and Platform Chemicals*, https://doi.org/10.1007/978-981-15-1804-1_14

Abstract

For the sustainable production of fuel, biomass pyrolysis processes have appeared as a promising alternative for the efficient utilization of biomass. The liquid product obtained from biomass pyrolysis, i.e., bio-oil has a very complex composition including large proportion of organic oxygenated hydrocarbon compounds and appreciable amount of water. The high oxygen content of the bio-oil creates many major drawbacks that hinder its vast application. The quality of bio-oil is not applicable for direct transportation purposes, and its upgrading is mandatory before it can be used practically in power engines. The abovementioned drawbacks can be overcome by the catalytic hydrodeoxygenation of bio-oil, in which the functional groups containing oxygen are removed and replaced by hydrogen atoms. For this purpose of selectively activating and breaking the C–O bonds while trying to maintain the C–C bonds intact, processing knowledge base of petroleum industry may not be utilized for hydrodeoxygenation of bio-oil. Therefore, novel catalysts, solvents, and processes need to be developed. This chapter reviews the state-of-the-art results obtained from the research and development of these technologies. The pros and cons, potential of future applications, challenges related to these technologies, and opportunities to maximize economic and environmental benefits while minimizing pollution are highlighted. Besides, the application of molecular modeling in the integration with experiment is highlighted in this chapter as a new paradigm for mechanistic studies, which could open new avenues to design and develop catalysts for a plethora of bio-oil upgrading processes that require high activity and selectivity. Finally, the breakthrough applications of novel sonochemical technique in biomass treatment and conversion are introduced in this chapter.

Keywords

Pyrolysis · Bio-oil upgrading · Hydrodeoxygenation · Density functional theory (DFT) · Molecular modeling · Sonochemical technique · Catalyst design

14.1 Introduction

Although conventional fossil fuels currently contribute to 80% of the world's energy consumption, their resources are diminishing. Therefore, there are strong incentives in finding alternative resources to reduce our reliance on fossil fuels (Amaniampong et al. 2015, 2017, 2018b, c; Trinh et al. 2015, 2016). Besides, fossil fuels also create several environmental problems such as increasing greenhouse gas emissions, accumulating CO_2, and pollution (release of fine particulate matter, NO_x and SO_x) of air and water systems. Among possible solutions, biofuel from renewable biomass is one of the potential alternatives to replace the fossil fuel.

Biomass is defined as materials produced by the growth of microorganisms, plants, or animals. It includes all the biological organic materials from living organisms produced by photosynthesis either indirectly or directly (Nachenius et al. 2013).

Biomass comprises animal-derived and mostly plant-derived materials such as crops, forestry and agricultural residues, wood, and industrial and municipal solid wastes. The estimation of biomass availability is reported as more than ten times higher than the energy demand of the world, which could be a promising resource for cleaner and sustainable energy production (Adams et al. 2018). A report from the International Energy Agency (IEA) shows that by 2035, approximately 15% of the global energy supply could be contributed from biomass and bioenergy, and by 2050, a large portion of world transportation fuel (~30%) could be replaced by biofuel (Nachenius et al. 2013; Trinh et al. 2016).

The first-generation biofuel raw materials are mostly food crop sources, such as sugar, starch, vegetable oil, and animal fats. These raw materials can generate a wide range of products, including biodiesel (via the transesterification), biogas, syngas, and bio-alcohols (via fermentation). However, the utilization of such feedstocks to produce biofuels is difficult and unsustainable globally due to the stress they put on the nutritional needs of the population. An alternative way to convert biomass to energy is the development of second-generation biofuels, which are produced from nonfood lignocellulosic biomass with many environmental and social benefits (Mushrif et al. 2015; Nanda et al. 2015). A major portion of biomass produced on earth is not utilized in the food and nonfood sectors such as animal feeds and papermaking, which could be utilized to produce next-generation biofuels. Among these nonfood biomasses, the renewable lignocellulosic biomass is the biggest carbon-containing source. It mainly consists of three basic components, i.e., cellulose (40–50 wt%), lignin (10–40 wt%), and hemicellulose (20–40 wt%), with the corresponding content amount of the individual components varying with the biomass type (Adams et al. 2018, Nachenius et al. 2013; Nanda et al. 2013).

To transform biomass into chemicals and fuels, three main technologies are available, such as thermal, physical/mechanical, and biochemical processes (Adams et al. 2018, Bridgwater 2012; Kabir and Hameed 2017; Nachenius et al. 2013; Nanda et al. 2014). Biochemical methods can only process the feedstock containing sugars and carbohydrates available in a relatively high purity. Additionally, they can barely breakdown some of the main biomass components like lignin. Hence, the thermochemical pyrolysis of biomass is the most promising method to convert cellulosic waste biomass into biofuel due to its convenience in scale-up and flexibility in controlling operation conditions to give a wide range of desired products and cost effectiveness. Primary processes in the thermal treatment of biomass include gasification (conversion of biomass to syngas and subsequently to liquid fuels via Fischer–Tropsch synthesis), combustion/incineration at high temperature (to recover heat, which is then used to produce electricity), and pyrolysis process (conversion of biomass to liquid fuels, gases, and solid char). Figure 14.1 summarizes various thermal processes along with the relevant primary products and their corresponding markets. Gasification and combustion processes are highly energy intensive. Therefore, pyrolysis is considered as the most promising method to make liquid fuels from the lignocellulosic biomass.

Technically, pyrolysis involves the decomposition of biomass in the absence of oxygen under high temperatures to produce solid char, liquid bio-oil, and

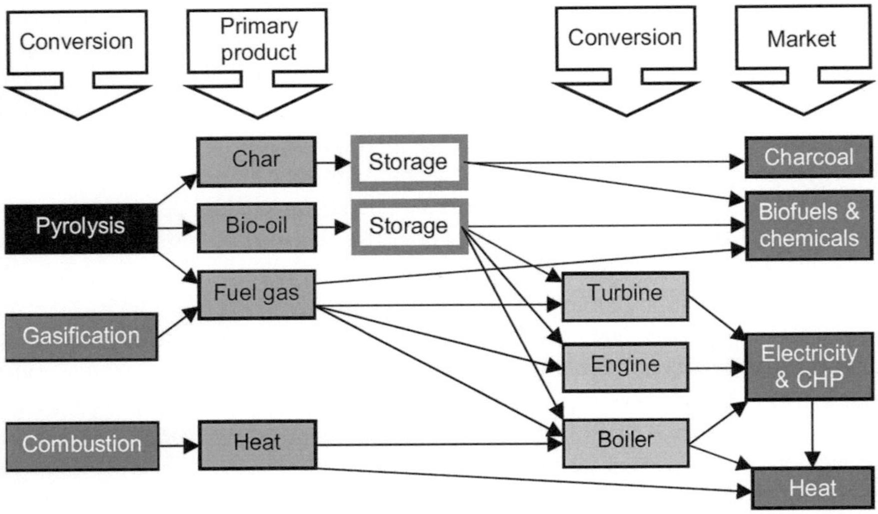

Fig. 14.1 Products from thermochemical biomass conversion (Bridgwater 2012)

noncondensable gases. There are three main types of pyrolysis processing, namely slow pyrolysis, fast pyrolysis, and flash pyrolysis, which are categorized based on the operation conditions (Adams et al. 2018; Bridgwater 2012; Bridgwater and Peacocke 2000; Mushrif et al. 2015; Nachenius et al. 2013). In slow pyrolysis process, biomass is heated for long residence times at low temperatures of 300–700 °C and slow heating rates to maximize the solid char product. In fast pyrolysis, a higher temperature range of 400–800 °C, coupled with short residence times and much faster heating rates (10–200 °C/s), promotes liquid product (bio-oil) formation. Very high temperatures in the range of 800–1000 °C along with extremely high heating rates (>1000 °C/s) and short residence times (<0.5 s) are characteristic of flash pyrolysis process with a high yield of bio-oil (up to 75 wt%). However, severe operating conditions along with the technical challenges related to its scalability make it difficult to commercialize the process. Fast pyrolysis is, therefore, still the most preferred method to convert biomass to liquid product.

14.2 Fast Pyrolysis Technology: Technoeconomic Assessment

As mentioned in the previous section, one of the most convenient technologies to convert biomass to fuels is the fast pyrolysis process. However, transportation of biomass appeared to be the major logistic issue for the conversion of cellulosic biomass to biofuel. It has been shown that the profitability of a power plant operated on biofuels could be affected to an extent of 30% due to transportation costs alone (Mushrif et al. 2015). Hence, to facilitate the commercialization of biomass to biofuel technology, the crucial step is to compromise between the transportation, handling, and processing of biomass, which is not the case for petroleum fuels. Fast pyrolysis process, on that front, is the only biomass processing technology, which offers the unique advantage of setting up processing plants at multiple geographic locations to convert biomass to bio-oil, which later can be stored/transported for the end use or for the upgradation in a much more economical fashion. In addition, an economical evaluation on a case study of a biomass processing plant with daily capacity of 2000 tons showed that fast pyrolysis process has the lowest capital and operating cost among other biomass platforms such as gasification and biochemical conversion (Anex et al. 2010; Swanson et al. 2010; Wright et al. 2010). The estimates of the product (gasoline equivalent) value for a pioneer plant and for the *n*th plant have also shown that the cost of pyrolysis fuel is almost half of that produced using biochemical methods.

Fast pyrolysis of biomass may appear to be an energy intensive process. However, analysis of the electricity consumption of a fast pyrolysis plant of the capacity of 20 tons of dry biomass per day has shown that if the product bio-oil is used to satisfy the plant's electricity requirement (with 40% efficiency), 18% of the total bio-oil produced would be used to run the plant. The key advantages that pyrolysis offers over the conventional incineration/combustion and gasification systems with respect to the generation of electricity from bio-oil are as follows:

1. The conversion of biomass to bio-oil can be decoupled from the electricity generation system, since bio-oil can be stored and transported. This not only provides flexibility in terms of electricity generation and distribution but also allows further production of high-grade fuels and chemicals from biomass. This decoupling becomes even more critical for geographically large countries.
2. The production cost of the pyrolysis-based electricity generation would be significantly lower than that of incinerators due to the decoupling.
3. The net system efficiency, in terms of energy in the electricity per unit heating value of the feedstock of the pyrolysis-generated electricity, is 25–50% more than that of incinerators, even with the existing pyrolysis technology. Further improvement can be achieved with improved bio-oil quality and yield.
4. The emission of furans and dioxins into the atmosphere and the generation of toxic fly ash and bottom ash in incinerators are hugely reduced.
5. Two side products generated by pyrolysis are char and gases. The gases formed can be utilized to heat the pyrolysis reactor, thus contributing to the economy of the process, and the char formed can be used as activated carbon for soil amendment.
6. The anthropogenic CO_2 emissions are significantly lower in the case of pyrolysis than that of conventional incineration/combustion.

The literature on the economic analysis of biomass pyrolysis to bio-oil for electricity and fuels/chemicals production suggests that the cost of the electricity generated from bio-oil and the cost of high-grade fuel produced from it are most sensitive to the yield and quality of bio-oil. To take a significant step forward to advance this technology, it is extremely crucial to be able to improve the pyrolysis process, to get higher bio-oil yields with better fuel quality.

14.3 Bio-oil Quality from Fast Pyrolysis and the Need for Upgrading

In fast pyrolysis, biomass is heated rapidly without the presence of oxygen at 400–1000 °C, leading to the formation of a short-lived condensed phase in which a variety of reactions occur to form volatile products and solid char. Most of the volatile products (except some light gases like CO, H_2, and light hydrocarbons) are later condensed to form a viscous fluid called as bio-oil. The quantity and quality of bio-oil depend strongly on the biomass feedstock, but other factors involving operating conditions and reactor design are also very important in varying the bio-oil properties. The operating temperature including heating and cooling rates, grinding particle size of the material, and pyrolysis residence times could significantly impact the distribution of the bio-oil liquid products and can even change the balance of bio-oil-to-char ratio (Adams et al. 2018; Bridgwater 2012; Kan et al. 2016; Nachenius et al. 2013).

Bio-oil obtained from fast pyrolysis process of biomass is a dark brown liquid with a characteristic acrid smoky smell due to high content of light aldehydes and acids. Its composition is very complex, including large proportion of organic compounds (oxygenated hydrocarbons like hydroxyaldehydes, hydroxyketones, hydroxyacetic acids, etc.) and appreciable amount of water. Bio-oil could be used as fuel for heating and electricity generation in diesel engines, combustors, furnaces, boilers, and gas turbines. Since there are many value-added chemical compounds in bio-oil, they can be extracted and used as the feedstocks for pharmaceutical, food, and fertilizer industries.

Some major drawbacks hinder the wider application of bio-oils. First, the bio-oil is highly acidic (pH ~2.5) due to its high content of organic acids, making it highly corrosive than fossil fuels. The corrosiveness of bio-oil is even much more severe at higher temperature and, therefore, limits the choice of material for bio-oil usage. Second, the presence of appreciable amount of water in the bio-oil (15–35 wt%) increases the ignition delay and reduces the heating value and the combustion rate of the bio-oil fuel. The high oxygen content of bio-oil resulting from the biomass is the main reason contributing to its low quality due to high polarity, low energy density, and low miscibility with other hydrocarbons components of the bio-oil. Besides, the oxygenated compounds (aldehyde and acids) are highly reactive and cause the bio-oil to become unstable due to the favorable condensation reactions occurring between these compounds at higher temperatures.

The drawbacks of bio-oil properties and their impact are summarized in Table 14.1. Therefore, it is generally accepted that the quality of bio-oil is not applicable for direct transportation purposes and significant upgrading of the bio-oils is required before it can be used practically as a transportation fuel (Elliott 2016). The upgrading of bio-oil can be done physically (e.g., extraction, filtration, and emulsion) or chemically with the help of a catalyst. The catalytic upgrading is the most widely applied technique, which involves different reaction pathways such as decarbonylation, decarboxylation, hydrodeoxygenation, and esterification for reduction of the oxygen content of the pyrolysis-derived bio-oil. The catalytic upgrading of bio-oil can be done in situ (co-introduction of the catalyst with biomass during the

Table 14.1 Drawback of bio-oils (Bridgwater 2012)

Characteristic	Cause	Effects
Acidity or low pH	• Organic acids from biopolymer degradation	• Corrosion of vessels and pipelines
Aging	• Continuation of secondary reactions including polymerization	• Slow increase in viscosity from secondary reactions such as condensation • Potential phase separation
Alkali metals	• Nearly all alkali metals retain in the char • High ash feed • Incomplete solid separation	• Catalyst poisoning • Deposition of solids in combustion • Erosion and corrosion • Slag formation • Damage to turbines
Char	• Incomplete char separation	• Aging of oil • Sedimentation • Filter blockage • Catalyst blockage • Engine injector blockage • Alkali metal poisoning
Chlorine	• Contaminants in biomass feed	• Catalyst poisoning in upgrading
Color	• Cracking of biopolymers and char	• Discoloration of some products such as resins
Contamination of feed	• Poor harvesting practice	• Contaminants notably soil act as catalysts and can increase particulate carry over
High oxygen content	• Biomass composition	• Poor stability • Nonmiscibility with hydrocarbons
High viscosity	• Organic and aqueous phase composition • Tar fraction	• Gives high-pressure drop increasing equipment cost • High pumping cost • Poor atomization
Low H/C ratio	• Biomass has low H/C ratio	• Upgrading to hydrocarbons is more difficult
Low miscibility with hydrocarbons	• Highly oxygenated nature of bio-oil	• Does not mix with any hydrocarbons so integration into a refinery is more difficult
Materials incompatibility	• Phenolic and aromatics	• Destruction of seals and gaskets
Nitrogen	• Contaminants in biomass feed • High nitrogen feed such as proteins in wastes	• Unpleasant smell • Catalyst poisoning in upgrading NO in combustion
Phase separation or nonhomogeneity	• High feed water • High ash in feed • Poor char separation	• Phase separation • Partial phase separation • Layering • Poor mixing • Inconsistency in handling • Storage and processing

(continued)

Table 14.1 (continued)

Characteristic	Cause	Effects
Poor distillation	• Reactive mixture of degradation products	• Bio-oil cannot be distilled completely (maximum 50% typically) • Liquid begins to react below 100 °C and substantially decomposes above 100 °C
Smell or odor	• Aldehydes and other volatile organics (mostly from cellulose and hemicellulose)	• While not toxic, the smell is often objectionable
Solids	• Includes char, particulates from reactor such as sand (fluidizing medium), particulates from feed contamination	• Sedimentation • Erosion and corrosion • Blockage
Sulfur	• Contaminants in biomass feed	• Catalyst poisoning in upgrading
Temperature sensitivity	• Incomplete reactions	• Irreversible decomposition of liquid into two phases above 100 °C • Irreversible viscosity increases above 60 °C • Potential phase separation above 60 °C
Toxicity	• Biopolymer degradation products	• Human toxicity is positive but small • Eco-toxicity is negligible
Water content	• Pyrolysis reactions • Feed water	• Complex effect on viscosity and stability • Increased water lowers heating value, density, stability, and increase pH • Affects catalysts

fast pyrolysis) or ex situ (catalytic upgrading of bio-oil vapors is performed after pyrolysis). The summary of catalysts being used for bio-oil upgrading and the strategies to design new catalysts with better efficiency and stability are described in later sections.

14.4 Influence of Heat and Mass Transfer on the Bio-oil Yield and Composition

It is difficult to systematically optimize a variety of factors including operating conditions of the pyrolysis process (i.e., particle size of feedstock and operating temperature) and the design of reactor for different types of biomass feedstocks to achieve the desired yield and appropriate product distribution of bio-oil. Most of the research is based on trying and testing different reactor configurations and operating conditions for different biomass feedstocks (Bridgwater 2012). Different temperatures, residence time, and particle sizes were tried and tested in several pyrolysis reactors without much insight on how these operating conditions could affect the transport phenomenon (i.e., heat and mass transport) inside the biomass particles. These operating conditions could also affect numerous reactions taking place in the condensed phase leading to the decomposition of biomass into char and bio-oil (Ansari et al. 2018; Paulsen et al. 2013).

Due to the high temperature's reactions coupled with heat and mass transport effects, the high complexity of biomass feedstock composition and the extremely fast reaction kinetics made it challenging to investigate fundamentals of pyrolysis. Because there is the lack of insights into pyrolysis reactions, most of the theoretical work, aimed at pyrolysis process design, used a lumped approach, where all the reactions in the condensed phase are represented in the form of a few pseudo reactions (Zhou et al. 2014). Kinetic parameters derived using such an approach were empirical, and the models were only applicable to specific biomass pyrolysis system and feedstock and did not have any predictive or interpretive capabilities for the variety of biomass (Mayes et al. 2014; Seshadri and Westmoreland 2012). Therefore, it is necessary to develop a comprehensive pyrolysis reaction network in the condensed phase and to combine it with heat and mass transport effects at particle level to be able to predict the effects of operating variables and the reactor design on cellulosic biomass decomposition at the particle level.

In principle, bio-oil derived from pyrolysis contains hundreds of compounds, which can be broadly categorized into five main groups as follows (Ansari et al. 2018):

1. Anhydrosugar compounds comprising levoglucosan (LGA), levoglucosenone (LGO), dianhydroglucopyranose (DAGP), and 1,6-anhydroglucofuranose (AGF).
2. Furanic compounds consisting of 2-methylfuran, furfural, 5-methylfurfural, 2,5-dimethylfuran (DMF), 5-hydroxymethylfurfural (HMF), 2-furanmethanol, furan, and furanone 2(5H).
3. Pyrans consisting of 2,3-dihydro-3,5-dihydroxy-6-methyl-4HPyran-4-one (DHMDHP) and 1,5-anhydro-4-deoxy-D-glycerohex-1-en-3-ulose (ADGH).
4. Light oxygenates compounds collectively formed from acetic acid (AA), 1,2-cyclopentanedione (CPD), methylglyoxal (MG), formic acid (FA), glyoxal (GL), hydroxyacetone (HAA), 2,3-butanedione (BD), and 2-hydroxy-3-methyl-2-cyclopenten-1-one (CPHM).

The percentage of each compound governs the bio-oil composition, and it varies significantly with the pyrolysis operating conditions, especially temperature and feedstock size. Paulsen et al. (2013) have systematically showed the changes in the yield of pyrolysis products, especially forming bio-oil, with respect to the biomass (or cellulose) particle size and operating temperature. In the subsequent study, Ansari et al. (2018) have succeed in decoupling the heat and mas transport in studying the biomass pyrolysis by taking glucose as a model biomass compound. By pyrolyzing two different sizes of glucose such as thin film, the influence of heat and mass transport was eliminated, and in powdered form, the heat and mass transport controlled the pyrolysis process. Fast pyrolysis of glucose thin film produced bio-oil with a lower yield than it was observed in the glucose powder at the temperature range between 300 °C and 500 °C, demonstrating the important role of heat/mass transport in this process. Generally, the yields of noncondensable gases were similar for both cases, but the observed amount of char was much higher in the case of glucose thin film. The higher yield of char was because of some extent of re-polymerization reactions taking place during glucose thin-film pyrolysis.

The yield of broader category of bio-oil compounds including furans, pyrans, anhydrosugars, and light oxygenates also changed with pyrolysis temperature. Within the bio-oil composition produced from the pyrolysis of glucose thin-film samples, the amount of anhydrosugars decreased when the operating temperature is increased, whereas an opposite increase was observed in the yield of light oxygenate, and the number of furans and pyrans altered only slightly. The trend of produced anhydrosugars and light oxygenate in derived bio-oil from the glucose powder pyrolysis was similar to that of glucose thin film. Furans showed a marginal decrease in their yields, while pyrans remained unchanged with an increase in the operating temperature. At higher operating temperatures, anhydrosugars tend to convert into lighter compounds via ring operating and fragmentation reactions. The uniform temperature inside glucose thin film assisted anhydrosugars in further decomposition, which in the case of glucose powders restricted because of the thermal gradient.

The insightful information gained from analyzing the product yields and composition of bio-oil obtained from powder and thin-film samples of glucose pyrolysis using the thin-plate spline interpolation method was helpful to optimize the operating conditions of biomass pyrolysis process (Ansari et al. 2018). Figure 14.2 shows the yields of char, bio-oil, noncondensable gases, and different bio-oil compounds (including levoglucosan, 5-hydroxymethylfurfural, furfural, furans, and anhydrosugars). The large size of glucose particles (460–500 μm) and the higher operating temperature (500 °C) maximized the noncondensable gases as shown in Fig. 14.2a. When the transport is limited, the bio-oil yield remained maximum with the glucose powders (500 μm) at 500 °C. The lowest yield of bio-oil was observed at 300 °C in the pyrolysis of thin-film glucose as shown in Fig. 14.2b. The yield of char yield exhibited an opposite trend to that of noncondensable gases/bio-oil and reached the minimum at pyrolysis temperature of 500 °C and particle size around 500 μm (Fig. 14.2c).

For determining the quality of bio-oil or better product distribution, the maximum yield of furans could be obtained at the operating conditions with the particle size of 300 μm and pyrolysis temperature of 500 °C (Fig. 14.2d). The amount of 5-hydroxymethylfurfural was maximized at 500 °C with particle size in the range of 200–400 μm, suggesting its production was competitive with other furans formation in glucose–pyrolysis as shown in Fig. 14.2e. Another furanic compound, i.e., furfural, became highest at 400 °C and with particle size of 200–300 μm (Fig. 14.2f). Finally, the formation of anhydrosugars and the major compound (e.g., levoglucosan) (Lin et al. 2009) reached the maximum at 500 °C with the biomass particle of 460–500 μm (Fig. 14.2g, h). In summary, it is demonstrated that by varying the size

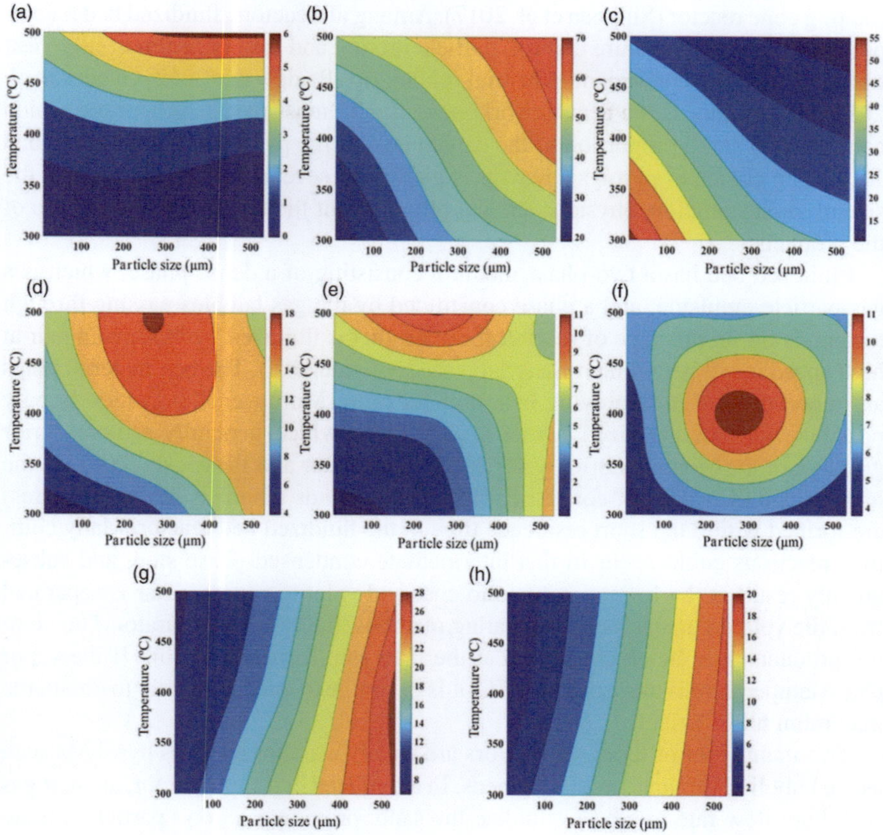

Fig. 14.2 Yields of bio-oil compositions versus pyrolysis temperature and feed particle size (Ansari et al. 2018): (**a**) gases yield (% C), (**b**) bio-oil yield (% C), (**c**) char yield (% C), (**d**) furans yield (% C), (**e**) HMF yield (% C), (**f**) furfural yield (% C), (**g**) anhydrosugars yield (% C), (**h**) levoglucosan yield (% C)

of biomass particles and the operating temperature range, the yield of bio-oil and its composition could be significantly changed. Such an approach for optimizing the operating conditions of biomass pyrolysis (i.e., particle size and operating temperature) could help in obtaining the highest yield together with achieving the desired composition of bio-oil produced from fast pyrolysis.

14.5 Fluidized Bed Reactor for Fast Pyrolysis Process

During the recent decades, numerous researches on biomass pyrolysis have focused on the design of reactors for biomass pyrolysis, resulting in a development of many different reactor designs such as fluidized bed reactor (Ly et al. 2019), fixed bed (Mohanty et al. 2013), transported bed, ablative reactor (Auersvald et al. 2019), and rotating cone reactor (Singbua et al. 2017). Among all reactors, fluidized bed reactor provides better temperature control, particle mixing and residence times, high heat transfer rates, and convenient industrial scaling-up (Bridgwater 2012; Guedes et al. 2018). The reactor is the central workhorse of a biomass pyrolysis process, which contributes a small proportion of the capital cost of the pyrolysis system (less than 15%). The choice of pyrolysis reactor is guided by several parameters such as the size of the installation, physicochemical properties of the feedstock, and the use of the products.

Fluidized bed has a two-phase medium consisting of a dense phase, which is a gas-particle emulsion, and a phase constituted by the gas bubbles passing through the emulsion. In this type of reactor, the solid fuel is fluidized by introducing air at high speed into a bed of inert materials (sand, olivine, etc.). The bed material plays an important role in improving heat transfer rates. Moreover, fluidization is only possible with smaller sized particles (2–5 mm), which generally requires prior grinding of the biomass (Bridgwater 2007; Bridgwater and Peacocke 2000). At the initial stage of pyrolysis, condensed-phase compounds (short-lived intermediates) are formed within the short residence time in the fluidized bed reactor. Many complex reactions could occur in this intermediate condensed-phase step, and subsequently result in the formation of solid char and volatile vapors. Char is separated from the volatile matter using separating methods like cyclone separator. The volatile products after being condensed at the later step form the bio-oil. If the vapor phase temperature is more than 400 °C, it is desirable to quench rapidly to obtain the maximum bio-oil yield.

Two main types of fluidized reactors are widely used for biomass pyrolysis such as the bubbling bed and circulating beds. In a fluidized bed-type reactor, an inert gas with high flow rate is used to fluidize the sand, providing a good particle mixing with excellent heat transfer. Heat can be transferred to the biomass particles by circulating hot gas around the reactor or by injecting hot inert gas through the reactor (Bridgwater 2007). Bubbling fluidized bed is the most popular design for biomass pyrolysis. Circulating fluidized beds also have the similar design to bubbling fluidized bed reactors. However, the gas flow rate in circulating fluidized beds is intentionally set high to carry all the particles out of the reactor compared to bubbling fluidized beds where the bed material remains suspended in the reactor (Treedet and Suntivarakorn 2018). The bed material is separated by gas cyclones, reheated in a combustion chamber, and returned to the bottom of the pyrolyzer.

Recently, many researchers have studied the pyrolysis process using fluidized bed reactors with different types of biomass feedstock (Eri et al. 2017; Park et al. 2019; Suntivarakorn et al. 2018; Upadhyay et al. 2018). These researches aim to investigate the effects of different parameters such as the biomass particle size, temperature, residence time, and flow rate of fluidizing medium. Bubbling fluidized bed reactors are mostly applied in those studies. In a study on the effect of fluidized bed materials in the tulip tree pyrolysis using a bubbling fluidized bed-type reactor, Ly et al. (2019) reported that the highest yield of bio-oil reached 49 wt% at 450 °C with nitrogen as the fluidizing medium and silica sand as the fluidizing bed material. The bio-oil yield reached 45 wt% at 400 °C using dolomite catalyst. Treedet and Suntivarakorn (2018) developed a system that was able to produce highest bio-oil yield of 78.1 wt% at 480 °C. The particle size effect in the fluidized bed reactor has also been reported (Septien et al. 2012; Shen et al. 2009). Shen et al. (2009) concluded that when the average size of biomass particle was greater, the bio-oil yield was lower at the bed temperature of 500 °C. It is generally accepted that to observe high biomass heating rates, small particle sizes should be used.

Industrial developments in bubbling fluidized beds have been chosen by several companies, including Dynamotive, which currently has units with a production capacity range from 450 kg/h up to 250 tons/day (Pattiya 2018). Figure 14.3 presents the RTI pilot process, which can process 200–250 tons/day of biomass in Guelph, Ontario, Canada, developed by Dynamotive Energy Systems (Radlein and Quignard 2013). The advantage of this process is that it achieves high bio-oils yields (72% for hardwoods and 65% for grasses). Many research units have also been developed by Union Fenosa (Jendoubi 2011), Dynamotive, and Wellman (Radlein and Quignard 2013). Some processes fail to scale up due to their low bio-oil yield, low particle mixing, heat transfer rates, and design complexities (Pinheiro 2008).

The pretreatment of biomass to reduce oxygen content in bio-oil is carried out at laboratory scales. Co-processing of bio-oil in fluid catalytic cracking unit (FCC) has been used in some refineries. Pinho et al. (2015) demonstrated that most of the oxygen in bio-oil was removed as water through the catalytic cracking, and the rest of oxygen was accumulated in liquid product as phenolic compounds. Bio-oil could be co-fed directly into a regular gasoil FCC feed up to 20 wt% without utilizing any type of hydrodeoxygenation (HDO). Both cases partially upgraded bio-oil and raw bio-oil with high oxygen content (50 wt%), and the co-processing would also be feasible. Furthermore, through ^{14}C isotopic analysis, Pinho et al. (2017) detected the presence of renewable carbon in gasoline and diesel cuts, clearly demonstrating that renewable carbon is not only transformed into coke, CO, and CO_2 but also to refining liquid products with higher value. The carbon efficiency (conversion of carbon to liquid products) was computed approximately to 30%, more than the values found in literature (15–20%).

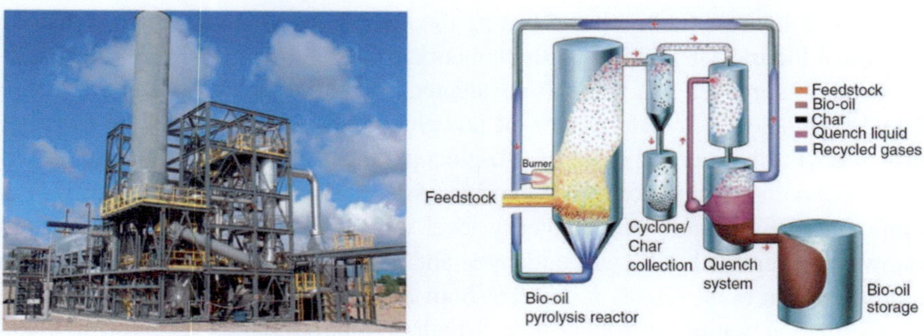

Fig. 14.3 One of the world's largest industrial-scale fast pyrolysis plant with a biomass input of 200 ton/day by Dynamotive Energy Systems located at Guelph, Ontario, Canada, and schematic representation of the RTI pilot process (Pattiya 2018)

Some issues must be considered when co-processing bio-oil in FCC unit:

1. To inject separately the fossil stream and the bio-oil at different reactor heights into the reactor.
2. To prevent corrosion of the materials used for feed lines and bio-oil storage in the FCC unit due to high acidity of the bio-oil.
3. The alkaline metals present in the bio-oil could destroy the structure and deactivate the zeolite in the FCC catalyst. Therefore, higher make-up rates might have to be utilized to ensure FCC equilibrium catalyst activity.

14.6 The Role of Molecular Modeling in Investigating Biomass Pyrolysis

Several studies on the biomass pyrolysis mechanism have been performed in the last decade. The state-of-the-art knowledge gained for the pyrolysis of lignocellulosic biomass has been well reviewed in terms of experimental and kinetic modeling (Shen et al. 2015; Wang et al. 2017). This section concerns an updated review of the most recent molecular modeling works using quantum chemistry tool. Quantum chemistry approaches offer a brilliant alternative way in providing insightful understanding into the chemistry of reactions occurring in the fast pyrolysis at the micro- and molecular-scale levels (Arora et al. 2018; Trinh et al. 2012, 2016, 2017; Varghese et al. 2016).

Since cellulose has a high polymerization structure, it is crucial to choose appropriate surrogate compounds for the modeling to study the mechanism of pyrolysis process using quantum chemistry. Glucose is chosen as a basic monomer unit of cellulose, and it has been popularly used for several mechanistic studies of glucose pyrolysis. Seshadri and Westmoreland (2012) developed elementary-reaction pathways, i.e., β-D-glucopyranose using the theory level of UB3LYP/6–311++G(d,p). In this work, different reactions pathways for the glucose ring-opening and reformation, retro-aldol condensation, ring contraction, tautomerization transformation between keto-enol structures, and dehydration were evaluated.

Mayes et al. (2014) investigated the conversion of glucose in fast pyrolysis at 500 °C to 5-hydroxymethylfurfural using theoretical calculations based on density functional theory (DFT) with the implementation of M06-2X/6-311+G(2df,p) theory details. Moreover, this work also evaluated the role of the unimolecular glucose dehydration and the transformation of levoglucosan to the pyranose and furanose structures. Furthermore, the use of β-D-glucopyranose (the close ring structure) as the model compound for cellulose pyrolysis was studied by Zhang et al. (2014) in employing DFT simulations at theory level of B3LYP/6-31G(d,p). The reaction pathways generated hydroxyacetaldehyde and acetol at around 126.8 °C and 1 atm conditions. Hu et al. (2018) found that both mannose and glucose have the same transformation pathways to produce 5-hydroxymethylfurfural and furfural and some key intermediates such as D-fructose and 3-deoxy-glucosone.

Although several works have chosen glucose as model compound representative of cellulose, this choice has serious limit concerning the fact that glycosidic bonds are not presented in glucose (Mettler et al. 2012). Besides, the yields of the pyrolytic products in glucose and their distribution are considerably different from those of cellulose. Therefore, cellobiose and cellotriose are found more suitable to be used as surrogate models in theoretical and modeling studies of cellulose fast pyrolysis. Zhang et al. (2015) employed cellobiose as a cellulose model compound to investigate the pyrolysis mechanism at a temperature range of 24.5–826.8 °C and use the theory level of M06-2X/6-31+G(d,p). Several primary products of the depolymerization reactions such as 5-hydroxymethylfurfural, hydroxyacetaldehyde, and levoglucosan were recognized. Furthermore, proton (H^+) played a crucial role in facilitating the glucosidic bonds cracking with the feasible reaction barrier of 146 kJ/mol at operation conditions of biomass pyrolysis.

Lu et al. (2018) performed fast pyrolysis of cellulose at 500 °C by DFT simulations at the theory level of M06-2X/6-31+G(d,p) using surrogate model compounds of both cellobiose and cellotriose. The first decomposition tends to produce important characteristic chain ends such as levoglucosan end, acyclic D-glucose end, reducing and nonreducing ends, unsaturated ends, and dehydrated units. In the second decomposition, these two types of model compounds showed different behavior in the final product formation. The authors also showed that levoglucosan is the main product formed from all three basic units of cellulose. Ascopyrone P and 1-hydroxyl-3,6-dioxabicyclo-octan-2-one are noncompetitive products and rely to the unsaturated end of cellulose. The acyclic D-glucose end consists in key component for the formation of 5-hydroxymethyl furan, furfural, and hydroxyacetaldehyde. The dehydrated unit tends to form a certain amount of hydroxyacetaldehyde. It is worthy to note that polymerization degree has significant influences on the pyrolytic product distribution. The larger the polymerization degree, greater is the amount of levoglucosan formed. The reason behind this is that the interior units enhance the levoglucosan formation. This provides a way to control the selectivity for the pyrolysis process to make specific products (Lu et al. 2018).

Lignocellulosic biomass usually contains from 15 to 35 wt% hemicelluloses (or polyoses). These heteropolysaccharides present in the hemicellulose are more complicated in structure than cellulose. The monosaccharide unit includes mainly pentoses (5-member aldoses such as xylose and arabinose) and hexoses (5-member aldoses such as mannose, glucose, and galactose) and other saccharides with lower content like fucose and rhamnose (Huang et al. 2016). There has been fewer studies on hemicellulose pyrolysis modeling at the molecular level compared to cellulose. Xylose is widely been chosen as a model compound for hemicellulose for mechanistic investigation of pyrolysis.

Huang et al. (2016) studied the pyrolysis of xylose using DFT calculations at the theory level of M06-2X/6-31++G(d,p). The authors found three major reaction pathways among eight studied pathways. The first pathway consisted of the conversion of xylose to acrylic containing carbonyl isomer, which could be able to process via the ring-opening reaction with a reaction barrier of 189 kJ/mol. In the second reaction route, xylose can be decomposed to generate ethanol with a much higher barrier of 300 kJ/mol. Finally, in the third proposed pathway, xylose is activated via the dehydration between the hydroxyl group attached to carbon C_4 and the hydrogen atom coordinated to carbon C_5, forming the product anhydroxylopyranose with computed reaction barrier energy of 295 kJ/mol. Overall, it is observed that the ring-opening reaction is the most favored to initiate the pyrolysis of xylose.

Lu et al. (2016) employed B3LYP/6-31+G(d,p) DFT simulations to explore fast pyrolysis mechanism for the formation of hydroxyacetaldehyde from β-D-mannopyranose, β-D-glucopyranose, *O*-acetyl-β-D-xylopyranose, and β-D-xylopyranose. The authors found that along with the pyrolysis of β-D-glucopyranose compound, hydroxyacetaldehyde is essentially derived from different possible positions such as C_3–C_4, C_1–C_2, and C_5–C_6, which are initiated by the dissociation of C–O_{ring} bond characterized by transition state activation barrier of 178 kJ/mol. Furthermore, the breaking of C_2–C_3 bond forming 1,2-ethenediol could be processed via the retro-aldol reaction, and subsequently undergoes the tautomerization into the hydroxyacetaldehyde with high selectivity from C_1–C_2. The β-D-mannopyranose and β-D-xylopyranose have similar reaction mechanism with the first studied component. The *O*-acetyl-β-D-xylopyranose has different behavior compared to the other three compounds, in which it is shown that the hydroxyacetaldehyde is mainly obtained from C_4–C_5 positions, which is processed together with the scission of the acetyl branch producing acetic acid (Lu et al. 2016).

Lignin is characterized by more complicated structures than the cellulose and hemicellulose. The polymerization of three main structures of monolignols including sinapyl alcohol, coniferyl, and *p*-coumaryl results in the formation of three corresponding structures of lignin units, i.e., syringyl lignin, guaiacyl, and *p*-hydroxyl phenyl. These units are characterized with the interlinkages such as α-*O*-4, β-*O*-4, β-1, and 5-5. Among those largest dominate interlinkages, β-*O*-4 and α-*O*-4 bonds are presented with the highest proportions, and their contributions in the total linkages in lignin is 48–60% and 6–8%, respectively (Huang and He 2015).

Huang et al. (2014) employed DFT simulations at B3LYP/6-31G(d,p) details to theoretically evaluate the pyrolysis of lignin containing the β-*O*-4 linkages using the dimer structure of 1-phenyl-2-phenoxy-1,3-propanediol as a model compound. They reported that the C_β–O bond homolytic cleavage reaction was the most dominating pathway with the lowest reaction barrier equal to 245 kJ/mol, whereas the one of C_α–C_β bond has the second lowest reaction barrier, i.e., 259 kJ/mol. Huang and He (2015) investigated α-*O*-4 linkage dimer model pyrolysis employing the theory level of B3LYP/6-31G(d,p) in their DFT simulations. The authors found that the hemolytic C_α–O bond dissociation reaction (reaction barrier of 177 kJ/mol) was the most preferred reaction pathway, whereas the O–CH_3 bond homolytic cleavage reaction (reaction barrier of 228 kJ/mol) and the one of C_α–C_β bond (reaction barrier of 267 kJ/mol) were less favorable being the weaker competitive routes in pyrolysis mechanism of the studied model compound.

It is evident that the quantum chemistry modeling still occurs as an important parameter allowing understanding the complicated reaction schemes of lignocellulosic biomass at the molecular level. Although an enormous number of quantum chemistry studies have been contributed to the biomass pyrolysis mechanism, there are still several challenges. Many attempts can be made for predicting the preponderance of

Table 14.2 Different catalytic processes for bio-oil upgrading

Methods	Catalyst	References
Hydrogenation	Ru/TiO_2	Wang et al. (2016)
	Ru/Fe_3O_4-SiO_2	Cherkasov et al. (2017)
Hydrodeoxygenation	• Metal sulfides (supported Co-MoS_2 and Ni-MoS) • Nobble metal catalyst (supported Pd, Pt, Rh, Ru) • Nonnoble metal catalysts (supported Fe, Ni, Co, and W)	Mortensen et al. (2011), Si et al. (2017)
Catalytic cracking	Zeolite (HZSM-5, HY, H-Mordenite, Silicalite, SAPO-5, SAPO-11, MgAPO-36, etc.)	Adjaye and Bakhshi (1995a), Adjaye and Bakhshi (1995b), Liao et al. (2014), Mortensen et al. (2011)
Catalytic esterification	Ion exchanged resin (732, NKC)	Wang et al. (2010)
	Zeolite (medium-pore ZSM-5, Faujasite)	Milina et al. (2014)
	Sulfated ZrO_2–TiO_2	Liu et al. (2015)
Catalytic steam reforming	Ni-Co/ and Ni-Cr/ MgO-La_2O_3-α-Al_2O_3	Garcia et al. (2000)
	Ni/Al_2O_3	Bimbela et al. (2013), Galdámez et al. (2005)
	Ni/MgO	Wu and Liu (2011)
	Supported Ni, Ru, or Rh catalysts	Trane et al. (2012)
	Ni-Co/MgO	Mei et al. (2016)
	Ni/zeolite L, Ni/CeO_2-zeolite L	Bizkarra et al. (2019)
	Ni/Al_2O_3, Ni/ZrO_2, and Ni/MgO	Santamaria et al. (2019)

reaction pathways, which form the representative oxygenated compounds from fast pyrolysis and the possible interactions between these pathways.

14.7 Catalytic Bio-oil Upgrading

The quality and stability of crude bio-oil can be improved by removal of oxygen content that is present with high concentration. Several methods such as hydrogenation, hydrodeoxygenation, catalytic cracking, esterification, emulsification, and steam reforming have been extensively studied for the upgrading of bio-oil (Table 14.2).

After the catalytic upgrading, the properties and quality of bio-oil can be significantly modified. In summary, each method of catalytic upgrading has its own advantages and disadvantages that could be described as below:

1. Hydrogenation of unsaturated compounds and hydrodeoxygenation of oxygen-rich organic compounds require a high pressure of hydrogen, special catalysts, complex facilities, and high cost of the operation.

2. Catalytic cracking using a solid acid catalyst with high mechanical and thermal stability such as zeolites converts oxygen-organic compounds to CO, CO_2, H_2O and a hydrocarbon-rich high-grade liquid fuel. The catalyst for the catalytic upgrading of bio-oils can have the following characteristics: (a) proper attrition strength; (b) good chemical stability to be used in high temperature and corrosive environment; and (c) suitable mesoporosity of zeolite crystals to allow the bio-oil molecules to access the catalytic acid sites located inside the pore and permit the products to diffuse back out of the zeolite crystals.
3. The catalytic esterification can be used to neutralize bio-oil with high content of carboxylic acids by alcohols (such as ethanol or methanol) and acid catalyst in reactive distillation (Xiu and Shahbazi 2012). By adding these polar solvents, the viscosity of the bio-oil also decreases that leads to better miscibility with diesel fuel and higher volatility and heating value. Catalytic esterification is also one of the most appropriate methods for the upgrading of bio-oil because of the simplicity of the method and the low cost of alcohols.
4. Catalytic steam reforming of bio-oil recently is also an important upgrading technology for converting into hydrogen. However, some oxygenated compounds such as phenol cannot be completely converted. Carbon deposition on nickel-supported catalysts, high reaction temperature that lead to the loss of catalyst, and reactor corrosion are some disadvantages of this method.
5. The emulsification is used to produce homogeneous and stable fuel obtained from bio-oil and diesel, which can be burned in existing engines. This a simple method, but the obtained fuel can cause the corrosion of the injector nozzle because of the presence of acidic compound.

Mortensen et al. (2011) and Tran et al. (2014) reported that the deoxygenation of bio-oil by hydrodeoxygenation and catalytic cracking might be the most prospective options due to its economically feasible. Bio-oil can be upgraded either by catalytic cracking over zeolite catalysts (oxygen removed as CO_2 and water) at atmospheric pressures and high temperatures (300–600 °C) (Cheng and Huber 2011; Jae et al. 2011; Mukarakate et al. 2014). This can also be done through catalytic hydrodeoxygenation over metal catalysts (oxygen removed as water) like Ru, Pt, Pd, Ni, Cu, Co, etc. in the presence of H_2 (Nolte and Shanks 2017; Vispute and Huber 2009; Vispute et al. 2010; Wildschut et al. 2009).

The changes in bio-oil properties before and after catalytic upgrading via hydrodeoxygenation and zeolite cracking processes are presented in Table 14.3. Although the quality of bio-oil can be improved, the catalytic cracking still suffers from a poor quality of the upgraded bio-oil with O/C ratios higher than 0.6 and significant coke formation due to the cleavage of C–C bonds. Due to the similarity between hydrodeoxygenation and conventional hydrotreating and hydrodesulfurization processes used in petroleum refineries, Mo-sulfide catalysts promoted with Co and Ni have been the most widely used hydrodeoxygenation catalysts. However, the sulfided catalysts suffer from rapid deactivation and require sulfur in the feed streams for activation, which could lead to bio-oil product contamination. Catalytic hydrodeoxygenation produces high H/C ratios with low oxygen content but requires high hydrogen pressures. Due

Table 14.3 The properties of bio-oil before and after catalytic upgrading (Mortensen et al. 2011)

	Bio-oil	Hydrodeoxygenation	Zeolite cracking	Crude oil
Upgraded bio-oil				
Y_{oil} (wt%)	100	21–65	12–28	–
$Y_{Waterphase}$ (wt%)	–	13–49	24–28	–
Y_{Gas} (wt%)	–	3–15	6–13	–
Y_{Carbon} (wt%)	–	4–26	26–39	–
Oil characteristics				
Water (wt%)	15–30	1.5	–	0.1
pH	2.8–3.8	5.8	–	–
ρ (kg/L)	1.05–1.25	1.2	–	0.86
$\mu_{50°C}$ (cP)	40–100	1–5	–	180
Higher heating value (MJ/kg)	16–19	42–45	21–36	44
C (wt%)	55–65	85–89	61–79	83–86
O (wt%)	28–40	<5	13–24	<1
H (wt%)	5–7	10–14	2–8	11–14
S (wt%)	<0.05	<0.005	–	<4
N (wt%)	<0.4	–	–	<1
Ash (wt%)	<0.2	–	–	0.1
Atomic H/C	0.9–1.5	1.3–2.0	0.3–1.8	1.5–2.0
Atomic O/C	0.3–0.5	<0.1	0.1–0.3	0

to the limitations of the catalytic cracking process, most of the researches on catalyst development have focused on the hydrodeoxygenation process.

A few other novel process is also being developed recently. A new method for bio-oil upgrading using methane as a cheaper reductant instead of hydrogen for the deoxygenation of guaiacol, a model compound in the lignin thermal degradation, on Pt/C and Pt-Bi/C catalyst has been reported (Varma and Xiao 2017, 2018). In these studies, Bi-promoter is used to prevent the coking or carbon deposition on the Pt/C catalyst. However, there is still a long way of developing these processes to be commercially applicable.

14.8 Mechanistic Study and State-of-the Art Computational DFT Approach in Catalyst Design for Bio-oil Upgrading

One of the most challenging issue in upgrading of bio-oil is to design and develop better catalysts, which can achieve the dual tasks. For example, the catalyst needs to be efficient in decreasing the oxygen content and in the meantime being able to facilitate the production of drop-in liquid fuels for blending as additives into conventional gasoline fuels or directly utilizing as transportation fuels. The catalyst should thus selectively break C–O bonds while preserving C–C bonds to form olefins/isoalkanes, which are some excellent fuel additives for gasoline. A good hydrodeoxygenation catalyst needs to have appropriate surface oxophilicity (to facilitate C–O bond cleavage) and moderate hydrogenation capability (to achieve selective hydrogenation) of the bio-oil derivatives to desired target products. Thus, hydrodeoxygenation catalyst design requires an in-depth understanding of the metal–adsorbate interaction. Bimetallic catalysts having dual functionalities or addition of promotors to single metal catalysts can alter the electronic structure and properties and act as possible candidates for hydrodeoxygenation catalysis. Additionally, an insightful understanding on the underlying reaction chemistry is essential for rational catalyst design and predicting product distributions and their corresponding selectivity.

The advent of supercomputers over the last decade has seen a significant improvement in the computing technology. This has resulted in the advancement of techniques and development of tools for understanding the chemical reaction at the micro-level, i.e., at atomistic scales. A computational tool like density functional theory (DFT) has been useful in establishing the electronic structure of catalyst surfaces and also in predicting reaction mechanisms, energetics, and chemical reactivity for different reactions on these model catalyst surfaces (Mondal et al. 2016; Nandula et al. 2015; Nihei et al. 2018; Sarkar et al. 2019; Singuru et al. 2016; Trinh et al. 2012, 2013, 2017, 2018; Varghese et al. 2016; Yang et al. 2015). The theoretical insights gained from numerous computational studies are highlighted in this section, which have thrown light on our understanding of the chemical reaction kinetics, metal–adsorbate interactions, and developed structure activity relationships for the catalytic hydrodeoxygenation of bio-oil derived from pyrolysis processes.

Although several experimental works on the catalytic upgradation of bio-oil have been reported, very few have focused on understanding the underlying mechanisms using theoretical tools. As discussed in the earlier sections, the bio-oil derived from biomass pyrolysis is a complex mixture of aldehydes, ketones, furanic, and phenolic compounds. The major challenge in developing metal–adsorbate interactions and detailed reaction mechanism for bio-oil upgradation on a catalyst surface is the choice of model compound, which could provide an accurate representation of the real bio-oil. Model compounds like furfural, guaiacol, phenol, acetone, propanol, *p*-cresol, *m*-cresol, ethanol, etc. have been widely used as representative compounds in evaluating the hydrodeoxygenation on different metal surfaces (Chiu et al. 2014; Gu et al. 2016; Mironenko and Vlachos 2016; Shi et al. 2015; Sitthisa et al. 2011b; Vorotnikov et al. 2012; Wan et al. 2018). Among these, furfural and guaiacol have gained the most attention due to the abundance of furanic and phenolic compounds observed in the pyrolysis-derived bio-oil. Although model compounds may not give a true representation of the hydrodeoxygenation of bio-oil, the insights gained could be useful for predicting catalyst behavior of actual bio-oil. The difference in activity of different metals toward hydrodeoxygenation can be broadly attributed to two factors, such as (a) nature of the metal–adsorbate interaction, i.e., orientation of adsorption and strength of adsorption, and (b) properties of the catalyst active site and energetics of the hydrodeoxygenation mechanism.

Multiple theoretical investigations have examined the adsorption of bio-oil model compounds on the transition metal surfaces. The most stable adsorption configuration of bio-oil surrogate model compound on the catalyst is also one of the important factor that control the selectivity/activity of the catalyst

(Banerjee and Mushrif 2018; Trinh et al. 2013, 2015, 2016, 2017). The DFT studies have shown that furfural adopts two different adsorption configurations such as (a) it is adsorbed on Cu surfaces via the O of the carbonyl group in a η^1(–O) structure (Shi et al. 2015; Sitthisa et al. 2011c), and (b) a η^2(–C–O) orientation, i.e., adsorption via both the ring and the carbonyl group favored on Mo_2C, Ni, Pt, Pd, and Ru surfaces (Banerjee and Mushrif 2017, 2018; Sitthisa et al. 2011a, b; Vorotnikov et al. 2012).

A comparison of the furfural binding energies computed over different metal surfaces in literature suggests that the strength of furfural adsorption follows the trend, $Mo_2C(001) > Ru(001) > Ni(111) > Pd(111) \sim Pt(111) > Cu(111)$, correlating to the corresponding activities of the catalysts. The η^1(–O) orientation of adsorption coupled with a low binding strength makes ring activation difficult on the Cu surfaces. In an opposite trend, the calculated furfural binding energy in the η^2(–C–O) configuration is a combination of the metal–ring and metal–carbonyl interaction and the dominant interaction between the two is estimated from the furan binding energies, which acts as a descriptor for metal–ring binding strength. The nature of the binding interaction can give a rough estimation of the observed product distributions. For example, the strong metal–ring and the metal–carbonyl interaction on the Ru surface would facilitate ring opened and C–O scission products, while the strong interaction of the carbonyl group on the Mo_2C surface results in very facile C–O scission barrier leading to high methylfuran selectivity.

The catalyst activity and selectivity toward desired products can be further tuned by the use of bimetallic catalysts, which either alter the adsorption strength (or surface oxophilicity) or modify the orientation of adsorption (Jain et al. 2018; Sitthisa et al. 2011a, b; Trinh et al. 2012; Wan et al. 2018). For example, the incorporation of an oxophilic metal like Fe in Ni catalysts was found to stabilize furfural binding on the NiFe surface, via a strong Fe–O interaction (Sitthisa et al. 2011a). The carbonyl C–O of furfural was elongated in the presence of Fe, which facilitated C–O bond scission to form methylfuran and suppressed the decarbonylation of furfural to furan. Theoretical studies have also shown the doping of boron onto Ni surfaces to form a corrugated structure, which enhanced the furfural binding energies by 0.66 eV (Banerjee and Mushrif 2018; Trinh et al. 2016, 2017).

A lowering of furan yields and increase in furfuryl alcohol yields was observed on Cu-modified Pd catalysts (Sitthisa et al. 2011b). The presence of Cu destabilized the η^2(–C–O) bonded furfural relative to the pure Pd surface, resulting in a decrease in stability of the acyl species, which is an intermediate for furan formation. Jalid et al. (2017) utilized a microkinetic modeling-based approach to determine the activity toward C–O cleavage for ethanol on a stepped (211) surface of different transition metals using carbon- and oxygen-binding energies as the descriptors. The catalysts such as Ru, Rh, and Ni had high C- and O-binding energies and were found to selectively produce methane via C–C cleavage, while Co, Ir, and Ni were found to be ideal for C–O cleavage with high TOF for ethane. The abovementioned studies show that adding suitable secondary adatoms to the monometallic catalyst surface could modify the metal–adsorbate interaction to achieve desired products. The principle of using bimetallic catalysts can be extended for replacing expensive monometallic catalysts like Ru, which are highly active for hydrodeoxygenation of bio-oil with cheaper catalyst formulations having similar metal–adsorbate interactions to achieve desired product selectivity. However, metal–adsorbate interactions may not be the only descriptor for catalyst reactivity, and the energetics of the reaction on these modified surfaces need to be explored fully, to comprehend the catalyst activity and selectivity.

Mechanistic investigations detailing the systematic progress from the reactant to the product are required to explain the multiple product distributions that are observed experimentally. Such an analysis would also be required for the design of new, novel catalysts which can selectively produce the desired target products (Trinh et al. 2012, 2017). The ability of computational tools like DFT in elucidating the reaction mechanism and its energetics has led to numerous insights into catalyst design and development. The reaction mechanism of model compounds like guaiacol, furfural, and phenol has been studied on several different metal surfaces to understand the activation behavior on the different surfaces.

The different bond-breaking or bond-forming pathways involved in the hydrodeoxygenation of furfural include C–O cleavage, C–H cleavage, C–O hydrogenation, C–C cleavage, C_{aryl} hydrogenation, and C_{aryl}–O cleavage as is illustrated in Fig. 14.4. On Pd(111) surfaces, furfural hydrogenation to furfuryl alcohol proceeds via hydrogenation of the O end of the carbonyl atom, with a highest barrier of 0.8 eV. On the

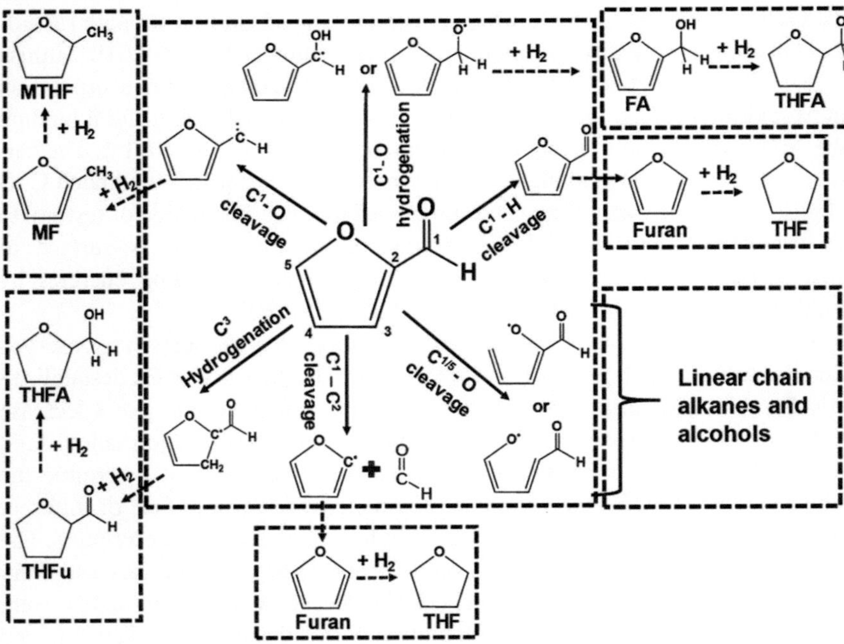

Fig. 14.4 Possible furfural activation pathways and possible products. Abbreviations: Methylfuran *MF*, furfuryl alcohol *FA*, tetrahydrofurfuryl alcohol *THFA*, tetrahydrofuran *THF*, tetrahydrofurfural *THFu*, methyltetrahydrofuran *MTHF* (Banerjee and Mushrif 2017)

other hand, furfural decomposition to furan proceeds via C–H cleavage followed by C–C cleavage with a highest barrier of 0.95 eV (Vorotnikov et al. 2012). CH_x–O ($x = 0$–2) scission of the carbonyl group had high activation barriers, thus making methylfuran formation kinetically unfavorable on Pd surfaces. The effect of surface H coverage and operating conditions on the reactivity of furfural was also explored with DFT calculations (Wang et al. 2015). Furfural-binding configuration changes from a flat to a "tilted" configuration with effective hydrogenation and decarbonylation barriers of 0.78 and 0.63 eV, respectively. This results in a reversal of the main product selectivity from furan (at low H coverage limit at low H_2 pressures) to furfuryl alcohol (at high H coverage limit at high H_2 pressures).

Cai et al. (2015) investigated the structure sensitivity of furfural activation on different sites of Pt, including flat terrace site and under-coordinated site such as step and corner sites. Both furfuryl alcohol and furan formation proceeded via a similar pathway as that on Pd surfaces. Furfuryl alcohol formation was kinetically favored on the terrace sites, i.e., Pt(111) surfaces with a highest barrier of 0.34 eV, while furan formation dominated at the corner sites (Pt_{55} cluster) with a highest barrier of 1.18 eV. On Cu(111) surfaces, hydrogenation of the carbonyl group proceeds via the C atom first, followed by the O atom to form furfuryl alcohol (Shi et al. 2015). However, CH_2–O scission (and subsequent hydrogenation to methylfuran) and hydrogenation to furfuryl alcohol were found to have competitive barriers of 1.18 and 1.17 eV, respectively. The experimentally observed furfuryl alcohol yields were attributed to the high H coverage present on the Cu surfaces, which could favor furfuryl alcohol formation and hinder methylfuran formation. The C–O scission always proceeded after hydrogenation of the carbonyl group, and the direct scission of the carbonyl C–O was kinetically limited on all the studied metal surfaces. All the above computational studies focused on the formation of selected products like furan, methylfuran, and furfuryl alcohol and did not provide a complete reaction pathway detailing all the possible products of furfural activation (see Fig. 14.4). Such a reaction pathway analysis is especially useful for developing micro-kinetic models, which evaluate surface coverages of the intermediates and predict product selectivity.

The first detailed mechanistic studies for the activation of furfural on transition metal Ni(111) and Ru(001) surfaces were performed recently in which the reaction pathways resulting in the formation of different products like furfuryl alcohol, tetrahydrofurfuryl alcohol, furan, tetrahydrofuran, methylfuran, methyltetrahydrofuran, diols, and alkanes were evaluated (Banerjee and Mushrif 2017, 2018). Nickel catalysts show a strong product selectivity dependence with temperature. At lower temperature, furans and furfuryl alcohol were obtained as the dominant products, while higher temperatures favored furan and ring opened products. Banerjee and Mushrif (2018) explained this unique behavior to be due to the surface H coverage on Ni at different temperatures. At low temperatures, the adsorption of H on the Ni surface is spontaneous, and therefore, a high surface H coverage is proposed, which favors furfuryl alcohol formation and hinders ring activation.

At the certain high level of reaction temperature, the spontaneous adsorption of H is no longer occurred, resulting in low H coverages, which favor furfural ring opening and furan formation. On Ru surfaces, ring opening via a C_{ring}–O cleavage was found to be very facile with a barrier of only 0.66 eV making it the kinetically favored first activation pathway for furfural (Banerjee and Mushrif 2017). The ring opened moiety then undergoes successive C–O scissions (with barriers between 0.4 and 0.7 eV) and subsequent hydrogenations to form saturated alkanes as the dominant products, in agreement with experimental observations. Except Ru, none of the catalyst surfaces studied were successful in removing all the O-atoms from furfural. The excellent capability of Ru toward hydrodeoxygenation of bio-oil as observed via experiments was thus explained by the mechanistic studies.

Besides furfural, the reaction mechanism for hydrodeoxygenation of guaiacol, *m*-cresol, *p*-cresol, and phenol has also been studied on Pt, Ru, and NiFe surfaces (Chiu et al. 2014; Gu et al. 2016; Nelson et al. 2015; Shetty et al. 2017). The vapor phase hydrodeoxygenation of guaiacol at low hydrogen pressures of ~1 bar leads to phenol and catechol on Pt and phenol formation on Ru catalysts. On Ru, guaiacol activation involves the step-wise dehydrogenation of the methoxy group followed by a C_{alkyl}–O and then C_{aryl}–O cleavage to form phenol. Prior to the C_{aryl}–O scission, the catecholate radical (15 in Fig. 14.5) can either undergo hydrogenation at the O end (to 14 followed by hydrogenation to catechol) or undergo a C_{aryl}–O scission (to 17 followed by hydrogenation to phenol), with comparable activation barriers of 99 and 106 kJ/mol. However, the low reverse barriers in the former suggest (14) to be an unstable species, which would dehydrogenate back to (15). As a result, catechol formation is kinetically and thermodynamically less favored on the Ru surfaces.

Lee et al. (2015) studied guaiacol activation on Pt surfaces and proposed a similar pathway as that on Ru surfaces to form a catecholate radical. However, hydrogenation of this radical was kinetically favored with a barrier of only 0.29 eV relative to the C_{aryl}–O cleavage, which had a barrier of over 2 eV. Thus, catechol formation was favored on the Pt surfaces. The phenol observed during hydrodeoxygenation of guaiacol on Pt was proposed to form via a dehydroxygenation of catechol molecule on the acidic Al_2O_3 sites of the Pt/alumina interface. Thus, hydrodeoxygenation of guaiacol proceeded via a dehydrogenation of the methoxy group followed by C_{aryl}–OH scission on Pt and Ru surfaces and not by a direct demethoxylation or demethylation step.

A similar reaction mechanism has also been proposed for hydrodeoxygenation of anisole on Pt, Ru, and Fe surfaces (Tan et al. 2017). Interestingly, mechanistic studies on the hydrodeoxygenation of *p*-cresol to toluene on Pt surfaces revealed the hydrodeoxygenation to proceed via 3–4 ring hydrogenations followed by a C_{aryl}–OH scission (Gu et al. 2016). Hydrogenation of the ring was found to lower the C_{aryl}–OH scission barriers. The hydrogenated complex then underwent a dehydrogenation step to form toluene as the dominant product. The importance of theoretical calculations in understanding the mechanism for hydrodeoxygenation is further highlighted as molecules with similar binding adsorbate geometries (*p*-cresol and guaiacol here) behave differently on the same metal surface leading to different product distributions.

Most of the above reaction pathway studies were evaluated for monometallic catalyst surfaces, resulting in at least two dominant products (except for Cu). As suggested earlier, the role of catalysts would be to produce the desired product/products with high selectivity. In case of hydrodeoxygenation of bio-oil, the target

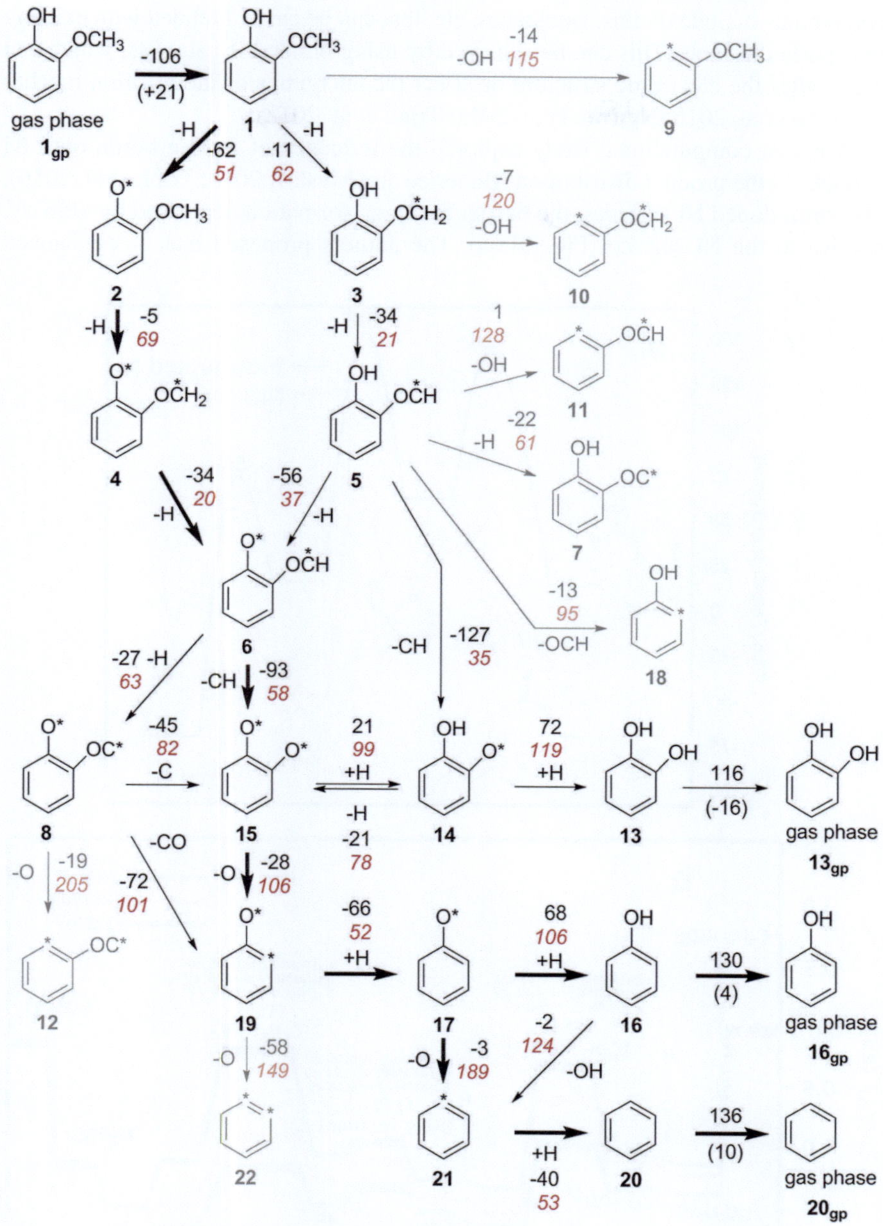

Fig. 14.5 Reaction network of hydrodeoxygenation of guaiacol on Ru(001) surface (Chiu et al. 2014)

compounds include olefins, isoalkanes, etc. that can be easily blended with existing transportation fuels. This can be achieved by using bimetallic catalysts, which can either alter the electronic structure or affect the energetics of the reaction mechanism (Le et al. 2016; Nguyen et al. 2014; Trinh et al. 2012).

A recent computational study explored the influence of doping boron onto Ni surfaces to the product distribution (Banerjee and Mushrif 2018; Trinh et al. 2016). On boron doped Ni surfaces, the barrier for furan formation decreased by ~0.6 eV, relative to the Ni surface (Fig. 14.6a). The authors proposed that this enhanced

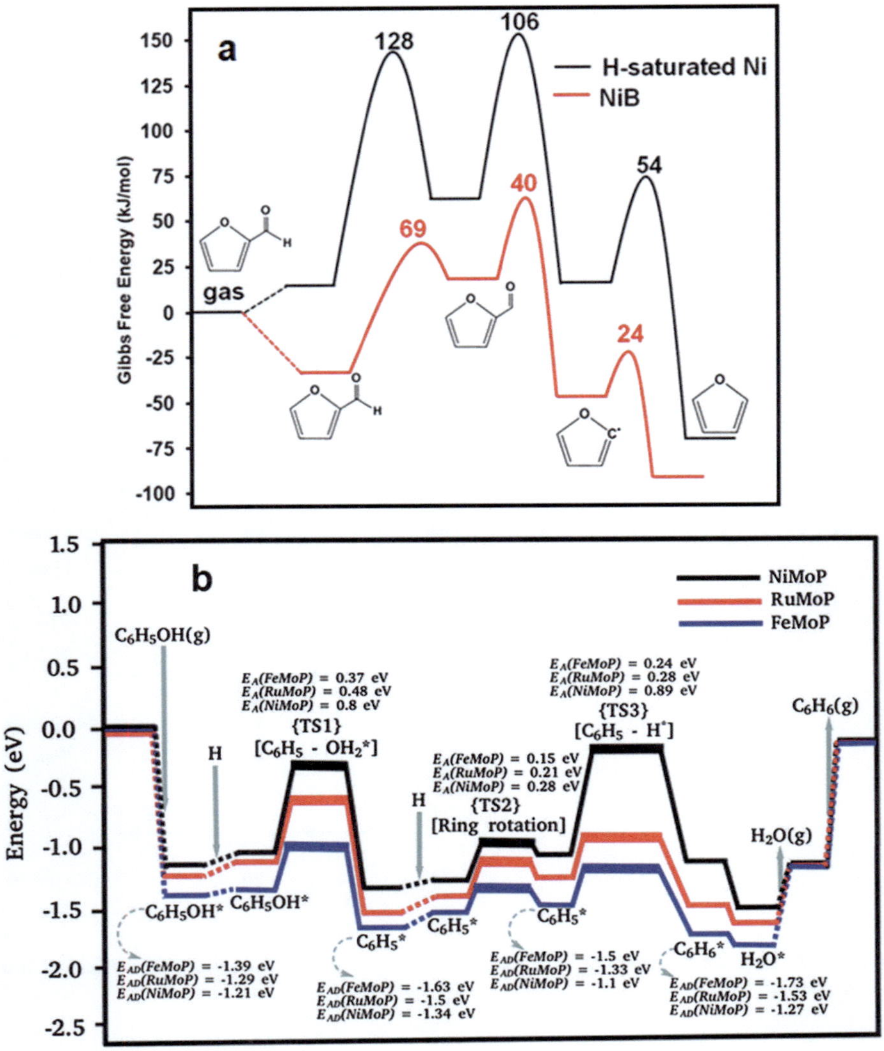

Fig. 14.6 Reaction energetics for (**a**) hydrodeoxygenation of furfural to furan on Ni and boron-doped Ni surfaces (Banerjee and Mushrif 2018), and (**b**) hydrodeoxygenation of phenol to benzene on NiMoP, FeMoP, and RuMoP (Jain et al. 2018)

catalytic activity could facilitate furan formation at lower reaction temperatures. A combined theoretical and experimental study evaluated the activity of bimetallic phosphides like FeMoP, RuMoP, and NiMoP for the hydrodeoxygenation of phenol (Jain et al. 2018).

Three reaction pathways were proposed to explain the benzene and cyclohexane/cyclohexene formation on FeMoP and NiMoP/RuMoP surfaces, respectively. The direct deoxygenation pathway (DDO) proceeds via an attack of surface H with the phenolic OH to form Ph-OH_2, followed by a C_{aryl}–OH_2 bond cleavage in a concerted step, while hydrogenolysis (RH-DO) and dehydration (RH-DEHYD) proceeded via ring hydrogenation to form a saturated ring followed by C–O hydrogenolysis or dehydration. The experimentally observed high benzene selectivity on FeMoP surfaces is due to the significantly low C_{aryl}–O cleavage barriers (0.37 eV via DDO) and high ring hydrogenation barriers (>1 eV via RH-DO) on the FeMoP surfaces making DDO the dominant pathway to form benzene (Fig. 14.6b). On NiMoP and RuMoP surfaces, ring hydrogenation is very facile with barriers <0.3 eV, suggesting the RH-DO/RH-DEHYD pathways to cyclohexane/cyclohexene to be the dominant pathways. Similar mechanistic studies on other bimetallic systems proposed in literature are presently lacking and need to be evaluated for a better understanding of the underlying chemistry. Such reaction mechanism analysis can help propose secondary metals/adatoms, which may either suppress the formation of side reactions or enhance the activity of the primary metal toward a route/pathway leading to highly selective desired products.

In summary, this section highlights the importance of theoretical calculations for understanding the catalyst reactivity on different metal surfaces. With the help of numerous examples, an attempt is made to emphasize the importance of evaluation of the metal–adsorbate interactions and mechanism and energetics of the reaction to understand the underlying chemistry for a rational design of catalysts. Computational studies for hydrodeoxygenation have, however, been largely limited to studies on stable facets of monometallic systems. Under reaction conditions, catalysts have defects and kink sites, which are likely higher active for the hydrodeoxygenation reaction. In addition, a high hydrogen surface coverage is expected to be presented on the catalyst surface under hydrodeoxygenation conditions. However, only a few studies have evaluated the influence of H coverage to the reaction mechanism (Banerjee and Mushrif 2018; Wang et al. 2015).

The development of novel catalysts would require a close collaboration between computational scientists and experimentalists. In situ experimental studies can identify the nature of the actual active phase along with the dominant surface species under reaction conditions and will help the computational scientists to model “realistic” catalyst surfaces under dynamic conditions. Such collaborations would help in developing structure–activity relationships for different metal facets for multimetallic catalyst systems.

14.9 Lignocellulosic Biomass Pretreatment and Reactions Promoted by Ultrasound

Recently, the scientific community has been faced with major scientific hurdles in the manufacturing of atom efficient chemical products from renewable bioresources in the drive toward a sustainable economy. As a result, breakthrough technologies can efficiently convert biomass to useful chemicals and fuels. In this section, the utilization of low- and high-intensity ultrasound reactors for biomass conversions will be reviewed and discussed. The advantages and disadvantages of employing this unconventional activation technique in biomass conversions will also be discussed with its potential and prospects.

Many breakthrough technologies capable of efficiently circulating renewable carbon-containing resources are an urgent task to produce atom efficient bio-based commodities. However, due to the low thermal conductivity of biomass materials that generates a barrier for mass and heat transfers, the conversion of concentrated feeds of biomass remains a significant scientific challenge (Yunus et al. 2010; Zhao et al. 2007). The use of ultrasound energy could provide an environment with unique physicochemical properties to process the conversion of crude biomass feedstock. Owing to the high corrosion and energy impact provided by the high-intensity ultrasound, biomass conversion coupling with ultrasound techniques can offer many advantages such as easier fractionation, pretreatment, and the chemical conversion of biomass materials. This could be facilitated under profoundly mild operation conditions to eventually enhance the catalytic activity and reaction efficiency (He et al. 2017; Shi et al. 2013).

The improvement of heat and mass transfer reactions is often observed during ultrasound applications in biomass conversions (Moholkar et al. 2004). In heterogeneous catalytic systems, the enhancement in substrate–catalyst contact and acceleration in reaction rate and kinetics are also observed (Amaniampong et al. 2018c; Suslick and Price 1999). Recent advances in the application of ultrasound for biomass conversions have focused on the sono-assisted lignocellulosic pretreatment, sonochemistry of carbohydrate compounds, extraction of natural products for pharmaceutical applications (Baxi and Pandit 2012), transesterification and esterification of fatty acids (Amaniampong et al. 2018c), and fermentation and pretreatment of organic wastes (Shirsath et al. 2012). Despite this crucial advancement in the application of unconventional activation techniques for biomass conversions, critical assessment of the reactions and pretreatments of biomass assisted by ultrasound techniques to produce value-added chemicals and biofuels are required. This section focuses on the judicious application of ultrasound for biomass conversions.

When ultrasound energy is applied to lignocellulosic biomass, the structural integrity of the biomass material is greatly compromised via depolymerization of the recalcitrant crystalline strong hydrogen bonding in the lignocellulosic network. Biomass particle size attrition and reduction is greatly enhanced under ultrasonic applications, leading to the ease of further conversions to other platform commodities. The application of high-intensity ultrasound with an average power rate of 3.0–10 W/mL to cellulosic materials can result in crushing of the particles and crystalline grains of the biomass materials having particle size from micro to even nano-ranges (Chen et al. 2011; Filson and Dawson-Andoh 2009). More interestingly, Pinjari and Pandit (2010) also reported the success on the formation of nanofibrils from the milling of natural cellulose via hydrodynamic and ultrasound cavitation. They observed that the particle size of the cellulosic grains was finally reduced to 301 nm from the initial size of 1360 nm after hydrodynamic cavitation under the process of 110-min sonication at the rate of 3.0 W/mL.

Sonication is also being used to obtain interesting results in the fractionation of crude biomass and in the extraction of value-added compounds from raw biomass such as polysaccharides and phenolic compounds (Filson and Dawson-Andoh 2009). The high-speed microjets and shockwaves generated from the ultrasonic cavitation have great impacts in loosing and destroying the chemical linkages in the lignocellulosic structure (Ashokkumar 2011). Amaniampong et al. (2017) reported the successful conversion of glucose, a bio-derived substrate to glucuronic acid with excellent yields (94%). This work was an important step in the production of rare platform chemicals via the application of unconventional activation techniques. In another work, Amaniampong et al. (2018a) investigated the glycosylation of concentrated feed of carbohydrates in alcoholic media. They reported an interesting pyrolysis-like mechanism occurring at the gas–liquid interface, leading to the formation of levoglucosan as intermediate for the oligomerization of carbohydrates.

Karnjanakom et al. (2015) reported a highest yield value of 80.3% for the hydrocarbon formation via the upgrading of bio-oil obtained from ultrasound pretreated cedar wood with 2.5 wt% Zn/Al_2O_3. A maximum bio-oil yield of 56 wt% was obtained over ultrasound pretreated cedar wood as compared to 45 wt% bio-oil from untreated cedar wood. Shi et al. (2013) utilized ultrasound as a pretreatment tool which resulted in an enhanced efficiency of bio-oil formation from pure cellulose liquefaction in hot-compressed water. Their results showed that the yield (42 wt%) of bio-oil was significantly increased using the ultrasonic pretreatment. In fact, emulsification and the mixing of bio-oil derived from lignocellulosic biomass via fast pyrolysis process with diesel fuel forms emulsified oil faster and easier under ultrasonic conditions, and this mixture fuel can be directly employed in existing engines (Li et al. 2010). This is because of improved phase contacts and mass transfer effects.

Ultrasound intensification demonstrated beneficial and integrated applications in biorefinery systems including the reactions and pretreatment of lignocellulosic materials and other biomass-related reactions and pretreatment, such as preemulsification of bio-oil to prepare input materials for catalytic upgrading, liquefaction of lignocellulosic biomass to crude bio-oil, and selective oxidation of biomass-related substrates (oxygenates hydrocarbons such as sugars). Nonetheless, the utilization of ultrasound in biomass conversions faces significant challenges that need to be investigated and addressed. The correct choice of ultrasound parameters and skillful use of the reactor is desired to maximize energy output during its utilization. A proper choice of solvent compatible to sonochemical reactions regarding biomass conversions is crucial. The integration of ultrasound energy into biomass conversion and downstream treatment has a bright and promising future, although further and dedicated research studies are required.

14.10 Conclusions

This chapter demonstrated the biomass fast pyrolysis as a very promising alternative way to produce cleaner and sustainable fuel and reduce the reliance on fuel from fossil resources. The liquid product obtained from the fast pyrolysis of biomass, called bio-oil, should undergo catalytic upgrading before it can be used for transportation purposes. Several methods to upgrade bio-oil are described along with their corresponding advantages and disadvantages. One of the most widely applied and most extensively studied approach for the catalytic upgrading of bio-oil is the hydrodeoxygenation in which the oxygen-containing functional groups are removed and replaced by hydrogen atoms. However, for this purpose of selectively activating and breaking the C–O bonds while trying to maintain the C–C bonds intact, processing knowledge base of petroleum industry may not be utilized for hydrodeoxygenation of bio-oil, and therefore novel catalysts, solvents, and processes need to be developed.

The importance of utilizing insights gained from theoretical calculations for a coherent catalysts design with desirable properties such as high selectivity and activity for the hydrodeoxygenation of bio-oil has been highlighted. The development of low-cost and nonprecious metal catalysts, which selectively cleave the C–O bonds while preserving the C–C bonds at low hydrogen pressures, could remain the major challenge for bio-oil upgradation. Using DFT calculations, suitable descriptors like bond lengths, adsorbate binding energies, etc. can be established for understanding chemical reactivity on different catalyst surfaces. Such descriptors can be useful for proposing new multimetallic catalyst systems and predicting their activity and/or selectivity toward hydrodeoxygenation of bio-oil.

The catalyst activity can be enhanced by the addition of dopants/promotors to the existing catalysts. Predicting the behavior of these doped catalyst systems could, however, require an in-depth understanding of the energetics and mechanism of hydrodeoxygenation of bio-oil. In addition, most of the catalytic activity is attributed to the presence of defect sites on the catalysts surface. A detailed in situ characterization study of the catalyst under working conditions could help to model the real catalytic surface under realistic surface coverages and accurately predict product selectivity. Future work should also focus on improving our understanding of hydrodeoxygenation by incorporating the effect of reaction conditions on the reactivity at the molecular level. Finally, the application of sonochemical technique in combination with catalyst design could make breakthrough developments in the catalytic upgrading of bio-oil.

Acknowledgments Q.T. Trinh acknowledges the financial support by the National Research Foundation (NRF), Prime Minister's Office, Singapore, under its Campus for Research Excellence and Technological Enterprise (CREATE) program. K.B. Ansari acknowledges the financial support from the Ministry of Singapore under the Academic Research Fund (AcRF) Tier-2 and Nanyang Technological University (NTU) for its research facility. D.Q. Dao thanks the funding from the Vietnam National Foundation for Science and Technology Development (NAFOSTED) under the grant number 103.03-2018.366. P.N. Amaniampong and A. Drif are grateful to the CNRS, the Ministry of Research and the Région Aquitaine-Limousin-Poitou-Charentes for their financial support. The funding from the International Consortium on Eco-conception and Renewable Resources (FR CNRS INCREASE 3707) and the Chair "TECHNOGREEN" is also acknowledged.

References

Adams P, Bridgwater T, Lea-Langton A, Ross A, Watson I (2018) Biomass conversion technologies. In: Thornley P, Adams P (eds) Greenhouse gas balances of bioenergy systems. Academic Press, London, pp 107–139

Adjaye JD, Bakhshi NN (1995a) Production of hydrocarbons by catalytic upgrading of a fast pyrolysis bio-oil. Part I: conversion over various catalysts. Fuel Process Technol 45:161–183

Adjaye JD, Bakhshi NN (1995b) Production of hydrocarbons by catalytic upgrading of a fast pyrolysis bio-oil. Part II: comparative catalyst performance and reaction pathways. Fuel Process Technol 45:185–202

Amaniampong PN, Trinh QT, Wang B, Borgna A, Yang Y, Mushrif SH (2015) Biomass oxidation: formyl C-H bond activation by the surface lattice oxygen of regenerative CuO nanoleaves. Angew Chem Int Ed 54:8928–8933

Amaniampong PN, Karam A, Trinh QT, Xu K, Hirao H, Jérôme F, Chatel G (2017) Selective and catalyst-free oxidation of d-glucose to D-glucuronic acid induced by high-frequency ultrasound. Sci Rep 7:40650

Amaniampong PN, Clément J-L, Gigmes D, Ortiz Mellet C, García Fernández JM, Blériot Y, Chatel G, De Oliveira Vigier K, Jérôme F (2018a) Catalyst-free synthesis of alkylpolyglycosides induced by high-frequency ultrasound. ChemSusChem 11:2673–2676

Amaniampong PN, Trinh QT, Li K, Mushrif SH, Hao Y, Yang Y (2018b) Porous structured CuO-CeO_2 nanospheres for the direct oxidation of cellobiose and glucose to gluconic acid. Catal Today 306:172–182

Amaniampong PN, Trinh QT, Varghese JJ, Behling R, Valange S, Mushrif SH, Jérôme F (2018c) Unraveling the mechanism of the oxidation of glycerol to dicarboxylic acids over a sonochemically synthesized copper oxide catalyst. Green Chem 20:2730–2741

Anex RP, Aden A, Kazi FK, Fortman J, Swanson RM, Wright MM, Satrio JA, Brown RC, Daugaard DE, Platon A, Kothandaraman G, Hsu DD, Dutta A (2010) Techno-economic comparison of biomass-to-transportation fuels via pyrolysis, gasification, and biochemical pathways. Fuel 89:S29–S35

Ansari KB, Arora JS, Chew JW, Dauenhauer PJ, Mushrif SH (2018) Effect of temperature and transport on the yield and composition of pyrolysis-derived bio-oil from glucose. Energy Fuel 32:6008–6021

Arora JS, Chew JW, Mushrif SH (2018) Influence of alkali and alkaline-earth metals on the cleavage of glycosidic bond in biomass pyrolysis: a DFT study using cellobiose as a model compound. J Phy Chem A 122:7646–7658

Ashokkumar M (2011) The characterization of acoustic cavitation bubbles—An overview. Ultrason Sonochem 18:864–872

Auersvald M, Shumeiko B, Vrtiška D, Straka P, Staš M, Šimáček P, Blažek J, Kubička D (2019) Hydrotreatment of straw bio-oil from ablative fast pyrolysis to produce suitable refinery intermediates. Fuel 238:98–110

Banerjee A, Mushrif SH (2017) Reaction pathways for the deoxygenation of biomass-pyrolysis-derived bio-oil on Ru: a DFT study using furfural as a model compound. ChemCatChem 9:2828–2838

Banerjee A, Mushrif SH (2018) Investigating reaction mechanisms for furfural hydrodeoxygenation on Ni and the effect of boron doping on the activity and selectivity of the catalyst. J Phys Chem C 122:18383–18394

Baxi PB, Pandit AB (2012) Using cavitation for delignification of wood. Bioresour Technol 110:697–700

Bimbela F, Oliva M, Ruiz J, García L, Arauzo J (2013) Hydrogen production via catalytic steam reforming of the aqueous fraction of bio-oil using nickel-based coprecipitated catalysts. Int J Hydrogen Energy 38:14476–14487

Bizkarra K, Barrio VL, Gartzia-Rivero L, Bañuelos J, López-Arbeloa I, Cambra JF (2019) Hydrogen production from a model bio-oil/bio-glycerol mixture through steam reforming using zeolite L supported catalysts. Int J Hydrogen Energy 44:1492–1504

Bridgwater AV (2007) The production of biofuels and renewable chemicals by fast pyrolysis of biomass. Int J Glob Energy Issues 27:160–203

Bridgwater AV (2012) Review of fast pyrolysis of biomass and product upgrading. Biomass Bioenergy 38:68–94

Bridgwater AV, Peacocke GVC (2000) Fast pyrolysis processes for biomass. Renew Sust Energ Rev 4:1–73

Cai Q-X, Wang J-G, Wang Y-G, Mei D (2015) Mechanistic insights into the structure-dependent selectivity of catalytic furfural conversion on platinum catalysts. AICHE J 61:3812–3824

Chen W, Yu H, Liu Y, Chen P, Zhang M, Hai Y (2011) Individualization of cellulose nanofibers from wood using high-intensity ultrasonication combined with chemical pretreatments. Carbohydr Polym 83:1804–1811

Cheng Y-T, Huber GW (2011) Chemistry of furan conversion into aromatics and olefins over HZSM-5: a model biomass conversion reaction. ACS Catal 1:611–628

Cherkasov N, Jadvani V, Mann J, Losovyj YB, Shifrina ZB, Bronstein LM, Rebrov EV (2017) Hydrogenation of bio-oil into higher alcohols over Ru/Fe_3O_4-SiO_2 catalysts. Fuel Process Technol 167:738–746

Chiu C-C, Genest A, Borgna A, Rösch N (2014) Hydrodeoxygenation of guaiacol over Ru(0001): a DFT study. ACS Catal 4:4178–4188

Elliott DC (2016) Production of biofuels via bio-oil upgrading and refining. In: Luque R, Lin CSK, Wilson K, Clark J (eds) Handbook of biofuels production, 2nd edn. Woodhead Publishing, Sawston, pp 595–613

Eri Q, Zhao X, Ranganathan P, Gu S (2017) Numerical simulations on the effect of potassium on the biomass fast pyrolysis in fluidized bed reactor. Fuel 197:290–297

Filson PB, Dawson-Andoh BE (2009) Sono-chemical preparation of cellulose nanocrystals from lignocellulose derived materials. Bioresour Technol 100:2259–2264

Galdámez JR, García L, Bilbao R (2005) Hydrogen production by steam reforming of bio-oil using coprecipitated Ni−Al catalysts. Acetic acid as a model compound. Energy Fuel 19:1133–1142

Garcia LA, French R, Czernik S, Chornet E (2000) Catalytic steam reforming of bio-oils for the production of hydrogen: effects of catalyst composition. Appl Catal A Gen 201:225–239

Gu GH, Mullen CA, Boateng AA, Vlachos DG (2016) Mechanism of dehydration of phenols on noble metals via first-principles microkinetic modeling. ACS Catal 6:3047–3055

Guedes RE, Luna AS, Torres AR (2018) Operating parameters for bio-oil production in biomass pyrolysis: a review. J Anal Appl Pyrolysis 129:134–149

He Z, Wang Z, Zhao Z, Yi S, Mu J, Wang X (2017) Influence of ultrasound pretreatment on wood physiochemical structure. Ultrason Sonochem 34:136–141

Hu B, Lu Q, Jiang X, Dong X, Cui M, Dong C, Yang Y (2018) Pyrolysis mechanism of glucose and mannose: the formation of 5-hydroxymethyl furfural and furfural. J Energy Chem 27:486–501

Huang J, He C (2015) Pyrolysis mechanism of α-O-4 linkage lignin dimer: a theoretical study. J Anal Appl Pyrolysis 113:655–664

Huang J, Liu C, Wu D, Tong H, Ren L (2014) Density functional theory studies on pyrolysis mechanism of β-O-4 type lignin dimer model compound. J Anal Appl Pyrolysis 109:98–108

Huang J, He C, Wu L, Tong H (2016) Thermal degradation reaction mechanism of xylose: a DFT study. Chem Phys Lett 658:114–124

Jae J, Tompsett GA, Foster AJ, Hammond KD, Auerbach SM, Lobo RF, Huber GW (2011) Investigation into the shape selectivity of zeolite catalysts for biomass conversion. J Catal 279:257–268

Jain V, Bonita Y, Brown A, Taconi A, Hicks JC, Rai N (2018) Mechanistic insights into hydrodeoxygenation of phenol on bimetallic phosphide catalysts. Cat Sci Technol 8:4083–4096

Jalid F, Khan TS, Mir FQ, Haider MA (2017) Understanding trends in hydrodeoxygenation reactivity of metal and bimetallic alloy catalysts from ethanol reaction on stepped surface. J Catal 353:265–273

Jendoubi N (2011) Transfer mechanisms of inorganics in biomass fast pyrolysis processes: impacts of lignocellulosic resources variability on bio-oils quality. Institut National Polytechnique de Lorraine

Kabir G, Hameed BH (2017) Recent progress on catalytic pyrolysis of lignocellulosic biomass to high-grade bio-oil and bio-chemicals. Renew Sust Energ Rev 70:945–967

Kan T, Strezov V, Evans TJ (2016) Lignocellulosic biomass pyrolysis: a review of product properties and effects of pyrolysis parameters. Renew Sust Energ Rev 57:1126–1140

Karnjanakom S, Guan G, Asep B, Du X, Hao X, Yang J, Samart C, Abudula A (2015) A green method to increase yield and quality of bio-oil: ultrasonic pretreatment of biomass and catalytic upgrading of bio-oil over metal (cu, Fe and/or Zn)/γ-Al_2O_3. RSC Adv 5:83494–83503

Le MT, Do VH, Truong DD, Bruneel E, Van Driessche I, Riisager A, Fehrmann R, Trinh QT (2016) Synergy effects of the mixture of bismuth molybdate catalysts with $SnO_2/ZrO_2/MgO$ in selective propene oxidation and the connection between conductivity and catalytic activity. Ind Eng Chem Res 55:4846–4855

Lee K, Gu GH, Mullen CA, Boateng AA, Vlachos DG (2015) Guaiacol hydrodeoxygenation mechanism on Pt(111): insights from density functional theory and linear free energy relations. ChemSusChem 8:315–322

Li Y, Wang T, Liang W, Wu C, Ma L, Zhang Q, Zhang X, Jiang T (2010) Ultrasonic preparation of emulsions derived from aqueous bio-oil fraction and 0# diesel and combustion characteristics in diesel generator. Energy Fuel 24:1987–1995

Liao HT, Ye XN, Lu Q, Dong CQ (2014) Overview of bio-oil upgrading via catalytic cracking. Adv Mater Res 827:25–29

Lin Y-C, Cho J, Tompsett GA, Westmoreland PR, Huber GW (2009) Kinetics and mechanism of cellulose pyrolysis. J Phys Chem C 113:20097–20107

Liu Y, Li Z, Leahy JJ, Kwapinski W (2015) Catalytically upgrading bio-oil via esterification. Energy Fuel 29:3691–3698

Lu Q, Tian H-Y, Hu B, Jiang X-Y, Dong C-Q, Yang Y-P (2016) Pyrolysis mechanism of holocellulose-based monosaccharides: the formation of hydroxyacetaldehyde. J Anal Appl Pyrolysis 120:15–26

Lu Q, Hu B, Zhang Z-X, Wu Y-T, Cui M-S, Liu D-J, Dong C-Q, Yang Y-P (2018) Mechanism of cellulose fast pyrolysis: the role of characteristic chain ends and dehydrated units. Combus Flame 198:267–277

Ly HV, Choi JH, Woo HC, Kim S-S, Kim J (2019) Upgrading bio-oil by catalytic fast pyrolysis of acid-washed Saccharina japonica alga in a fluidized-bed reactor. Renew Energy 133:11–22

Mayes HB, Nolte MW, Beckham GT, Shanks BH, Broadbelt LJ (2014) The alpha–bet(a) of glucose pyrolysis: computational and experimental investigations of 5-hydroxymethylfurfural and levoglucosan formation reveal implications for cellulose pyrolysis. ACS Sustain Chem Eng 2:1461–1473

Mei Y, Wu C, Liu R (2016) Hydrogen production from steam reforming of bio-oil model compound and byproducts elimination. Int J Hydrogen Energy 41:9145–9152

Mettler MS, Paulsen AD, Vlachos DG, Dauenhauer PJ (2012) The chain length effect in pyrolysis: bridging the gap between glucose and cellulose. Green Chem 14:1284–1288

Milina M, Mitchell S, Pérez-Ramírez J (2014) Prospectives for bio-oil upgrading via esterification over zeolite catalysts. Catal Today 235:176–183

Mironenko AV, Vlachos DG (2016) Conjugation-driven "reverse Mars–van Krevelen"-type radical mechanism for low-temperature C–O bond activation. J Am Chem Soc 138:8104–8113

Mohanty P, Nanda S, Pant KK, Naik S, Kozinski JA, Dalai AK (2013) Evaluation of the physiochemical development of biochars obtained from pyrolysis of wheat straw, timothy grass and pinewood: effects of heating rate. J Anal Appl Pyrolysis 104:485–493

Moholkar VS, Warmoeskerken MMCG, Ohl CD, Prosperetti A (2004) Mechanism of mass-transfer enhancement in textiles by ultrasound. AICHE J 50:58–64

Mondal J, Trinh QT, Jana A, Ng WKH, Borah P, Hirao H, Zhao Y (2016) Size-dependent catalytic activity of palladium nanoparticles fabricated in porous organic polymers for alkene hydrogenation at room temperature. ACS Appl Mater Interfaces 8:15307–15319

Mortensen PM, Grunwaldt JD, Jensen PA, Knudsen KG, Jensen AD (2011) A review of catalytic upgrading of bio-oil to engine fuels. Appl Catal A Gen 407:1–19

Mukarakate C, Watson MJ, ten Dam J, Baucherel X, Budhi S, Yung MM, Ben H, Iisa K, Baldwin RM, Nimlos MR (2014) Upgrading biomass pyrolysis vapors over β-zeolites: role of silica-to-alumina ratio. Green Chem 16:4891–4905

Mushrif SH, Vasudevan V, Krishnamurthy CB, Venkatesh B (2015) Multiscale molecular modeling can be an effective tool to aid the development of biomass conversion technology: a perspective. Chem Eng Sci 121:217–235

Nachenius RW, Ronsse F, Venderbosch RH, Prins W (2013) Biomass pyrolysis. In: Murzin DY (ed) Advances in chemical engineering. Academic Press, London, pp 75–139

Nanda S, Mohanty P, Pant KK, Naik S, Kozinski JA, Dalai AK (2013) Characterization of north American lignocellulosic biomass and biochars in terms of their candidacy for alternate renewable fuels. Bioenergy Res 6:663–677

Nanda S, Mohammad J, Reddy SN, Kozinski JA, Dalai AK (2014) Pathways of lignocellulosic biomass conversion to renewable fuels. Biomass Conv Bioref 4:157–191

Nanda S, Azargohar R, Dalai AK, Kozinski JA (2015) An assessment on the sustainability of lignocellulosic biomass for biorefining. Renew Sust Energ Rev 50:925–941

Nandula A, Trinh QT, Saeys M, Alexandrova AN (2015) Origin of extraordinary stability of square-planar carbon atoms in surface carbides of cobalt and nickel. Angew Chem Int Ed 54:5312–5316

Nelson RC, Baek B, Ruiz P, Goundie B, Brooks A, Wheeler MC, Frederick BG, Grabow LC, Austin RN (2015) Experimental and theoretical insights into the hydrogen-efficient direct hydrodeoxygenation mechanism of phenol over Ru/TiO_2. ACS Catal 5:6509–6523

Nguyen QB, Nguyen VB, Lim CYH, Trinh QT, Sankaranarayanan S, Zhang YW, Gupta M (2014) Effect of impact angle and testing time on erosion of stainless steel at higher velocities. Wear 321:87–93

Nihei M, Ida H, Nibe T, Moeljadi AMP, Trinh QT, Hirao H, Ishizaki M, Kurihara M, Shiga T, Oshio H (2018) Ferrihydrite particle encapsulated within a molecular organic cage. J Am Chem Soc 140:17753–17759

Nolte MW, Shanks BH (2017) A perspective on catalytic strategies for deoxygenation in biomass pyrolysis. Energ Technol 5:7–18

Park JY, Kim J-K, Oh C-H, Park J-W, Kwon EE (2019) Production of bio-oil from fast pyrolysis of biomass using a pilot-scale circulating fluidized bed reactor and its characterization. J Environ Manag 234:138–144

Pattiya A (2018) Fast pyrolysis. In: Rosendahl L (ed) Direct thermochemical liquefaction for energy applications. Woodhead Publishing, Sawston, pp 3–28

Paulsen AD, Mettler MS, Dauenhauer PJ (2013) The role of sample dimension and temperature in cellulose pyrolysis. Energy Fuel 27:2126–2134

Pinheiro A (2008) Impact of oxygenated compounds from lignocellulosic biomass and of their hydrodeoxygenation reaction products on the kinetics of gas oil hydrotreating. Université Claude Bernard—Lyon I

Pinho AR, de Almeida MBB, Mendes FL, Ximenes VL, Casavechia LC (2015) Co-processing raw bio-oil and gasoil in an FCC unit. Fuel Process Technol 131:159–166

Pinho AR, de Almeida MBB, Mendes FL, Casavechia LC, Talmadge MS, Kinchin CM, Chum HL (2017) Fast pyrolysis oil from pinewood chips co-processing with vacuum gas oil in an FCC unit for second generation fuel production. Fuel 188:462–473

Pinjari DV, Pandit AB (2010) Cavitation milling of natural cellulose to nanofibrils. Ultrason Sonochem 17:845–852

Radlein D, Quignard A (2013) A short historical review of fast pyrolysis of biomass. Oil Gas Sci Technol 68:765–783

Santamaria L, Lopez G, Arregi A, Amutio M, Artetxe M, Bilbao J, Olazar M (2019) Stability of different Ni supported catalysts in the in-line steam reforming of biomass fast pyrolysis volatiles. Appl Catal B Environ 242:109–120

Sarkar C, Pendem S, Shrotri A, Dao DQ, Mai PPT, Ngoc TN, Chandaka DR, Rao TV, Trinh QT, Sherburne MP, Mondal J (2019) Interface engineering of graphene-supported cu nanoparticles encapsulated by mesoporous silica for size-dependent catalytic oxidative coupling of aromatic amines. ACS Appl Mater Interfaces 11:11722–11735

Septien S, Valin S, Dupont C, Peyrot M, Salvador S (2012) Effect of particle size and temperature on woody biomass fast pyrolysis at high temperature (1000–1400°C). Fuel 97:202–210

Seshadri V, Westmoreland PR (2012) Concerted reactions and mechanism of glucose pyrolysis and implications for cellulose kinetics. J Phys Chem A 116:11997–12013

Shen J, Wang X-S, Garcia-Perez M, Mourant D, Rhodes MJ, Li C-Z (2009) Effects of particle size on the fast pyrolysis of oil mallee woody biomass. Fuel 88:1810–1817

Shen D, Jin W, Hu J, Xiao R, Luo K (2015) An overview on fast pyrolysis of the main constituents in lignocellulosic biomass to valued-added chemicals: structures, pathways and interactions. Renew Sust Energ Rev 51:761–774

Shetty M, Buesser B, Román-Leshkov Y, Green WH (2017) Computational investigation on hydrodeoxygenation (HDO) of acetone to propylene on α-MoO_3 (010) surface. J Phys Chem C 121:17848–17855

Shi W, Jia J, Gao Y, Zhao Y (2013) Influence of ultrasonic pretreatment on the yield of bio-oil prepared by thermo-chemical conversion of rice husk in hot-compressed water. Bioresour Technol 146:355–362

Shi Y, Zhu Y, Yang Y, Li Y-W, Jiao H (2015) Exploring furfural catalytic conversion on cu(111) from computation. ACS Catal 5:4020–4032

Shirsath SR, Sonawane SH, Gogate PR (2012) Intensification of extraction of natural products using ultrasonic irradiations—a review of current status. Chem Eng Process Process Intensif 53:10–23

Si Z, Zhang X, Wang C, Ma L, Dong R (2017) An overview on catalytic hydrodeoxygenation of pyrolysis oil and its model compounds. Catalysts 7:169

Singbua P, Treedet W, Duangthong P, Seithtanabutara V, Suntivarakorn R (2017) Bio-oil production of napier grass by using rotating cylindrical reactor. Energy Procedia 138:641–645

Singuru R, Trinh QT, Banerjee B, Rao BG, Bai L, Bhaumik A, Reddy BM, Hirao H, Mondal J (2016) Integrated experimental and theoretical study of shape-controlled catalytic oxidative coupling of aromatic amines over CuO nanostructures. ACS Omega 1:1121–1138

Sitthisa S, An W, Resasco DE (2011a) Selective conversion of furfural to methylfuran over silica-supported NiFe bimetallic catalysts. J Catal 284:90–101

Sitthisa S, Pham T, Prasomsri T, Sooknoi T, Mallinson RG, Resasco DE (2011b) Conversion of furfural and 2-methylpentanal on Pd/SiO_2 and Pd–cu/SiO_2 catalysts. J Catal 280:17–27

Sitthisa S, Sooknoi T, Ma Y, Balbuena PB, Resasco DE (2011c) Kinetics and mechanism of hydrogenation of furfural on cu/SiO_2 catalysts. J Catal 277:1–13

Suntivarakorn R, Treedet W, Singbua P, Teeramaetawat N (2018) Fast pyrolysis from Napier grass for pyrolysis oil production by using circulating fluidized bed reactor: improvement of pyrolysis system and production cost. Energy Rep 4:565–575

Suslick KS, Price GJ (1999) Applications of ultrasound to materials chemistry. Annu Rev Mater Sci 29:295–326

Swanson RM, Platon A, Satrio JA, Brown RC (2010) Techno-economic analysis of biomass-to-liquids production based on gasification. Fuel 89:S11–S19

Tan Q, Wang G, Long A, Dinse A, Buda C, Shabaker J, Resasco DE (2017) Mechanistic analysis of the role of metal oxophilicity in the hydrodeoxygenation of anisole. J Catal 347:102–115

Tran N, Uemura Y, Chowdhury S, Ramli A (2014) A review of bio-oil upgrading by catalytic hydrodeoxygenation. Appl Mech Mater 625:255–258

Trane R, Dahl S, Skjøth-Rasmussen MS, Jensen AD (2012) Catalytic steam reforming of bio-oil. Int J Hydrogen Energy 37:6447–6472

Treedet W, Suntivarakorn R (2018) Design and operation of a low cost bio-oil fast pyrolysis from sugarcane bagasse on circulating fluidized bed reactor in a pilot plant. Fuel Process Technol 179:17–31

Trinh QT, Yang J, Lee JY, Saeys M (2012) Computational and experimental study of the volcano behavior of the oxygen reduction activity of PdM@PdPt/C (M=Pt, Ni, co, Fe, and Cr) core–shell electrocatalysts. J Catal 291:26–35

Trinh QT, Tan KF, Borgna A, Saeys M (2013) Evaluating the structure of catalysts using core-level binding energies calculated from first principles. J Phys Chem C 117:1684–1691

Trinh QT, Chethana BK, Mushrif SH (2015) Adsorption and reactivity of cellulosic aldoses on transition metals. J Phys Chem C 119:17137–17145

Trinh QT, Nguyen AV, Huynh DC, Pham TH, Mushrif SH (2016) Mechanistic insights into the catalytic elimination of tar and the promotional effect of boron on it: first-principles study using toluene as a model compound. Cat Sci Technol 6:5871–5883

Trinh QT, Banerjee A, Yang Y, Mushrif SH (2017) Sub-surface boron-doped copper for methane activation and coupling: first-principles investigation of the structure, activity, and selectivity of the catalyst. J Phys Chem C 121:1099–1112

Trinh QT, Bhola K, Amaniampong PN, Jérôme F, Mushrif SH (2018) Synergistic application of xps and dft to investigate metal oxide surface catalysis. J Phys Chem C 122:22397–22406

Upadhyay M, Park HC, Choi HS (2018) Multiphase fluid dynamics coupled fast pyrolysis of biomass in a rectangular bubbling fluidized bed reactor: process intensification. Chem Eng Process 128:180–187

Varghese JJ, Trinh QT, Mushrif SH (2016) Insights into the synergistic role of metal–lattice oxygen site pairs in four-centered C–H bond activation of methane: the case of CuO. Cat Sci Technol 6:3984–3996

Varma A, Xiao Y (2017) Catalytic deoxygenation of bio-oils using methane. United States Patent 20170029712

Varma A, Xiao Y (2018) Catalytic deoxygenation of bio-oils using methane. United States Patent 10023809

Vispute TP, Huber GW (2009) Production of hydrogen, alkanes and polyols by aqueous phase processing of wood-derived pyrolysis oils. Green Chem 11:1433–1445

Vispute TP, Zhang H, Sanna A, Xiao R, Huber GW (2010) Renewable chemical commodity feedstocks from integrated catalytic processing of pyrolysis oils. Science 330:1222–1227

Vorotnikov V, Mpourmpakis G, Vlachos DG (2012) DFT study of furfural conversion to furan, furfuryl alcohol, and 2-methylfuran on Pd(111). ACS Catal 2:2496–2504

Wan W, Ammal SC, Lin Z, You K-E, Heyden A, Chen JG (2018) Controlling reaction pathways of selective C–O bond cleavage of glycerol. Nature Comm 9:4612

Wang J-J, Chang J, Fan J (2010) Catalytic esterification of bio-oil by ion exchange resins. J Fuel Chem Technol 38:560–564

Wang S, Vorotnikov V, Vlachos DG (2015) Coverage-induced conformational effects on activity and selectivity: hydrogenation and decarbonylation of furfural on Pd(111). ACS Catal 5:104–112

Wang H, Lee S-J, Olarte MV, Zacher AH (2016) Bio-oil stabilization by hydrogenation over reduced metal catalysts at low temperatures. ACS Sustain Chem Eng 4:5533–5545

Wang S, Dai G, Yang H, Luo Z (2017) Lignocellulosic biomass pyrolysis mechanism: a state-of-the-art review. Prog Energy Combust Sci 62:33–86

Wildschut J, Mahfud FH, Venderbosch RH, Heeres HJ (2009) Hydrotreatment of fast pyrolysis oil using heterogeneous noble-metal catalysts. Ind Eng Chem Res 48:10324–10334

Wright MM, Daugaard DE, Satrio JA, Brown RC (2010) Techno-economic analysis of biomass fast pyrolysis to transportation fuels. Fuel 89:S2–S10

Wu C, Liu R (2011) Sustainable hydrogen production from steam reforming of bio-oil model compound based on carbon deposition/elimination. Int J Hydrogen Energy 36:2860–2868

Xiu S, Shahbazi A (2012) Bio-oil production and upgrading research: a review. Renew Sust Energ Rev 16:4406–4414

Yang C, Trinh QT, Wang X, Tang Y, Wang K, Huang S, Chen X, Mushrif SH, Wang M (2015) Crystallization-induced red emission of a facilely synthesized biodegradable indigo derivative. Chem Commun 51:3375–3378

Yunus R, Salleh SF, Abdullah N, Biak DRA (2010) Effect of ultrasonic pre-treatment on low temperature acid hydrolysis of oil palm empty fruit bunch. Bioresour Technol 101:9792–9796

Zhang Y, Liu C, Xie H (2014) Mechanism studies on β-d-glucopyranose pyrolysis by density functional theory methods. J Anal Appl Pyrolysis 105:23–34

Zhang Y, Liu C, Chen X (2015) Unveiling the initial pyrolytic mechanisms of cellulose by DFT study. J Anal Appl Pyrolysis 113:621–629

Zhao W, Zong Z-M, Lin J, Song Y-M, Guo X-F, Yao Z-S, Zhang L-N, He R-L, Cao J-P, Wei X-Y (2007) Dewaxing from stalks with petroleum ether by different methods. Energy Fuel 21:1165–1168

Zhou X, Nolte MW, Mayes HB, Shanks BH, Broadbelt LJ (2014) Experimental and mechanistic modeling of fast pyrolysis of neat glucose-based carbohydrates. 1. Experiments and development of a detailed mechanistic model. Ind Eng Chem Res 53:13274–13289

Heterogeneous Catalysis in Hydroxymethylfurfural Conversion to Fuels and Chemicals

15

Chanatip Samart, Thi Tuong Vi Tran,
Suwadee Kongparakul, Surachai Karnjanakom,
and Prasert Reubroycharoen

Abstract

Hydroxymethylfurfural is an important intermediate compound to produce biomass-based fine chemicals. The conventional organic reactions including oxidation, hydrogenation, reduction, and condensation can be used to transform hydroxymethylfurfural to different intermediates of fine chemicals where heterogeneous catalysts play significant roles. Furandicarboxylic acid, 2,5-dihydroxymethylfuran, 2,5-furandicarboxyamide, and 5-(hydroxymethyl) furfurylamine are a few important value-added products obtained through oxidation, selective hydrogenation, oxidative amidation, and reductive amination of hydroxymethylfurfural, respectively. In addition, hydroxymethylfurfural is a precursor to synthesize monomers of polyester and polyurethane.

Keywords

Hydroxymethylfurfural · Heterogeneous catalysis · Biomass · Fine chemicals · Biofuels

C. Samart (✉) · S. Kongparakul
Department of Chemistry, Faculty of Science and Technology, Thammasat University, Pathumthani, Thailand

Bioenergy and Biochemical Refinery Technology Program, Faculty of Science and Technology, Thammasat University, Pathumthani, Thailand
e-mail: samart.chanatip22@gmail.com

T. T. V. Tran
Department of Chemistry, Faculty of Science and Technology, Thammasat University, Pathumthani, Thailand

S. Karnjanakom
Department of Chemistry, Rangsit University, Pathumthani, Thailand

P. Reubroycharoen
Department of Chemical Technology, Chulalongkorn University, Bangkok, Thailand

S. Nanda et al. (eds.), *Biorefinery of Alternative Resources: Targeting Green Fuels and Platform Chemicals*, https://doi.org/10.1007/978-981-15-1804-1_15

15.1 Introduction

Conventional fuels and petrochemical products are mostly derived from fossil resources, which are on a verge of depletion. In addition, massive greenhouse gas emissions and environmental pollution are also associated with their utilization. Accordingly, the use of sustainable green chemistry approach has been considered for the design of environmentally benign processes. Green carbon science was introduced to the chemical industries related with carbon processing, utilization, and recycling, including the fuel and petrochemical industries (He et al. 2013). Waste plant biomass is an interesting carbon source based on its renewable resource that consists mainly of the three different components such as cellulose, hemicellulose, and lignin. Cellulose, a macromolecule of glucose units, is the main component of lignocellulosic biomass.

Glucose is mainly used as a building block for biomass-derived chemicals in biorefinery process. On the other hand, hydroxymethylfurfural (HMF) and levulinic acid (LA) are the key compounds in the biomass-based chemical platforms that can be derived through the dehydration of glucose (Kucherov et al. 2018). To convert glucose to chemical intermediates, resins, polymers, solvents, fuels, and other value-added products, it is important to discover several catalytic thermochemical and biochemical pathways. Many studies on HMF utilization have been published, including fuel synthesis, chemicals, and polymers, while several organic reactions related with HMF conversion have been presented in Fig. 15.1.

15.2 Conversion of Hydroxymethylfurfural to Biofuels and Fine Chemicals

Lignocellulosic biomass can be divided into the two groups of woody and agricultural biomass, both differing in their chemical and structural composition (Nanda et al. 2018). The non-woody biomass such as agricultural crop residues have gained the most attention in HMF production due to the economic consideration of their abundance and low cost. The cellulose content in agricultural biomass is in the range of 30–50 wt%, in which the total annual cellulose availability is recently estimated at over 700 million ton (Tye et al. 2016).

HMF is derived from lignocellulose biomass through a three-step mechanism, as shown in Fig. 15.2, which includes (a) hydrolysis of cellulose to glucose, (b) isomerization of glucose to fructose, and (c) dehydration of fructose or glucose to HMF (Li et al. 2019). Many catalytic systems can produce HMF with yields up to 70%. The ionic liquid, especially metal halide system, uses Lewis acid catalysts for the dehydration of glucose to HMF, which can be performed under mild conditions. In such systems, higher HMF yields are generally achieved compared to autocatalytic systems, which suffer from low HMF yield due to the formation of by-products such as levulinic acid and formic acid. This catalytic system was initially developed to increase the HMF yield with a biphasic aqueous and organic mixture system where the reaction takes place in an aqueous phase. The HMF product is later dissolved in the organic layer resulting in a shift in the chemical equilibrium forward to produce more HMF with yields approaching 79–81% (Binder and Raines 2009).

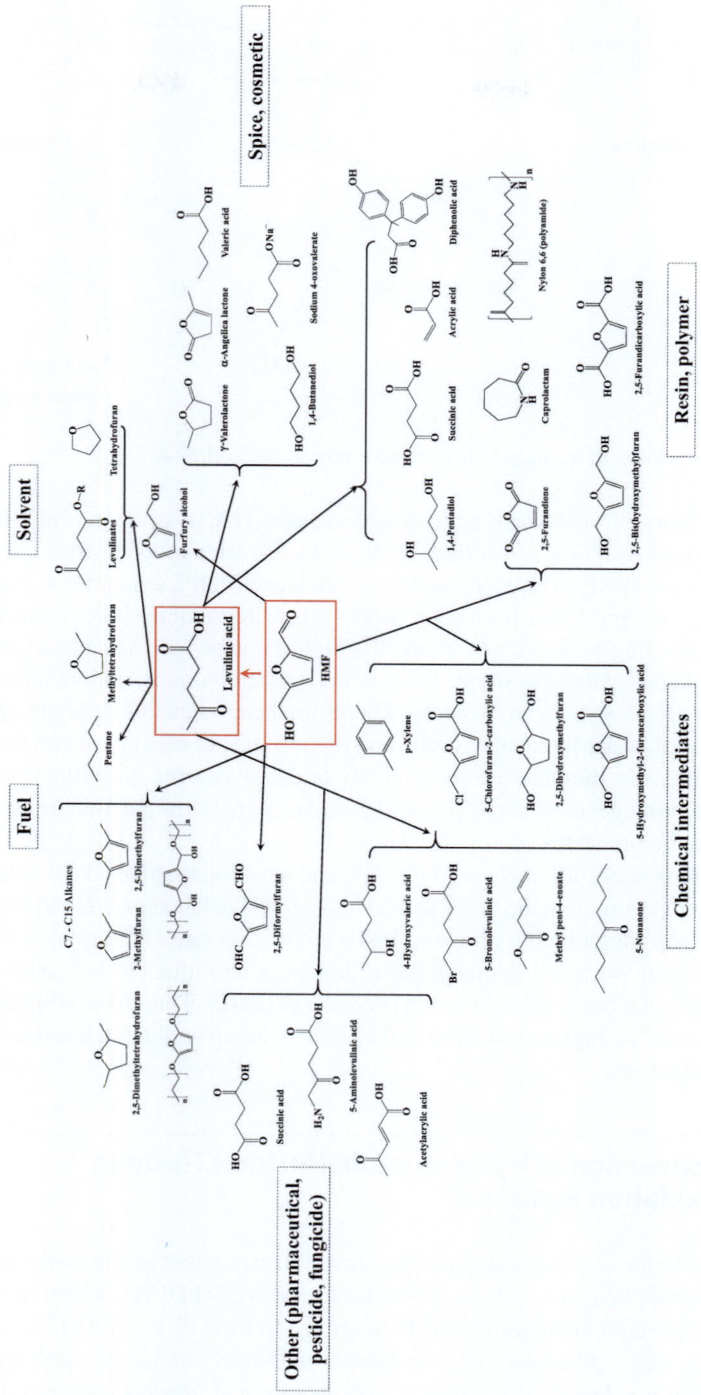

Fig. 15.1 Value-added derivatives and compounds from hydroxymethylfurfural

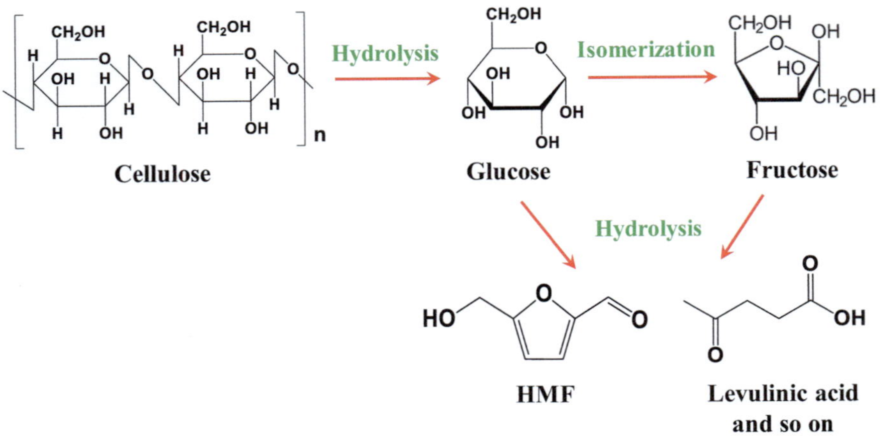

Fig. 15.2 Mechanism of hydroxymethylfurfural synthesis from cellulose

Various types of acids such as hydrochloric acid (HCl), sulfuric acid (H_2SO_4), phosphoric acid (H_3PO_4), and formic acid (HCOOH) are applied in the conversion of biomass into HMF in aqueous solutions. However, large amounts of corrosive acidic wastes are produced from such processes, which require proper waste handling and recycling technologies. Since the biphasic aqueous and organic mixture are homogenous catalytic systems, the product separation, purification, and recovery of acids are costly and challenging. Therefore, heterogeneous catalytic systems using acid catalyst are interesting according to their ease of separation and recovery process. Moreover, the heterogeneous catalysts can overcome the difficulty in the recycling of homogeneous catalysts including the decrease in the amount of acidic wastes from the processes.

Mesoporous materials such as Al-KCC-1 and acid-functionalized SBA-15 have also been used to improve the HMF selectivity (>90%) (Shahangi et al. 2018; Wang et al. 2019). Although highly efficient, HMF production rate is reported to be only 13% of the total world oil demand. Nevertheless, a less efficient but much faster reaction could overcome this limitation (Perez et al. 2019). Therefore, HMF synthesis still presents the highest potential and is already being applied in biomass-based chemical industries.

15.3 Conversion of Hydroxymethylfurfural Through Oxidation Reactions

Oxidation reaction is a technique for HMF transformation that results in the production of dicarboxylic acids such as 2,5-furandicarboxylic (FDCA), which is formed through the selective oxidation of HMF as shown in Eq. (15.1). The FDCA is then used as a new precursor of bio-based polymers including polyethylene 2,5-furandicarboxylate (PEF), which possesses physical, mechanical, and thermal properties superior to polyethylene terephthalate (PET) and acts as a potential material for PET replacement in plastic industry. PEF is synthesized by the polymerization of the dimethyl ester of FDCA and ethylene glycol (EG), as shown in Eq. (15.2). From an economic consideration, copper-based catalysts are used for the oxidation of HMF. However, 2,5-diformylfuran (DFF) is the main product from such reactions (Kucherov et al. 2018). A palladium-gold (Pd-Au) catalyst was introduced for the oxidation of HMF to FDCA and achieved nearly 90% yield. The synergistic interaction of the bimetallic catalyst played an important effect on HMF oxidation activity (Xia et al. 2019).

(15.1)

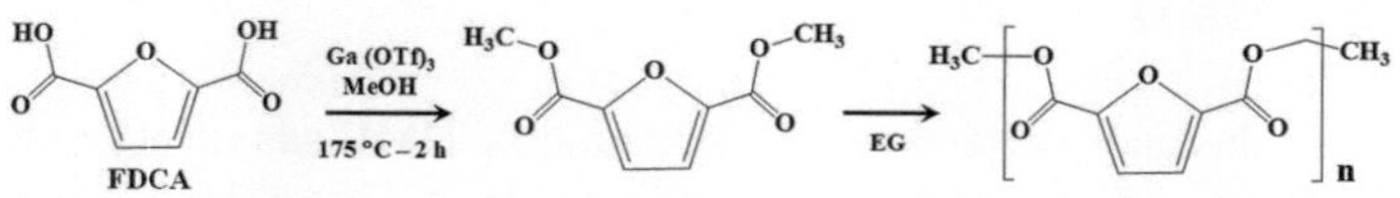

(15.2)

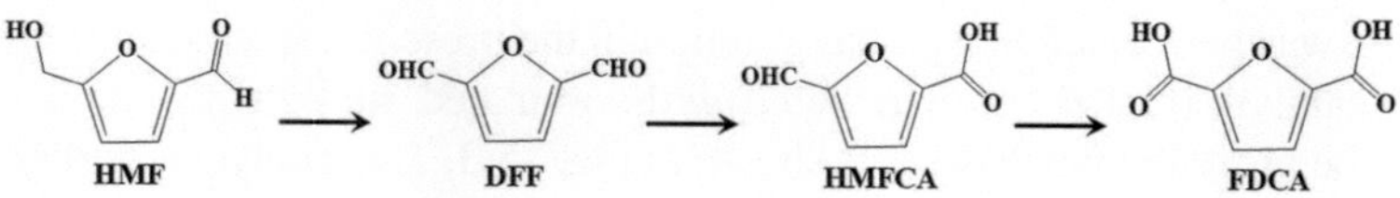

(15.3)

For Au-based catalysts, the catalyst support also presented a significant effect on the activity and selectivity in HMF conversion to FDCA due to the interaction between the Au-based catalyst and catalyst support (Albonetti et al. 2015). Moreover, the morphology of the catalyst support affected the growth of Au nanoparticles during the reaction condition. The Au nanoparticles grew from 2.4 to 10.1 nm on the surface of Aerosil (silica) (Masoud et al. 2018). Even though the precise metal catalyst can achieve high yield and selectivity of HMF conversion to FDCA, researchers have tried to develop low-cost catalysts such as copper or manganese with efficient catalytic properties according to the economic concern. For example, the oxidation pair of CuCl/tertiary-butanoic acid gave 50% yield of FDCA (Hansen et al. 2013), while the FDCA yield was improved up to 98.6% by the use of needlelike CuO/Al_2O_3 fibers. The conversion pathway of HMF to FDCA was through the oxidation of aldehyde and alcohol functional groups present in HMF to DFF and 5-hydroxymethyl-2-furandicarboxylic acid (HMFCA), respectively, as shown in Eq. (15.3) (Tirsoaga et al. 2018; Zhou et al. 2019).

On the other hand, the oxides of manganese acted as highly selective catalysts for DFF formation with 97% selectivity (Chen et al. 2019; Tong et al. 2017). DFF can be used as a substrate for various compounds in pharmaceuticals and as an antifungal agent (Kucherov et al. 2018). The combination of Mn with Co as bimetallic catalyst can convert HMF via oxidation to provide 71% yield of FDCA (Liu and Zhang 2016). Among heterogeneous catalyst systems, to achieve an efficient PEF production, the limitation of catalyst dispersion in the mixture is observed, which results in low productivity and high energy consumption during the process (Kucherov et al. 2017).

To overcome the limitation of catalyst dispersion, an efficient shortcut route for PEF synthesis from HMF-acetal with 1,3-propanediol (PD-HMF), which is stable in concentrated aqueous solution (10–20 wt%), has been studied in an aerobic oxidation system. The homogeneous and heterogeneous catalysts (e.g., PdCoBi/C catalysts and CeO_2-supported Au catalyst) have been used for anaerobic oxidative esterification of HMF and HMF-cyclic acetal for FDCA synthesis (Kim et al. 2018; Li et al. 2018a). A one-pot reaction system by aerobic oxidative esterification from PD-HMF with methanol and ethylene glycol can be successfully converted to methyl furan-2,5-dicarboxylate (MFDC) and bis(2-hydroxyethyl)furan-2,5-dicarboxylate (HEFDC), which can further undergo self-condensation for high-quality PEF synthesis.

15.4 Conversion of Hydroxymethylfurfural Through Photocatalytic and Electrochemical Oxidation Reactions

In addition to thermal conversion, the oxidation of HMF can also be attained by electrochemical and photocatalytic-assisted oxidations. The selective oxidation of HMF to FDCA was performed in aqueous solution in the presence of a cobalt thio-porphyrazine catalyst (CoPz) when radiated with simulated sunlight to attain a 96.1% yield. The proposed mechanism is shown in Fig. 15.3. The catalyst adsorbed sunlight and produced $^1O_2$, which is a powerful oxidant that can oxidize HMF to FDCA (Xu et al. 2017). In the case of HMF oxidation to DFF, polymeric carbon nitride-hydrogen peroxide adducts were highly selective for DFF formation with an obtained yield of 40–50% in aqueous media and natural sunlight. The mechanism of HMF oxidation to DFF is shown in Eqs. (15.4)–(15.8) (Ilkaeva et al. 2018a, b).

$$O_2 + e^- \longrightarrow O_2^- \quad (15.4)$$

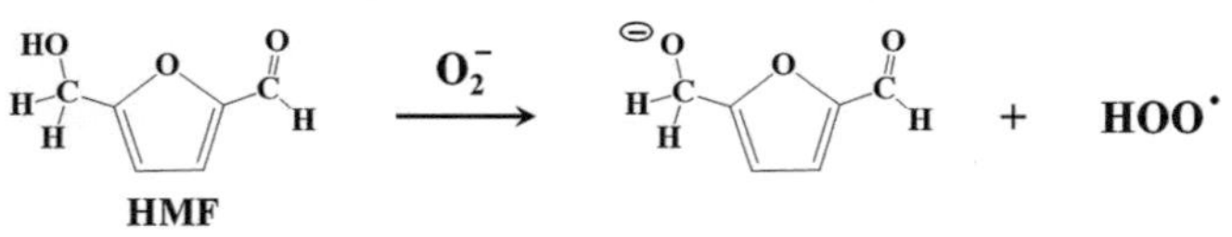

(15.5)

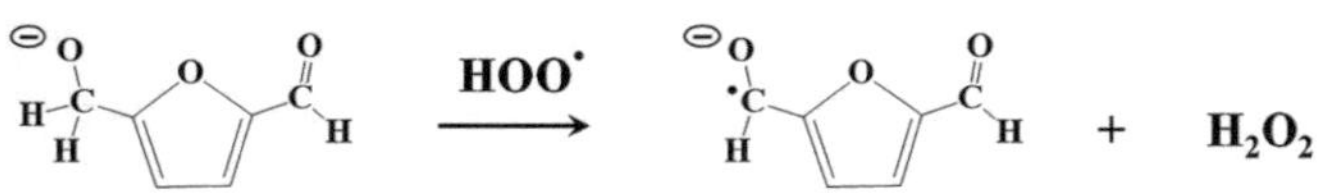

(15.6)

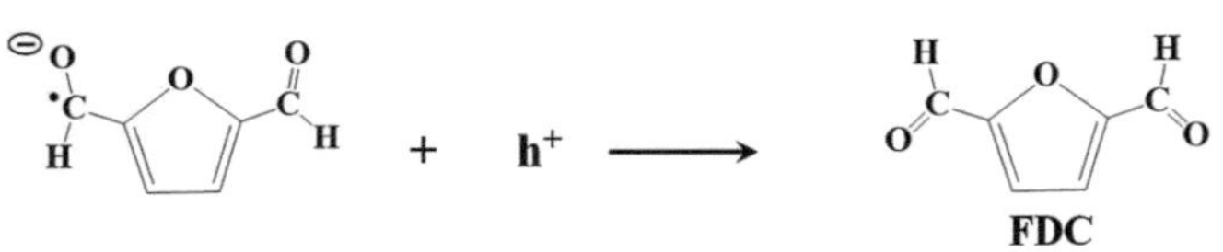

(15.7)

$$H_2O_2 + H^+ + e^- \longrightarrow {}^{\bullet}OH + H_2O \quad (15.8)$$

For a green chemistry approach, electrochemical oxidation appears to be a promising technique for the conversion of HMF to FDCA, because it can be performed at ambient temperature and avoids the use of hazardous chemicals. The electrochemical cell consists of the two compartments of anode and cathode sides. The oxidation of HMF is performed at the anode side, whereas the cathode

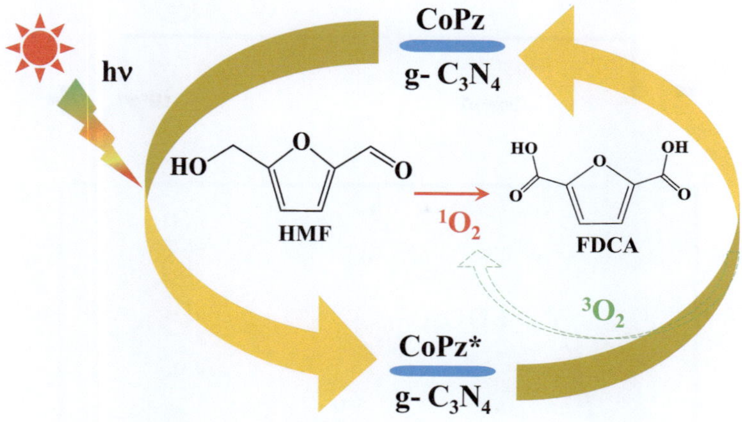

Fig. 15.3 Mechanism of photocatalytic oxidation of hydroxymethylfurfural

side presents hydrogen evolution as shown in Fig. 15.4. The product distribution was significantly affected by the applied potential and the surface composition of electrode. The oxyhydroxide on the surface of nickel (Ni) and cobalt (Co) presented a highly efficient oxidation of HMF to FDCA. The proposed mechanism was that HMF was oxidized with NiOOH at the electrode during the open circuit, while NiOOH was reduced to $Ni(OH)_2$. The $Ni(OH)_2$ was then regenerated by linear sweep voltammetry to NiOOH. In the case of CoOOH, it approached the same mechanism as NiOOH (Latsuzbaia et al. 2018; Taitt et al. 2019). The reaction pathway proceeded through the formation of HMFCA intermediate before undergoing oxidation to obtain FDCA (Gao et al. 2018).

The electrochemical oxidation of HMF could be combined with hydrogen production unit because the reduction of water occurred to produce hydrogen while HMF was oxidized (Cha and Choi 2015). In addition, a Cu-based electrode was introduced in the oxidation of HMF, which did not catalyze water oxidation as a side reaction. Therefore, the Cu-based electrode performed with a wide potential window to oxidize HMF without the side effect of water oxidation (Nam et al. 2018). The electrochemical oxidation was performed in an alkaline media as the FDCA product can be separated in an acidic condition. Accordingly, the separation cost needs to be considered for any large-scale process. Manganese oxide (MnO) could oxidize HMF under a low pH to overcome the problem of FDCA separation where the FDCA could be formed and precipitated simultaneously in an acidic medium (Kubota and Choi 2018). The oxidation of HMF proceeds to FDCA as a major product that has the potential to replace conventional terephthalic acid in PET polymerization. The FDCA can be formed through thermal catalytic, photocatalytic, and electrochemical oxidations. Although thermochemical conversion is still a cost-effective technology, the photocatalytic and electrochemical conversion have the advantage of being energy efficient or using less hazardous chemicals, respectively.

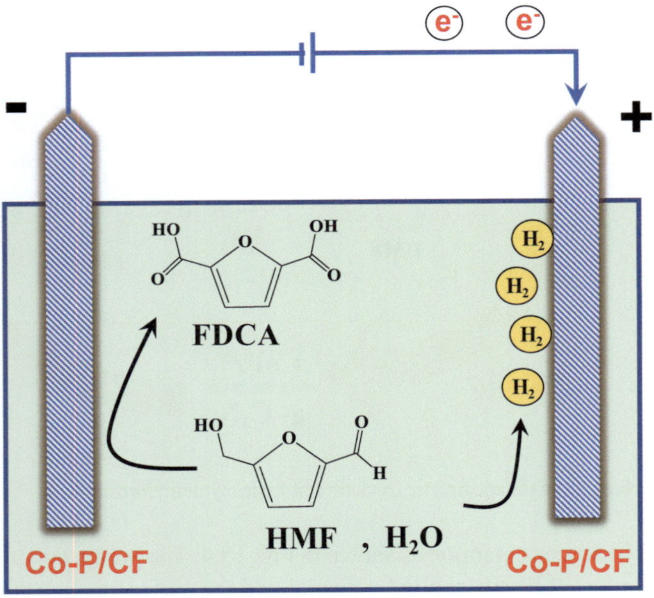

Fig. 15.4 Mechanism of electrochemical oxidation of hydroxymethylfurfural

15.5 Selective Hydrogenation of Hydroxymethylfurfural

The selective hydrogenation of HMF to 2,5-dihydroxymethylfuran (DHMF) is an attractive and important chemical platform that can be further transformed to other derivatives such as 1,2,6-hexanetriol for medicament and cosmetics, 1-hydroxyhexane-2,5-dione for polymer and solvent, and 3-hydroxymethylcyclopentanone for the synthesis of fragrances, pesticides, and polymers (Hu et al. 2017). In addition, DHMF can further reduce to diol tetrahydrofurandimethanol (THFDM), which was possibly used as a bio-based monomer for polyesters or polyurethanes as shown in Eq. (15.9).

The selective hydrogenation of HMF has been reported using both homogeneous and heterogeneous catalytic process. The hydrogenation catalyst should have a high selectivity to minimize the formation of undesired products. Since HMF molecule consists of a furan ring containing both aldehyde (HC=O) and alcohol (–OH) functional groups, the selection of metal catalyst and ligand is a key role for catalytic cycle in homogeneous hydrogenation of HMF. The metal salts of Ru, Rh, Ir, Re, and Ni have been used as hydrogenation catalysts coupled with various types of ligands such as N-heterocyclic carbene (NHC) ligands, phosphorus-based ligands, and auxiliary ligands (Cadu et al. 2018). By tuning the parameters, Ru-based catalysts with nitrogen-containing ligands have shown the most active and selectivity for HMF conversion to DHMF. Carbene ligands offered moderate activity and selectivity with a preference for unsaturated backbones (for selectivity) and for bulky aromatic substitution pattern (for conversion). However, there is some limitation for upscaling to industrial production due to high catalyst amount required.

Heterogeneous catalysts including Pt, Pd, Ru, Ni, and Ir can perform selective hydrogenation of HMF to DHMF where the metal species and support characteristic significantly affected the catalyst activity. The DHMF yield was in the range of 34–89.3% (Cai et al. 2014). A zirconia-based inorganic-organic coordination polymer was introduced in the catalytic transfer-based hydrogenation of HMF to DHMF with a high catalytic efficiency giving a DHMF yield of 94%. The reaction pathway followed the Meerwein-Ponndorf-Verley reaction (Hu et al. 2018) as shown in Eq. (15.9). Chen et al. (2015) reported on the development of Pd supported on amine-functionalized metal-organic frameworks (MOFs), high surface area, tunable pore sizes, and controllable structures for metal supports.

The presence of free amine (–NH_2) moieties in the frameworks played a key role for the formation of uniform and well-dispersed palladium nanoparticles on the support. The maximum DHMF yield achieved up to 96% with a full conversion of HMF using 3 wt% Pd loading over MOF support and operated at a 30 °C in aqueous medium. The recycling of the fine particle of heterogeneous catalyst eases to separate by filtration or centrifugation. However, the catalyst needs to form the certain shape for industrial application to prevent the loss of solid catalyst during the process. To overcome separation issues of homogeneous and heterogeneous catalysts, the use of magnetic catalysts is one of an interesting concept, which efficiently removed catalyst from reaction mixtures by applying an external magnetic field (Liu and Zhang 2016). The catalytic reaction could be selective to the type of metal loading over magnetic nanoparticles.

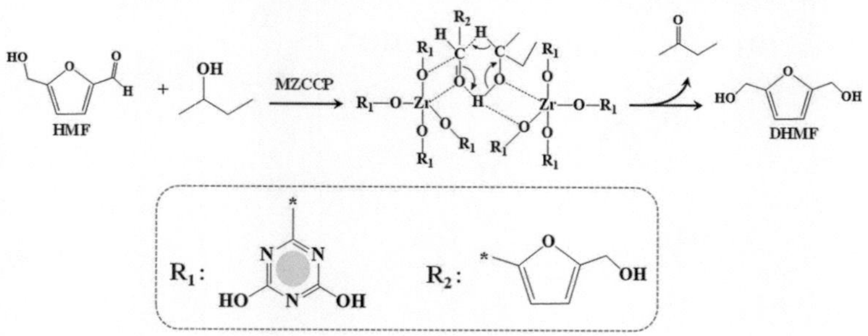

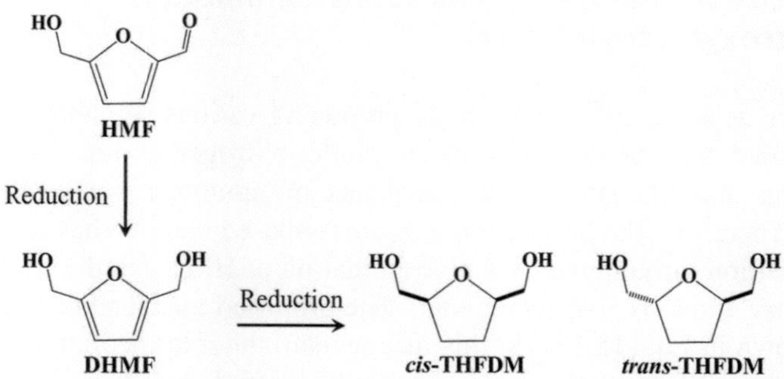

(15.9)

Several authors have reported on the conversion of HMF to DHMF or further reduction to THFDM by operating at pressurized hydrogen system (28–350 bar) operated at high temperatures more than 100 °C using various precious metal catalysts. However, there is another opportunity to convert HMF to DHMF by electrochemical reduction with no hydrogen required. The reaction occurs in the cathode compartment as shown in Fig. 15.5, where the electrode is covered by HMF and its reduction intermediate. The electron transfer for the reduction of the intermediate to DHMF is shown in Eq. (15.10). This gives a 100% DHMF selectivity and a Faradic efficiency of 100% (Roylance et al. 2016).

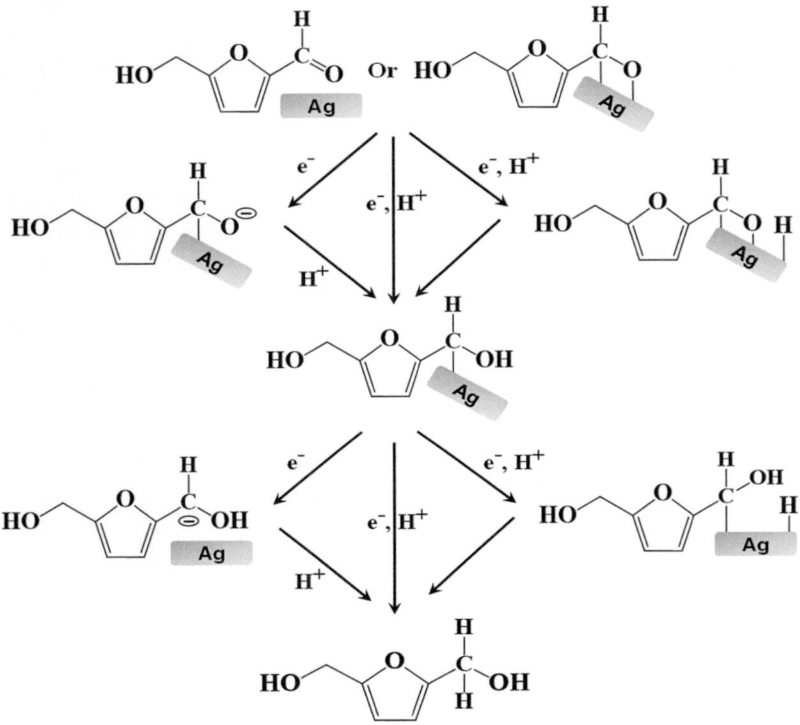

(15.10)

15.6 Oxidative Amidation and Reductive Amination of Hydroxymethylfurfural

HMF could serve as a versatile platform for producing various valuable chemicals by the reaction with or formation of an amide. Nitrogen-containing compounds including diamide are key intermediates of monomer synthesis and pharmaceutics. Therefore, the production of biomass-based diamide has recently gained more attention through the oxidative amidation of HMF, but the selectivity for the primary amide is still a challenge. The proposed mechanism of HMF amidation is shown in Eq. (15.11). In this mechanism, the intermediate product of 5-hydroxymethyl-2-furancarbonitrile is initially formed, which is then converted to 2,5-dicyanofuran and 5-cyano-2-furancarboxamide, respectively, forming 2,5-furandicarboxyamide.

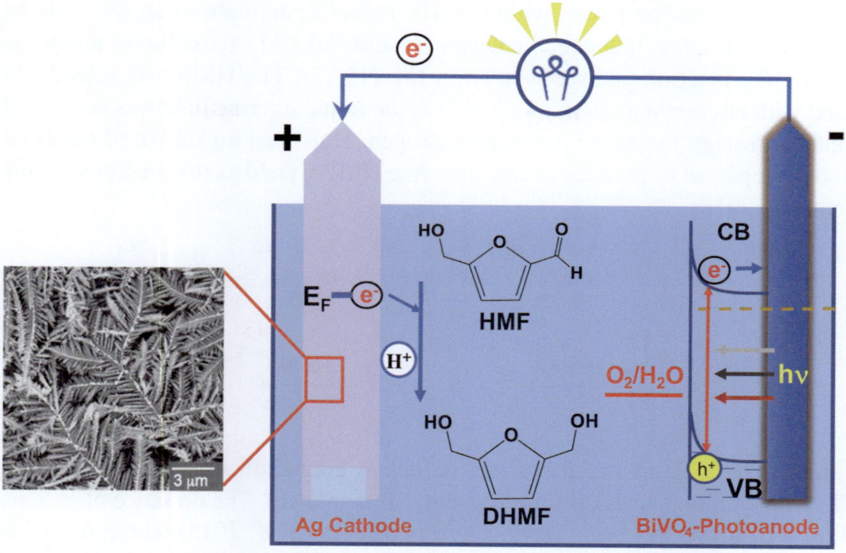

Fig. 15.5 Photo-electrochemical reduction of hydroxymethylfurfural (HMF) to 2,5-dihydroxymethylfuran (DHMF)

Alkali manganese oxide (α-MnO_2/Na_xMnO_2) was introduced as a catalyst for the oxidative amidation of HMF demonstrating nearly 85% yield of 2,5-dicyanofuran. From a kinetic study, the hydration of 5-cyano-2-furancarboxamide was found to be the controlling step (Li et al. 2017). The doping of cryptomelane with different metals was investigated for the catalytic oxidative amidation of HMF. Both the metal species and doping method played an important role on the HMF amidation efficiency, and the yield of 2,5-dicyanofuran could be improved up to 97% (Li et al. 2018b).

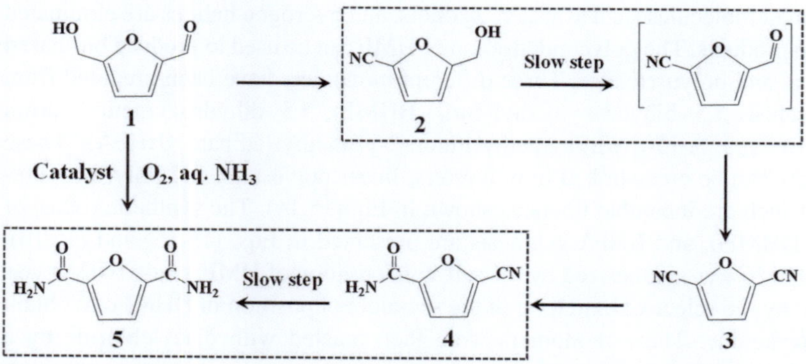

(15.11)

Nitrogen-containing carbon compounds were not only derived from the oxidative amidation of HMF, but could also be prepared by reductive amination. Here, the 5-(hydroxymethyl)furfurylamine (HMFA) product is a key compound in pharmaceuticals and polymers synthesis. The reductive amination can generally be performed with a precious metal catalyst (Pt and Pd) and excess liquid ammonia following the reaction pathway shown in Eq. (15.12). The HMF was selectively reacted with ammonia to an imine, and then the imine intermediate was converted to HMFA through the reduction with hydrogen. However, nickel-based catalysts have been reported to produce nearly the same HMFA yield as the precious metals (Chen et al. 2018).

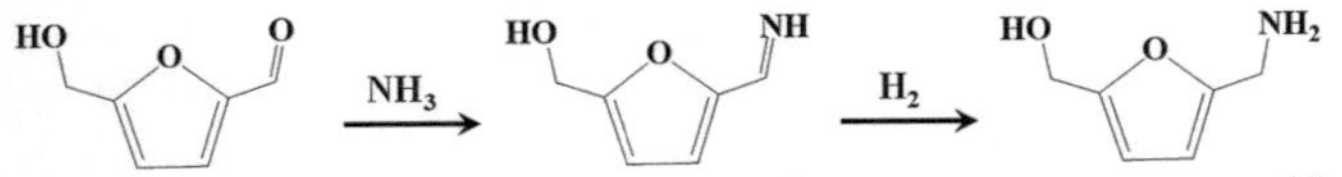

(15.12)

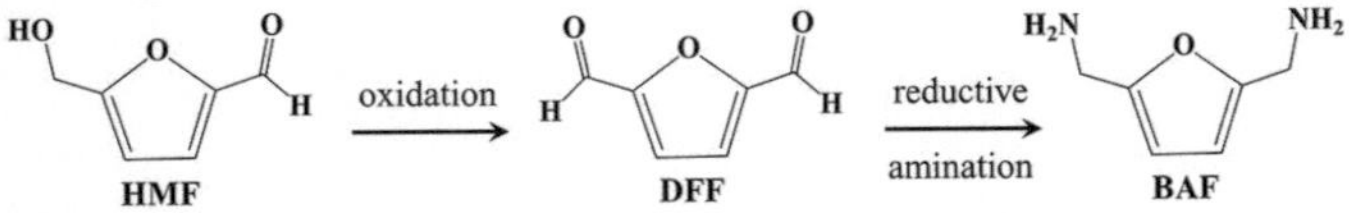

(15.13)

Equation (15.13) shows the direct reductive amination of 2,5-diformylfuran (DFF) with ammonia to 2,5-bis(aminomethyl)furan (BAF). The presence of various Ni-Raney catalysts has been previously reported (Le et al. 2015) where the acid-treated Ni-Raney catalysts yielded 42.6% BAF in tetrahydrofuran/water mixture at optimal reaction conditions. The electrochemical reductive amination is of interest in the HMF conversion to amine derivatives of furan compounds because it requires fewer chemicals in the system. Ethanolamine and water were used as the nitrogen source and the reducing agent, respectively. The catalytic efficiency depends on the electrode material and applied voltage. The electrical potential showed strong effect on the desired product (Roylance and Choi 2016).

15.7 Polycondensation of Hydroxymethylfurfural

Polycondensation is the formation of complex esters via a condensation reaction where small molecules such as water, alcohols, and hydrogen halides are eliminated as the by-products. The polycondensation of HMF can be used to produce bio-based polyesters and polyurethanes. Three different monomers have been prepared from HMF such as 5,5′-bihydroxymethyl furil (BHMF), 5,5′-dihydroxymethyl furoin (DHMFo), and bis[5-(hydroxymethyl)furan-2-yl)methyl]adipate (BHFA). These monomers can be cross-linked to polyesters, linear polyesters, and poly(ester urethane), which are insoluble fibers as shown in Eq. (15.14). The synthetic routes of BHMF, DHMFo, and BHFA synthesis are presented in Eqs. (15.15) and (15.16), where DHMF was synthesized by the self-condensation of HMF, while BHMF was prepared by the selective oxidation of the sec-alcohol position in DHMFo to obtain α-hydroxyketone. These monomers were then reacted with diacylchloride by a polycondensation reaction to obtain the cross-linked polyester and linear polyester as shown in Eqs. (15.17)–(15.19).

The BHFA could react with diisocyanates to produce polyurethane as shown in Eq. (15.20) (Mou and Chen 2016). HMF is also a precursor of a bifunctional furan monomer in the thermoplastics preparation. The degree of completion of polycondensation and the mean length of the macromolecules are limited by the equilibrium concentration of the reagents and reaction products (called as reversible equilibrium polycondensation). Therefore, the polycondensation by-products must be removed to achieve an interesting chain length for its commercial production.

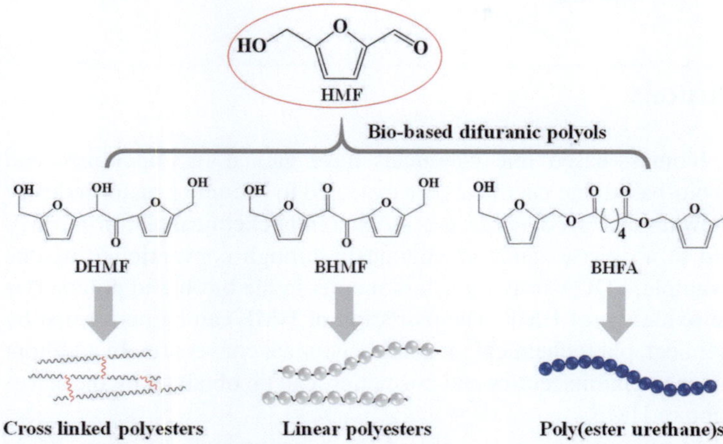

(15.14)

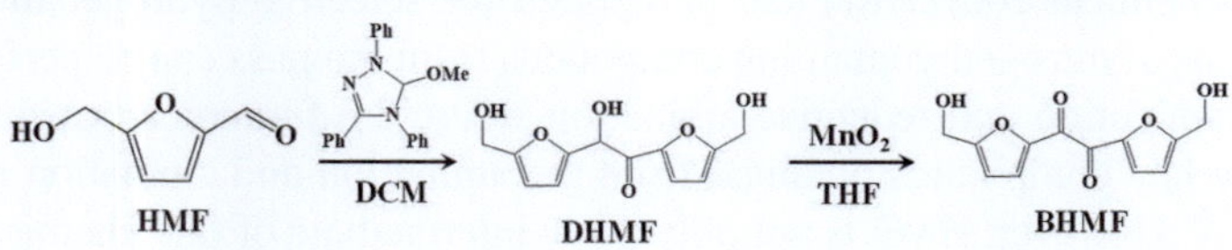

(15.15)

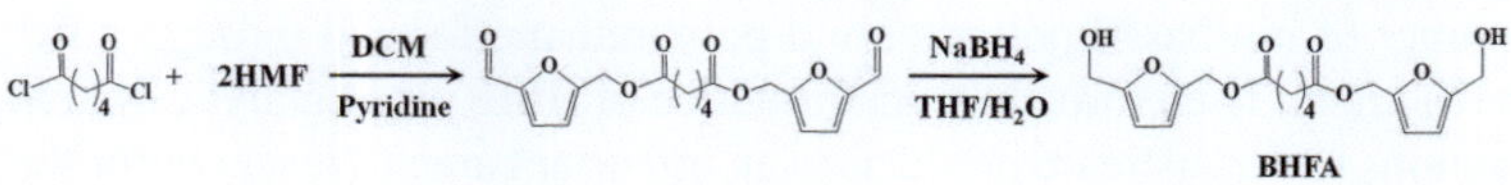

(15.16)

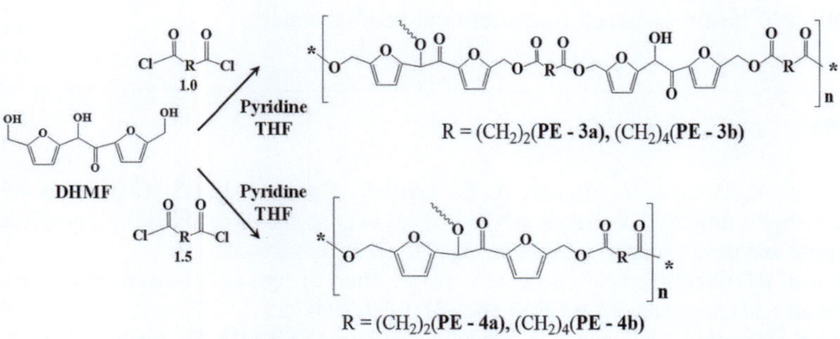

(15.17)

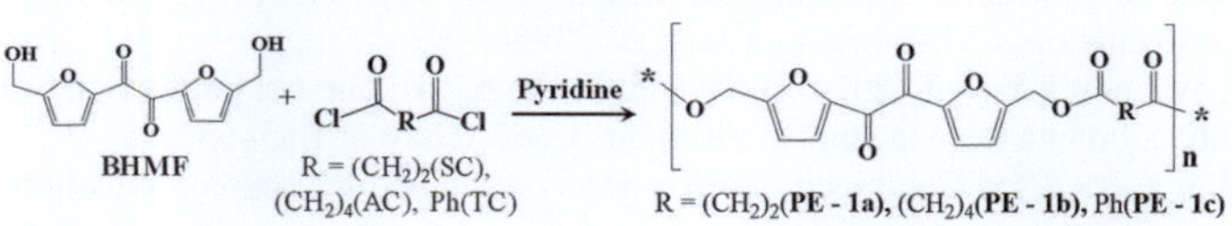

(15.18)

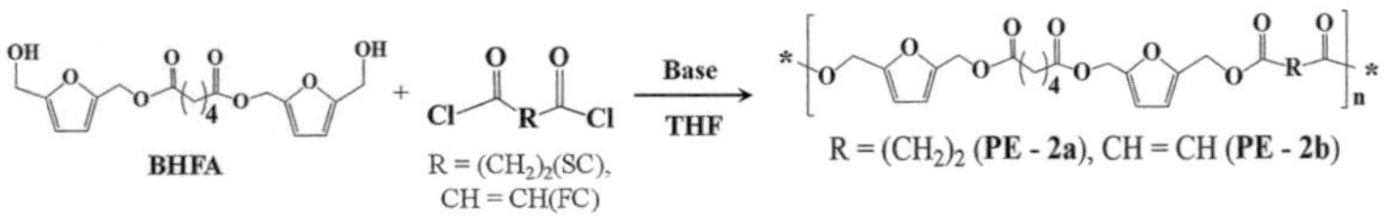

(15.19)

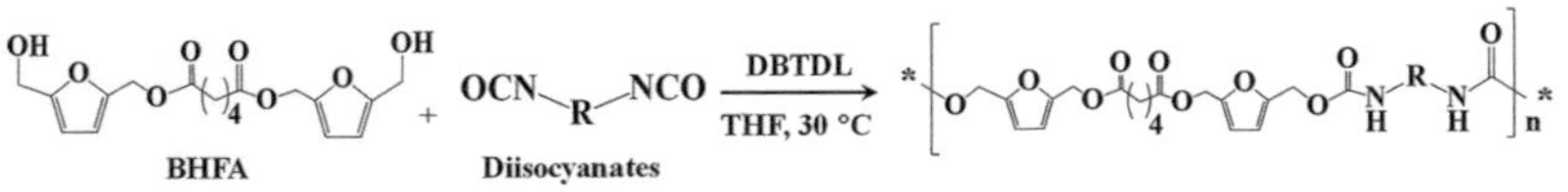

(15.20)

15.8 Conclusions

Lignocellulosic biomass-based fine chemicals have gained in importance and acceptance. The bio-based fine chemicals are expected to become a major industry in the near future where HMF could serve as an important chemical platform. HMF can be converted to a diverse range of chemicals through conventional organic reactions. For example, FDCA formation, a monomer in the bio-based polymer, is produced by the oxidation of HMF. The oxidation of HMF can be performed by either thermochemical, photochemical, or electrochemical conversion. In addition, DHMF, a precursor in pharmaceutics and cosmetics, can be obtained by the selective hydrogenation of HMF.

Electrochemical conversion can be applied for selective hydrogenation, while the synthesis of nitrogen-containing compounds from biomass can be performed by oxidative amidation and reductive amination where 2,5-furandicarboxyamide and HMFA are key compounds obtained from the amidation and amination reactions, respectively. However, HMF is not only a key intermediate of fine chemicals, but it is also a monomer of bio-based polyester and polyurethane formed through polycondensation reactions. Therefore, the transformation of HMF via a catalytic system in organic reactions to downstream products is an important area of research for the next-generation of biomass-based fine chemical production.

References

Albonetti S, Lolli A, Morandi V, Migliori A, Lucarelli C, Cavani F (2015) Conversion of 5-hydroxymethylfurfural to 2,5-furandicarboxylic acid over au-based catalysts: optimization of active phase and metal–support interaction. Appl Catal B 163:520–530

Binder JB, Raines RT (2009) Simple chemical transformation of lignocellulosic biomass into furans for fuels and chemicals. J Am Chem Soc 131:1979–1985

Cadu A, Sekine K, Mormul J, Ohlmann DM, Schaub T, Hashmi ASK (2018) Homogeneous catalysed hydrogenation of HMF. Green Chem 20:3386–3393

Cai H, Li C, Wang A, Zhang T (2014) Biomass into chemicals: one-pot production of furan-based diols from carbohydrates via tandem reactions. Catal Today 234:59–65

Cha HG, Choi K-S (2015) Combined biomass valorization and hydrogen production in a photoelectrochemical cell. Nat Chem 7:328

Chen J, Liu R, Guo Y, Chen L, Gao H (2015) Selective hydrogenation of biomass-based 5-hydroxymethylfurfural over catalyst of palladium immobilized on amine-functionalized metal–organic frameworks. ACS Catal 5:722–733

Chen W, Sun Y, Du J, Si Z, Tang X, Zeng X, Lin L, Liu S, Lei T (2018) Preparation of 5-(aminomethyl)-2-furanmethanol by direct reductive amination of 5-hydroxymethylfurfural with aqueous ammonia over the Ni/SBA-15 catalyst. J Chem Technol Biotechnol 93:3028–3034

Chen L, Yang W, Gui Z, Saravanamurugan S, Riisager A, Cao W, Qi Z (2019) MnO_x/P25 with tuned surface structures of anatase-rutile phase for aerobic oxidation of 5-hydroxymethylfurfural into 2,5-diformylfuran. Catal Today 319:105–112

Gao L, Bao Y, Gan S, Sun Z, Song Z, Han D, Li F, Niu L (2018) Hierarchical Ni-co based transition metal oxide catalysts for electrochemical conversion of biomass into valuable chemicals. ChemSusChem 11:2547–2553

Hansen TS, Sádaba I, García-Suárez EJ, Riisager A (2013) Cu catalyzed oxidation of 5-hydroxymethylfurfural to 2,5-diformylfuran and 2,5-furandicarboxylic acid under benign reaction conditions. Appl Catal A 456:44–50

He M, Sun Y, Han B (2013) Green carbon science: scientific basis for integrating carbon resource processing, utilization, and recycling. Angew Chem Int Ed 52:9620–9633

Hu L, Lin L, Wu Z, Zhou S, Liu S (2017) Recent advances in catalytic transformation of biomass-derived 5-hydroxymethylfurfural into the innovative fuels and chemicals. Renew Sust Energ Rev 74:230–257

Hu L, Li T, Xu J, He A, Tang X, Chu X, Xu J (2018) Catalytic transfer hydrogenation of biomass-derived 5-hydroxymethylfurfural into 2,5-dihydroxymethylfuran over magnetic zirconium-based coordination polymer. Chem Eng J 352:110–119

Ilkaeva M, Krivtsov I, García-López EI, Marcì G, Khainakova O, García JR, Palmisano L, Díaz E, Ordóñez S (2018a) Selective photocatalytic oxidation of 5-hydroxymethylfurfural to 2,5-furandicarboxaldehyde by polymeric carbon nitride-hydrogen peroxide adduct. J Catal 359:212–222

Ilkaeva M, Krivtsov I, García JR, Díaz E, Ordóñez S, García-López EI, Marcì G, Palmisano L, Maldonado MI, Malato S (2018b) Selective photocatalytic oxidation of 5-hydroxymethyl-2-furfural in aqueous suspension of polymeric carbon nitride and its adduct with H_2O_2 in a solar pilot plant. Catal Today 315:138–148

Kim M, Su Y, Fukuoka A, Hensen EJM, Nakajima K (2018) Aerobic oxidation of 5-(hydroxymethyl)furfural cyclic acetal enables selective furan-2,5-dicarboxylic acid formation with CeO_2-supported gold catalyst. Angew Chem Int Ed 57:8235–8239

Kubota SR, Choi K-S (2018) Electrochemical oxidation of 5-hydroxymethylfurfural to 2,5-furandicarboxylic acid (FDCA) in acidic media enabling spontaneous FDCA separation. ChemSusChem 11:2138–2145

Kucherov FA, Gordeev EG, Kashin AS, Ananikov VP (2017) Three-dimensional printing with biomass-derived PEF for carbon-neutral manufacturing. Angew Chem Int Ed 56:15931–15935

Kucherov FA, Romashov LV, Galkin KI, Ananikov VP (2018) Chemical transformations of biomass-derived c6-furanic platform chemicals for sustainable energy research, materials science, and synthetic building blocks. ACS Sustain Chem Eng 6:8064–8092

Latsuzbaia R, Bisselink R, Anastasopol A, Van der Meer H, Van Heck R, Yagüe MS, Zijlstra M, Roelands M, Crockatt M, Goetheer E, Giling E (2018) Continuous electrochemical oxidation of biomass derived 5-(hydroxymethyl)furfural into 2,5-furandicarboxylic acid. J Appl Electrochem 48:611–626

Le N-T, Byun A, Han Y, Lee K-I, Kim H (2015) Preparation of 2, 5-bis (aminomethyl) furan by direct reductive amination of 2,5-diformylfuran over nickel-raney catalysts. Green Sustain Chem 5:115

Li X, Jia X, Ma J, Xu Y, Huang Y, Xu J (2017) Catalytic amidation of 5-hydroxymethylfurfural to 2,5-furandicarboxamide over alkali manganese oxides. Chin J Chem 35:984–990

Li F, Li X-L, Li C, Shi J, Fu Y (2018a) Aerobic oxidative esterification of 5-hydroxymethylfurfural to dimethyl furan-2,5-dicarboxylate by using homogeneous and heterogeneous PdCoBi/C catalysts under atmospheric oxygen. Green Chem 20:3050–3058

Li X, Ma J, Jia X, Xia F, Huang Y, Xu Y, Xu J (2018b) Al-doping promoted aerobic amidation of 5-hydroxymethylfurfural to 2,5-furandicarboxamide over cryptomelane. ACS Sustain Chem Eng 6:8048–8054

Li X, Xu R, Yang J, Nie S, Liu D, Liu Y, Si C (2019) Production of 5-hydroxymethylfurfural and levulinic acid from lignocellulosic biomass and catalytic upgradation. Ind Crop Prod 130:184–197

Liu B, Zhang Z (2016) Catalytic conversion of biomass into chemicals and fuels over magnetic catalysts. ACS Catal 6:326–338

Masoud N, Donoeva B, de Jongh PE (2018) Stability of gold nanocatalysts supported on mesoporous silica for the oxidation of 5-hydroxymethyl furfural to furan-2,5-dicarboxylic acid. Appl Catal A 561:150–157

Mou Z, Chen EYX (2016) Polyesters and poly(ester-urethane)s from biobased difuranic polyols. ACS Sustain Chem Eng 4:7118–7129

Nam D-H, Taitt BJ, Choi K-S (2018) Copper-based catalytic anodes to produce 2,5-furandicarboxylic acid, a biomass-derived alternative to terephthalic acid. ACS Catal 8:1197–1206

Nanda S, Rana R, Sarangi PK, Dalai AK, Kozinski JA (2018) A broad introduction to first, second and third generation biofuels. In: Sarangi PK, Nanda S, Mohanty P (eds) Recent advancements in biofuels and bioenergy utilization. Springer Nature, Singapore, pp 1–25

Perez GP, Mukherjee A, Dumont M-J (2019) Insights into HMF catalysis. J Ind Eng Chem 70:1–34

Roylance JJ, Choi K-S (2016) Electrochemical reductive amination of furfural-based biomass intermediates. Green Chem 18:5412–5417

Roylance JJ, Kim TW, Choi K-S (2016) Efficient and selective electrochemical and photoelectrochemical reduction of 5-hydroxymethylfurfural to 2,5-bis(hydroxymethyl)furan using water as the hydrogen source. ACS Catal 6:1840–1847

Shahangi F, Chermahini AN, Saraji M (2018) Dehydration of fructose and glucose to 5-hydroxymethylfurfural over Al-KCC-1 silica. J Energ Chem 27:769–780

Taitt BJ, Nam D-H, Choi K-S (2019) A comparative study of nickel, cobalt, and iron oxyhydroxide anodes for the electrochemical oxidation of 5-hydroxymethylfurfural to 2,5-furandicarboxylic acid. ACS Catal 9:660–670

Tirsoaga A, El Fergani M, Parvulescu VI, Coman SM (2018) Upgrade of 5-hydroxymethylfurfural to dicarboxylic acids onto multifunctional-based Fe_3O_4@SiO_2 magnetic catalysts. ACS Sustain Chem Eng 6:14292–14301

Tong X, Yu L, Chen H, Zhuang X, Liao S, Cui H (2017) Highly efficient and selective oxidation of 5-hydroxymethylfurfural by molecular oxygen in the presence of cu-MnO_2 catalyst. Catal Commun 90:91–94

Tye YY, Lee KT, Wan Abdullah WN, Leh CP (2016) The world availability of non-wood lignocellulosic biomass for the production of cellulosic ethanol and potential pretreatments for the enhancement of enzymatic saccharification. Renew Sust Energ Rev 60:155–172

Wang L, Zhang L, Li H, Ma Y, Zhang R (2019) High selective production of 5-hydroxymethylfurfural from fructose by sulfonic acid functionalized SBA-15 catalyst. Compos Part B 156:88–94

Xia H, An J, Hong M, Xu S, Zhang L, Zuo S (2019) Aerobic oxidation of 5-hydroxymethylfurfural to 2,5-difurancarboxylic acid over Pd-au nanoparticles supported on mg-Al hydrotalcite. Catal Today 319:113–120

Xu S, Zhou P, Zhang Z, Yang C, Zhang B, Deng K, Bottle S, Zhu H (2017) Selective oxidation of 5-hydroxymethylfurfural to 2,5-furandicarboxylic acid using O_2 and a photocatalyst of Co-thioporphyrazine bonded to g-C_3N_4. J Am Chem Soc 139:14775–14782

Zhou X-H, Song K-H, Li Z-H, Kang W-M, Ren H-R, Su K-M, Zhang M-L, Cheng B-W (2019) The excellent catalyst support of Al_2O_3 fibers with needle-like mullite structure and HMF oxidation into FDCA over CuO/Al_2O_3 fibers. Ceram Int 45:2330–2337

Conversion of Glycerol to Value-Added Products

16

Parmila Devi and Ajay K. Dalai

Abstract

The dramatic increase in biodiesel production has increased the oversupply of glycerol as a by-product. Low cost, ready availability, and surplus supply of glycerol is a spectacular opportunity for chemical industry to utilize glycerol as feedstock for production of high-value products. This chapter provides the information about different pathways available for conversion of glycerol into high specialty chemicals. Various approaches and strategies have been developed by researchers to investigate the effects of reaction parameters, i.e., nature and type of catalyst, reaction time, temperature, pressure, and type of solvent, on glycerol conversion and product selectivity. Based on the information provided in this chapter, researchers can select a viable process for the conversion of waste glycerol into valuable commodities.

Keywords

Glycerol · Value-added products · Biodiesel · Esterification · Transesterification

16.1 Introduction

The increasing global energy demand, declining fossil fuel reserves, and associated environmental concerns have stimulated the global interest in renewable sources of energy. Biofuels and biomass are emerging as popular alternatives due to their vast availability, nontoxic nature, and biodegradable properties. Biodiesel is conventionally produced from saponification and transesterification of vegetable oil (Ciriminna

P. Devi · A. K. Dalai (✉)
Department of Chemical and Biological Engineering, University of Saskatchewan, Saskatoon, Saskatchewan, Canada
e-mail: ajay.dalai@usask.ca

S. Nanda et al. (eds.), *Biorefinery of Alternative Resources: Targeting Green Fuels and Platform Chemicals*, https://doi.org/10.1007/978-981-15-1804-1_16

et al. (2014) (Fig. 16.1)). In the saponification process, soap and glycerol are generated from hydrolysis of oil in presence of alkali salt. In the transesterification process, methyl ester and glycerol are produced from reaction of triglycerides with methanol or ethanol. Both of these processes concurrently generate large amounts of glycerol (Gholami et al. 2014).

About 1 kg of glycerol is generated during the production of 10 kg of biodiesel (Devi et al. 2018). In the last two decades, production of biodiesel has significantly grown due to various governmental initiatives. Rapid development in the biodiesel industry has led to the oversupply of crude glycerol in the market. The unavailability of suitable method for the processing of crude glycerol has led to the drop in the prices of crude glycerol. At the same time, the surplus amount of glycerol is generated as a by-product (Gholami et al. 2014; Devi et al. 2018). Owing to its limited commercial applications, mainly in cosmetics, food, and pharmaceutical industries, glycerol price dropped as low as \$0.04–0.09/lb (Ciriminna et al. 2014). Therefore, researchers across the globe are attempting to find ways to turn waste glycerol into a profit.

Based on the predicted high growth rate of biodiesel production, it is expected that the crude glycerol prices will further decline if the current market situation continues. The economic feasibility of glycerol conversion to value-added product is entirely dependent on the purity of glycerol, market price, and utility of the final products. The current prices of pure glycerol are high compared to the crude glycerol. Thus, the utilization of crude glycerol in glycerol catalytic conversion processes is vital for attaining a sustainable and cost-effective production of products with high market value. Given the high market value of chemicals obtained from

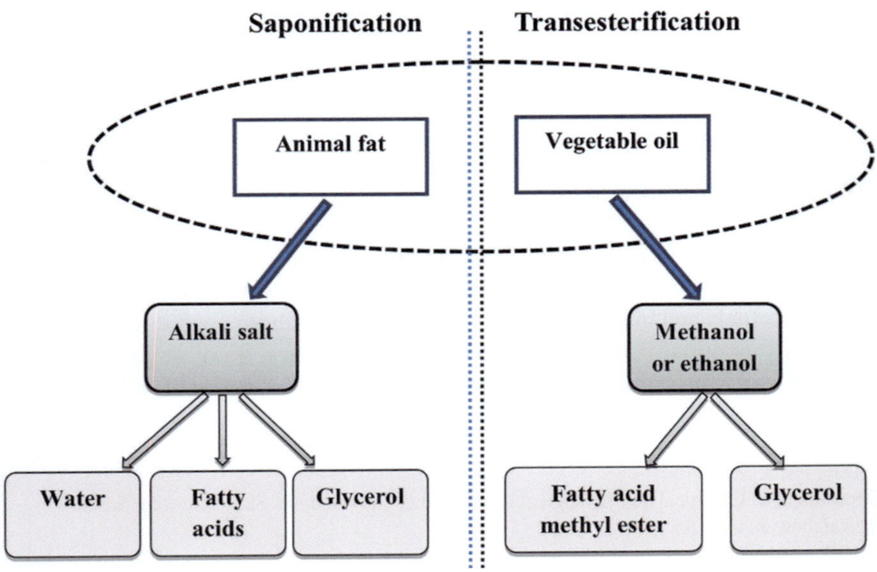

Fig. 16.1 Schematic process flow for the production of biodiesel

Table 16.1 Current market prices of glycerol-based value-added products

Compound	Chemical formula	Chemical structure	Approximate price ($/lbs)	Estimated US capacity (MMlbs)
Glycerol	$C_3H_8O_3$	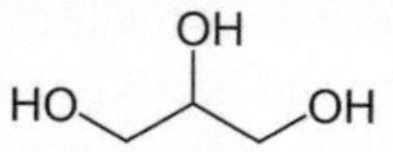	0.04–0.5	251
Propionaldehyde	C_3H_6O	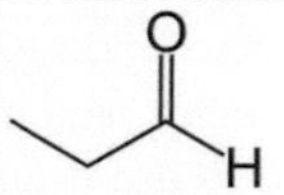	0.5	401
Propionic acid	$C_3H_6O_2$	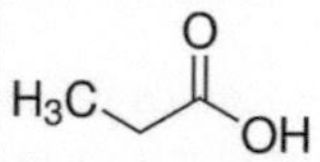	0.5–0.6	439
Glycidol	$C_3H_6O_2$	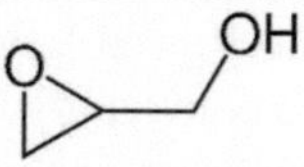	>$10,500	–
Allyl alcohol	C_3H_6O	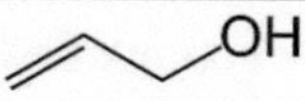	1.1	65
Acrylic acid	$C_3H_4O_2$	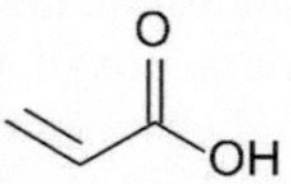	0.5–1.1	2779
Glycerol carbonate	$C_4H_6O_4$	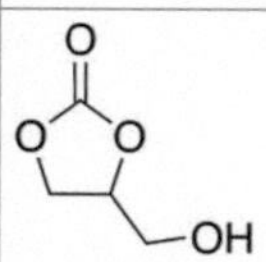	406	–
Solketal	$C_6H_{12}O_3$	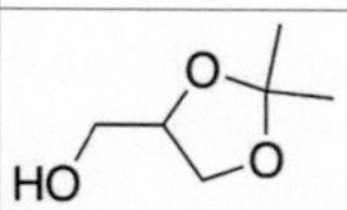	–	–
Tri-tert-butyl glycerol	$C_{15}H_{32}O_3$	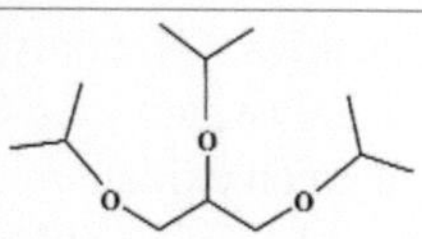	–	–

catalytic conversion of glycerol, glycerol offers tremendous potential (Table 16.1). Therefore, glycerol can be utilized as a feedstock for the manufacture of value-added chemicals or products.

Until 2008, glycerol was mainly consumed in refined form in the chemical and soap preparation industries. During 2009–2010, a sudden boom in biodiesel industry resulted in an increased production of glycerol as by-product. The biodiesel production market was at 37 billion gallons in 2016. Annually, the biodiesel market is increasing by the growth of 42% indirectly generating about 4.1 billion gallons of crude glycerol. Based on current market trends, the annual production of glycerol will reach 41.9 billion liters by 2020 (Talebian-Kiakalaieh et al. 2018). The following factors are triggering to the growth of global glycerol market:

1. Growth in food and beverage industries
2. Flourishing biodiesel industry

3. Optimistic approach of administrators toward environmentally friendly products

Crude glycerol is mainly used in animal feed, dust suppression, and deicing. Recently, glycerol is reported to be used as an additive material and binder for biomass pellets inducing hydrophobicity and heating value (Azargohar et al. 2019). The chemical composition of crude glycerol obtained from biodiesel industry is given in Table 16.2 (Ciriminna et al. 2014). Crude glycerol has substantial color (yellow to dark brown), high salt and fatty acids, and high contents of toxic methanol. Current technologies are not designed to use it directly and require expensive refining for its further use.

Various researchers have developed processes for the transformation of glycerol into marketable high-value chemicals/products using different approaches and methodologies. In order to manage surplus glycerol glut and make biodiesel industries more cost-effective, it is necessary to convert glycerol into chemicals or value-added products of higher price and larger market. In this chapter, several methods of converting glycerol into value-added products are described, which include oxidation, hydrogenolysis, dehydration, acetylation, etherification, esterification, polymerization, ammonification, chlorination, and carboxylation. The various intermediate and final products during processing of glycerol using the abovementioned processes are shown in Fig. 16.2.

16.2 Oxidation of Glycerol

Oxidation is a commonly used technique for the conversion of glycerol into dihydroxyacetone, hydroxypyruvic acid, formic acid, glycolic acid, etc. The glycerol oxidation products have various applications in cosmetics, pharmaceutical, food, detergents, dyeing, and chemical industries. Generally, heterogeneous noble metal catalysts like Au-, Pt-, and Pd-based catalysts are used for the oxidation of glycerol. Coupling of these catalysts with some green oxidant makes more sustainable system for glycerol oxidation. Photocatalysts like TiO_2, ZnO_2, SiC, and CdS are another most commonly used class of catalysts due to their low cost, high efficiency, and nontoxic nature. Various researchers have investigated the effects of reaction conditions and type of catalyst on glycerol conversion and product selectivity (Table 16.3).

Table 16.2 Typical composition of crude glycerol obtained from biodiesel industry

Components	Typical composition (wt%)
Glycerol	66–84
Methanol	24–38
Ash	3–5
Water	0.9–3.1
Glycerides	0.4–2
Potassium	~0.2
FFAs	0.5–1

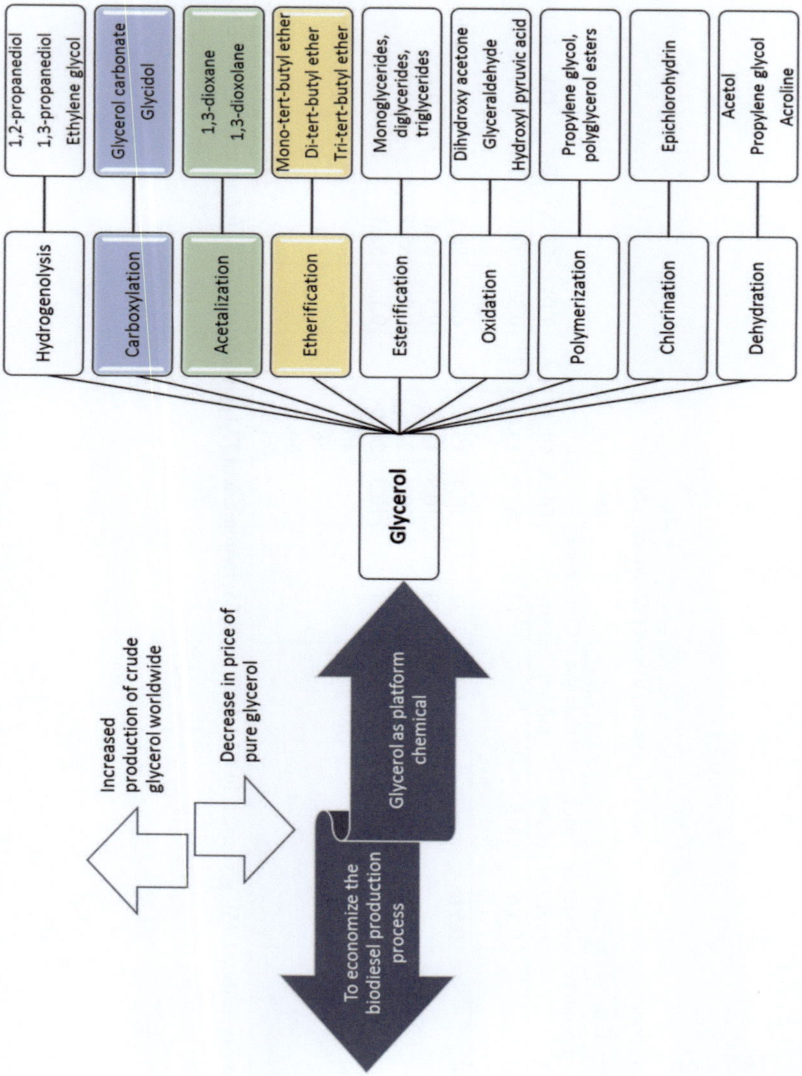

Fig. 16.2 Various pathways for the transformation of glycerol into value-added chemicals and products

Table 16.3 Reaction conditions for the oxidation of glycerol to several by-products

Catalyst	Reaction parameters				Conversion (%)	Selectivity (%)							References
	Temperature (°C)	Reaction time (h)	Molar ratio	Catalyst loading		DHA	GLA	LA	TTA	GCA	FA	OA	
Au-based catalysts	80	0.5 h	315		13.8	67.3	13.4	0.2	0.3	8.6	1.7	2.4	Yuan et al. (2015)
Pt/CNTs	60	0.5 h		10 wt%	22	10.4	19.5	–	0.2	58.8	–	0.2	Ning et al. (2015)
Pt/O-CNTs					13	10.8	13.6		0.3	65.0		0.1	
Pt/NCNTs					32	10.2	24.6		0.2	55.7		0.1	
Pt/O-NCNTs					34	10.3	26.7		0.1	58.3		0.2	
Ag/Al_2O_3	60	3 h	4	0.5 g		–	46.8				0		Skrzyńska et al. (2015)
Au/Al_2O_3							58.3				27.7		
Pd/Al_2O_3							84.1				0.2		
Pt/Al_2O_3							75.3				14.5		

Abbreviations: dihydroxyacetone *DHA*, glyceric acid *GLA*, lactic acid *LA*, tartronic acid *TTA*, glycolic acid *GCA*, formic acid *FA*, oxalic acid *OA*, carbon nanotubes *CNT*

It was found that the impurities present in crude glycerol could affect oxidation process. Therefore, Skrzyńska et al. (2015) studied the synergetic and antagonistic effects of various impurities (methanol, mineral salts, and organic sulfur derivatives) on the oxidation process. Au catalysts show good activity in pure glycerol oxidation, but the conversion rate was low for crude glycerol due to their sensitivity toward mineral salts and organic sulfur derivatives. Pd was found to be the most resistant toward methanol, mineral salts, and organic sulfur derivatives.

The effectiveness and efficiency of the catalyst can be changed by manipulating the surface electronic properties, functionalities distribution, and deposition of metal particles. Ning et al. (2015) studied the role of nitrogen and oxygen functionality of nitrogen-doped carbon nanotubes in oxidation of glycerol. It was found that graphitic nitrogen interacts with Pt nanoparticles by electron transfer mechanism and enhances the dispersion of Pt nanoparticles. However, the presence of oxygen-containing groups reduces the donor–acceptor interaction.

Yuan et al. (2015) studied the oxidation of glycerol in the presence of Au-based catalyst supported on $MgO\text{-}Al_2O_3$ materials using NaOH as the activation agent. The acid–base property of the catalyst is reported to have strong effect on product distribution depending on the molar ratio of Mg/Al in the catalyst. The high selectivity of dihydroxyacetone was observed in acidic environment, while the selectivity of glyceric acid was found to be increased in basic environment. Similarly, other researchers reported a significant catalytic activity (13–34%) of Pd/TiO_2-, Pd-Ni/C-, and Pd-Co/Au-based catalysts in alkaline solutions in comparison to methanol and ethylene glycol (Su et al. 2009; Simoes et al. 2010; Rostami et al. 2015). Pd-Co/Au catalyst was found very stable even after 200 cycles (Rostami et al. 2015).

16.3 Hydrogenolysis of Glycerol

1,2-Propanediol and ethylene glycol are the two major products produced from the hydrogenolysis of glycerol. 1,2-Propanediol is prepared through selective hydrogenation of glycerol by the insertion of hydrogen molecule without attacking C–C bond in glycerol. 1,2-Propanediol is mainly used in the production of pharmaceutical products, polymeric resins, cosmetic products, and paints. Previously, different types of reaction schemes (gas or liquid phases) and pathways were reported for the hydrogenation of glycerol to 1,2-propanediol. Casale and Gomez (1993) patented a technique for hydrogenation of glycerol using copper and zinc catalysts in a temperature range of 240–270 °C and at a pressure of 150 bar. The requirement of high temperature and pressure conditions, low production efficiency, and low selectivity to 1,2-propanediol limits the application of this method.

Several studies have been reported in the literature on usage of various transition metal catalysts such as Ni, Pt, Rh, Ru, Cu, and bimetallic systems (Pt-Ru). Ruthenium has been reported as the most active catalyst for hydrogenation of glycerol. The design of a bifunctional catalyst containing acidic/basic and metal active sites is a crucial step for hydrogenation of glycerol to 1,2-propanediol. Platinum catalysts with hydrotalcite and MgO as support exhibited good conversion and higher 1,2-propanediol selectivity than Al_2O_3, H-ZSM5, and H-beta (Yuan et al. 2010).

Hamzah et al. (2012) prepared a catalyst by the combination of Ru/TiO_2 using bentonite–TiO_2 as the support. The activity of Ru/TiO_2 catalyst was found to be improved by the addition of bentonite due to proper dispersion of Ru particles on catalyst surface that resulted in high activity of Ru/TiO_2/bentonite for hydrogenolysis of glycerol. The mechanism of hydrogenolysis of glycerol to 1,2-propanediol depends on the reaction conditions (acidic/alkaline). In acidic conditions, hydrogenolysis of glycerol occurs in two steps, such as (a) dehydration of glycerol to acetol and (b) hydrogenation of acetol to 1,2-propanediol (Alhanash et al. 2008; Furikado et al. 2007; Miyazawa et al. 2007; Wang and Liu 2007). In alkaline conditions, the reaction proceeds in three steps, such as (a) formation of glyceraldehyde from dehydrogenation of glycerol, (b) formation of 2-hydroxyacrolein from dehydration of glyceraldehyde, and (c) formation of 1,2-propanediol from hydrogenation of 2-hydroxyacrolein (Feng et al. 2007, 2008; Maris and Davis 2007).

16.4 Dehydration of Glycerol

Acrolein is a key intermediate in the chemical synthesis industry and can be manufactured from the dehydration of glycerol. The transformation of glycerol to acrolein is a double dehydration reaction that proceeds via the formation of 3-hydroxypropionaldehyde followed by 1-hydroxyacetone. Generally, liquid-phase dehydration and gas-phase dehydration are used for the transformation of glycerol to acrolein. The poor conversion of glycerol (15–25%) and low catalytic activity in the liquid-phase processes limit its practical applications. Various studies have been reported on the application of subcritical and supercritical conditions to improve the catalytic activity in liquid-phase reactions. However, the application of severe reaction conditions during subcritical and supercritical reactions leads to an increase in corrosiveness of the reactor material and creates safety concerns (Ginjupalli et al. 2014; Dalil et al. 2015; Alhanash et al. 2010). The gas-phase reactions are progressively used for the conversion of glycerol to acrolein due to their high conversion efficiency (65–99%) and high selectivity (Gholami et al. 2014).

Various solid acid catalysts including metal sulfates, metal oxides, metal phosphates, heteropolyacids (HPAs), and zeolites have been studied for gas-phase reactions. The acid strength and textural properties of these catalysts determine their efficiency and selectivity toward acrolein synthesis. Ginjupalli et al. (2014) found that the catalytic activity in dehydration process was greatly affected by surface Lewis/Brønsted acid sites of the catalysts. Lewis acid catalyst sites show high selectivity for acetol due to the participation of –OH terminal of glycerol, while Brønsted acid sites show higher selectivity for acrolein due to the participation of the secondary –OH group of glycerol (Alhanash et al. 2010). Furthermore, strong acidity of the catalyst promotes catalyst deactivation due to the deposition of carbonaceous species in its micropores resulting in low activity on the catalyst surface (Ginjupalli et al. 2014).

The presence of oxygen promotes the formation of oxidation products like formic acid, acetic acid, propionic acid, and carbon oxides (Dalil et al. 2015). The reaction mechanism of glycerol dehydration in presence of WO_3-TiO_2 catalyst involves various steps as shown in Fig. 16.3 such as:

1. Production of hydroxyacetone and 3-hydroxypropanal through loss of water
2. Formation of acrolein from dehydration of 3-hydroxypropanal
3. Retroaldol reaction of 3-hydroxypropanal leading to the production of formaldehyde and acetaldehyde
4. Formation of cyclic ethers from the reaction of formaldehyde with oxygen

16.5 Carboxylation of Glycerol

Carboxylation of glycerol produces glycerol carbonate and glycidol as the main reaction product. Glycerol carbonate has various applications in chemical, pharmaceutical, agricultural, semiconductor, and cosmetic industry. The synthesis pathways and applications of glycerol carbonate are summarized in Fig. 16.4. It can be produced from glycerol using direct and indirect pathways. Direct pathway involves

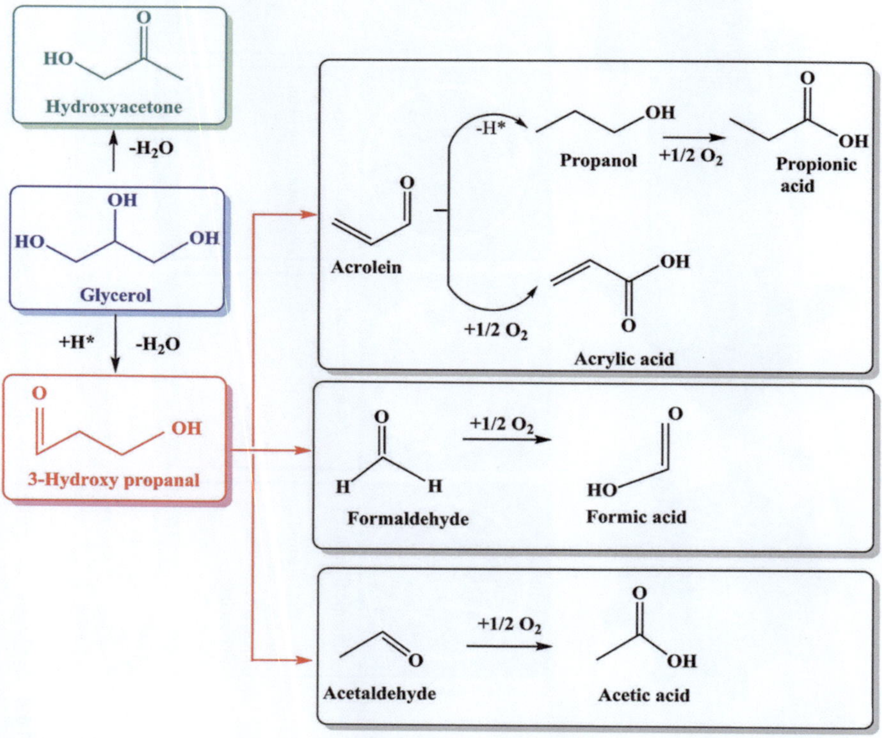

Fig. 16.3 Reaction mechanism for dehydration of glycerol using WO_3-TiO_2 catalyst

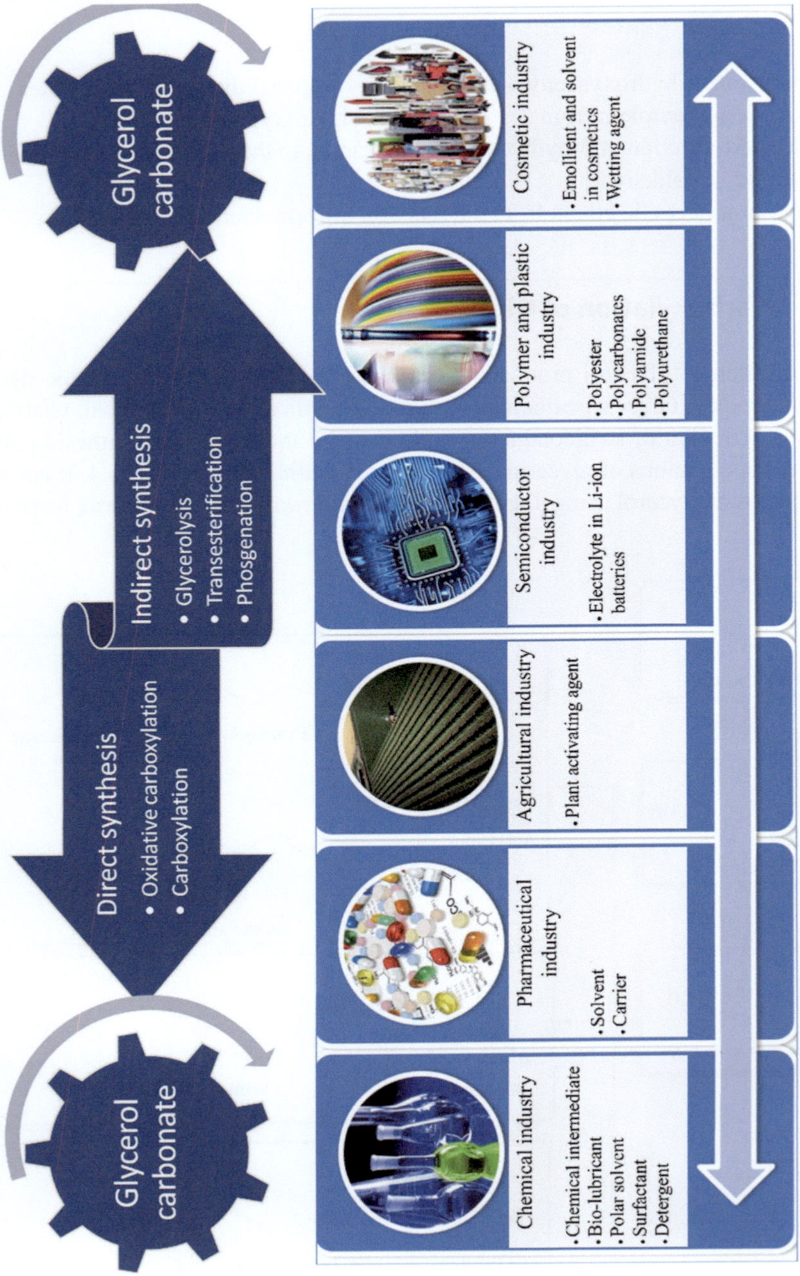

Fig. 16.4 Synthesis and applications of glycerol carbonate

carboxylation of glycerol using alkyl carbonates, urea, or CO_2 as the carbonate source, while indirect pathway includes glycerolysis, transesterification, and phosgenation reaction.

Carboxylation of glycerol to glycerol carbonate using CO_2 is carried out at elevated temperature and pressure conditions in presence of catalyst (Aresta et al. 2006). The direct reaction of glycerol with urea is another method, but the requirement of strict reaction conditions (use of vacuum to evolve ammonia to start reaction) and the production of side products (isocyanic acid and biuret) are the main disadvantages associated with this process (Aresta et al. 2009). Many studies have reported dimethyl carbonate as an alternative carboxylating agent (Table 16.4). Carboxylation with dimethyl carbonate can be performed at milder conditions without the production of problematic side products (Simanjuntak et al. 2011). The synthesis of glycerol carbonate using indirect pathways like phosgenation and transesterification is reported as the most industrially viable routes to achieve high glycerol carbonate yield.

Several studies have reported the transformation of glycerol to glycerol carbonate via transesterification using dialkyl carbonate (Kumar et al. 2012; Pan et al. 2012; Liu et al. 2013). A wide range of catalysts (homogeneous or heterogeneous and acidic or alkaline) has been investigated for transesterification of glycerol. Based on the literature, low catalytic activities for glycerol transesterification were achieved using homogeneous and heterogeneous acidic catalysts due to the mass transport limitations. On the other hand, homogeneous and heterogeneous basic catalysts show high activity for transformation of glycerol to glycerol carbonate and more than 90% glycerol carbonate yield. The main limitation of transesterification process is the formation of glycidol due to decarboxylation of glycerol carbonate (Devi et al. 2018; Algoufi et al. 2014). In addition, the maintenance of optimum reaction conditions and design of suitable catalysts are the key parameters to achieve high glycerol carbonate yield.

16.6 Acetylation of Glycerol

Acetylation of glycerol using acetic acid is gaining special attention due to the high industrial importance of the final products. The products formed from glycerol acetylation are monoacetylglycerol (monoacetin or MAG), diacetylglycerol (diacetin or DAG), and triacetylglycerol (triacetin or TAG). Monoacetin is basically used in the preparation of explosives, smokeless powder, and cosmetics. The mixture on monoacetin, diacetin, and triacetin can be used in printing inks, plasticizers, and softening agents. Triacetin is of most industrial importance due to its application as an antiknock agent to improve the viscosity and cold flow properties of the biodiesel (Khayoon et al. 2014). Additionally, triacetin meets the standards specification (EN14214 and ASTM D6751) for flash point and oxidation stability, which makes it product of high industrial importance (Gracia et al. 2009).

It is observed from the literature that strong acid catalysts perform best in acetylation of glycerol using acetic acid. Generally, the homogeneous catalysts like sulfuric acid, hydrofluoric acid, or *p*-toluenesulfonic acid are used in acetylation

Table 16.4 Reaction conditions for conversion of glycerol to glycerol carbonate

Catalyst	Reaction parameters				Conversion (%)	Solvents	References
	Temperature (°C)	Reaction time (h)	Molar ratio	Catalyst loading			
Amberlyst 39 wet	75	1.5	5:1	6 wt%	6.2	DMC	Ochoa-Gómez et al. (2009)
CaO	75	1.5	5:1	10 mol%	91.1		
$CaCO_3$	75	1.5	5:1	10 mol%	90.6		
CaO	60	2	1:1	2 wt%	69	DMC	Li and Wang (2011)
Calcium complex $Ca(C_3H_7O_3)_2$	60	3	2.5:1	8 mol%	95		
CaO	75	0.5	2:1	3 wt%	90.2	–	Simanjuntak et al. (2011)
Na_2O	75	0.5	2:1	3 mol%	92.6		
ZnO	75	0.5	2:1		0.5		
MgO	75	3	2:1		10.2		
CaO	35	1	2:1	0.5 wt%	81	EC	Climent et al. (2010)
Al/Mg hydrotalcite	35	1	2:1	0.5 wt%	57	–	
Al/Li hydrotalcite	35	1	2:1	0.5 wt%	85	–	
Al/Ca hydrotalcite	35	1	2:1	0.5 wt%	87		
K_2CO_3	73–75	3	3:1	4.5 wt%	97	DMC	Rokicki et al. (2005)
Mg/Zr/Sr mixed oxides	90	1.5	5:1	0.1 wt%	94		Malyaadri et al. (2011)
Mg/Al/Zr mixed oxides	75	1.5	5:1	0.1 wt%	94		
Mg/Al hydrotalcite	100	1	5:1	54 wt%	75	DMF	Takagaki et al. (2010)
Mg/ Al hydrotalcite–hydromagnesium	100	1.16	5:1	54 wt%	88	–	Kumar et al. (2012)
Mg/Al hydrotalcite	100	2	3:1	10 wt%	65	–	Liu et al. (2013)
Mg/Al hydrotalcite	130	8	21:1	2.16 g/g	84	DMSO	Álvarez et al. (2012)
Zeolite (NaY)	70	4	3:1	54 wt%	77	DMF	Pan et al. (2012)
KF/hydroxyapatite	78	0.83	2:1	3 wt%	99	–	Bai et al. (2011)
K_2CO_3/MgO	80	2	2.5:1	1 wt%	99	–	Du et al. (2012)

Abbreviations: dimethyl carbonate *DMC*, ethylene carbonate *EC*, dimethyl sulfoxide *DMSO*, dimethylformamide *DMF* reactions (Kale et al. 2015). However, there are certain limitations associated with the use of such catalysts due to their hazardous and corrosive properties. Recently, many studies have reported the use of solid acid catalyst for acetylation of glycerol, namely sulfated zirconia, sulfated activated carbon, sulfated mesoporous silica, ionic liquids, heteropolyacids, and ion exchange resins (Table 16.5). However, certain limitations on the wide industrial application of these catalysts are related to the functionalization of silicates catalysts, which is difficult and costly, and they have low surface area, low thermal stability of zeolites and solubility of heteropolyacids in polar media (Khayoon et al. 2014; Melero et al. 2007).

It is widely reported that acetylation of glycerol is an equilibrium reaction. In addition, acetic acid is required in excess quantities to maximize the selectivity for triacetin formation. Most of the studies reported low triacetin selectivity, and the reaction mixture obtained from acetylation reaction contains all three types of acetin (monoacetin, diacetin, and triacetin) (Melero et al. 2007; Zhu et al. 2013). Rezayat and Ghaziaskar (2009) achieved 100% selectivity for triacetin using Amberlyst-15 as the catalyst at very high glycerol-to-acetic acid molar ratio (1:24) and 200 bar pressure. However, such high-pressure conditions with extremely high molar ratio are not economically viable for industrial applications. On the contrary, Liao et al. (2009) performed the acetylation in the presence of excess of glycerol (as glycerol is a low-value product) using ion exchange resins as the catalysts. Several authors worked on the hypothesis that equilibrium of the reaction shifts toward the products by using an excess of glycerol, instead of an excess of acetic acid. Kale et al. (2015) suggested that the addition of some external component (entrainer) could be helpful to increase the selectivity of triacetin as the stoichiometric reaction product. It was found that the removal of water using azeotropic distillation aids to achieve high yield of triacetin.

Several studies have claimed better selectivity of triacetin in presence of acetic anhydride as acylating reagent for acetylation of glycerol. Liao et al. (2009) reported 100% selectivity for triacetin with the Amberlyst-35 in the presence of acetic anhydride. Similarly, Silva et al. (2011) reported 100% selectivity for triacetin using glycerol-to-acetic anhydride molar ratio of (1:4) over zeolite, Amberlyst-15, and niobium phosphate at 60 °C after 2 h. However, the use of acetic anhydride as the acylating agent is not viable for industrial applications due to its higher cost and associated health hazards (Silva et al. 2011).

16.7 Etherification of Glycerol

In etherification process, branched oxygen-containing compounds are formed from the stepwise reaction occurring between the hydroxyl groups of glycerol and the functional reactant groups such as alcohols (e.g., tert-butyl alcohol) or alkenes (isobutene). The product obtained from etherification reaction contains a mixture of monoglyceride, diglyceride, and triglyceride due to the variation in reactivity of hydroxyl groups of glycerol (Fig. 16.5). The oxygenate compounds obtained from etherification of glycerol can be used as fuel additive. Glycerol tert-butyl ethers can

Table 16.5 Reaction conditions for acetylation of glycerol

Catalyst	Reaction parameters				Conversion (%)	Selectivity (%)			References
	Temperature (°C)	Reaction time (h)	Molar ratio	Catalyst loading		Monoacetin	Diacetin	Triacetin	
Amberlyst-36	105	10	1:8	0.25 g	95.6	70.3	4.5		Dosuna-Rodríguez and Gaigneaux (2012)
Amberlyst-15					95.3	70.3	2.5		
Dowex resin-2					95.2	80.8	5.1		
Dowex resin-4					94.8	71.6	4.2		
Dowex resin-8					94.7	72.9	4.7		
Ar-SBA-15	125	4	9:1	0.2 g	96	15	47	38	Melero et al. (2007)
F-SBA-15					90	14	50	36	
Pr-SBA-15					80	17	44	39	
Amberlyst-15	80	8	6:1	0.64 g	100	88.5	11.2	0.3	Kim et al. (2014)
$HPMo/Nb_2O_5$					87	81.8	17.5	0.7	
SCZ					81	84.7	14.8	0.5	
SO_3H-SBA15					100	11.1	61.9	27.0	
SO_3H-cell					100	37.6	55.0	13.4	
TPA_3/ZrO_2	100	6	6:1	0.15 g	80	60	36	4	Patel and Singh (2014)
TPA_3/ MCM-41					87	25	60	15	
Amberlyst-15	105	10	6:1	5 wt%	100	0	12.3	83.9	Kale et al. (2015)
Amberlyst-70					100	0	7.5	87.6	
STA/S11					100	1	55.5	35.8	
Amberlyst-15	110	0.5	3:1	0.47 g	97	31	54	13	Gonçalves et al. (2008)
K-10				4.0 g	96	44	49	5	
Niobic acid				6.3 g	30	83	-	-	
HZSM-5				1.6 g	30	83	10	0	
PW2_AC	120	3	16:1	0.2 g	86	25	63	11	Ferreira et al. (2010)

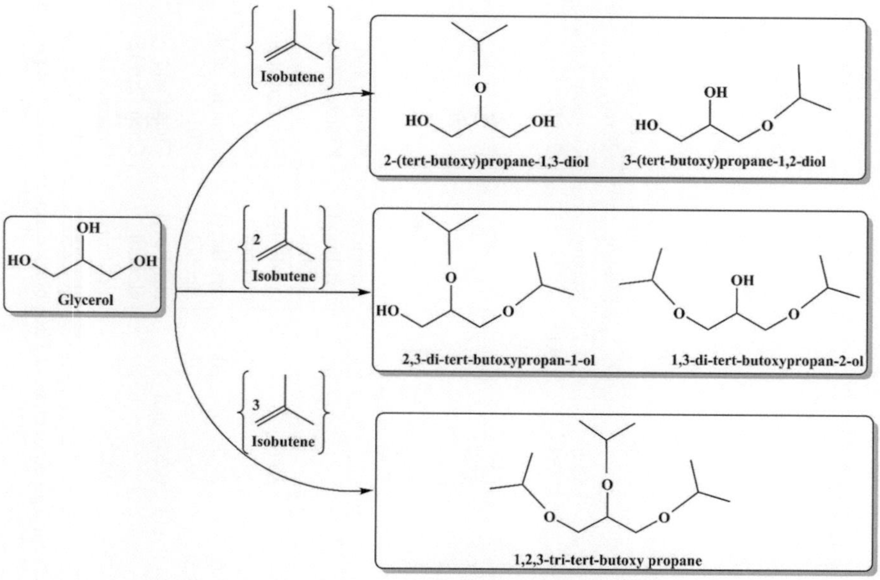

Fig. 16.5 Etherification of glycerol using isobutene

be used as an alternative to ethyl tert-butyl ether as it helps to improve octane properties of gasoline. Other reaction products like di-tert-butylglycerols (DTBG) and tri-tert-butylglycerol (TTBG) are reported to improve cold flow properties of the fuel as valuable fuel additives to reduce the toxic emissions.

Homogeneous and heterogeneous catalysts (e.g., zeolites, mesoporous silica, and metal oxides) can catalyze glycerol etherification. Generally, homogeneous catalysts are not preferred due to their corrosive nature and difficulty to impossible in recovery from reaction mixture and hence heterogeneous catalysts are preferred. Heterogeneous acid catalysts including zeolites, ion exchange resins, and mesoporous silicates are reported to be more active for etherification of glycerol (Izquierdo et al. 2012). However, the application of heterogeneous catalysts is limited due to their low selectivity and formation of secondary products. On the contrary, the high product selectivity and higher etherification activity can be achieved using basic heterogeneous catalysts (Table 16.6). Therefore, basic catalysts are usually preferred compared to the acidic catalyst (Melero et al. 2008; Klepacova et al. 2005).

The effects of different solvents (e.g., isobutene, dioxane, sulfolane, and dimethyl sulfoxide) were investigated on etherification reaction, and highest conversion was achieved using isobutene as solvent. Although isobutene is efficient in the etherification of glycerol, its high cost and low selectivity and formation of secondary products (e.g., isobutene oligomerization) are some disadvantages of this process. Etherification of glycerol using tert-butyl alcohol is widely reported in the literature (Frusteri et al. 2009; Chang and Chen 2011; Ozbay et al. 2010). In addition, oligomerization of isobutene can be prevented by the addition of tert-butyl alcohol to the reaction mixture (Karinen and Krause 2006; Melero et al. 2008). However,

Table 16.6 Reaction conditions for etherification of glycerol

Catalyst	Reaction parameters				Conversion (%)	Selectivity (%)			Solvents	References
	Temperature (°C)	Reaction time (h)	Molar ratio	Catalyst loading		X_{MTBG}	X_{DTBG}	X_{TTBG}		
Amberlyst-15	75	4	0.25	0.5 g	73	65	32	3	Isobutene	González et al. (2012)
SBA-15					41	87	13	0		
SBA-15-Cs					98	39	56	5		
SBA-15-Mw-S					99	9	55	36		
BCC-S	393 K	4	4:1	5 wt%	70	42	20 (+)		tert-Butyl alcohol	Gonçalves et al. (2016)
Amberlyst-15	90	8	4:1	7.5 wt%	100	8.5	59.6	21.2	Isobutylene	Klepacova et al. (2005)
Amberlyst-35					100	7.6	59.3	29.8		
Zeolite H-Y					94.8	22.5	64.9	7.1		
Zeolite H-beta					100	12.7	72.9	–		
Amberlyst-15	60	8	4:1	7.5 wt%	95.9	13.2	55.8	16.6		
Amberlyst-35					80.1	23.4	46.9	6.8		
Zeolite H-Y					100	16.2	57.3	14.3		
Zeolite H-beta					18.0	6.9	16.3	–		
Amberlyst-35[a]	80	7	3	1 g	95				Isobutene	Karinen and Krause (2006)
	60	8	4	7.5 wt%	86				tert-Butyl alcohol	
Spherical silica-supported perfluorosulfonic acid	70	6	4	5.0 wt%	99		85 (+)		tert-Butyl alcohol	Beatrice et al. (2013)
A-15	70	6	4	7.5 wt%	78.9	71.6	27.7	0.5	tert-Butyl alcohol	Cannilla et al. (2014)
H-730/ES70Y					34.1	79.8	19.3	0.9		
Amberlyst-15	89.4	8	4	7.3 wt%	72.3	0.509 mol	0.205 mol		tert-Butyl alcohol	Chang and Chen (2011)
Propyl-SO_3H/SiO_2	85-95	3	1:10	10 wt%	90	60–70	20–30		tert-Butyl alcohol	Drago et al. (2013)

Abbreviation: mono-tert-butyl glycerol ethers *MTBG*, di-tert-butyl glycerol ethers *DTBG*, tri-tert-butyl glycerol ethers *TTBG*, black carbon from coffee ground *BCC* treated with sulfuric acid

[a]Reaction was performed at different reaction conditions using the same catalyst

tert-butyl alcohol-based etherification process has low selectivity for triethers when compared to isobutene-based etherification process.

16.8 Acetalization of Glycerol

Acetalization of glycerol to ketal or acetal can be performed with aldehyde and ketones. Highly branched oxygenated compounds with five-membered ring ketal (2,2-dimethyl-[1,3]-dioxane-4-yl)-methanol and six-membered ring acetal (2,2-dimethyl-dioxane-5-ol) are generally produced from acetalization of glycerol using ketones. Acetals can be used as base in surfactant and as a precursor for manufacturing of 1,3-dihydroxyacetone and 1,3-propanediol (Umbarkar et al. 2009). It is observed from the literature that glycerol acetalization reaction mixture contains both five- and six-membered ring acetals. The fraction of five- and six-membered acetals in acetalization reaction mixture can be controlled by varying the several reaction parameters like temperature, aldehyde-to-glycerol molar ratio or ketone-to-glycerol molar ratio, and the type of solvent used. Cesar et al. (2002) reported that 100% selectivity for five- and six-membered ring acetals is not achieved even under supercritical conditions using acetone as the acetalization reagent. In contrast, very high selectivity for six-membered ring acetals can be achieved using aldehyde-based acetalization at higher temperatures (Pagliaro et al. 2007).

Acetalization of glycerol can be performed in the presence of homogeneous catalysts like strong acids (sulfuric acid, hydrofluoric acid, hydrochloric acid, and *p*-toluenesulfonic acid) (Frusteri et al. 2009). Kaufhold and El-Chahawi (1996) patented a procedure for acetal production using homogeneous catalyst in the presence of water-soluble entrainer (hexane or pentane) in the temperature range of 25–75 °C. Acetalization of glycerol is a reversible process; therefore, water is continuously removed from reaction mixture to ensure the shift of reaction in the desired direction. However, this process is not economically and environmentally feasible due to the corrosive nature of acid catalyst.

Recently, many studies have reported the application of heterogeneous acid catalysts to overcome the limitations associated with homogeneous catalysts (Khayoo and Hameed 2013). Umbarkar et al. (2009) investigated the application of MoO_3/SiO_2 catalyst prepared by sol–gel techniques for acetalization of glycerol using aldehydes and obtained 78% conversion with very high selectivity (up to 100%) for six-membered acetal. Güemez et al. (2013) reported the conversion of glycerol using *n*-butyraldehyde in the presence of Amberlyst-47 catalyst and obtained mixtures of cyclic acetals as the main product. The selectivity of acetalization process mainly depends on the actualization position of hydroxyl group in glycerol molecule, which in turn depends on electronic and steric factors (Ferreira et al. 2010). The typical reaction mechanism for acetalization of glycerol involves the following steps: (a) formation of hemiketal, (b) removal of water molecule through dehydration reaction, (c) formation of tertiary carbonium ion, and (d) formation of ketal or acetal.

16.9 Polymerization of Glycerol

Polymerization of glycerol to polyglycerol is one way to expand the industrial applications of glycerol. Polyglycerols are viscous, water-soluble, highly branched compounds and can be synthesized from glycerol using two routes, especially (a) etherification of glycerol and (b) glycerol condensation in the presence of excess water (Gholami et al. 2014; Vila et al. 2012). The properties of polyglycerol esters depend on the degree of esterification, length of polyglycerol chain, and molecular weight of fatty acid (Clacens et al. 2002). Generally, it is difficult to control the selectivity of polymerization process as the products contain mixture of polyglycerols (monomers to hexamers). Diglycerol is a simplest polyglycerol, which is formed by the condensation of two glycerol molecules. Furthermore, progressive condensation of glycerol molecules can lead to the formation of triglycerol, tetraglycerol, pentaglycerol, and long-chain polyglycerol. Polyglycerols have many industrial applications from the cosmetic industry to drug delivery. Polyglycerol esters can be used in cosmetics, lubricants, food additive and as a surfactant (Guerrero-Urbaneja et al. 2014). The polymerization of glycerol to polyglycerols using different catalysts is widely reported in literature (Table 16.7).

Traditionally, epichlorohydrin is used as a starting material for the preparation of polyglycerols in industrial processes, but the time-consuming multiple purification steps make it an expensive process (Pérez-Barrado et al. 2015). Therefore, researchers are moving toward the alternative processes like catalytic etherification of glycerol. Many studies have been reported in literature on polyglycerol synthesis using different homogeneous (alkaline carbonates, Cs_2CO_3 and Na_2CO_3, and hydroxides, NaOH, CsOH, and H_2SO_4) and heterogeneous catalysts (metal-impregnated MCM and zeolites and mixed oxides). Catalytic activity of Na_2CO_3 in glycerol is better than hydroxides due to its better solubility and high basicity (Bookong et al. 2015; Barrault et al. 2004). Although homogeneous catalysts are more active than heterogeneous catalyst, low selectivity, corrosive nature, and difficulty in separation are the factors that limit their applications.

The heterogeneous catalysts (mesoporous materials) have high selectivity for low molecular weight oligomers like diglycerol and triglycerol due to their smaller pore size. Several studies have been reported by the Barrault group on the use of zeolitic and mesoporous catalysts for etherification of glycerol to polyglycerols. Polyglycerols production from glycerol etherification involves complex reaction, which makes difficult to control the degree of polymerization and leads to the formation of undesirable side products. Generally, the formation of toxic acrolein is reported at temperatures >200 °C due to double dehydration reactions (Ruppert et al. 2008; Guerrero-Urbaneja et al. 2014). The mechanism of glycerol etherification comprises the following steps: (a) protonation of –OH group of first glycerol molecule, (b) protonated glycerol molecule nucleophilic attack hydroxyl group on other glycerol molecules; (c) protonated glycerol molecule combined with other glycerol molecules, and (d) release of water molecule.

Table 16.7 Reaction conditions for polymerization of glycerol

Catalyst	Reaction parameters				Selectivity (%)		References
	Temperature (°C)	Reaction time (h)	Catalyst loading	Conversion (%)	Diglycerol	Triglycerol	
MgAl-Na	220	24	300 mg	50	85	15	García-Sancho et al. (2011)
Cs (exchanged)	260	8		51	83	17	Clacens et al. (2002)
CsX (impregnated)				36	100	0	
$Cs_{25}Al(20)$				80	55	25	
Cs-impregnated MCM-41	260		2 wt%	80	75	25	Clacens et al. (2000)
La-incorporated MCM-41				90	40	23	
Cs-exchanged X-zeolite				80	65	20	
Amberlyst-15	60	8		68			Tseng and Wang (2011)
Amberlyst-35				70.9			
Zeolite H-beta				64.9			
Zeolite H-Y				88			

16.10 Esterification of Glycerol

The production of polyglycerides using esterification of glycerol offers many promising industrial applications (Table 16.8). Particularly, monoglycerides are used as food additive in food industry, personal care formulations, and pharmaceutical and chemical industries. Generally, esterification reactions are performed out in homogeneous liquefied phase in a batch reactor. Acid catalysts (sulfuric acid, hydrochloric acid, and orthophosphoric acid) are used for the direct esterification of glycerol with fatty acids at mild reaction temperature (Molinero et al. 2014).

Several reports are available on the use of microporous zeolites, mesoporous silica, and porous carbon materials as catalysts for esterification of glycerol. Mesoporous materials are found to be more efficient compared to microporous zeolites in esterification of glycerol because the pores of mesoporous materials are easily assessable by fatty acids and their esters (Sanchez et al. 2011). The porous carbon materials produced from starch and polysaccharides are proven to be more recalcitrant to the structural changes compared to mesoporous silica. Generally, mesoporous are treated with alkyl sulfonic acid to impart acidic characters to the catalyst. The presence of acidic R–SO_3H groups is reported to increase the catalytic activity of mesoporous silicates for esterification of glycerol (Jeenpadiphat et al. 2015). The concentration of methyl groups can affect the performance of the catalyst for each fatty acid, and higher catalytic activity can be achieved by tuning the methyl-to-sulfonic acid ratio (Díaz et al. 2000). The optimum balance between different parameters like catalyst acidity, pore size, specific surface area, and

Table 16.8 Value-added products from esterification of glycerol and their application

Products	Structure	Applications
Monoglycerol ester	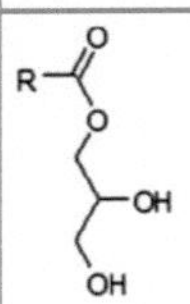 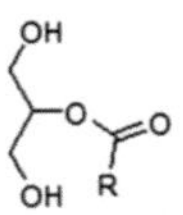	• Food industry • Cosmetic industry • Pharmaceutical and chemical industries
Diglycerol ester	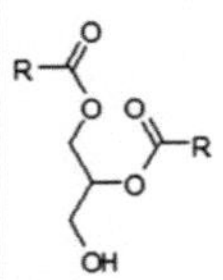 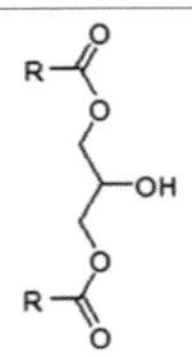	• Personal care formulations and fragrances • Plasticizer in polyvinyl alcohol films • Manufacture of polyurethanes and polyesters
Triglycerol ester	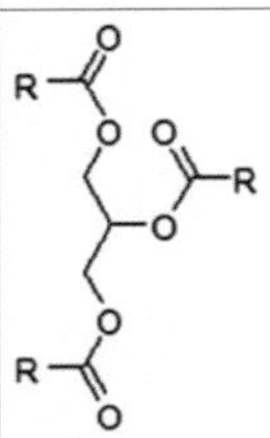	• Cosmetic industry • Food industry

methyl-to-sulfonic group ratio is necessary to obtain catalysts with improved catalytic activity and selectivity for the desired products.

Various factors like surface area, adsorption, glycerol/fatty acid molar ratio, swelling ratio, and catalyst acidity have strong effects on the catalyst activity and selectivity toward the desired product (Pouilloux et al. 1999; Kotwal et al. 2011). Water is generally considered as a limiting factor in esterification reaction. Therefore, molecular sieve can be used as a water trap to increase the rate of reaction. Moreover, adsorption and swelling pattern are also found to influence the reaction rate. Zhou et al. (2013) reported that the length of fatty acid chain and the steric effect influence the selectivity of the process and rate of reaction.

16.11 Chlorination of Glycerol

Chlorination of glycerol to dichlorohydrins and epichlorohydrin is an interesting alternative. Epichlorohydrin has many industrial applications like chemical synthesis of resins and synthetic elastomers, water purification process, and, in reinforcement, intermediate for the synthesis of oligoglycerols and sizing agent in paper industry. Several studies have reported the chlorination of glycerol to dichlorohydrin and subsequently to epichlorohydrin with gaseous hydrochloric acid (Santacesaria et al. 2010). The products obtained from chlorination reaction are 1,3-chloro-2-propanol, 1,2-chloro-3-propanol, 1-chloro-2,3-pro-panediol, and 2-chloro-1,3-propanediol (Tesser et al. 2012). The mechanism of epichlorohydrin synthesis involves two chlorination steps such as (a) chlorination of glycerol to 1-chloro-2,3-propanediol and 2-chloro-1,3-propanediol and (b) chlorination reaction leading to the formation of 1,3-chloro-2-propanol and 1,2-chloro-3-propanol.

The chlorination reaction of glycerol and hydrochloric acid is traditionally catalyzed by acetic acid, but the use of acetic acid as a catalyst has some disadvantages. Acetic acid is volatile under high temperatures. Therefore, acetic acid is required to be replaced by less volatile catalysts (e.g., glycolic acid, amino acid, and chlorotrimethylsilane) to perform reactions at a higher temperature. Vitiello et al. (2014) used a series of glycolic acid catalysts and amino acid series catalysts for chlorination of glycerol. The performance and selectivity of catalysts were found to depend on pK_a value and molecular structure (number of carboxylic groups) of the catalyst. The glycolic acid series catalyst performed better than amino acid series catalysts in terms of activity and selectivity. Escribà et al. (2009) investigated the production of chlorohydrin esters using esterification–chlorination of glycerol using chlorotrimethylsilane as the catalyst in the presence of classical or microwave heating. The results showed that 2-chloro-1-(chloromethyl)ethyl esters can be obtained in high quantity even at lower reaction temperatures without using any solvent addition.

16.12 Ammoxidation of Glycerol

Ammoxidation of glycerol is performed to produce acrolein and acrylonitrile. Based on the number of steps involved, ammoxidation can be divided into two types, such as direct ammoxidation and indirect ammoxidation. Acrolein can be directly produced from ammoxidation of glycerol. However, acrylonitrile is indirectly produced from glycerol dehydration of acrolein followed by ammoxidation of acrolein to acrylonitrile (Liebig et al. 2013). Usually, high temperature (270 °C) is required for ammoxidation process.

Mono-oxides and mixed metal oxide-based catalysts (e.g., Mo, Sn, W, Zr, Al, P, Ti, and Sn) are mainly used for the manufacture of acrylonitrile via ammoxidation of glycerol (Bagheri et al. 2015). Liebig et al. (2013) performed the indirect ammoxidation of glycerol to acrylonitrile using dehydration and ammoxidation steps. WO_3/TiO_2 catalyst was used for dehydration step, while Sb/V and Sb/Fe catalysts were used for ammoxidation step. Sb/Fe catalysts were found to be highly selective for acrylonitrile. The selectivity of acrylonitrile was found to rise with an increase in reaction time due to the formation of $FeSbO_4$ during the reaction. Similarly, alumina-supported Sb/V catalyst was found to be highly active for the conversion of glycerol into acrylonitrile under solvent-free microwave activation.

The acrylonitrile selectivity of more than 80% can be achieved using Sb/V catalyst in microwave activation within short reaction times compared to conventional thermal activation (Calvino-Casilda et al. 2009). It is reported that various kinds of Sb–V–O structures can be obtained with the variation in Sb–V–O loading and the method of impregnation. The oxidation state of the vanadium species in these Sb–V–O structures significantly affects the catalytic behavior and selectivity toward acrylonitrile (Golinska et al. 2010). The metal loading, chemical composition, and morphology of the support are the key parameters affecting catalytic activity in ammoxidation process.

16.13 Conclusions

Glycerol has emerged as a low-cost biodiesel industry by-product that has great potential to be transformed into marketable value-added products. This chapter highlights different potential reaction pathways for the catalytic conversion of glycerol into value-added marketable fine chemicals. Interestingly, different kinds of catalysts (homogeneous and heterogeneous) are explored for the conversion of glycerol in a wide range of studies. However, few catalysts have shown long-term stability without severe deactivation. Many researchers have provided information about the optimization of experimental conditions to obtain high glycerol conversion and good product selectivity, but majority of these experiments were limited to batch reaction systems.

The time and temperature profiles of the catalytic reaction and product yield can be significantly improved by the application of advanced reactors like ultrasonic, microwave, membrane reactor, etc. A few studies are available on the usage of crude glycerol for the manufacture of high market value chemicals. The utilization of crude glycerol can provide new prospects on glycerol waste minimization and can help in potential commercialization and industrialization of these processes.

Acknowledgments The authors are thankful for the financial support provided by the Agriculture Development Fund (ADF), Saskatchewan and Canada Research Chair program.

References

Algoufi YT, Akpan UG, Asif M, Hameed BH (2014) One-pot synthesis of glycidol from glycerol and dimethyl carbonate over KF/sepiolite catalyst. Appl Catal A 487:181–188

Alhanash A, Kozhevnikova EF, Kozhevnikov IV (2008) Hydrogenolysis of glycerol to propanediol over Ru: polyoxometalate bifunctional catalyst. Catal Lett 120:307–311

Alhanash A, Kozhevnikova EF, Kozhevnikov IV (2010) Gas-phase dehydration of glycerol to acrolein catalysed by caesium heteropoly salt. Appl Catal A Gen 378:11–18

Álvarez MG, Plíšková M, Segarra AM, Medina F, Figueras F (2012) Synthesis of glycerol carbonates by transesterification of glycerol in a continuous system using supported hydrotalcites as catalysts. Appl Catal B Environ 113–114:212–220

Aresta M, Dibenedetto A, Nocito F, Pastore C (2006) A study on the carboxylation of glycerol to glycerol carbonate with carbon dioxide: the role of the catalyst, solvent and reaction conditions. Atmos Environ 41:407–416

Aresta M, Dibenedetto A, Nocito F, Ferragina C (2009) Valorization of bio-glycerol: new catalytic materials for the synthesis of glycerol carbonate via glycerolysis of urea. J Catal 268:106–114

Azargohar R, Nanda S, Kang K, Bond T, Karunakaran C, Dalai AK, Kozinski JA (2019) Effects of bio-additives on the physicochemical properties and mechanical behavior of canola hull fuel pellets. Renew Energy 132:296–307

Bagheri S, Julkapli NM, Yehye WA (2015) Catalytic conversion of biodiesel derived raw glycerol to value added products. Renew Sust Energ Rev 41:113–127

Bai R, Wang S, Mei F, Li T, Li G (2011) Synthesis of glycerol carbonate from glycerol and dimethyl carbonate catalyzed by KF modified hydroxyapatite. J Ind Eng Chem 17:777–781

Barrault J, Clacens J, Pouilloux Y (2004) Selective oligomerization of glycerol over mesoporous catalysts. Top Catal 27:137–142

Beatrice C, Di Blasio G, Lazzaro M, Cannilla C, Bonura G, Frusteri F, Asdrubali F, Baldinelli G, Presciutti A, Fantozzi F, Bidini G, Bartocci P (2013) Technologies for energetic exploitation of biodiesel chain derived glycerol: oxy-fuels production by catalytic conversion. Appl Energy 102:63–71

Bookong P, Ruchirawat S, Boonyarattanakalin S (2015) Optimization of microwave-assisted etherification of glycerol to polyglycerols by sodium carbonate as catalyst. Chem Eng J 275:253–261

Calcens JM, Pouilloux Y, Barrault J (2002) Selective etherification of glycerol to polyglycerols over impregnated basic MCM-41 type mesoporous catalysts. Appl Catal A Gen 227:181–190

Calvino-Casilda V, Guerrero-Pérez MO, Bañares MA (2009) Efficient microwave-promoted acrylonitrile sustainable synthesis from glycerol. Green Chem 11:939–941

Cannilla C, Bonura G, Frusteri L, Frusteri F (2014) Glycerol etherification with TBA: high yield to poly-ethers using a membrane assisted batch reactor. Environ Sci Technol 48:6019–6026

Casale B, Gomez AM (1993) Method of hydrogenating glycerol. US Patent 5:214–219

Cesar A, Ferreira DS, Barbe JC, Bertrand A (2002) Heterocyclic acetals from glycerol and acetaldehyde in port wines: evolution with aging. J Agric Food Chem 50:2560–2564

Chang JS, Chen DH (2011) Optimization on the etherification of glycerol with tert-butyl alcohol. J Taiwan Inst Chem Eng 42:760–767

Ciriminna R, Pina CD, Rossi M, Pagliaro M (2014) Understanding the glycerol market. Eur J Lipid Sci Technol 116:1432–1439

Clacens JM, Pouilloux Y, Barrault J (2000) Synthesis and modification of basic mesoporous materials for the selective etherification of glycerol. Stud Surf Sci Catal 143:687–695

Climent MJ, Corma A, Frutos PD, Iborra S, Noy M, Velty A, Concepción P (2010) Chemicals from biomass: synthesis of glycerol carbonate by transesterification and carbonylation with urea with hydrotalcite catalysts The role of acid–base pairs. J Catal 269:140–149

Dalil M, Carnevali D, Dubois JL, Patience GS (2015) Transient acrolein selectivity and carbon deposition study of glycerol dehydration over WO3/TiO_2 catalyst. Chem Eng J 270:557–563

Devi P, Das U, Dalai AK (2018) Production of glycerol carbonate using a novel Ti-SBA-15 catalyst. Chem Eng J 346:477–488

Díaz I, Márquez-Alvarez C, Mohino F, Pérez-Pariente J, Sastre E (2000) Combined alkyl and sulfonic acid functionalization of MCM-41-type silica. Part II Esterification of glycerol with fatty acids. J Catal 193:295–302

Dosuna-Rodríguez I, Gaigneaux EM (2012) Glycerol acetylation catalysed by ion exchange resins. Catal Today 195:14–21

Drago C, Liotta LF, La Parola V, Testa ML, Nicolosi G (2013) One-pot microwave assisted catalytic transformation of vegetable oil into glycerol-free biodiesel. Fuel 113:707–711

Du M, Li Q, Dong W, Geng T, Jiang Y (2012) Synthesis of glycerol carbonate from glycerol and dimethyl carbonate catalyzed by K_2CO_3/MgO. Res Chem Intermed 38:1069–1077

Escribà M, Eras J, Duran M, Simon S, Butchosa C, Villorbina G, Balcells M, Canela R (2009) From glycerol to chlorohydrin esters using a solvent-free system. Microwave irradiation versus conventional heating. Tetrahedron 65:10370–10376

Feng J, Wang J, Zhou Y, Fu H, Chen H, Li X (2007) Effect of base additives on the selective Hydrogenolysis of glycerol over Ru/TiO_2 catalyst. Chem Lett 36:1274–1278

Feng J, Fu H, Wang J, Li R, Chen H, Li X (2008) Hydrogenolysis of glycerol to glycols over ruthenium catalysts: effect of support and catalyst reduction temperature. Catal Commun 9:1458–1464

Ferreira P, Fonseca IM, Ramos AM, Vital J, Castanheiro JE (2010) Valorisation of glycerol by condensation with acetone over silica-included heteropolyacids. Appl Catal B Environ 98:94–99

Frusteri F, Arena F, Bonura G, Cannilla C, Spadaro L, Di Blasi O (2009) Catalytic etherification of glycerol by tert-butyl alcohol to produce oxygenated additives for diesel fuel. Appl Catal A General 367:77–83

Furikado I, Miyazawa T, Koso S, Shimao A, Kunimori K, Tomishige K (2007) Catalytic performance of Rh/SiO_2 in glycerol reaction under hydrogen. Green Chem 9:582–588

García-Sancho C, Moreno-Tost R, Mérida-Robles JM, Santamaría-González J, Jiménez-López A, Torres PM (2011) Etherification of glycerol to polyglycerols over MgAl mixed oxides. Catal Today 167:84–90

Gholami Z, Abdullah AZ, Lee KT (2014) Dealing with the surplus of glycerol production from biodiesel industry through catalytic upgrading to polyglycerols and other value-added products. Renew Sust Energ Rev 39:327–341

Ginjupalli SR, Mugawar S, Rajan N, Kumar Balla P, Chary Komandur VR (2014) Vapour phase dehydration of glycerol to acrolein over tungstated zirconia catalysts. Appl Surf Sci 309:153–159

Golinska H, Rojas E, López-Medina R, Calvino-Casilda V, Ziolek M, Bañares MA (2010) Designing new V–Sb–O based catalysts on mesoporous supports for nitriles production. Appl Catal A 380:95–104

Gonçalves VLC, Pinto BP, Silva JC, Mota CJA (2008) Acetylation of glycerol catalyzed by different solid acids. Catal Today 133–135:673–677

Gonçalves M, Soler FC, Isoda N, Carvalho WA, Mandelli D, Sepúlveda J (2016) Glycerol conversion into value-added products in presence of a green recyclable catalyst: acid black carbon obtained from coffee ground wastes. J Taiwan Inst Chem Eng 60:294–301

González MD, Cesteros Y, Llorca J, Salagre P (2012) Boosted selectivity toward high glycerol tertiary butyl ethers by microwave-assisted sulfonic acid-functionalization of SBA-15 and beta zeolite. J Catal 290:202–209

Gracia MD, Balu AM, Campelo JM, Luque R, Marinas JM, Romero AA (2009) Evidences of the in situ generation of highly active Lewis acid species on Zr-SBA-15. Appl Catal A 37:85–91
Güemez MB, Requies J, Agirre I, Arias PL, Barrio VL, Cambra JF (2013) Acetalization reaction between glycerol and n-butyraldehyde using an acidic ion exchange resin. Kinetic modelling. Chem Eng J 228:300–307
Guerrero-Urbaneja P, García-Sancho C, Moreno-Tost R, Mérida-Robles J, Santamaría-González J, Jiménez-López A, Maireles-Torres P (2014) Glycerol valorization by etherification to polyglycerols by using metal oxides derived from MgFe hydrotalcites. Appl Catal A Gen 470:199–207
Hamzah N, Nordin NM, Nadzri AHA, Nik YA, Kassim MB, Yarmo MA (2012) Enhanced activity of Ru/TiO_2 catalyst using bisupport, bentonite-TiO_2 for hydrogenolysis of glycerol in aqueous media. Appl Catal A Gen 419-420:133–141
Izquierdo JF, Montiel M, Palés I, Outón PR, Galán M, Jutglar L, Villarrubia M, Izquierdo M, Hermo MP, Ariza X (2012) Fuel additives from glycerol etherification with light olefins: state of the art. Renew Sust Energ Rev 16:6717–6724
Jeenpadiphat S, Björk EM, Odén M, Tungasmita DN (2015) Propylsulfonic acid functionalized mesoporous silica catalysts for esterification of fatty acids. J Mol Catal A Chem 410:253–259
Kale S, Umbarkar SB, Dongare MK, Eckelt R, Armbruster U, Martin A (2015) Selective formation of triacetin by glycerol acetylation using acidic ion-exchange resins as catalyst and toluene as an entrainer. Appl Catal A Gen 490:10–16
Karinen RS, Krause AOI (2006) New biocomponents from glycerol. Appl Catal A Gen 306:128–133
Kaufhold M, El-Chahawi M (1996) Process for preparing acetaldehyde diethyl acetal. US Patent 5527969
Khayoo MS, Hameed BH (2013) Solventless acetalization of glycerol with acetone to fuel oxygenates over Ni-Zr supported on mesoporous activated carbon catalyst. Appl Catal A Gen 464–465:191–199
Khayoon MS, Triwahyono S, Hameed BH, Jalil AA (2014) Improved production of fuel oxygenates via glycerol acetylation with acetic acid. Chem Eng J 243:473–484
Kim I, Kim J, Lee D (2014) A comparative study on catalytic properties of solid acid catalysts for glycerol acetylation at low temperatures. Appl Catal B Environ 148-149:295–303
Klepacova K, Mravec D, Bajus M (2005) Brønsted acidic ionic liquids: the dependence on water of the Fischer esterification of acetic acid and ethanol. Appl Catal A Gen 294:141–147
Kotwal M, Deshpande SS, Srinivas D (2011) Esterification of fatty acids with glycerol over Fe–Zn double-metal cyanide catalyst. Catal Commun 12:1302–1306
Kumar A, Iwatani K, Nishimura S, Takagaki A, Ebitani K (2012) Promotion effect of coexistent hydromagnesite in a highly active solid base hydrotalcite catalyst for transesterifications of glycols into cyclic carbonates. Catal Today 185:241–246
Li J, Wang T (2011) Chemical equilibrium of glycerol carbonate synthesis from glycerol. J Chem Thermodyn 43:731–736
Liao X, Zhu Y, Wang SG, Li Y (2009) Producing triacetylglycerol with glycerol by two steps: esterification and acetylation. Fuel Process Technol 90:988–993
Liebig C, Paul S, Katryniok B, Guillon C, Couturier JL, Dubois JL (2013) Glycerol conversion to acrylonitrile by consecutive dehydration over WO_3/TiO_2 and ammoxidation over Sb-(Fe, V)-O. Appl Catal B Environ 132–133:170–182
Liu P, Derchi M, Hensen EJM (2013) Synthesis of glycerol carbonate by transesterification of glycerol with dimethyl carbonate over MgAl mixed oxide catalysts. Appl Catal A Gen 467:124–131
Malyaadri M, Jagadeeswaraiah K, Sai Prasad PS, Lingaiah N (2011) Synthesis of glycerol carbonate by transesterification of glycerol with dimethyl carbonate over mg/Al/Zr catalysts. Appl Catal A Gen 401:153–157
Maris EP, Davis RJ (2007) Hydrogenolysis of glycerol over carbon-supported Ru and Pt catalysts. J Catal 249:328–337
Melero JA, Van Grieken R, Morales G, Paniagua M (2007) Acidic mesoporous silica for the acetylation of glycerol: synthesis of bioadditives to petrol fuel. Energy Fuel 21:1782–1791
Melero JA, Vicente G, Morales G (2008) Acid-catalyzed etherification of bio-glycerol and isobutylene over sulfonic mesostructured silicas. Appl Catal A 346:44–51

Miyazawa T, Koso S, Kunimori K, Tomishige K (2007) Development of a Ru/C catalyst for glycerol hydrogenolysis in combination with an ion-exchange resin. Appl Catal A 318:244–251
Molinero L, Ladero M, Tamayo JJ, Garcia-Ochoa F (2014) Homogeneous catalytic esterification of glycerol with cinnamic and methoxycinnamic acids to cinnamate glycerides in solventless medium: kinetic modelling. Chem Eng J 247:174–182
Ning X, Yu H, Peng F, Wang H (2015) Pt nanoparticles interacting with graphitic nitrogen of N-doped carbon nanotubes: effect of electronic properties on activity for aerobic oxidation of glycerol and electro-oxidation of CO. J Catal 325:136–144
Ochoa-Gómez JR, Gómez-Jiménez-Aberasturi O, Maestro-Madurga B, Pesquera-Rodríguez A, Ramírez-López C, Lorenzo-Ibarreta L, Torrecilla-Soria J, Villarán-Velasco MC (2009) Synthesis of glycerol carbonate from glycerol and dimethyl carbonate by transesterification: catalyst screening and reaction optimization. Appl Catal A Gen 366:315–324
Ozbay N, Oktar N, Dogu G, Dogu T (2010) Conversion of biodiesel by-product glycerol to fuel ethers over different solid acid catalysts. Int J Chem React Eng 8:1–10
Pagliaro M, Ciriminna R, Kimura H, Rossi M, Della Pina C (2007) From glycerol to value-added products. Angew Chem Int Ed 46:4434–4439
Pan S, Zheng L, Nie R, Xia S, Chen P, Hou Z (2012) Transesterification of glycerol with dimethyl carbonate to glycerol carbonate over Na–based zeolites. Chin J Catal 33:1772–1777
Patel A, Singh S (2014) A green and sustainable approach for esterification of glycerol using 12-tungstophosphoric acid anchored to different supports: kinetics and effect of support. Fuel 118:358–364
Pérez-Barrado E, Pujol MC, Aguiló M, Llorca J, Cesteros Y, Díaz F, Pallarès J, Marsal LF, Salagre P (2015) Influence of acid–base properties of calcined MgAl and CaAl layered double hydroxides on the catalytic glycerol etherification to short-chain polyglycerols. Chem Eng J 264:547–556
Pouilloux Y, Abro S, Vanhove C, Barrault J (1999) Reaction of glycerol with fatty acids in the presence of ion-exchange resins: preparation of monoglycerides. J Mol Catal A 149:243–254
Rezayat M, Ghaziaskar HS (2009) Continuous synthesis of glycerol acetates in super- critical carbon dioxide using Amberlyst 15. Green Chem 11:710–715
Rokicki G, Rakoczy P, Parzuchowski P, Sobiecki M (2005) Hyperbranched aliphatic polyethers obtained from environmentally benign monomer: glycerol carbonate. Green Chem 7:529–539
Rostami H, Omrani A, Rostami AA (2015) On the role of electrodeposited nanostructured Pd–co alloy on au for the electrocatalytic oxidation of glycerol in alkaline media. Int J Hydrog Energy 40:9444–9451
Ruppert AM, Weckhuysen BM, Meeldijk JD, Kuipers BWM, Erné BH (2008) Glycerol etherification over highly active CaO-based materials: new mechanistic aspects and related colloidal particle formation. Chem Eur J 14:2016–2020
Sanchez JA, Hernandez DL, Moreno JA, Mondragon F, Fernandez JJ (2011) Alternative carbon based acid catalyst for selective esterification of glycerol to acetylglycerols. Appl Catal A Gen 405:55–60
Santacesaria E, Tesser R, Serio MD, Casale L, Verde D (2010) New process for producing epichlorohydrin via glycerol chlorination. Ind Eng Chem Res 49:964–970
Silva LN, Valter LC, Mota CJA (2011) Catalytic acetylation of glycerol with acetic anhydride. Catal Commun 11:1036–1039
Simanjuntak FSH, Kim TK, Lee SD, Ahn BS, Kim HS, Lee H (2011) CaO-catalyzed synthesis of glycerol carbonate from glycerol and dimethyl carbonate: isolation and characterization of an active ca species. Appl Catal A Gen 401:220–225
Simoes M, Baranton S, Coutanceau C (2010) Electro-oxidation of glycerol at Pd based nanocatalysts for an application in alkaline fuel cells for chemicals and energy cogeneration. Appl Catal B Environ 93:354–362
Skrzyńska E, Zaid S, Girardon JS, Capron M, Dumeignil F (2015) Catalytic behaviour of four different supported noble metals in the crude glycerol oxidation. Appl Catal A Gen 499:89–100
Su L, Jia W, Schempf A, Lei Y (2009) Palladium/titanium dioxide nanofibers for glycerol electrooxidation in alkaline medium. Electrochem Commun 11:2199–2202

Takagaki A, Iwatani K, Nishimura S, Ebitani K (2010) Synthesis of glycerol carbonate from glycerol and dialkyl carbonates using hydrotalcite as a reusable heterogeneous base catalyst. Green Chem 12:578–581

Talebian-Kiakalaieh A, Amin NAS, Najaafi N, Tarighi S (2018) A review on the catalytic acetalization of bio-renewable glycerol to fuel additives. Front Chem 6:1–25

Tesser R, Di Serio M, Vitiello R, Russo V, Ranieri E, Speranza E, Santacesaria E (2012) Glycerol chlorination in gas–liquid semibatch reactor: an alternative route for chlorohydrins production. Ind Eng Chem Res 51:8768–8776

Tseng YH, Wang ML (2011) Kinetics and biphasic distribution of active intermediate of phase transfer catalytic etherification. J Taiwan Inst Chem Eng 42:129–131

Umbarkar SB, Kotbagi TV, Biradar AV, Pasricha R, Chanale J, Dongare MK, Mamede AS, Lancelot C, Payen E (2009) Acetalization of glycerol using mesoporous MoO_3/SiO_2 solid acid catalyst. J Mol Catal A Chem 310:150–158

Vila F, López Granados M, Ojeda M, Fierro JLG, Mariscal R (2012) Glycerol hydrogenolysis to 1,2-propanediol with cu/γ-Al_2O_3: effect of the activation process. Catal Today 187:122–128

Vitiello R, Russo V, Turco R, Tesser R, Di Serio M, Santacesaria E (2014) Glycerol chlorination in a gas-liquid semibatch reactor: new catalysts for chlorohydrin production. Chin J Catal 35:663–669

Wang S, Liu H (2007) Selective hydrogenolysis of glycerol to propylene glycol on cu-ZnO catalysts. Catal Lett 117:62–67

Yuan Z, Wang J, Wang L, Xie W, Chen P, Hou Z, Zheng X (2010) Biodiesel derived glycerol hydrogenolysis to 1,2-propanediol on cu/MgO catalysts. Bioresour Technol 101:7088–7092

Yuan Z, Gao Z, Xu BQ (2015) Acid-base property of the supporting material controls the selectivity of au catalyst for glycerol oxidation in base-free water. Chin J Catal 36:1543–1551

Zhou L, Al-Zaini E, Adesina AA (2013) Catalytic characteristics and parameters optimization of the glycerol acetylation over solid acid catalysts. Fuel 103:617–625

Zhu S, Zhu Y, Gao X, Mo T, Zhu Y, Li Y (2013) Production of bioadditives from glycerol esterification over zirconia supported heteropolyacids. Bioresour Technol 130:45–51

Recent Advances in Steam Reforming of Glycerol for Syngas Production

17

Tan Ji Siang, Nurul Asmawati Roslan, Herma Dina Setiabudi, Sumaiya Zainal Abidin, Trinh Duy Nguyen, Chin Kui Cheng, Aishah Abdul Jalil, Minh Thang Le, Prakash K. Sarangi, Sonil Nanda, and Dai-Viet N. Vo

Abstract

Syngas, a mixture of carbon monoxide and hydrogen, has recently emerged as an important intermediate feedstock in petrochemical industry and an efficient and eco-friendly energy carrier to substitute petroleum-based fuels. Although methane is conventionally employed as a main feedstock for syngas production via indus-

T. J. Siang
School of Chemical and Energy Engineering, Faculty of Engineering, Universiti Teknologi Malaysia, Johor, Malaysia

N. A. Roslan · H. D. Setiabudi · S. Z. Abidin · C. K. Cheng
Faculty of Chemical and Natural Resources Engineering, Universiti Malaysia Pahang, Kuantan, Pahang, Malaysia

T. D. Nguyen · D.-V. N. Vo (✉)
Center of Excellence for Green Energy and Environmental Nanomaterials, Nguyễn Tất Thành University, Hồ Chí Minh City, Vietnam
e-mail: vndviet@ntt.edu.vn

A. A. Jalil
School of Chemical and Energy Engineering, Faculty of Engineering, Universiti Teknologi Malaysia, Johor, Malaysia

Centre of Hydrogen Energy, Institute of Future Energy, Universiti Teknologi Malaysia, Johor, Malaysia

M. T. Le
School of Chemical Engineering, Hanoi University of Science and Technology, Hai Bà Trưng, Hà Nội, Vietnam

P. K. Sarangi
Directorate of Research, Central Agricultural University, Imphal, Manipur, India

S. Nanda
Department of Chemical and Biological Engineering, University of Saskatchewan, Saskatoon, Saskatchewan, Canada

S. Nanda et al. (eds.), *Biorefinery of Alternative Resources: Targeting Green Fuels and Platform Chemicals*, https://doi.org/10.1007/978-981-15-1804-1_17

trial methane steam reforming process, a rising interest about the implementation of glycerol for generating syngas is widely reported in literature due to its great abundance, low cost, and hydrogen-rich content. This chapter summarizes the recent progress in catalytic steam reforming of glycerol for syngas yield in terms of catalytic design using various supports and promoters and the manipulation of operating variables. The mechanistic pathways and their fundamentally derived kinetic models for expressing glycerol reaction rate and estimating associated kinetic parameters are also comprehensively reviewed throughout this chapter.

Keywords
Glycerol · Glycerol steam reforming · Syngas · Hydrogen · Methane · Carbon dioxide

17.1 Introduction

Since currently all aspects of the modern life, transportation and industrial production, are heavily relying on petroleum-based energy, the fast and unavoidable depletion of these main and nonrenewable fossil fuels is predictable in coming decades (Nanda et al. 2015). In addition, the excessive combustion of oil-based fuels leads to a significant increase in CO_2 emissions, especially in the power and transport industries contributing to around 266.7 metric tons per year in 2016 (EDGAR 2017). Hence, minimizing the undesirable CO_2 emissions is crucial for mitigating global warming mainly induced by the anthropogenic greenhouse gases. In fact, the concerns on the security of energy supply and environmental issues have driven the dependency of energy system from solids to liquid fuels and subsequently toward other alternative gas sources. As seen in Fig. 17.1, about 90% of the world energy supply could be eventually derived mainly from hydrogen by 2080.

Based on Fig. 17.1, hydrogen could emerge as one of the main energy sources to substitute crude oil in the next centuries. Hydrogen is industrially separated from syngas (a mixture of H_2 and CO) produced from various processes through reforming (Singh et al. 2018), gasification (Reddy et al. 2014; Nanda et al. 2016a; Okolie et al. 2019), and electrochemical, photochemical, and biological routes (Nanda et al. 2017b). Therefore, apart from utilizing other renewable energy sources, namely solar energy, wind power, and biofuels, the global mitigation strategies are currently focusing on the use of syngas as an alternative and sustainable source of energy. In fact, syngas is extensively employed as important building blocks for numerous petrochemical purposes, including crucial feedstocks for the production of synthetic hydrocarbon fuels (Vo and Adesina 2011; Vo et al. 2012; Nanda et al. 2016b), methyl tert-butyl ether (Siang et al. 2018), methanol (Pham et al. 2019), and ammonia (Schwengber et al. 2016; Inayat et al. 2019), while hydrogen is widely used for high-temperature fuel cells and as an efficient energy carrier in hydrogen cars (Holladay et al. 2009; Nanda et al. 2017a).

The current large-scale production processes of syngas are steam reforming of methane (Arcotumapathy et al. 2015), autothermal reforming of methane, and

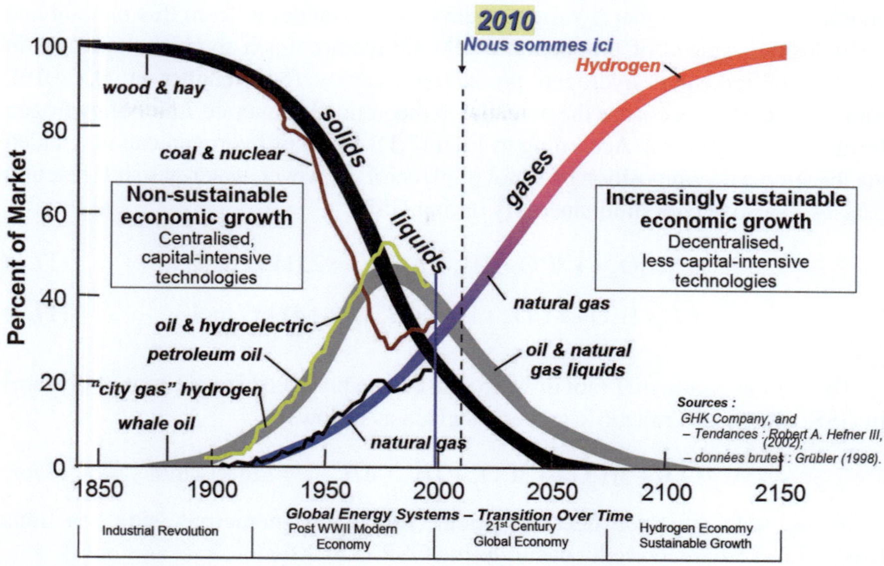

Fig. 17.1 Global energy system transition (1850–2150). (Adapted from Hefner III (1995) with permission. Copyright 1995, Elsevier)

partial oxidation of methane (Minh et al. 2018; Siang et al. 2019). However, methane, a main constituent of natural gas, is also considered as a non-replenished energy and can be used directly as fuel. Thus, the implementation of other biomass-derived feedstocks such as ethanol (Bahari et al. 2016; Fayaz et al. 2019), ethylene glycol (Jun et al. 2018), and glycerol (Arif et al. 2016) in the reforming processes has been broadly considered in academic and industrial R&D to yield an eco-friendly syngas product. Among biomass-based feedstocks, glycerol ($C_3H_8O_3$), which is a highly available by-product from industrial biodiesel production, has become increasingly attractive as a hydrogen-containing feedstock for syngas production. In fact, the global biodiesel production rate has substantially increased about 74.9 million metric tons from 2000 to 2017 (Statista 2018), and the amount of unwanted glycerol by-product is around 10 wt% of biodiesel productivity (Huang et al. 2018). Additionally, the Food and Agriculture Organization predicted that as biodiesel production attains 41 billion liters in 2020, approximately three million metric tons of discarded glycerol would be generated (FAO 2015). The large amount of low-cost glycerol co-generation is one of the main concerns in biodiesel industry although glycerol is nonhazardous and easily handled and stored (Reddy et al. 2016). Thus, the conversion of abundantly undesired glycerol to syngas is considered as an efficient waste-to-wealth approach.

Syngas can be effectively produced from glycerol via various pathways such as autothermal reforming (Hajjaji et al. 2014), partial oxidation (Wang 2010), dry reforming (Tavanarad et al. 2018; Arif et al. 2017), steam reforming (Ghasemzadeh et al. 2019), and supercritical water gasification (Nanda et al. 2018). Nevertheless, glycerol steam reforming (GSR) is more preferred routes among the abovementioned approaches since a higher hydrogen yield could be achieved from this method and the industrial scale GSR production would not require much significant changes in the current industrial hydrogen production process (Schwengber et al. 2016). Moreover, GSR process has the potential to theoretically generate 7 mol of hydrogen from 1 mol of glycerol. According to Eq. (17.1), 4 mol of hydrogen can be yielded via the simple decomposition of 1 mol of glycerol. However, water-gas shift reaction (Eq. 17.2) also occurs simultaneously during GSR:

$$C_3H_8O_3 \leftrightarrow 3CO + 4H_2 \quad \Delta H^0_{298\,K} = 251\,kJ/mol \tag{17.1}$$

$$CO + H_2O \leftrightarrow CO_2 + H_2 \quad \Delta H^0_{298\,K} = -41\,kJ/mol \tag{17.2}$$

Thus, a maximum of 7 mol of hydrogen can be produced from 1 mol of glycerol in GSR, and the overall GSR process is given as follows:

$$C_3H_8O_3 + 3H_2O \leftrightarrow 3CO_2 + 7H_2 \quad \Delta H^0_{298\,K} = 128\,kJ/mol \tag{17.3}$$

Along with glycerol decomposition (Eq. 17.1), numerous side reactions (Eqs. 17.4–17.11) reportedly occur during GSR (Lin 2013):

$$3H_2 + CO \leftrightarrow H_2O + CH_4 \quad \Delta H^0_{298\,K} = -206\,kJ/mol \tag{17.4}$$

$$4H_2 + CO_2 \leftrightarrow 2H_2O + CH_4 \quad \Delta H^0_{298\,K} = -165\,kJ/mol \tag{17.5}$$

$$CH_4 + CO_2 \rightarrow 2H_2 + 2CO \quad \Delta H^0_{298\,K} = +247\,kJ/mol \tag{17.6}$$

$$H_2 + CO_2 \leftrightarrow H_2O + CO \quad \Delta H^0_{298\,K} = +41\,kJ/mol \tag{17.7}$$

$$2CO \leftrightarrow C_{(s)} + CO_2 \quad \Delta H^0_{298\,K} = -172\,kJ/mol \tag{17.8}$$

$$CH_4 \rightarrow C_{(s)} + 2H_2 \quad \Delta H^0_{298\,K} = +75\,kJ/mol \tag{17.9}$$

$$H_2 + CO \leftrightarrow C_{(s)} + H_2O \quad \Delta H^0_{298\,K} = -131\,kJ/mol \tag{17.10}$$

$$2H_2 + CO_2 \leftrightarrow C_{(s)} + 2H_2O \quad \Delta H^0_{298\,K} = +306\,kJ/mol \tag{17.11}$$

In order to generate renewable syngas or hydrogen using glycerol as a main feedstock, the efficient scission of C–H and C–C bonds as well as maintenance of the C–O bond to minimize carbon deposition is currently one of the main challenges in large-scale production. From the above reasons, the right selection and development of catalysts are extremely important in GSR in order to optimize glycerol conversion, catalytic stability, and hydrogen yield. Thus, this chapter intensively discusses various types of catalysts used in GSR. In addition, this article provides a clear picture on catalyst design in GSR by examining the role of different active metals, supports, and promoters as well as their synergic effects on catalytic performance and stability. Besides, this chapter comprehensively reviews the influence of varied operating conditions including reaction temperature, gas hourly space velocity, and reactant feed composition in GSR. Finally, the current progress about kinetics and mechanisms of GSR are also discussed in details.

17.2 Glycerol Steam Reforming Catalysts

Catalysts play a vital role in controlling the mechanistic path and product distribution of GSR via the cleavage of C–H, C–C, and O–H bonds amidst the glycerol molecules (without changing C–O bonds) (Charisiou et al. 2018). Therefore, various types of catalysts were suggested by researchers including monometallic, bimetallic, and trimetallic systems. The summary of recent catalysts used for GSR is summarized in Table 17.1 and thoroughly reviewed in this section.

17.2.1 Active Metal Catalysts

Active metals play a vital role in the GSR whereby the dissociation of glycerol takes place on these sites. Several metals have been used as active metals for metal-based catalysts including the noble metals (Rh, Ru, Pt, and Ir) (Senseni et al. 2017), transition metals (Ni, Co, and Cu) (Papageridis et al. 2016; Kousi et al. 2016; Ramesh and Venkatesha 2017; Demsash et al. 2018; Carrero et al. 2017; Dobosz et al. 2018; Wang et al. 2018), and bi- or trimetallic catalysts (Ramesh and Venkatesha 2017; Wang et al. 2018; Shejale and Yadav 2018).

Senseni et al. (2017) studied the GSR reaction over noble metal nanocatalysts (Rh, Ru, Pt, and Ir) adopted on MgO-promoted Al_2O_3. At 300–600 °C, the catalytic activity of GSR over the catalysts followed a sequence of Rh > Pt > Ru > Ir, while hydrogen selectivity followed a sequence of Rh > Ir > Ru > Pt. Additionally, Rh exhibited superior catalytic activity and stability with glycerol conversion of 100%, hydrogen selectivity of 100%, and good longevity evaluation up to 20-h time-on-stream (TOS). Superior catalytic performance of Rh catalyst was owing to its smallest metal crystal size (1.28 nm), higher BET (Brunauer, Emmett, and Teller) surface area (159 m^2/g), and easier metal reduction as compared to the other catalysts. Although noble metals exhibit higher activity and stability toward GSR, their high expenditures may limit their commercial applications especially in industrial fields (Papageridis et al. 2016). Thus, substantial efforts have been devoted to develop transition metal-based catalysts.

Papageridis et al. (2016) used a constant loading (8 wt%) for Co, Cu, and Ni adopted on γ-alumina (Al_2O_3) and tested for GSR. Results demonstrated the behavior and stability of GSR reaction following the sequence of $Co/Al_2O_3 > Cu/Al_2O_3 > Ni/Al_2O_3$. The varying catalytic activity achieved by the catalysts was related to the acid-base natures of the catalysts owing to the importance of a balanced acid-base distribution in GSR. Ni/Al_2O_3 displayed relatively poor catalytic activity and stability owing to the occurrence of dehydration, cracking, and polymerization reactions over the strong acid

Table 17.1 Summary of catalytic performance for recently employed catalysts in glycerol steam reforming

Catalysts	Steam/glycerol ratio	T (°C)	GHSV ($L/g_{cat}/h$)	Initial performance				TOS (h)	Final performance				D^a (%)	References
				Glycerol conversion (%)	H_2 selectivity (%)	CO selectivity (%)	CO_2 selectivity (%)		Glycerol conversion (%)	H_2 selectivity (%)	CO selectivity (%)	CO_2 selectivity (%)		
Supported catalysts														
8%Ni/γ-Al_2O_3	4/1	600	50	78.2	61.0	57.9	34.5	20	10.8	85.1	56.0	43.7	86.2	Papageridis et al. (2016)
8%Co/γ-Al_2O_3	4/1	600	50	63.3	55.1	64.8	31.3	20	23.9	68.5	82.7	16.8	62.2	Papageridis et al. (2016)
8%Cu/γ-Al_2O_3	4/1	600	50	54.3	54.6	73.9	23.7	20	21.0	71.9	77.8	22.3	61.3	Papageridis et al. (2016)
10%Ni/γ-Al_2O_3	4/1	600	57	95.1	93.8	8.4	86.1	55	29.5	27.6	7.4	22.3	68.9	Kousi et al. (2016)
Ni/γ-Al_2O_3	3/1	650	10000[b]	58.7	66.7	4.8	25.1	24	54.3	69.4	4.2	25.0	7.5	Ramesh and Venkatesha (2017)
10%Ni/γ-Al_2O_3	12/1	650	10[b]	99.9	58.7	13.5	21.9	24	78.0	37.9	36.2	14.0	21.9	Demsash et al. (2018)
7%Co/SBA-15	6/1	600	7.7	92.3	61.7	–	–	50	91.3	–	–	–	1.1	Carrero et al. (2017)
7.5%Co/HAp	9/1	800	6	99.1	74.8	–	–	6	94.2	51.5	–	–	4.9	Dobosz et al. (2018)
10%Ni/ATP	9/1	600	9619[b]	67.0	61.9	49.4	45.7	30	19.8	84.3	53.1	34.2	70.4	Wang et al. (2018)

10%Ni/Al_2O_3/CeO_2	12/1	650	10[b]	99.9	59.9	11.5	22.8	24	94.4	49.9	22.6	17.1	5.5	Demsash et al. (2018)
Promoted catalysts														
16.8%La-promoted Ni/γ-Al_2O_3	4/1	600	57	64.8	66.3	6.9	57.0	20	36.2	36.6	7.9	27.3	44.1	Kousi et al. (2016)
10%Mo7%CeAl	15/1	500	9.2	64.9	57.8	18.2	21.6	10	62.4	54.3	23.3	19.1	3.8	Mitran et al. (2016)
8.5%Ce-promoted Co/SBA-15	6/1	600	7.7	99.8	70.7	–	–	50	49.2	–	–	–	50.7	Carrero et al. (2017)
10%Ce-promoted Co/HAp	9/1	800	6	99.1	79.5	–	–	6	98.8	56.2	–	–	0.3	Dobosz et al. (2018)
1%Ru-promoted Ni/Al_2O_3/CeO_2	12/1	650	10[b]	100	65.0	9.2	23.9	24	99.5	65.2	11.1	21.8	0.5	Demsash et al. (2018)
Bimetallic or trimetallic catalysts														
$LaNiO_3$	3/1	650	10000[b]	59.9	65.3	3.1	27.9	24	59.2	65.2	4.1	27.1	1.2	Ramesh and Venkatesha (2017)
$LaNiO_3$ (E)[c]	3/1	650	10000[b]	68.8	68.0	3.6	26.4	24	66.7	67.5	4.6	25.2	3.1	Ramesh and Venkatesha (2017)
5%Ni-5%Co/ATP	9/1	600	9619[b]	70.5	58.2	61.6	53.8	30	60.0	71.2	42.7	54.3	14.9	Wang et al. (2018)
5%Ni-5%Cu/ATP	9/1	600	9619[b]	75.2	61.6	56.3	47.4	30	48.0	74.9	55.7	45.9	36.2	Wang et al. (2018)

(continued)

Table 17.1 (continued)

Catalysts	Steam/glycerol ratio	T (°C)	GHSV ($L/g_{cat}/h$)	Initial performance: Glycerol conversion (%)	H_2 selectivity (%)	CO selectivity (%)	CO_2 selectivity (%)	TOS (h)	Final performance: Glycerol conversion (%)	H_2 selectivity (%)	CO selectivity (%)	CO_2 selectivity (%)	D^a (%)	References
5%Ni-5%Zn/ATP	9/1	600	9619[b]	67.9	62.0	40.2	60.0	30	62.6	67.2	41.0	60.9	7.8	Wang et al. (2018)
10%Ni-10%Cu/ La_2O_3-MgO (CP)[d]	6/1	500	3120[b]	66.2	96.1	–	–	15	80.7	80.1	–	–	–	Shejale and Yadav (2018)
10%Ni-10%Co/ La_2O_3-MgO (CP)[d]	6/1	500	3120[b]	60.0	88.8	–	–	15	66.6	68.7	–	–	–	Shejale and Yadav (2018)

[a]Degree of catalyst deactivation, D (%) = [1 – (final glycerol conversion/initial glycerol conversion)] × 100%
[b]GHSV (gas hourly space velocity) unit is h^{-1}
[c]Catalyst was prepared with ethylenediaminetetraacetic acid
[d]CP and IP represent the corresponding coprecipitation and impregnation methods

sites-rich catalyst's surface. On the contrary, the relatively good performance of Co/Al_2O_3 was probably due to the enhanced acidity nature of the catalyst upon the addition of Co via the transformation of intrinsic weaker acid sites to stronger acid sites. Regarding the product distribution, Ni/Al_2O_3 has the highest potential in producing the gaseous products via GSR than Co/Al_2O_3 and Cu/Al_2O_3, which might be owing to its higher ability in promoting the rupture of C–C bond. The application of Ni/Al_2O_3 in GSR was also reported by other researchers (Ramesh and Venkatesha 2017; Demsash et al. 2018; Kousi et al. 2016) at different operating conditions as listed in Table 17.1.

For overcoming the drawbacks of Ni-derived catalysts, several studies have reported the combination of Ni-based catalysts with the second and/or third metals to form bimetallic or trimetallic catalysts, respectively. Wang et al. (2018) studied the performance of monometallic Ni- and bimetallic Ni-M (M = Zn, Co, or Cu)-based attapulgite (ATP) catalysts in GSR. The catalytic results revealed relatively better hydrogen yields and glycerol conversions for all bimetallic catalysts than monometallic Ni/ATP catalyst, which significantly associated with the improved strength of metal-support interaction upon the inclusion of the second metals. Stronger metal-support interactions in bimetallic catalysts inhibited the coke deposition and metal sintering. As compared, Ni-Cu/ATP showed superiority in terms of hydrogen yield and glycerol conversion in GSR than Ni supported on Co/ATP and Zn/ATP catalysts. This was owing to the greater particle size of 12.2 nm and higher reducibility for Ni-Cu/ATP, which resulted in its outstanding performance. In terms of catalytic stability, deactivation resistance followed the order of Ni-Zn/ATP > Ni-Co/ATP > Ni-Cu/ATP > Ni/ATP. Ni-Zn/ATP exhibited long-term stability for GSR, which can be credited to its outstanding anti-sintering ability proven from the smallest crystal size of spent catalyst.

On the contrary, Shejale and Yadav (2018) compared the catalytic performance of trimetallic catalysts, Ni-Cu/La_2O_3-MgO and Ni-Co/La_2O_3-MgO in GSR. Both catalysts demonstrated satisfactorily stability over 15-h TOS, and Ni-Cu/La_2O_3-MgO marked higher glycerol conversion and hydrogen selectivity than Ni-Co/La_2O_3-MgO. The high performance of Ni-Cu/La_2O_3-MgO can be explained by the presence of Cu, which triggered dehydrogenation reaction in parallel with high water-gas shift reaction, subsequently resulting in higher hydrogen production. In addition, the combination of both Mg and Cu triggered the formation of an electronic effect on the active Ni phase and therein limited the formation of coke deposition. Moreover, the larger surface area and more homogeneous Ni distribution also contributed to the higher hydrogen production in Ni-Cu/La_2O_3-MgO.

In addition to Ni-based catalyst, perovskite-type oxide catalysts have attracted great attention for GSR owing to their special characteristics. Ramesh and Venkatesha (2017) compared the performance of Ni/Al_2O_3 with Ni-based perovskite prepared with and without template, denoted as $LaNiO_3$(E) and $LaNiO_3$, respectively. As compared, $LaNiO_3$ and $LaNiO_3$(E) perovskite catalysts performed better catalytic performance and stability than Ni/Al_2O_3. The superiority of perovskite catalysts ($LaNiO_3$(E) and $LaNiO_3$) compared with Ni/Al_2O_3 was due to the smaller NiO nanoparticle sizes formation, easier NiO reduction, and moderately weak acid sites, which prevented the filamentous carbon deposition, metal sintering, and deactivation.

17.2.2 Catalyst Supports

Catalyst support plays a significant role in the GSR since it facilitates the dissociation of water into –OH groups, thus triggering the migration of the species to the active metal sites where CO_x and H_2 are formed (Charisiou et al. 2018). In addition, support enhances the stabilization of the metal particles at elevated reaction temperatures (Charisiou et al. 2018). Several types of supports have been used in GSR in an effort to determine the most favorable physicochemical properties including acid-base properties and metal-support interaction. Alumina (Al_2O_3) is one of the most widely adopted catalyst supports in GSR owing to its excellent physicochemical attributes and thermal endurance. According to literature, Ni/Al_2O_3 was mainly populated with strong acid sites, and these sites play a significant factor in the GSR (Papageridis et al. 2016; Kousi et al. 2016; Ramesh and Venkatesha 2017; Demsash et al. 2018). However, the adoption of Al_2O_3 as support triggered catalyst deactivation as a result of coke deposition and catalyst sintering. At strong acid sites of Al_2O_3, the coke formed can be related with cracking, dehydration, and polymerization reactions, while sintering can be explained by the gradual transformation of Al_2O_3 to crystalline phase within the reaction.

To overcome the limitation related with the Al_2O_3-adopted catalysts, several researchers reported on the potential of mesoporous silica as catalyst support for GSR. Carrero et al. (2017) explored the performance of Co/SBA-15 in GSR. SBA-15 has several interesting characteristics including significantly high surface area and uniform pore size dispersion, which may obstruct the agglomeration of large metal and catalyst deactivation by metal sintering. 7%Co/SBA-15 exhibited good activity in GSR with glycerol conversion of 92.3% and 1.08% degree of catalyst deactivation during 50-h TOS.

Dobosz et al. (2018) examined the catalytic behavior of Co-based calcium hydroxyapatite (HAp) in GSR. HAp with a chemical formula of $Ca_{10}(PO_4)_6(OH)_2$ possesses a high thermal stability, mesoporous structure, and acid-based characteristics. The catalytic activity study showed that an efficient transformation of glycerol into reforming products was achieved by 7.5%Co/HAp with glycerol conversion of 99.1% and hydrogen selectivity up to 74.8% at 800 °C.

Wang et al. (2018) discovered the potential of attapulgite (ATP) as supports for Ni-based catalyst. ATP, commonly known as polygorskite, is a kind of hydrated magnesium aluminum silicate clay mineral ($(H_2O)_4(Mg,Al,Fe)_5(OH)_2Si_8O_{20}\cdot 4H_2O$) covered with reactive hydroxy groups. Its lath or fibrous morphology is formed as its upper and lower parts are integrated by octahedral magnesium/aluminum/iron atoms layer in sixfold coordination structure and its structure was constituted by the Si-O tetrahedra parallel double-strand. ATP possesses excellent properties, including high surface area, superb thermal/mechanical properties, and silanol-based chemistry surface. Ni/ATP displayed good activity in GSR with glycerol conversion of 67% and hydrogen selectivity of 61.9% at 800 °C.

Several researchers have reported the potential of mixed oxides as support materials. Demsash et al. (2018) explored the catalytic activity of GSR over 10%Ni/Al_2O_3/CeO_2. The results showed that Ni supported on Al_2O_3/CeO_2 exhibited good activity and stability owing to the greater enhancement in catalytic properties than that of single oxide, Al_2O_3. In fact, 10%Ni/Al_2O_3/CeO_2 exhibited glycerol conversion of 99.9% and hydrogen selectivity of 59.9% at 650 °C.

17.2.3 Promoters

Various oxides, which are mostly alkaline earth or lanthanide oxides, have been used as promoters for improving the resistance of Ni-based catalysts toward metal sintering and coking. The presence of promoter is anticipated to promote the high longevity of Ni particles at elevated reaction temperature and increase NiO particles reducibility. Kousi et al. (2016) explored the potential of lanthanide oxide (La_2O_3) as a promoter for Ni/γ-Al_2O_3. They found that the addition of La_2O_3 facilitated the reactant conversions to gas-phase products. In addition, La_2O_3 played an important role in hindering the carbonaceous species formation by interacting La_2O_3 with CO_2 to form intermediate $La_2O_2CO_3$ species. However, Ni/La_2O_3/γ-Al_2O_3 did not offer a higher stability than Ni/γ-Al_2O_3 owing to its poorer textural properties. This circumstance can be claimed on the formation of $LaAlO_3$ phase as a result of the detrimental impact by the addition of La_2O_3 onto Al_2O_3 support and severe calcination at 900 °C.

Mitran et al. (2016) explored the combination of two metal oxides (CeO_2 and MoO_3) on Al_2O_3 for GSR. The combination of 7%Ce with 10%Mo resulted in the most selective catalyst toward GSR owing to the synergetic interaction between Ce and Mo. The high activity of GSR over MoCe7Al was closely related to Ce addition that was able to alter the phase of Mo species from tetrahedral molybdena to octahedral molybdena, thus contributing to the enhancement of both activity and hydrogen selectivity for MoCe7Al. In contrast, Carrero et al. (2017) investigated Ce addition as a promoter for Co/SBA-15 in GSR. Within 5-h TOS, the introduction of Ce induced better glycerol conversion compared with Co/SBA-15, especially at 600 °C. The superiority of Co-based Ce/SBA-15 over Co/SBA-15 was owing to the higher strength interactions amidst Co and Ce/SBA-15. At 600 °C, Co-Ce/SBA-15 yielded higher H_2 and CO_2 products than Co/SBA-15 as high redox properties of Ce ameliorated water reactivity and oxygen mobility and thus facilitated water-gas shift reaction.

Dobosz et al. (2018) compared the hydrogen production via GSR by using the calcium hydroxyapatite (HAp)-supported cobalt (Co/HAp) and cobalt-cerium catalysts (Ce-promoted Co/HAp) at 800 °C. During the reaction, the individual selectivity of hydrogen and glycerol conversion declined along the TOS for Co/Hap, while the glycerol conversion for Co-Ce/HAp remained stable within 6-h TOS. The Ce addition improved the glycerol conversion, hydrogen selectivity, and stability of Co-Ce/HAp by suppressing the growth or sintering of Co particles. The lower susceptibility of Co-Ce/HAp to sintering was relatively associated with strong interaction amidst Ce and Co species.

Demsash et al. (2018) performed GSR activities using Ni/Al_2O_3, Ni/Al_2O_3/CeO_2, and Ru-promoted Ni/Al_2O_3/CeO_2. As evidenced, Ru-promoted Ni/Al_2O_3/CeO_2 demonstrated better and more stable catalytic behavior than Ni/Al_2O_3 and Ni/Al_2O_3/CeO_2. Ni/

Al_2O_3, $Ni/Al_2O_3/CeO_2$, and Ru-promoted $Ni/Al_2O_3/CeO_2$ maintained their stabilities for 7 h, 17 h, and 24 h, respectively. Highest hydrogen selectivity (88.6%) versus thermodynamic selectivity (95.9%) was marked by Ru-promoted $Ni/Al_2O_3/CeO_2$, and it was found to restrict the formation of by-products such as CO and CH_4. This phenomenon can be credited to the parallel water-gas shift side reaction ($CO + H_2O(g) \leftrightarrow CO_2 + H_2$) and methane reforming processes ($CO_2 + CH_4 \rightarrow 2H_2 + 2CO$; $CH_4 + H_2O \rightleftharpoons CO + 3H_2$) that consumed the corresponding CO and CH_4. Moreover, it can be noted that Ru-promoted $Ni/Al_2O_3/CeO_2$ catalyst enhances metal dispersion and NiO reducibility, which are favorable for GSR.

17.3 Effects of Operating Conditions on the Performance of Glycerol Steam Reforming

A systematic understanding of the impacts of process variables on catalytic performance is enormously essential for catalyst development (Wang et al. 2018); optimization of GSR in terms of activity, yield, and selectivity (He et al. 2010); as well as the derivation of comprehensive kinetic models (Bobadilla et al. 2015; Menezes et al. 2018). As a result, a wide range of studies regarding the role and contribution of process parameters (viz., temperature, gas hourly space velocity (GHSV), and steam/glycerol feed ratio) on GSR performance have been extensively conducted to determine their correlations with reaction activity and stability. The following sections discuss the detailed and recent findings about the effect of process variables on catalytic GSR performance and desired product distribution. The influence of steam/glycerol ratio, reaction temperature, and GHSV on glycerol conversion, selectivity, and H_2/CO ratio over various supported catalysts is also summarized in Table 17.2 for comparison purpose.

17.3.1 Effect of Reaction Temperature

Based on the stoichiometric GSR reaction (see Eq. 17.3), every mole of glycerol reactant consumed theoretically yields 7 mol of hydrogen as the product. However, the presence of other parallel side reactions such as glycerol dehydrogenation, hydrogenolysis of glycerol, CO methanation, and CO_2 methanation could result in the loss of H_2 selectivity and yield. As GSR possesses endothermic nature, its catalytic activity is favored by rising reaction temperature, while the degree of undesired side reactions could be controlled effectively by the manipulation of reaction temperature. From the above reasons, the selection of suitable reaction temperature is a fundamental and important factor requiring detailed investigation to limit unfavorably concomitant side reactions and concurrently improve the GSR activity.

In the evaluation of catalytic GSR stability at various temperature regions for Ni supported on Al_2O_3, CeO_2, and SiC with the corresponding acidic, basic, and neutral attributes, Kim and Woo (2012) observed that increasing temperature from 300 °C to 500 °C could lead to an improvement in glycerol conversion around 76–83% and

Table 17.2 Influence of process variables on the performance of glycerol steam reforming over different catalysts

Catalyst	Operating conditions: Steam/glycerol ratio	Temperature (°C)	GHSV (L/g_{cat}/h/)	Glycerol conversion (%)	Selectivity or yield (%): CH_4	H_2	CO	CO_2	H_2/CO ratio	References
Ni/Al_2O_3	6/1	300–500	33.3[a]	20.1–92.7	–	–	–	–	1.3–4.9	Kim and Woo (2012)
	(3–9)/1	400	33.3[a]	46.4–73.0	–	–	–	–	2.7–12.7	
Ni/CeO_2	6/1	300–500	33.3[a]	20.3–87.3	–	–	–	–	1.0–4.8	Kim and Woo (2012)
	(3–9)/1	400	33.3[a]	50.5–87.7	–	–	–	–	2.5–6.4	
Ni/SiC	6/1	300–500	33.3[a]	15.9–93.7	–	–	–	–	1.0–1.7	Kim and Woo (2012)
	(3–9)/1	400	33.3[a]	65.7–95.2	–	–	–	–	1.3–1.8	
Ni/ATP	9/1	500–700	9619[b]	45.0–65.3	15.3–8.8	20.8–61.8	45.5–53.0	39.2–38.2	1.8–2.2	Wang et al. (2018)
Ni-Co/ATP	9/1	500–700	9619[b]	50.8–78.0	5.1–2.5	50.9–65.5	30.0–43.0	64.9–54.5	2.9–3.2	
Ni-Cu/ATP	9/1	500–700	9619[b]	60.6–73.4	1.9–0.6	44.4–64.0	35.3–45.1	62.8–54.3	3.3–3.0	
Ni-Zn/ATP	9/1	500–700	9619[b]	63.2–73.1	3.8–0.8	39.8–63.0	30.3–39.7	65.9–59.5	2.6–3.6	
NiNbAl	1/2	450–650	300,000[b]	7.37–81.9	0.8–2.9[c]	4.3–44.3[c]	6.2–41.9[c]	2.5–36.9[c]	–	Menezes et al. (2018)
	(5–2)/1	450	500,000[b]	73.0–6.8	0.3–1.0[c]	6.2–2.7[c]	4.4–3.6[c]	5.0–0.8[c]	–	
	1/3	500	200,000–500,000[b]	97.0–26.0	2.9–0.3[c]	33.9–6.4[c]	29.1–4.4[c]	28.7–5.9[c]	–	
MoCe7Al	15/1	400–500	9.2[d]	15.5–65.5	–	22.4–58.3	33.9–16.5	14.0–21.7	–	Mitran et al. (2016)
	(9–20)/1	500	9.2[d]	32.8–83.9	–	29.0–65.6	24.6–18.8	17.9–12.3	–	
	15/1	500	6.4–12.9[d]	67.7–57.5	1.1–2.1	60.0–55.5	14.5–19.3	23.3–22.7	–	

(continued)

Table 17.2 (continued)

Catalyst	Operating conditions: Steam/glycerol ratio	Temperature (°C)	GHSV (L/g_{cat}/h/)	Glycerol conversion (%)	Selectivity or yield (%): CH_4	H_2	CO	CO_2	H_2/CO ratio	References
3%Pt/C	46/1	225	10,000–30,000[b]	45.2–22.6	3.3–2.4	41.5–51.8	73.6–85.0	3.9–5.5	–	Wei et al. (2015)
3%Pt–3%Re/C	46/1	225	25,000–153,000[b]	59.4–18.8	1.8-0	45.2–56.4	72.9–83.4	6.1–5.8	–	
Ni/CeZrO	24/1	500–700	36	94.0–100	5.1–13.4	76.7–33.9	29.2–59.7	66.4–28.8	–	Shao et al. (2014)
	(6–18)/1	600	36	80.9–92.8	4.7–4.8	62.7–71.9	44.1–35.7	51.3–61.4	–	
NiSn/AlMgCe	12/1	600–750	100	63.0–82.0	–	18.0–49.0[c]	–	–	2.0-4.5	Bobadilla et al. (2015)
	(20–8)/1	750	100	91.0–76.0	–	63.1–39.9[c]	–	–	–	
	12/1	750	50–100	–	7.8–7.2	60.2–56.2	8.7–19.6	25.3–15.7	–	
Co–Ni/HTls	4/1	500–600	100	–	2.1–1.7	98.0–95.7	0.1–3.1	0–1.4	–	He et al. (2010)
	(3–9)/1	575	100	–	66.9–7.9	97.7–98.9	43.2–18.8	35.5–30.7	–	

[a]LHSV (liquid hourly space velocity) unit is h^{-1}
[b]GHSV unit is h^{-1}
[c]Yield value (%)
[d]GHSV based on glycerol feed

enhance H_2/CO ratio by up to 41–73% as shown in Table 17.2. This observation was ascribed to the endothermic nature of GSR inducing the facilitation of C–H, C–C, and C–O bonds cleavage at rising reaction temperature (Kim and Woo 2012). Additionally, in a thermodynamic study of GSR, Chen et al. (2011) found that CH_4 production decreased, while the amount of H_2 and CO enhanced with increasing temperature, suggesting the excessive H_2 formation may be generated not only from GSR and water-gas shift reaction but also from methane steam reforming (MSR), in which methane intermediate product is converted to valuable hydrogen.

Wang et al. (2018) synthesized bimetallic Ni-M (M:Co, Cu, or Zn) supported on attapulgite (ATP) and examined their catalytic performance for GSR at varying temperature from 500 °C to 700 °C with steam/glycerol ratio of 9 and atmospheric pressure. Similar findings from Kim and Woo (2012) were reported for the role of reaction temperature. Increasing temperature from 500 °C to 700 °C has a positive effect on the bond cleavage of glycerol molecule, hence enhancing GSR activity. Interestingly, it is also reported that the selectivity of methane possessed an opposite trend with hydrogen selectivity, suggesting that methane by-product originated from glycerol decomposition during GSR process was converted to hydrogen via MSR.

17.3.2 Effect of Gas Hourly Space Velocity

As one of the most important process variables in heterogeneous catalytic system, the impact of GHSV on catalytic performance of GSR is rigorously assessed to achieve the independence of GSR activity from mass and heat transfer limitations and optimize the performance of catalysts with respect to conversion, H_2 yield, and H_2/CO ratio. Bobadilla et al. (2015) prepared a bimetallic NiSn/AlMgCe catalyst for catalyzing the GSR reaction at various GHSV values and 750 °C. They reported that the growth in GHSV from 50 to 100 $L/g_{cat}/h$ adversely decreased hydrogen selectivity from 60.2% to 56.2% and decreased catalytic stability reasonably owing to the thermal cracking of glycerol to carbonaceous species clogging the active sites.

According to Wang et al. (1997), the glycerol decomposition reportedly occurs before reaching catalyst bed, and few glycerol compounds (without thermal cracking) successfully access active sites for catalytic surface reaction. Hence, Bobadilla et al. (2015) proposed that GSR is a catalytic system in which glycerol is thermally decomposed to carbon deposition on active metal sites and the steam gasification of these species to gaseous products (viz., CH_4, C_2H_4, CO, H_2, and CO_2) subsequently occurs to prevent catalyst from deactivation. From the abovementioned reasons, high GHSV could not provide enough residence time for thoroughly gasifying deposited carbon from glycerol decomposition to the desired H_2 and CO products as well as protect active sites from carbonaceous species.

Menezes et al. (2018) examined the impacts of operating parameters on GSR by Ni-supported Al_2O_3 and Nb_2O_5. They found that rising GHSV values led to a significant decline in glycerol conversion from 97% to 26% in agreement with findings of Bobadilla et al. (2015). As reported by Menezes et al. (2018), coke formation was the main factor accounted for the drastic drop in GSR activity. They also suggested that the formation of carbonaceous species preferred at low GHSV led to the suppression of hydrogen formation, while the liquid by-products were dominant at high GHSV since the feed rate of glycerol was too excessive on catalytic bed to be converted to gaseous products.

17.3.3 Effect of Reactant Partial Pressure

At high reaction temperature, GSR could inevitably encounter glycerol thermal cracking side reaction leading to the blockage of active sites with deposited carbon. Therefore, scrutinizing the impact of reactant partial pressure (or water/glycerol ratio) on GSR activity, stability and hydrogen selectivity is appreciably vital for kinetic investigation and industrial applications. Shao et al. (2014) prepared Ni/CeZrO via coprecipitation method and assessed its catalytic performance for GSR at 600 °C and varied water/glycerol ratios ranging from 6 to 18. They observed that glycerol conversion increased from 80.9% to 92.8% with increasing water/glycerol ratio from 6:1 to 18:1 owing to decreasing glycerol feed composition. In addition, increasing water/glycerol ratio reportedly favored water-gas shift reaction because a substantial rise in H_2 and CO_2 selectivity was evident while CO selectivity experienced a considerable drop. Interestingly, a comparative stability with desired catalytic performance was achieved when a water/glycerol ratio of 6 was employed, indicating this ratio could be appropriate for carbon removal and enhanced the GSR activity by retaining catalytic active sites (Shao et al. 2014).

A similar behavior was also observed by Mitran et al. (2016) for their assessment of GSR over MoCeAl catalyst at water/glycerol ratios of 9–20 and temperature of 500 °C. Glycerol conversion was considerably increased up to 83.9% with growing water/glycerol ratios, rationally because of the reducing glycerol concentration in feed. With increasing water/glycerol ratio, the selectivity of both CO and CO_2 declined substantially, whereas the rise in H_2 selectivity was observed. Although hydrogen formation rate could be facilitated with an excessive steam content in feedstock, the employment of extreme water/glycerol ratios could induce an extensive energy prerequisite for evaporating a large amount of water into steam. Thus, from an economic perspective, it is crucial to achieve a balance between GSR performance and operating cost for sufficient energy supply (Bobadilla et al. 2015).

17.4 Mechanisms and Kinetics of Glycerol Steam Reforming

The comprehensive reaction rate expressions for heterogeneous catalytic reactions are fundamentally essential for the prediction of consumption rate, reactor design, and scale-up in commercialized industrial applications. Indeed, the intrinsic catalytic features play a key role in the establishment of the mechanism-derived reaction rate expressions. Thus, kinetic studies of glycerol steam reforming have arisen to be one of the main concerns in both industrial and academic realms in recent years, and several studies about GSR mechanism and kinetics have been lately reported in literature. In general, GSR kinetics reportedly include empirical power law model and mechanistic-based models, namely Langmuir-Hinshelwood-Hougen-Watson (LHHW) and Eley-Rideal (ER) models.

The implementation of power law model for the rough calculation of reaction order and apparent activation energy for reactant consumption and product formation is widely conducted owing to its simplicity. In fact, the experimental GSR reaction rates measured at varied reaction temperature and reactant partial pressure are conventionally fitted to the simple power law model (see Eq. 17.12) without the prerequisite of determining mechanistic pathways and rate-controlling step in order to estimate the associated model parameters:

$$-r_{C_3H_8O_3} = kP^{\alpha}_{C_3H_8O_3} P^{\beta}_{H_2O} \tag{17.12}$$

where $-r_{C_3H_8O_3}$ and k are the corresponding glycerol reaction rate and reaction rate coefficient, whereas P_i (with i: $C_3H_8O_3$ or H_2O) denotes the reactant partial pressure. In addition, α and β signify the reaction order of $C_3H_8O_3$ and H_2O, respectively.

Table 17.3 summarizes the calculated activation energy of glycerol consumption and reaction orders corresponding to glycerol and steam reactants from the GSR power law models. Adhikari et al. (2009) conducted GSR on Ni/CeO_2 with the excessive presence of water in comparison with glycerol concentration. As a result, they could use a simplified power law model excluding the effect of steam concentration to capture experimental glycerol consumption rates at varying reaction temperature of 600–650 °C and reported glycerol reaction order and activation energy were around 0.23 and 103.4 kJ/mol, correspondingly (Table 17.3). An analogous power law expression was also proposed by Dave and Pant (2011) for GSR over $Ni\text{-}ZrO_2/CeO_2$. A lower activation energy for glycerol conversion of 43.4 kJ/mol was reported, whereas the glycerol reaction order was observed as 0.3.

Additionally, in the kinetic investigation of GSR over Ni-Mg-Al-based catalysts, Wang et al. (2013) developed a GSR pseudo-n^{th} order model, which was only associated with the concentration of glycerol, for capturing the consumption rate of glycerol. They also reported that depending on catalysts employed, glycerol activation energy could vary from 37.8 to 131.6 kJ/mol. In addition, glycerol reaction order of unity was verified with the highest accuracy and best fitness for describing glycerol reaction rate.

Unlike aforementioned literature, the undeniably significant role of steam in GSR was also taken into account for power law models in other studies. Cheng et al. (2010a) employed $Co\text{-}Ni/Al_2O_3$ for GSR at varied operating conditions of 500–550 °C and 1 atm. The fitting of their experimental data for product formation rates including H_2, CO_2, CO, and CH_4 into a power law equation using nonlinear regression based on Levenberg-Marquardt algorithm revealed that partial pressure of steam possessed a great effect on CH_4, H_2, CO, and CO_2 formation rates. Apart from CO, the production rates of CH_4, H_2, and CO_2 increased with rising steam partial pressure since positive reaction orders with respect to steam partial pressure were observed for the formation of these species. The negative value of steam reaction order toward CO formation possibly suggested the competitive H_2O adsorption occupying active sites for generating CO product or loss of CO via simultaneous water-gas shift side reaction, particularly at steam-rich environment. Cheng et al. (2010a) also reported that for glycerol reaction rate, the fractional orders were estimated as 0.25 and 0.36 for the corresponding glycerol and steam reactants, while the activation energy for glycerol conversion was computed as about 63.3 kJ/mol on Co-Ni/Al_2O_3 catalyst. In the study of GSR over Ni/Al_2O_3, Cheng et al. (2011) found relatively similar activation energy values for H_2 and CO_2 product formation as well as glycerol reactant consumption ranging from 60 to 61.8 kJ/mol. These results could indicate that the direct GSR pathway involving the reaction between glycerol and water was the predominant step accounting for H_2 and CO_2 production.

Although the empirical power law model is able to provide an initiate conjecture of overall activation energy and reaction order for each reactant involved, this simple model fails to yield the accurate estimation of associated model parameters over a wide-ranging feedstock composition. Additionally, the power law model only complies with statistical criteria and does not obey or reflect the inherent GSR mechanisms. Hence, other fundamentally mechanism-derived reaction rate expressions, namely ER and LHHW models, are widely used for verifying the best fitness of obtained experimental data to the proposed kinetic models. Based on the degree of fitness and thermodynamic criteria, the most appropriate kinetic model could be determined and hence contributing to the confirmation of intrinsic GSR reaction steps on catalyst surface.

In the inspection of GSR over Ni/CeO_2 synthesized via precipitation deposition method, Pant et al. (2011) proposed an ER kinetic model, in which the molecularly adsorbed glycerol on catalyst surface could react with water in gas phase followed by decomposition into various intermediates as the rate-determining step (see Eq. 17.13) in Table 17.4. By fitting experimental GSR consumption rates to the proposed ER kinetic model, the overall GSR activation energy was estimated at around 36.5 kJ/mol.

Table 17.3 Summary of kinetic parameters computed from GSR power law models on several catalysts

		Activation energy, E_a (kJ/mol)	Reaction order		
			α	β	
Catalyst	Temperature (°C)	Glycerol	Glycerol	Water	References
Ni/CeO_2	600–650	103.4	0.23	0	Adhikari et al. (2009)
Ni/Al_2O_3	450–550	60.0	0.48	0.34	Cheng et al. (2011)
Co/Al_2O_3	450–550	67.2	0.08	0.40	Cheng et al. (2010b)
Co-Ni/Al_2O_3	500–550	63.3	0.25	0.36	Cheng et al. (2010a)
Ni-ZrO_2/CeO_2	600–700	43.4	0.30	0	Dave and Pant (2011)
Ni-Mg-Al	400–600	131.6	1	0	Wang et al. (2013)
Ce-Ni/Al_2O_3	550–650	62.9	0.45	0	Demsash and Mohan (2016)

Cheng et al. (2011) used Ni/Al_2O_3 catalyst for GSR and detected the existence of both basic and acidic sites on catalyst surface via the corresponding CO_2 and NH_3 temperature-programmed desorption runs. Thus, they proposed a dual-site GSR mechanism (see Fig. 17.2) involving molecular glycerol adsorption and dissociative water adsorption on two various active sites. Additionally, the surface reaction between these adsorbed glycerol and water is assumed as a rate-determining step. Based on this suggested mechanism, a LHHW model was derived as given in Eq. (17.14) of Table 17.4 and adequately captured the consumption rate of glycerol as well as satisfied all associated statistical and thermodynamic standards.

However, it is well-known that heterogeneous catalytic reactions could alter its mechanistic pathway to different routes depending on the type and attributes of catalysts employed. For example, in the study of bimetallic $Co\text{-}Ni/Al_2O_3$ for GSR, Cheng et al. (2010a) proposed 12 different GSR mechanisms based on single-site or dual-site ER and LHHW mechanism as they found the existence of acidic and basic Brönsted sites on catalyst. The statistical and thermodynamic discrimination for these 12 derived GSR kinetic models was conducted and proved that the LHHW kinetic model (Eq. 17.15) in Table 17.4, involving the molecular adsorption of both glycerol and steam reactants on 2 distinct active sites, was the most suitable model with the best fitness to the experimental glycerol reaction rate data. Thus, the activation energy for glycerol consumption was further computed from this model as about 69.36 kJ/mol.

17.5 Conclusions

In general, GSR reaction depends on the nature of the catalyst that consists of active metal and support. Active metals are responsible for the dissociation of glycerol molecules. Meanwhile, supports are crucial in activating the dissociation of water into –OH groups as well as promoting the species to migrate toward the active metal species whereby the CO_x and hydrogen are formed. In addition, supports stabilize the metal particles at elevated reaction temperature under steam.

For active metal selection, transition metals have attracted considerable attention owing to its availability and low price. Nickel is the most popular active metal catalyst in GSR among transition metals, even though it is prone to coking issues. In order to address the challenge of coke issues from nickel, bimetallic catalysts are a suggestive solution. In the bimetallic system, a ratio of the constituents' metals is considered as a crucial factor in order to produce more active bimetallic catalyst as compared to the corresponding monometallic catalyst. Additionally, reported literatures indicated that the adoption of alkaline earth or lanthanide oxides aided in dispersing the active metal particles reinforced the metal-support interactions, balanced acid-base strength and density distribution, and thus improved the performance of catalysts (glycerol conversion, hydrogen production, and stability) toward GSR.

In addition, appropriate support selection must be done considering its textural and chemical attributes to boost the active metal particles dispersion, facilitate the metal-support interaction, minimize metal sintering, stimulate the reduction of the

Table 17.4 LHHW and ER kinetic expressions for catalytic glycerol steam reforming

Catalyst	Assumption	Mechanism	Kinetic model	Eq. no.	Ref.
Ni/ CeO_2	Decomposition of molecularly adsorbed glycerol with water forming intermediates as RDS	$C_3H_8O_3 + S_1 \overset{K_1}{\leftrightarrow} C_3H_8O_3 - S_1$	$-r_{C_3H_8O_3} = \frac{k_1 k_2 P_{C_3H_8O_3} P_{H_2O}}{k_{-1} + k_1 P_{C_3H_8O_3} + k_2 P_{H_2O} + \left(\frac{k_1 k_2 P_{C_3H_8O_3}}{k_3}\right)}$	(17.13)	Pant et al. (2011)
		$C_3H_8O_3 - S_1 + H_2O \overset{K_2}{\leftrightarrow} C_3H_{10}O_4 - S_1$			
		$C_3H_{10}O_4 - S_1 \overset{k_3}{\rightarrow} \text{Intermediates}\left(\text{RDS}^{a}\right)$			
		$\text{Intermediates} \overset{K_4}{\leftrightarrow} H_2 + CO_2 + S_1$	$K_1 = \frac{k_1^{b}}{k_{-1}}; K_2 = \frac{k_2}{k_{-2}}$		
Ni/ Al_2O_3	Surface reaction of dual-site molecularly adsorbed glycerol and dissociatively adsorbed water as RDS	$C_3H_8O_3 + S_1 \leftrightarrow C_3H_8O_3 - S_1$	$-r_{C_3H_8O_3} = \frac{k_{rxn} P_{C_3H_8O_3} \sqrt{P_{H_2O}}}{\left(1 + K_{C_3H_8O_3} P_{C_3H_8O_3}\right)\left(1 + \sqrt{K_{H_2O} P_{H_2O}}\right)}$ $k_{rxn} = 1.33 \times 10^{-7}$ mol/s/m²/kPa $K_{H_2O} = 0.043\ \text{kPa}^{-1}$ $K_{C_3H_8O_3} = 0.00056\ \text{kPa}^{-1}$	(17.14)	Cheng et al. (2011)
		$H_2O + 2S_2 \leftrightarrow OH - S_2 + H - S_2$			
		$C_3H_8O_3 - S_1 + H - S_2 \rightarrow CH_2OHCHOH - S_1 + CHOH - S_2 + 2H_2$ (RDS)			
		$CHOH - S_2 \rightarrow CO - S_2 + H_2$			
		$CH_2OHCHOH - S_1 + H - S_2 \rightarrow CH_2OH - S_1 + CH_3O - S_2$			
		$CH_2OH - S_1 + S_2 \rightarrow CH_2 - S_1 + OH - S_2$			

Catalyst	Assumption	Mechanism	Kinetic model	Eq. no.	Ref.
		$CH_2-S_1+H-S_2 \rightarrow S_2+CH_3-S_1$			
		$CH_3-S_1+H-S_2 \rightarrow S_2+CH_4+S_1$			
		$CH_3O-S_1+S_2 \rightarrow H-S_2+CH_2O-S_1$			
		$CH_2O-S_1+S_2 \rightarrow H-S_2+HCO-S_1$			
		$HCO-S_1+S_2 \rightarrow H-S_2+CO-S_1$			
		$CO-S_1 \leftrightarrow S_1+CO$			
		$CO-S_1+OH-S_2 \leftrightarrow CO_2+H-S_2+S_1$			
		$H-S_2+H-S_2 \leftrightarrow H_2+2S_2$			
Co-Ni/Al_2O_3	Dual-site molecular glycerol and water adsorption with surface reaction of adsorbed reactants as RDS	$C_3H_8O_3+S_1 \leftrightarrow C_3H_8O_3-S_1$	$-r_{C_3H_8O_3}=\frac{k_{rxn}P_{C_3H_8O_3}P_{H_2O}}{\left(1+K_{C_3H_8O_3}P_{C_3H_8O_3}\right)\left(1+K_{H_2O}P_{H_2O}\right)}$ $k_{rxn}=5.57\times10^{-7}$ mol/s/m²/kPa $K_{H_2O}=0.0369\,kPa^{-1}$ $K_{C_3H_8O_3}=0.283\,kPa^{-1}$	(17.15)	Cheng et al. (2010a)
		$H_2O+S_2 \leftrightarrow H_2O-S_2$			
		$C_3H_8O_3-S_1+H_2O-S_2 \rightarrow CH_2OHCHOH-S_1$ $+HCOO-S_2+2H_2$ (RDS)			
		$COO-S_2 \rightarrow H-S_2+CO_2$			
		$CH_2OHCHOH-S_1+H-S_2 \rightarrow CH_3O-S_2$ $+CH_2OH-S_1$			
		$H_2OH-S_1+S_2 \rightarrow OH-S_2+CH_2-S_1$			
		$CH_2-S_1+H-S_2 \rightarrow S_2+CH_3-S_1$			

(continued)

Table 17.4 (continued)

Catalyst	Assumption	Mechanism	Kinetic model	Eq. no.	Ref.
		$CH_3-S_1+H-S_2 \rightarrow S_2+CH_4+S_1$			
		$CH_3O-S_1+S_2 \rightarrow H-S_2+CH_2O-S_1$			
		$CH_2O-S_1+S_2 \rightarrow H-S_2+HCO-S_1$			
		$HCO-S_1+S_2 \rightarrow H-S_2+CO-S_1$			
		$CO-S_1 \leftrightarrow S_1+CO$			
		$CO-S_1+OH-S_2 \leftrightarrow H-S_2+S_1+CO_2$			
		$2H-S_2 \leftrightarrow H_2+2S_2$			

[a]RDS represents the rate-determining step, while S is the available active site on catalyst surface. S_1 and S_2 indicate two different active sites of catalysts
[b]k_i and k_{-i} are the corresponding adsorption and desorption rate constants, while K_i is the adsorption equilibrium constant and k_{rxn} is GSR reaction rate constant

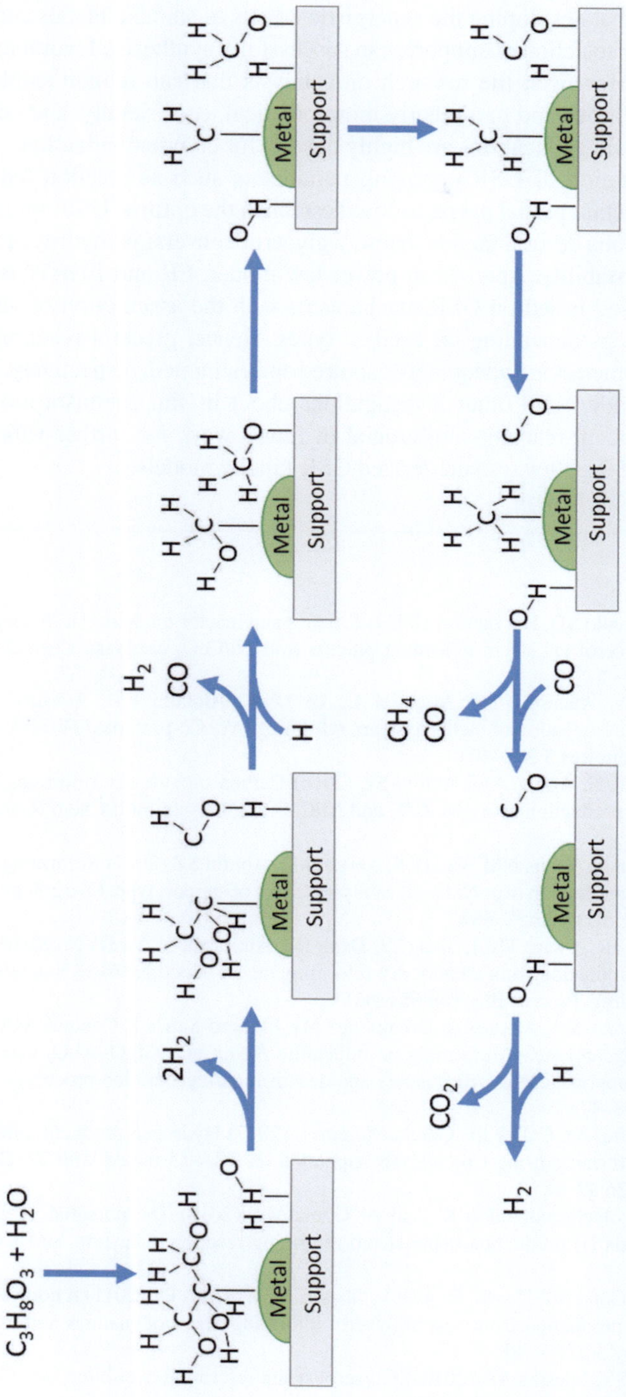

Fig. 17.2 Mechanistic steps in glycerol steam reforming reaction

catalyst, and, most importantly, reduce or resist the carbonaceous species formation. It is proposed that developing the synergistic effects of suitable metals composition with the proper selection of supports can successfully synthesize favorable catalysts toward GSR. Moreover, the research on catalysts that can remain stable at GSR operating conditions and particularly more practical, eco-friendly, and economical in industrial scale applications are highly needed for commercialization.

The examination of GSR operating parameters such as reaction temperature, GHSV, and reactant partial pressure could establish the optimal GSR operation conditions for various desired targets, namely glycerol conversion, hydrogen yield, H_2/CO ratio, and stability. Apart from power law model, ER and LHHW models are broadly proposed based on GSR mechanisms with the assumption of varied rate-determining steps depending on catalyst types. Hence, glycerol reaction rate and Arrhenius parameters are adequately captured and estimated, respectively. However, the isotopic studies and other investigations about in situ chemisorption of both glycerol and steam reactants are crucial in future work for further validating the mechanistic GSR pathways and derived GSR kinetic models.

References

Adhikari S, Fernando SD, Haryanto A (2009) Kinetics and reactor modeling of hydrogen production from glycerol via steam reforming process over Ni/CeO_2 catalysts. Chem Eng Technol 32:541–547

Arcotumapathy V, Alenazey FS, Al-otaibi L, Vo DVN, Alotaibi FM, Adesina AA (2015) Mechanistic investigation of methane steam reforming over Ce-promoted Ni/SBA-15 catalyst. Appl Petrochem Res 5:393–404

Arif NNM, Vo DVN, Azizan MT, Abidin SZ (2016) Carbon dioxide dry reforming of glycerol for hydrogen production using Ni/ZrO_2 and Ni/CaO as catalysts. Bull Chem React Eng Catal 11:200–209

Arif NNM, Harun N, Yunus NM, Vo DVN, Azizan MT, Abidin SZ (2017) Reforming of glycerol for hydrogen production over Ni based catalysts: effect of support type. Energ Sour Part A: Rec Util Environ Effect 39:657–663

Bahari MB, Goo BC, Pham TLM, Siang TJ, Danh HT, Ainirazali N, Vo DVN (2016) Hydrogen-rich syngas production from ethanol dry reforming on La-doped Ni/Al_2O_3 catalysts: effect of promoter loading. Process Eng 148:654–661

Bobadilla LF, Penkova A, Álvarez A, Domínguez MI, Romero-Sarria F, Centeno MA, Odriozola JA (2015) Glycerol steam reforming on bimetallic NiSn/CeO_2–MgO–Al_2O_3 catalysts: influence of the support, reaction parameters and deactivation/regeneration processes. Appl Catal A–Gen 492:38–47

Carrero A, Vizcaíno AJ, Calles JA, García-Moreno L (2017) Hydrogen production through glycerol steam reforming using Co catalysts supported on SBA-15 doped with Zr, Ce and La. J Energ Chem 26:42–48

Charisiou ND, Polychronopoulou K, Asif A, Goula MA (2018) The potential of glycerol and phenol towards H_2 production using steam reforming reaction: a review. Surf Coat Technol 352:92–111

Chen H, Ding Y, Cong NT, Dou B, Dupont V, Ghadiri M, Williams PT (2011) A comparative study on hydrogen production from steam-glycerol reforming: thermodynamics and experimental. Renew Energy 36:779–788

Cheng CK, Foo SY, Adesina AA (2010a) Glycerol steam reforming over bimetallic Co-Ni/Al_2O_3. Ind Eng Chem Res 49:10804–10817

Cheng CK, Foo SY, Adesina AA (2010b) H_2-rich synthesis gas production over Co/Al_2O_3 catalyst via glycerol steam reforming. Catal Commun 12:292–298

Cheng CK, Foo SY, Adesina AA (2011) Steam reforming of glycerol over Ni/Al_2O_3 catalyst. Catal Today 178:25–33

Dave CD, Pant KK (2011) Renewable hydrogen generation by steam reforming of glycerol over zirconia promoted ceria supported catalyst. Renew Energy 36:3195–3202

Demsash HD, Mohan R (2016) Steam reforming of glycerol to hydrogen over ceria promoted nickel–alumina catalysts. Int J Hydrogen Energ 41:22732–22742

Demsash HD, Kondamudi KVK, Upadhyayula S, Mohan R (2018) Ruthenium doped nickel-alumina-ceria catalyst in glycerol steam reforming. Fuel Process Technol 169:150–156

Dobosz J, Cichy M, Zawadzki M, Borowiecki T (2018) Glycerol steam reforming over calcium hydroxyapatite supported cobalt and cobalt-cerium catalysts. J Energy Chem 27:404–412

EDGAR, Emissions Database for Global Atmospheric Research (2017) European Commission. http://edgar.jrc.ec.europa.eu. Accessed 3 June 2019

FAO, Food and Agriculture Organization (2015) OECD-FAO agricultural outlook 2015. OECD Publishing, Paris

Fayaz F, Bach LG, Bahari MB, Nguyen TD, Vu KB, Kanthasamy R, Samart C, Nguyen-Huy C, Vo DVN (2019) Stability evaluation of ethanol dry reforming on Lanthania-doped cobalt-based catalysts for hydrogen-rich syngas generation. Int J Energy Res 43:405–416

Ghasemzadeh K, Ghahremani M, Amiri TY, Basile A (2019) Performance evaluation of Pd-Ag membrane reactor in glycerol steam reforming process: development of the CFD model. Int J Hydrogen Energ 44:1000–1009

Hajjaji N, Baccar I, Pons M-N (2014) Energy and exergy analysis as tools for optimization of hydrogen production by glycerol autothermal reforming. Renew Energy 71:368–380

He L, Parra JMS, Blekkan EA, Chen D (2010) Towards efficient hydrogen production from glycerol by sorption enhanced steam reforming. Energy Environ Sci 3:1046–1056

Hefner RA III (1995) Toward sustainable economic growth: the age of energy gases. Int J Hydrogen Energ 20:945–948

Holladay JD, Hu J, King DL, Wang Y (2009) An overview of hydrogen production technologies. Catal Today 139:244–260

Huang C, Xu C, Wang B, Hu X, Li J, Liu J, Jie L, Li C (2018) High production of syngas from catalytic steam reforming of biomass glycerol in the presence of methane. Biomass Bioenergy 119:173–178

Inayat M, Sulaiman SA, Kurnia JC, Shahbaz M (2019) Effect of various blended fuels on syngas quality and performance in catalytic co-gasification: a review. Renew Sust Energ Rev 105:252–267

Jun LN, Bahari MB, Phuong PTT, Phuc NHH, Samart C, Abdullah B, Setiabudi HD, Vo DVN (2018) Ethylene glycol dry reforming for syngas generation on Ce-promoted Co/Al_2O_3 catalysts. Appl Petrochem Res 8:253–261

Kim SM, Woo SI (2012) Sustainable production of syngas from biomass–derived glycerol by steam reforming over highly stable Ni/SiC. ChemSusChem 5:1513–1522

Kousi K, Chourdakis N, Matralis H, Kontarides D, Papadopoulou C, Verykios X (2016) Glycerol steam reforming over modified Ni-based catalysts. Appl Catal A Gen 518:129–141

Lin YC (2013) Catalytic valorization of glycerol to hydrogen and syngas. Int J Hydrogen Energ 38:2678–2700

Menezes JPDSQ, Manfro RL, Souza MM (2018) Hydrogen production from glycerol steam reforming over nickel catalysts supported on alumina and niobia: deactivation process, effect of reaction conditions and kinetic modeling. In J Hydrogen Energ 43:15064–15082

Minh DP, Siang TJ, Vo DVN, Phan TS, Ridart C, Nzihou A, Grouset D (2018) Hydrogen production from biogas reforming: an overview of steam reforming, dry reforming, dual reforming, and tri-reforming of methane. In: Azzaro-Pantel C (ed) Hydrogen supply chains design, deployment and operation. Academic Press, Waltham, MA, pp 111–166

Mitran G, Pavel OD, Mieritz DG, Seo DK, Florea M (2016) Effect of Mo/Ce ratio in Mo–Ce–Al catalysts on the hydrogen production by steam reforming of glycerol. Cat Sci Technol 6:7902–7912

Nanda S, Azargohar R, Dalai AK, Kozinski JA (2015) An assessment on the sustainability of lignocellulosic biomass for biorefining. Renew Sust Energ Rev 50:925–941

Nanda S, Isen J, Dalai AK, Kozinski JA (2016a) Gasification of fruit wastes and agro-food residues in supercritical water. Energ Convers Manage 110:296–306

Nanda S, Kozinski JA, Dalai AK (2016b) Lignocellulosic biomass: a review of conversion technologies and fuel products. Curr Biochem Eng 3:24–36

Nanda S, Li K, Abatzoglou N, Dalai AK, Kozinski JA (2017a) Advancements and confinements in hydrogen production technologies. In: Dalena F, Basile A, Rossi C (eds) Bioenergy systems for the future. Woodhead Publishing, Elsevier, Sawston, pp 373–418

Nanda S, Rana R, Zheng Y, Kozinski JA, Dalai AK (2017b) Insights on pathways for hydrogen generation from ethanol. Sustain Energ Fuel 1:1232–1245

Nanda S, Reddy SN, Vo DVN, Sahoo BN, Kozinski JA (2018) Catalytic gasification of wheat straw in hot compressed (subcritical and supercritical) water for hydrogen production. Energ Sci Eng 6:448–459

Okolie JA, Rana R, Nanda S, Dalai AK, Kozinski JA (2019) Supercritical water gasification of biomass: a state-of-the-art review of process parameters, reaction mechanisms and catalysis. Sustain Energ Fuel 3:578–598

Pant KK, Jain R, Jain S (2011) Renewable hydrogen production by steam reforming of glycerol over Ni/CeO_2 catalyst prepared by precipitation deposition method. Korean J Chem Eng 28:1859–1866

Papageridis KN, Siakavelas G, Charisiou ND, Avraam DG, Tzounis L, Kousi K, Goula MA (2016) Comparative study of Ni, Co, Cu supported on γ-alumina catalysts for hydrogen production via the glycerol steam reforming reaction. Fuel Process Technol 152:156–175

Pham TLM, Vo DVN, Nguyen HNT, Pham-Tran NN (2019) CH versus OH bond scission in methanol decomposition on Pt (111): role of the dispersion interaction. Appl Surf Sci 481:1327–1334

Ramesh S, Venkatesha NJ (2017) Template free synthesis of Ni-perovskite: an efficient catalyst for hydrogen production by steam reforming of bioglycerol. ACS Sustain Chem Eng 5:1339–1346

Reddy SN, Nanda S, Dalai AK, Kozinski JA (2014) Supercritical water gasification of biomass for hydrogen production. Int J Hydrogen Energ 39:6912–6926

Reddy SN, Nanda S, Kozinski JA (2016) Supercritical water gasification of glycerol and methanol mixtures as model waste residues from biodiesel refinery. Chem Eng Res Des 113:17–27

Schwengber CA, Alves HJ, Schnaffner RA, Silva FA, Sequinel R, Bach VR, Ferracin RJ (2016) Overview of glycerol reforming for hydrogen production. Renew Sust Energ Rev 58:259–266

Senseni AZ, Rezaei M, Meshkani F (2017) Glycerol steam reforming over noble metal nanocatalysts. Chem Eng Res Des 123:360–366

Shao S, Shi AW, Liu CL, Yang RZ, Dong WS (2014) Hydrogen production from steam reforming of glycerol over Ni/CeZrO catalysts. Fuel Process Technol 125:1–7

Shejale AD, Yadav GD (2018) Ni–cu and Ni–Co supported on La–Mg based metal oxides prepared by coprecipitation and impregnation for superior hydrogen production via steam reforming of glycerol. Ind Eng Chem Res 57:4785–4797

Siang TJ, Singh S, Omoregbe O, Bach LG, Phuc NHH, Vo DVN (2018) Hydrogen production from CH_4 dry reforming over bimetallic Ni–Co/Al_2O_3 catalyst. J Energy Inst 91:683–694

Siang TJ, Minh DP, Singh S, Setiabudi HD, Vo DVN (2019) Recent advances in hydrogen production through bi-reforming of biogas. In: Nanda S, Sarangi PK, Vo DVN (eds) Fuel processing and energy utilization. CRC Press, New York, pp 69–90

Singh S, Kumar R, Setiabudi HD, Nanda S, Vo DVN (2018) Advanced synthesis strategies of mesoporous SBA-15 supported catalysts for catalytic reforming applications: a state-of-the-art review. Appl Catal A Gen 559:57–74

Statista (2018) Global biofuel production by select country 2017. https://www.statista.com/statistics/274168/biofuel-production-in-leading-countries-in-oil-equivalent. Accessed 3 June 2019

Tavanarad M, Meshkani F, Rezaei M (2018) Production of syngas via glycerol dry reforming on Ni catalysts supported on mesoporous nanocrystalline Al_2O_3. J CO2 Util 24:298–305

Vo DVN, Adesina AA (2011) Kinetics of the carbothermal synthesis of Mo carbide catalyst supported on various semiconductor oxides. Fuel Process Technol 92:1249–1260

Vo DVN, Cooper CG, Nguyen TH, Adesina AA, Bukur DB (2012) Evaluation of alumina-supported Mo carbide produced via propane carburization for the Fischer–Tropsch synthesis. Fuel 93:105–116

Wang W (2010) Thermodynamic analysis of glycerol partial oxidation for hydrogen production. Fuel Process Technol 91:1401–1408

Wang D, Czernik S, Montane D, Mann M, Chornet E (1997) Biomass to hydrogen via fast pyrolysis and catalytic steam reforming of the pyrolysis oil or its fractions. Ind Eng Chem Res 36:1507–1518

Wang C, Dou B, Chen H, Song Y, Xu Y, Du X, Luo T, Tan C (2013) Hydrogen production from steam reforming of glycerol by Ni–Mg–Al based catalysts in a fixed-bed reactor. Chem Eng J 220:133–142

Wang Y, Chen M, Yang Z, Liang T, Liu S, Zhou Z, Li X (2018) Bimetallic Ni-M (M= Co, Cu and Zn) supported on attapulgite as catalysts for hydrogen production from glycerol steam reforming. Appl Catal A Gen 550:214–227

Wei Z, Karim AM, Li Y, King DL, Wang Y (2015) Elucidation of the roles of Re in steam reforming of glycerol over Pt–Re/C catalysts. J Catal 322:49–59

Conversion of Biogas to Syngas via Catalytic Carbon Dioxide Reforming Reactions: An Overview of Thermodynamic Aspects, Catalytic Design, and Reaction Kinetics

18

Doan Pham Minh, Ahimee Hernandez Torres, Bruna Rego de Vasconcelos, Tan Ji Siang, and Dai-Viet N. Vo

Abstract

Biogas production has continuously increased worldwide during the last decades. Nowadays, heat, electricity, and biomethane production are the main utilization of biogas at large-scale industrial processes. The research and development on biogas valorization is currently related to synthesis gas production via reforming process, since syngas allows obtaining various chemicals and fuels of high-added value. However, biogas reforming is a complex process, which implies various reactions in parallel, and needs high temperature (>800 °C) to obtain high methane conversion. The development of a highly-performing catalyst, which must be active, selective, thermally stable, and resistant to solid carbon formation on its surface, is crucial. This chapter is devoted to an update of the thermodynamic aspect of biogas reforming under different conditions. This chapter

D. Pham Minh (✉) · A. H. Torres
Université de Toulouse, IMT Mines Albi, Centre RAPSODEE, UMR CNRS 5302, Albi, France
e-mail: doan.phamminh@mines-albi.fr

B. Rego de Vasconcelos
Department of Chemical and Biotechnological Engineering, Université de Sherbrooke, Sherbrooke, Québec, Canada

T. J. Siang
School of Chemical and Energy Engineering, Universiti Teknologi Malaysia, Johor, Malaysia

D.-V. N. Vo
Center of Excellence for Green Energy and Environmental Nanomaterials, Nguyễn Tất Thành University, Hồ Chí Minh City, Vietnam

S. Nanda et al. (eds.), *Biorefinery of Alternative Resources: Targeting Green Fuels and Platform Chemicals*, https://doi.org/10.1007/978-981-15-1804-1_18

also reviews recent significant works related to catalyst design as well as kinetic and mechanistic studies of biogas reforming processes.

Keywords
Biogas · Dry reforming of methane · Tri-reforming of methane · Catalyst · Kinetic study

18.1 Introduction

The green alternatives to fossil fuel have been constantly developed in order to reduce the greenhouse gas emissions. One of the possible solutions is related to biogas, which is considered as a renewable energy source to replace conventional fuels to produce heat, electricity, and chemicals. Biogas is obtained from the decomposition of organic matter under anaerobic conditions (absence of oxygen). The main feedstocks for biogas production include energy crops (e.g., sunflower, wheat grain, sugar beet, etc.), agricultural residues, animal manure, municipal solid waste (MSW), industrial wastes, wastewater, and sewage sludge. These feedstocks contain mostly complex organic compounds, which are degraded by microorganisms under the absence of air through four main steps, such as hydrolysis, acidogenesis, acetogenesis or dehydrogenation and methanation (Jaffrin et al. 2003; Weiland 2010; Mudhoo 2012; Pullen 2015). The hydrolysis of complex polymers into oligomers and monomers (e.g., sugars, amino acids, fatty acids) takes place in the first step. These oligomers and monomers are then degraded into volatile fatty acids ($>C_2$), acetate, and hydrogen. The volatile fatty acids are later converted into acetate, H_2, and CO_2 by acetogenic bacteria. Finally, acetate, H_2, and CO_2 produce CH_4 through methanation. This multi-step transformation needs specific conditions including the absence of air (anaerobic condition), uniform temperature, optimum nutrient supply, optimum and uniform pH (Kaparaju and Rintala 2013).

From the technical point of view, large-scale biogas production can be done in an anaerobic digestion plant or in a landfill site, leading to the formation of digested gas and landfill gas, respectively. The size of these bioreactors varies from some cubic meters to several thousand cubic meters (Persson et al. 2009; Mudhoo 2012; Pullen 2015; Di Maio et al. 2018). Depending on the reactor design, feedstock used, and operational conditions, the composition of biogas can vary. However, biogas typically contains mostly CH_4, CO_2, H_2, N_2, O_2, light hydrocarbons, water vapor, and some pollutants (hydrogen sulfide, ammoniac, siloxanes, and organometallic complexes) (Persson et al. 2009; Pullen 2015; Mudhoo 2012; Kvist and Aryal 2019; Hoo et al. 2018; Eklund et al. 1998; Biogas Renewable Energy 2019). The digested gas generally contains more CH_4 (50–75 vol%) and less CO_2 (30–40 vol%) than landfill gas (CH_4: 35–65 vol% and CO_2: 15–50 vol%). Landfill gas contains varying amounts of N_2 (5–40 vol%) and O_2 (0–5 vol%) while the digested gas generally contains <2 vol% of these gases. This is due to the difficulty of sealing control in landfill sites during landfill gas recovery. Thus, the lower heating value (LHV) of the digested gas (5.5–7.5 kwh/Nm^3) is higher than that of landfill gas (around 4.4 kwh/Nm^3) (Okonkwo et al. 2018; Persson et al. 2009).

Biogas production worldwide continuously increases during the last decades. According to IRENA (2017), the worldwide installed capacity for biogas production increased from 6323 MW in 2007 to 15,752 MW in 2016. Europe represents the highest biogas production capacity with 10,533 MW in 2016 or 66.9% followed by North America with 2639 MW or 16.8% in the same year. In Europe, the top-five biogas producers are, in descending order, Germany, UK, Italy, France, and Netherlands (Achinas et al. 2017). In addition, Scarlat et al. (2018) predicted that biogas production in Europe would still increase with the similar rate during the next years.

Among different feedstocks, MSW is particularly interesting because of its specific characteristics. MSW is generated by households and includes various types of organic and inorganic wastes, which are mostly food waste, waste paper, cardboard, wood, plastic, rubber, and glass. However, it should be noted that the following wastes are not considered as MSW sources, e.g., municipal wastewater sludge, industrial wastes, automobile bodies, combustion ash, or construction and demolition debris. According to Kaza et al. (2018), the worldwide MSW production is predicted to increase from 2.01 billion tons in 2016 to 3.04 billion tons in 2030, mostly due to population explosion and urbanization. MSW composition largely depends on the income of each country. In high-income countries, MSW contains less food waste but more paper and cardboard wastes (from packaging), more rubber and leather (from used tires, clothes, and accessories) compared to low-income, lower-middle, and upper-middle income countries (Kaza et al. 2018). An example of the typical MSW composition in the United States is presented in Table 18.1. The total organic biomass content represents 69.5 wt% illustrating the high potential of biogas generation from MSW.

Up-to-date, the MSW collection is different from each country. In high-income countries, most of the MSW is collected (~100% for both urban and rural zones). This collection rate is lower in other countries, around 85%, 71%, and 45% from urban municipalities in upper-middle, lower-middle, and low-income countries, respectively. In contrast, 45%, 33%, and 26% of MSW are obtained from the rural municipalities in the upper-middle, lower-middle, and low-income countries, respectively. Thus, the effort is still needed for improving MSW collection in these countries. The MSW management by landfill technique is also different from each country. In several developing countries of low and lower-middle income, and

Table 18.1 MSW composition in the United States (Themelis and Ulloa 2007)

Biomass components	Weight %	Petrochemical components	Weight %
Paper/board	36.2	Plastics	11.3
Wood	5.8	Rubber, nylon, etc.	3.7
Yard trimmings	12.1	–	
Food scraps	11.7	–	
Cotton, wool, and leather	3.7	–	
Total biomass	69.5	Total man-made	15

particularly in rural zone, MSW is still burnt in open atmosphere without considering the pollution impacts, technical specification of waste treatment. MSW in some developing countries is treated in open dumping sites without leachate treatment and biogas recovery. In developed countries, sanitary landfill sites are used, which are equipped with lining systems for leachate and biogas recovery, geo-membrane for securing underground water, covering and drainage systems (Youcai and Ziyang 2017). More research is invited to improve the valorization of biogas generated by landfill sites in the world.

18.2 Biogas Utilization and Valorization

18.2.1 Heat and Electricity Production

Biogas mainly contains CH_4 and CO_2 together with other compounds (e.g., water vapor, N_2, O_2, H_2S, NH_3, and siloxanes). One of the primary valorization techniques is the production of heat and electricity. Biogas production in rural areas can be directly used for cooking, household heating, and street lightening without purification or upgrading. This is largely applied in developing countries with the financial incentives by local governments to boost cost-effective waste recycling, local income, and simple utilization (Yasmin and Grundmann 2019; Clemens et al. 2018; Pradhan and Limmeechokchai 2017; Aziz et al. 2019; Khalil et al. 2019; Mittal et al. 2018).

In large-scale urban regions, biogas can be used as a fuel for producing heat and electricity. Biogas utilization is found in the following aspects (Kaparaju and Rintala 2013):

1. Production of heat and/or steam using a boiler.
2. Production of electricity combined with waste heat recovery (combined heat power).
3. Production of heat, steam and/or electricity and cooling using a micro-turbine.
4. Production of electricity using a fuel cell.

It is worth mentioning that raw biogas must be purified before its utilization in an engine to removal pollutants like sulfur-, nitrogen- and chlorine-containing compounds, siloxanes, etc. (Kaparaju and Rintala 2013). In general, the requirement for biogas purification can be classified as follows: fuel cell > gas engine, gas turbine and micro-turbine > stirling engine > boiler (Kaparaju and Rintala 2013).

18.2.2 Biomethane Production

Biogas can also be upgraded into biomethane having the quality of natural gas. This biomethane can be then injected to the gas grid. This valorization way is nowadays commercialized in several countries. The pollutants and CO_2 must be removed from biogas to increase CH_4 content to at least 80 vol% (Elfattah et al. 2016). Different methods are available for biogas upgrading including water scrubbing, chemical scrubbing, biological absorption, pressure swing adsorption, membrane separation, and cryogenic technology, although each technology has its own advantages and disadvantages (Khan et al. 2017; Sahota et al. 2018; Angelidaki et al. 2018).

Water scrubbing is a simple method with low operation cost and allows CH_4 recovering of more than 80%. However, it requires high energy and water consumption, and has high corrosion risk because of the dissolution of acid gas like hydrogen sulfide. Chemical scrubbing is faster and more efficient, but also more expensive than water scrubbing. Adsorption needs specific operational conditions of pressure, temperature, and adsorbent to reach high efficiency (Pertiwiningrum et al. 2018). According to Angelidaki et al. (2018), the applied commercial technologies for biogas upgrading are as follows: water scrubbing > chemical scrubbing ~ membrane separation > pressure swing adsorption > other technologies. As previously presented, France is in the top-five biogas producers in Europe. The number of biomethane injection stations to the gas grid in France was 37 in June 2017, which has increased to 87 in May 2019 with a total production of 1382 GWh/year (GRDF 2019).

18.2.3 Syngas Production

Synthetic gas (or syngas—a gas mixture rich in CO and H_2) is well known as a versatile platform gas to produce various chemicals and fuels, e.g., methanol, light hydrocarbons, gasoline, diesel, waxes, and alcohols (Brown 2011; Puigjaner 2011; Abdoulmoumine et al. 2015; Liu et al. 2010; Shah 2017; Asimakopoulos et al. 2018; Abdulrasheed et al. 2019). The commercial syngas is produced from fossil resources, e.g., coal, natural gas and heavy hydrocarbons by catalytic steam reforming process (Schildhauer and Biollaz 2016; Liu et al. 2010; Shah 2017). Biogas, mainly containing CH_4 and CO_2, can be used as a renewable feedstock for syngas production.

Syngas production by catalytic steam reforming of biogas has been deployed at large-scale facilities. In the case of the VABHYOGAZ project in France, purified landfill gas was reformed with a molar ratio of steam/water around 4:1 to produce syngas, which is similar to the conditions of natural gas reforming, according to Eq. (18.1). This syngas was later transformed into hydrogen by the water-gas shift process (Grouset and Ridart 2018; Nanda et al. 2017). However, the large excess of steam makes the process highly intensive in energy because the reforming takes place around 900 °C. Current research on biogas reforming into syngas is focused on catalytic dry reforming and tri-reforming processes.

As previously mentioned, biogas mainly contains CH_4 and CO_2. This gas mixture can be transformed into syngas according to Eq. (18.2) which is the chemical reaction of dry reforming of methane (DRM). In this process, CO_2 is the principal oxidizing agent of the reforming of the CH_4 molecule. However, biogas generally contains more CH_4 than CO_2. In addition, water vapor and oxygen are generally present in biogas and can play the role of oxidizing agents in the steam reforming (Eq. 18.1) and partial oxidation (Eq. 18.3) of methane. The combination of three reactions, steam reforming, dry reforming, and partial oxidation of methane constitutes the process of tri-reforming of methane (TRM) (Eq. 18.4) (Table 18.2).

Table 18.2 Methane reforming reactions and side reactions

Reaction	Equation	
Steam reforming	$CH_4 + H_2O \leftrightarrow 3H_2 + CO$ ($\Delta H^\circ_{298} = 206.3$ kJ/mol)	Eq. (18.1)
Dry reforming	$CH_4 + CO_2 \leftrightarrow 2H_2 + 2CO$ ($\Delta H^\circ_{298} = 247$ kJ/mol)	Eq. (18.2)
Partial oxidation	$CH_4 + 0.5\ O_2 \leftrightarrow 2H_2 + CO$ ($\Delta H^\circ_{298} = -30.6$ kJ/mol)	Eq. (18.3)
Tri-reforming	$xCH_4 + yH_2O + zCO_2 + tO_2 \leftrightarrow uH_2 + vCO$	Eq. (18.4)
Water-gas shift	$CO_2 + H_2 \leftrightarrow CO + H_2O$ ($\Delta H^\circ_{298} = 41$ kJ/mol)	Eq. (18.5)
Methane cracking	$CH_4 \leftrightarrow C_{(s)} + 2H_2$ ($\Delta H^\circ_{298} = 75$ kJ/mol)	Eq. (18.6)
Boudouard reaction	$2CO \leftrightarrow C_{(s)} + CO_2$ ($\Delta H^\circ_{298} = -175$ kJ/mol)	Eq. (18.7)
Carbon gasification	$C_{(s)} + H_2O \leftrightarrow CO + H_2$ ($\Delta H^\circ_{298} = 131$ kJ/mol)	Eq. (18.8)
Carbon partial oxidation	$C_{(s)} + 0.5\ O_2 \leftrightarrow CO$ ($\Delta H^\circ_{298} = -110.5$ kJ/mol)	Eq. (18.9)
Carbon combustion	$C_{(s)} + O_2 \leftrightarrow CO_2$ ($\Delta H^\circ_{298} = -393.5$ kJ/mol)	Eq. (18.10)

Biogas reforming into syngas is accompanied by several side reactions (Eqs. 18.4–18.7), which have an important impact on methane conversion and syngas selectivity as well as catalyst stability. This chapter is focused on the thermodynamic aspects, catalytic design, and reaction kinetics of dry- and tri-reforming of methane. The study of the thermodynamic equilibrium allows determining the limit of a given chemical process under defined conditions of temperature, pressure, and initial composition. The common method of the thermodynamic equilibrium calculation is based on the minimization of Gibbs free energy (Faungnawakij et al. 2007; Nikoo and Amin 2011; Néron et al. 2012).

Nikoo and Amin (2011) have detailed the basis of this method for the dry reforming of methane. Chemical compounds at both gas and solid phases have been considered. The solid considered in this process is related to solid carbon formed during the reaction. The gaseous molecules considered include CH_4, CO_2, CO, H_2, water vapor, methanol, ethylene, ethane, and dimethyl ether (Nikoo and Amin 2011). In the literature, work has been devoted to the thermodynamic study of dry- and tri-reforming of methane (Song and Pan 2004; Faungnawakij et al. 2007; Nikoo and Amin 2011; Jafarbegloo et al. 2015; Abdullah et al. 2017; Pham Minh et al. 2018a, b; Cao et al. 2018; Chein et al. 2017; Chein and Hsu (2018); Rahnama et al. 2014; Diez-Ramirez et al. 2016; Rego de Vasconcelos (2016)). Similar conclusions have been obtained by these studies. This section summarizes the main characteristics of each process.

18.3 Dry Reforming of Methane

For dry reforming of the methane process, Fig. 18.1 shows the amounts of different species in equilibrium from an equimolar mixture of CH_4 and CO_2 (1 mole for each molecule) at 1 and 16 atm. These results were obtained by ASPEN Plus software on the principle of Gibbs free energy minimization described elsewhere (Nikoo and Amin 2011). When the gas mixture is set at 1 atm, solid carbon ($C_{(s)}$) and water vapor are strongly favored below 700 °C. On the other hand, H_2 and CO are predominant above 700 °C. The conversion of CH_4 and CO_2 as well as the selectivity

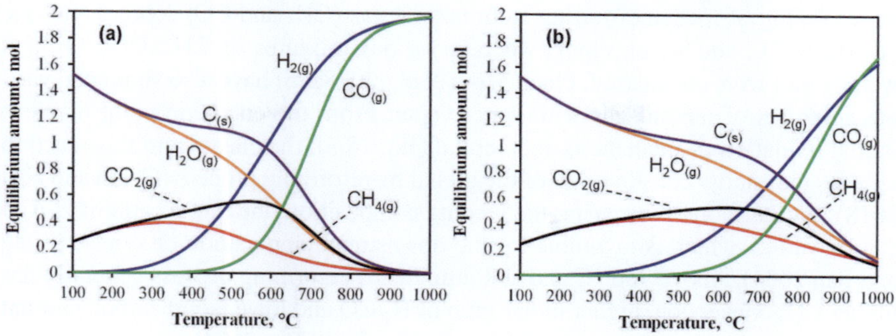

Fig. 18.1 Thermodynamic equilibrium of a gas mixture initially composed of 1 mol of CH_4 and 1 mol of CO_2 at (**a**) 1 atm and (**b**) 16 atm

of H_2 and CO are significantly favored above 900 °C. However, catalyst thermal sintering can occur at this high temperature, which is considered as irreversible catalyst deactivation. Therefore, the design of a thermally stable catalyst is a main challenge of dry reforming of methane. Another characteristic of the dry reforming process is related to solid carbon formation (Eqs. 18.6 and 18.7).

Wang and Lu (1996) have shown that the upper limit of the Boudouard reaction is equal to 700 °C while the lower limit of the methane cracking reaction is equal to 557 °C. Theoretically, there is no temperature range wherein solid carbon formation could be absent. Therefore, dry reforming catalyst must be not only thermally stable but also resistant to solid carbon formation. The formation of solid carbon can be reduced by increasing the molar ratio of CO_2/CH_4. By keeping the temperature at 700 °C and the total pressure at 1 atm, Rego de Vasconcelos (2016) showed that no solid carbon is formed when the molar ratio of CO_2/CH_4 is equal to 2.5/1. However, increasing this ratio also favors the formation of water vapor and slightly decrease the H_2 amount.

Dry reforming of methane produces more gaseous molecules than the initial gaseous reactants (Eq. 18.2). Thus, increasing pressure does not favor this reaction. This decreases CH_4 and CO_2 conversion as well as H_2 and CO selectivity while increases the selectivity in water vapor and solid carbon (Nikoo and Amin 2011; Rego de Vasconcelos 2016; Abdullah et al. 2017; Jafarbegloo et al. 2015). Figure 18.1b illustrates the equilibrium amount of an equimolar mixture of CH_4 and CO_2 when the total pressure is set at 16 atm. At 1000 °C and 16 atm, the unreacted CH_4 and CO_2 amounts are 11.1 and 8.8%, respectively, which are much higher than those at 1000 °C and 1 atm (<1%).

18.4 Tri-reforming of Methane

Tri-reforming process combines steam, dry reforming, and partial oxidation of methane. According to Eq. (18.4), CH_4 can be reformed by varying amounts of water vapor, CO_2, and O_2. Song and Pan (2004) have proposed a concept of tri-reforming of methane starting from natural gas (CH_4 and CO_2 source) and flue gas (CO_2, O_2, and water vapor) wherein various mixtures of CH_4, CO_2, O_2, and water vapor were considered. Pham Minh et al. (2018a, b) have also simulated various mixtures of landfill gas with water vapor. From the energetic point of view, partial oxidation of methane is exothermal (Eq. 18.3), thus heat from this reaction reduces the energy consumption of the global tri-reforming process (Chein and Hsu 2018). Furthermore, by varying the feeding composition, the molar ratio of H_2/CO can be varied, which is favorable for the downstream application of syngas (Song and Pan 2004). Above 750 °C and at 1 atm, the tri-reforming process generally has higher CH_4 conversion, higher molar ratio of H_2/CO and lower solid carbon amount compared to simple dry or steam reforming (Song and Pan 2004).

Figure 18.2 shows the thermodynamic equilibrium of a mixture initially composed of 2 mol of CH_4, 1 mol of CO_2, 1 mol of water vapor, and 0.1 mol of O_2 at 1 and 16 atm. The sum of oxidizing agents (O_2, CO_2, and water) is in excess for completely converting methane into syngas according to Eq. (18.1–18.3). The composition of this mixture is obtained from the addition of water vapor to a purified landfill gas having the molar ratio of $CH_4/CO_2/O_2$ of 2/1/0.1. At the total pressure of 1 atm, high temperature of at least 800 °C is required to selectively convert this mixture to syngas. Under these conditions, solid carbon could be prevented. In addition, increasing the pressure to 16 bar has negative impacts on reaction conversion and selectivity. At this pressure and 1000 °C, CH_4 conversion only reaches 86.3%.

From the thermodynamic point of view, both dry- and tri-reforming of methane are strongly influenced by the initial feeding composition, temperature, and total pressure. The gas mixture containing more oxidizing agents (e.g., water, CO_2, and O_2) than the stoichiometric amounts favors CH_4 conversion. High temperature also favors CH_4 conversion and syngas selectivity because of the endothermicity of steam and dry reforming of methane. On the other hand, high pressure is unfavorable considering the stoichiometric coefficients of methane reforming reactions.

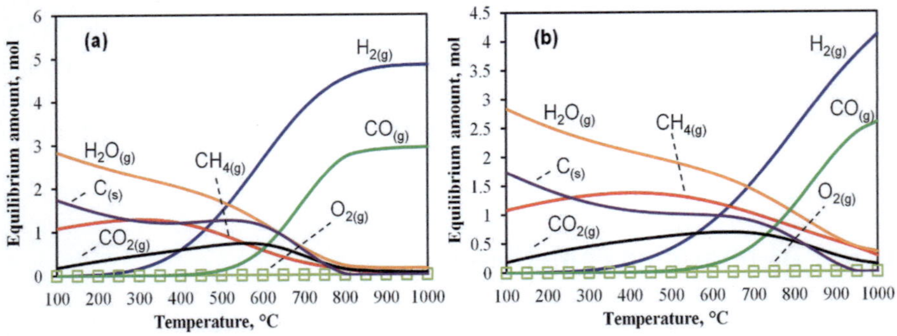

Fig. 18.2 Thermodynamic equilibrium of a gas mixture initially composed of 2 mol of CH_4, 1 mol of CO_2, 1 mol of water vapor, and 0.1 mol of O_2 at (**a**) 1 atm and (**b**) 16 atm

18.5 Catalysts Development

The development of an active and stable catalyst is one of the key points necessary for the implantation of biogas reforming technologies at large-scale operations. Catalyst deactivation related to carbon deposition, sulfur poisoning, and catalyst sintering is one of the biggest issues related to catalytic performance. Therefore, researches have focused on the development of more resistant catalysts using, for example, combination of different active metals, new supports, and promoters. Studies developed through the years on this area have shown that the performance of the catalysts mainly depends on the active metal used on the nature of the support, on the metal-support interaction as well as on other features, such as metal particle size and catalyst surface area. The following topics will discuss the recent advances in catalyst design and application in dry- and tri-reforming of biogas.

18.5.1 Catalysts in Dry Reforming of Methane

Noble metal-based catalysts have been reported as highly active and stable catalysts for dry reforming of biogas with only low amounts of coke formation (Papadopoulou et al. 2012). The main reasons related to the coke resistance of these catalysts are the small equilibrium constants for the decomposition of CH_4 as well as the carbon dissolution into their lattices (Papadopoulou et al. 2012). These findings were reported by Rostrup-Nielsen and Hansen (1993) when they compared the performance of different noble metals catalysts (Ru, Ir, Pd, Pt, and Rh) on dry reforming of methane at temperatures varying between 500 °C and 650 °C. Their results showed that ruthenium (Ru) and rhodium (Rd) had the best catalytic performances with high selectivity for carbon-free operation, thus proving the high coking resistance. Similarly, Hou et al. (2006) evaluated the performance of several noble metals (Ru, Pd, Pt, and Ir) doped on alumina (α-Al_2O_3) catalysts in the dry reforming of methane reaction at 950 °C. The results proved the coking resistance and the stability of these noble metal catalysts with CH_4 and CO_2 conversions higher than 84%. The following order of performance was observed: Rh > Ru > Ir > Pd > Pt. The less efficient performance of the Pd- and Pt-based catalysts was related to the metal particles sintering at high temperature.

Despite the high activity and stability of noble metal-based catalysts, their high cost and low availability render them unattractive for industrial-scale uses. Transition metals have become a suitable alternative due to their low cost, high availability, and good activity in the dry reforming reaction (Pakhare and Spivey 2014). Nickel-based catalysts are often studied since they are already used for advanced reforming process at industrial scale, such as steam reforming (Eq. 18.1). Goula et al. (2015) investigated the performance of Ni/Al_2O_3 catalysts with nickel (Ni) loadings varying between 8 and 16 wt% on the biogas reforming reaction. The authors reported that the catalysts with lower nickel loading exhibited higher specific surface area and low nickel particles than the catalysts with high nickel loadings, which was related to the coverage of the internal surface area by the nickel species adsorbed on the alumina active sites. However, the authors reported that the nickel loading had no direct effect on the catalytic performance. In this case, the preparation method showed to be the most influencing parameter since it had a direct impact on the reducibility of the nickel species and thus on the overall performance of the catalysts. The 16 wt% Ni/Al_2O_3 catalyst prepared by incipient wetness impregnation method showed the best catalytic performance at 750 °C with CH_4 and CO_2 conversions equaled to 69.1% and 88.3%, respectively.

Despite their low cost and proven catalytic activity, transition metal-based catalysts have been repeatedly reported as more prone to deactivation due to carbon deposition than noble metal-based catalysts (Pakhare and Spivey 2014). Kroll et al. (1997) showed Ni/SiO_2 as a better performing catalyst in the dry reforming reaction at 700 °C showing CH_4 conversion of 93% during the first hours of reaction. However, the catalyst quickly deactivated after less than 10 h of reaction due to carbon deposition. Hence, recent studies have focused on improving the catalytic properties of transition metal-based catalysts.

Bi-metallic catalysts using a combination of two transition metals or between transition metals and noble metals have been tested to improve the catalytic properties of the catalysts. Among the various combinations of transition metals used, nickel (Ni) and cobalt (Co) are by far the most investigated combination due to the synergy between these two metals as well as to the property of cobalt of decreasing the rate of coke formation (Estephane et al. 2015). The influence of cobalt addition to the performance of Ni/ZSM5 catalyst on dry reforming reaction at temperatures ranging from 600 °C to 800 °C was investigated by Estephane et al. (2015). The authors suggested that the cobalt presence could help oxidizing the carbon species close to the catalytic sites, thus increasing the stability of the catalyst. The best catalytic result was obtained using a Ni/Co ratio equal to 1:2. This catalyst exhibited CH_4 and CO_2 conversion equal to 80% and 85%, respectively, at 800 °C, proving the good catalytic performance of this bi-metallic catalyst.

Similar results were achieved by Xu et al. (2009). The addition of cobalt significantly increased the catalytic performance of the Ni/La_2O_3-Al_2O_3 catalyst. The authors reported CH_4 and CO_2 conversions equal to 93.7% and 94%, respectively, in biogas reforming at 800 °C and 290 h when a bi-metallic catalyst with a Ni/Co molar ratio equal to 7:3 was used. Besides carbon deposition, sulfur poisoning also leads to catalyst deactivation during biogas reforming. Recent studies have shown that the use of bi-metallic catalysts could improve the sulfur resistance of the catalysts. Sapountzi et al. (2018) showed that doping Ni/GDC (gadolinia-doped ceria) catalyst with 2.3 wt% of gold (Au) hindered the formation of strongly bonded sulfur components by producing an Au-Ni alloy.

The textural and chemical promoters have also been used to improve the catalytic performance of the catalysts. According to Jang et al. (2019), promoters are usually used to enhance the textural properties of the catalysts, which helps preventing or delaying sintering of the active phase. Chemical promoters provide new active sites or enhance chemical properties, such as basicity and redox property, which prevent carbon formation (Jang et al. 2019). San José-Alonso et al. (2011) investigated the use of potassium (K) and strontium (Sr) on the performance of Co/Al_2O_3 on CO_2 reforming of CH_4 at 700 °C. The authors reported that the Co/Al_2O_3 catalyst showed CH_4 and CO_2 conversions of 73% and 82%, respectively. The percentage of deposited carbon was 0.53%. When this catalyst was doped with 1 wt% K or 1 wt% Sr, CH_4 and CO_2 conversions decreased to about 15%. The percentage of deposited carbon also decreased, achieving 0.08% when K was used. The main reason for the decrease in the CH_4 conversion was related to the partial coverage of the active sites for CH_4 decomposition and to the ability of the catalyst to gasify the carbon species.

Ceria (Ce) has also been reported to improve the coke resistance of the catalysts. Daza et al. (2010) investigated the effect of Ce addition and loading on Ni/Mg-Al catalyst on dry reforming at 700 °C. The addition of 1 wt% of Ce to the catalyst was responsible for increasing CH_4 conversion from 34.9% to 88.3%. The authors reported that Ce addition increased the reducibility and the total basicity of the catalyst, which was responsible for the increase in the catalytic activity. However, increasing Ce content from 1 to 10 wt% did not have an influence on the catalyst activity but it greatly hindered the formation of coke deposits. The catalyst doped with Ce was stable in the reaction conditions during 100 h of reaction.

18.5.2 Catalysts in Tri-reforming of Methane

As presented in the previous sections, tri-reforming is a synergetic combination between dry reforming, steam reforming, and partial oxidation. Hence, it is not surprising that the catalytic systems used in the tri-reforming process are similar to those used in the other advanced reforming process and face similar problems, such as low stability due to coke deposition and metal sintering. However, the tri-reforming process is a much more complex process due to the presence of three oxidizing agents (e.g., O_2, CO_2, and H_2O), requiring a catalyst capable of adsorbing and activating all three reactants efficiently as well as to withstand the highly oxidizing atmosphere. The following topics will briefly present the recent advances in catalyst development for this process as well as the efforts for scaling up the technology.

Nickel supported on conventional catalytic supports, such as Al_2O_3, ZrO_2, SiO_2, and CeO_2 is the most studied active metal. However, these catalysts may behave differently in tri-reforming conditions mainly due to their re-oxidation by the oxygen in the feed, which leads to their deactivation (Amin et al. 2015). Majewski and Wood (2014) investigated the performance of Ni/SiO_2 core shell catalyst in steam reforming of methane in tri-reforming conditions at 750 °C. The authors reported that changing the percentage of each oxidizing agent (O_2, CO_2, and H_2O) in the feed mixture had a different impact on the performance of the catalyst. Increasing the amount of oxygen in the feed mixture increased the CH_4 conversion to 90% but did not improve the H_2/CO ratio in the produced syngas. When steam was not present in the feed, CO_2 conversion increased to 90%, but CH_4 conversion decreased. When steam was added with a CH_4/H_2O ratio of 1:0.5, the H_2/CO ratio of the syngas decreased from 2.5 to 1.5 and coke deposition increased from 5 mg $g_{cat}{}^{-1}$ up to 99 mg $g_{cat}{}^{-1}$. These results show the potential of Ni/SiO_2 core-shell catalyst in the tri-reforming process. However, it also highlights the complexity to maintain high reactants' conversion and good catalyst stability.

The role of the support in tri-reforming is also not yet defined and further studies to correlate the physicochemical properties of the catalysts to their performance are still needed. Different supports have been used to prepare catalysts for tri-reforming reaction, including CeO_2 (Pino et al. 2011), TiO_2 (Jiang et al. 2007), ZrO_2 (Si et al. 2012; Singha et al. 2016; Anchieta et al. 2019), $MgAl_2O_4$ (Lino et al. 2019), and SiC (García-Vargas et al. 2014a, 2014b, 2015b). In attempt to investigate the influence of different support materials on the performance of nickel-based catalysts, Kumar et al. (2019) prepared different catalysts using Al_2O_3, ZrO_2, MgO, CeO_2-ZrO_2, SBA-15, and TiO_2 as the supports and evaluated their performance in tri-reforming at 800 °C and with a feed composition of CH_4/CO_2/H_2O/O_2/N_2 of 1:0.23:0.46:0.07:0.28. The authors reported that Ni/Al_2O_3 had the best catalytic performance in the test conditions used with CH_4, H_2O, and CO_2 conversion rates equal to 8.72×10^{-1} mmol g_{cat}^{-1} s, 4.2×10^{-2} mmol g_{cat}^{-1} s, and 2.31×10^{-2} mmol g_{cat}^{-1} s, respectively. However, Ni/CeO_2-ZrO_2 and Ni/ZrO_2 showed better stability against oxidation, due to their redox and oxophilic properties, respectively. On the other hand, Ni/MgO and Ni/TiO_2 catalysts showed only very low CH_4 and CO_2 conversion rates since they presented a very low degree of reducibility. The authors concluded that small Ni particles with high dispersion, high degree of reducibility, strong metal-support interaction as well as high concentration of basic sites on the surface of the catalyst are crucial for the good catalytic performance of the catalysts.

García-Vargas et al. (2014b) investigated the performance of Ni/Al_2O_3 and Ni/CeO_2 catalysts in tri-reforming conditions at 800 °C. The best catalytic result was obtained using Ni/CeO_2 catalyst, showing CH_4 and CO_2 consumption rates around 8 and 3 mol s^{-1} g_{Ni}^{-1}, respectively, which was explained by its basic properties. Ni/Al_2O_3 showed the lowest CH_4 and CO_2 conversion rates due to the formation of $NiAl_2O_4$ phase, thus decreasing the reducibility of the catalyst. Pino et al. (2011) have found that Ni/CeO_2 (5 wt%) catalysts prepared by flaming combustion at 350 °C were highly active and selective in tri-reforming of methane at 800 °C with methane conversion up to 93%. The addition of La slightly increased the catalytic performance of these catalysts.

Singha et al. (2016) investigated Ni/ZrO_2 catalysts synthesized by hydrothermal method in the reforming of different mixtures containing CH_4, CO_2, water vapor, and oxygen at 500–800 °C. In most of the case, CH_4 conversion likely linearly increased with the reaction temperature while the molar ratio of H_2/CO varied in narrow range of 1.7–2.0. The optimal Ni content was found at 4.8 wt%, explained by the dependency of Ni particle size on Ni content. At 800 °C, this catalyst was found very active and stable with reactants' conversion close to 96–98% for 100 h of reaction time. This catalytic performance was mostly explained by the formation of metal nanoparticles strongly interacted with the support.

During the last decade, much work has been devoted to the design of a performing catalyst for biogas reforming, e.g., dry- and tri-reforming. Nickel has been found as the most appropriate active metal for these processes. Such a catalyst is generally constituted of (a) a thermally stable support at 700–1000 °C, (b) highly dispersed metal particles (<40 nm), (c) strong metal-support interaction, and (d) controlled basicity. The addition of promoters could also improve catalyst performance. Despite encouraging results, large pilot application of dry- and tri-reforming is still limited.

18.5.3 Catalyst Supports

Besides the active metal and promoters, the nature of the support also plays a significant role in the performance of the catalysts. Catalyst supports provide mechanical and thermal stability to the catalysts and confer high specific surface area, leading to high metal dispersion, which favors the CH_4 adsorption and dissociation (Papadopoulou et al. 2012). Moreover, strong metal-support interaction is responsible for the high stability of the metal particles during the reaction, which will prevent sintering and coking (Jang et al. 2019). Other support features, such as basic-acid characteristics and oxygen storage capacity also play an important role in the overall performance of the catalysts (Mohamedali et al. 2018).

Al_2O_3 is one of the support materials most used for reforming applications due to its low cost and high mechanical and thermal stability (Papadopoulou et al. 2012). Mo et al. (2015) recently investigated the application of Ni/Al_2O_3 porous catalysts in dry reforming of CH_4. The catalyst synthesized by hydrolysis-deposition method had high specific surface area of 477 m^2/g and small nickel particles of about 7 nm, which ensured high CH_4 and CO_2 conversions of around 95% during 10 h of time on stream. However, depending on the synthesis method used, large metal particles (>10 nm) or even phases difficult to reduce, such as nickel aluminate can be formed, thus decreasing the performance of the catalyst. The acid sites present in this support rendered the alumina-based catalysts more prone to deactivation by carbon deposition.

Hydrotalcite-like materials containing aluminum (Al) and magnesium (Mg) have been intensively investigated for dry reforming applications mainly due to their basic character, which improve coke gasification to their high surface area and to the presence of di and trivalent cations that leads to a homogeneous distribution of the active phase in the matrix (Dębek et al. 2015). de Vasconcelos et al. (2018a) recently investigated the performance of MgAl hydrotalcite-based catalysts doped with Ni and with MgO loads varying from 0 wt% to 70 wt% in the dry reforming reaction at 700 °C. The authors reported that the catalyst with an MgO loading of 70 wt% presented the best catalytic performance with CH_4 and CO_2 conversions around 80% during 50 h of time on stream as well as the lowest deactivation rate of about 0.004% h^{-1}. The catalytic performance of this catalyst was related to the presence of strong basic sites that favored the carbon gasification during the reaction. Moreover, a NiO-MgO solid solution was formed, which stabilized the active phase and prevented its sintering.

Rare earth-based supports, specially ceria, have been also been studied due to its ability to enhance the availability of hydroxyl (OH_{ads}) and surface oxygen (O_{ads}) species, which controls the formation of carbon deposits (Papadopoulou et al. 2012). Wolfbeisser et al. (2016) studied the dry reforming reaction over ceria-zirconia-supported Ni catalysts at 600 °C. The authors reported that the presence of Ce did not improve the catalytic stability of the catalyst. However, it considerably increased its stability but hindering the formation of filamentous coke. The authors further reported that the influence of ceria strongly depends on the catalyst synthesis method, which can affect the presence of surface oxygen species and of oxygen vacancies. These factors are crucial for the carbon removal at the metal/support surface.

Ceria/zirconia-based supports have been investigated by many authors (Djinovic et al. 2012, 2015; Aw et al. 2014a, b, 2015; Djinovic and Pintar 2017). Solid solutions of ceria/zirconia with high specific surface area, high oxygen storage capacity, and high interaction with nanoparticles of nickel and cobalt as active metals have been found as the main factors responsible for high catalytic performance in dry reforming of methane. These catalysts strongly limited the formation of coke during dry reforming reaction. However, water has been usually found as byproduct with high selectivity (up to 20% in the gas mixture at the reactor outlet) (Aw et al. 2015), suggesting that the designed catalyst favors reversed water-gas shift reaction.

Finally, hydroxyapatite (HAP, chemical formula: $Ca_{10}(PO_4)_6(OH)_2$) has also been found as a new promising catalyst support for the dry reforming of methane (Boukha et al. 2007, 2019; de Vasconcelos 2016; de Vasconcelos et al. 2018a, c; Phan et al. 2018). Hydroxyapatite-supported metals catalysts (e.g., Pt, Ni, and Co) have been found active, selective, and stable in dry reforming reactions and are competitive compared to catalysts prepared from conventional supports like Al_2O_3 and $MgAl_2O_4$ (de Vasconcelos et al. 2018a). The performance of hydroxyapatite in this reaction could be explained by a high thermal stability, a controlled acido-basicity via the molar ratio of Ca/P, and a good cation exchange capacity allowing a strong support-metal interaction between metal nanoparticles and hydroxyapatite surface (de Vasconcelos 2016; Boukha et al. 2019).

18.6 Kinetic Models and Mechanism Aspects

In this section, an overview of reaction mechanisms and kinetics for dry- and tri-reforming of methane are systematically discussed. Understanding the inherently catalytic mechanisms and establishing comprehensive kinetic models for precisely predicting the consumption and formation rates of the corresponding reactants and products are crucial for catalyst design, process optimization, and reactor fabrication and scale-up.

18.6.1 Kinetics of Dry Reforming of Methane

Dry reforming of methane is widely known as a highly endothermic reaction, which is thermodynamically favored at high reaction temperature (Aramouni et al. 2018). However, at such high reaction temperature, dry reforming is also accompanied by several parallel side reactions such as CH_4 decomposition, which provoke carbonaceous deposits leading to catalytic deactivation (Pakhare and Spivey 2014;

Kawi et al. 2015; Bian et al. 2017). Therefore, the understanding of fundamental dry reforming mechanism is crucial for kinetically resisting carbonaceous formation and simultaneously enhancing syngas yield as well as H_2 selectivity. Generally, the mechanistic pathways for syngas production from dry reforming of methane can be summarized into four major mechanisms, such as: (a) dissociative CH_4 adsorption, (b) dissociative CO_2 adsorption, (c) formation of hydroxyl groups, and (d) oxidation and desorption of intermediate species. The comprehensive review for these main dry reforming mechanisms is provided as follows:

1. Dissociative CH_4 adsorption: The dissociation energy for CH_x-H bonds in CH_4 reactant on the catalyst surface is mainly reliant on the intrinsic attributes of active phases and CH_4 dissociation step is widely recognized as the rate-determining step in dry reforming of methane (Tang et al. 2014; Horn and Schlögl 2015). In particular, CH_3 species are adsorbed on the top of active metal atoms, whereas CH_2 species are adsorbed in the middle of two active metal atoms also referred as bridged adsorption (Tang et al. 2014).
2. Dissociative CO_2 adsorption: CO_2 adsorption and dissociation are dependent on surface structure and structural defects of catalysts (Aramouni et al. 2018). In fact, this process can occur through three routes including C-only coordination, O-only coordination (in which an active metal surface is bonded with two O-atoms), as well as C and O coordination (wherein C and O atoms are adsorbed on catalyst surface meanwhile another O atom is left to expose) (Papadopoulou et al. 2012; Guharoy et al. 2018).
3. Formation of hydroxyl groups: Numerous dry reforming kinetic studies reveal that the water-gas shift (WGS) reaction nearly reaches equilibrium, indicating the rapid rate of associated surface reaction. The adsorbed H species from H_2 spillover on active metal surface reportedly migrate to support and subsequently react with adsorbed O-species to form hydroxyl groups. However, this process is unfavorable at high temperature beyond 800 °C (Papadopoulou et al. 2012; Bobadilla et al. 2017).
4. Oxidation and desorption of intermediates: The adsorbed CH_x species on active sites, denoted as S-CH_x, could react with surface oxygen leading to the formation of S-CH_xO and/or S-CO groups. Several proposed pathways for CO formation reportedly include the generation of intermediate S-CH_xO species and reduction of carbonates (formed by adsorbed CO_2) by surface carbon species. In general, the formation and/or decomposition of S-CH_xO species to H_2 and CO gases are recognized as the rate-determining step (Papadopoulou et al. 2012; Jiang et al. 2017; Das et al. 2018).

In general, the kinetic models typically employed in dry reforming of methane can be mainly categorized into empirical power law, Langmuir Hinshelwood-Hougen Watson (LHHW), and Eley Rideal (ER) models. Amongst these kinetic models, the practical and most simple power law model, as provided in Eq. (18.11), is commonly used since it is regarded as one of the easiest computational models for estimating the pertinent reaction parameters.

$$-r_{CH_4} = kP_{CH_4}^{m} P_{CO_2}^{n} \quad (18.11)$$

where k and P_i (i: CO_2 or CH_4) are the corresponding apparent rate constant of CH_4 and partial pressure of reactants, while m and n represent the reaction orders for the corresponding CH_4 and CO_2 reactants, and $-r_{CH4}$ is the CH_4 reaction rate.

The simplicity of power law model, which does not require the full understanding of elementary dry reforming reaction steps since this model is not derived from mechanistic pathways has made it to be extensively employed in many studies for approximate kinetic parameters estimation. As the power law model is not based on the surface mechanism of dry reforming of methane, this model is slightly inaccurate and inappropriate for the investigation of wide-ranging feedstock compositions in comparison with other mechanism-based kinetic models. Nevertheless, this empirical model is advantageous for providing an initial guess of reactants' power constants and apparent activation energy, E_a (kJ mol^{-1}). These useful parameters can be used for the justification of catalytic activity and comparison purpose among reported catalysts in literature.

Table 18.3 lists the computed reaction orders of CH_4 and CO_2 reactants and CH_4 activation energy obtained via power law model for numerous catalysts in recent dry reforming studies. Ayodele et al. (2016) found that CH_4 and CO_2 reaction orders were 3.66 and 0.35, respectively, with the apparent CH_4 activation energy of 96.4 kJ/mol on the La_2O_3-supported cobalt catalyst in dry reforming of methane. The higher CH_4 reaction order than that of CO_2 could indicate the larger dependence on CH_4 partial pressure for CH_4 dissociative adsorption. In addition, they found that the great oxygen storage capacity of La_2O_3 support supplied mobile lattice oxygen to enhance coke gasification from the catalyst surface. This evolution reasonably explains the low CH_4 activation energy (96.4 kJ/mol) on Co/La_2O_3 compared to other catalysts such as Pt/ZrO_2 (Wei and Iglesia 2004) and Ni/α-Al_2O_3 (Cui et al. 2007) with E_a of 100.56 and 106.84 kJ/mol, correspondingly.

In the kinetic study of dry reforming of methane over $SmCoO_3$ perovskite catalyst, Osazuwa et al. (2017) found similar activation energy values for CH_4 and H_2 whereas CO_2 activation energy was reportedly close to that of CO. Thus, they suggested that H_2 and CO formation rates were mainly affected by the corresponding CH_4 and CO_2 consumption rates. Karam and Hassan (2018) used SBA-15 supported Ni catalysts for dry reforming of methane and found that the activation energies of reactants (e.g., CH_4 and CO_2) were smaller than those of products (e.g., H_2 and CO), indicating that the formations of H_2 and CO via surface reaction between adsorbed species are the rate-determining step.

As previously mentioned, the empirical power law models cannot elucidate the involvement of inherent elementary steps on surface reaction during dry reforming of methane since these models are mainly reliant on statistical criteria and could yield untrue kinetic parameter values. Hence, ER and LHHW models are broadly implemented to capture reactant consumption rates and product distribution in dry reforming as they are derived from the intrinsic mechanistic reaction steps. In the dry reforming study of Ru or Pt loaded TiO_2, Singh and Madras (2016) found that the typical peaks belonging to CO_2 adsorption were not detected but CO_2 in the gas

Table 18.3 Kinetic parameters obtained via power law models for dry reforming on various catalysts

		Apparent CH_4 activation energy, E_a (kJ/Mol)	Reaction order		
			m	n	
Catalyst	Temperature (°C)	CH_4	CH_4	CO_2	References
Co/La_2O_3	650–750	96.4	3.66	0.35	Ayodele et al. (2016)
Ir-Ni/SBA-15	580–620	26.4	1.54	0.69	Karam and Hassan (2018)
Ni/SBA-15	580–620	35.1	0.99	0.97	Karam and Hassan (2018)
$Ni_{0.07}Mg_{0.93}O$	500–700	84.0	1	–	Zhou et al. (2018)
$Ru_{0.003}Ni_{0.067}Mg_{0.93}O$	500–700	92.0	1	–	Zhou et al. (2018)
$Ru_{0.07}Mg_{0.93}O$	500–700	74.0	1	–	Zhou et al. (2018)
$SmCoO_3$	700–800	41.0	1.15	0.33	Osazuwa et al. (2017)

Table 18.4 Representatives of LHHW and ER kinetic expressions for dry reforming of methane

Catalyst	Assumption	Mechanistic pathways	Kinetic model	Eq. No.	Refs.
Pt/TiO_2	Molecularly adsorbed CH_4 with the dissociation of CH_4 as RDS	$CH_4 + S^* \overset{K_1}{\leftrightarrow} CH_4 - S$ $CH_4 - S \overset{k_2}{\rightarrow} 2H_2 + C - S(RDS^*)$ $C - S + CO_2 + S \overset{K_3}{\leftrightarrow} 2CO - S$ $2CO - S \overset{K_4}{\leftrightarrow} 2S + 2CO$	$-r_{CH_4} = \frac{k_2 K_1 K_3 K_4 P_{CO_2} P_{CH_4}}{K_3 \sqrt{K_4} P_{CO} + P_{CO}^2 + \left(1 + K_1 P_{CH_4}\right) K_3 K_4 P_{CO_2}}$	(18.12)	Singh and Madras (2016)
NiCeMgAl	CO_2 and CH_4 dissociative adsorption and the removal of adsorbed C species by adsorbed OH as RDS	$CH_4 + S \overset{K_1}{\leftrightarrow} CH_4 - S$ $CH_4 - S + 4S \overset{K_2}{\leftrightarrow} C - S + 4H - S$ $CO_2 + S \overset{K_3}{\leftrightarrow} CO_2 - S$ $CO_2 - S + H - S \overset{K_4}{\leftrightarrow} CO - S + OH - S$ $C - S + OH - S \overset{k_5}{\rightarrow} CO - S + H - S(RDS)$ $2H - S \overset{K_6}{\leftrightarrow} H_2 + 2S$ $CO - S \overset{K_7}{\leftrightarrow} CO + S$	$-r_{CH_4} = k_5 \frac{K_1^4 K_2^4 K_3^4 K_4^4 K_6^4 K_7^4 P_{CH_4}^4 P_{CO_2}^4}{P_{CO_2}^4 P_{H_2}^6} \left(\frac{C_1}{DEN}\right)^8$ $DEN = 1 + K_1 P_{CH_4} + \frac{K_1 K_2 K_6^2 P_{CH_4}}{P_{H_2}^2} + \frac{P_{H_2}^{0.5}}{K_6^{0.5}} + K_3 P_{CO_2} + \frac{P_{CO}}{K_7} + \frac{K_3 K_4 K_7 P_{CO_2} P_{H_2}^{0.5}}{K_6^{0.5} P_{CO}}$	(18.13)	Bao et al. (2017)

$La_2RhZr_2O_7$	CO_2 and CH_4 dissociative adsorption and CH_4 dissociation as RDS	$CH_4 + S \overset{K_1}{\leftrightarrow} CH_4 - S$ $CH_4 - S \overset{k_2}{\rightarrow} C - S + 2H_2\,(RDS)$ $CO_2 + S \overset{K_3}{\leftrightarrow} CO_2 - S$ $CO_2 - S \overset{K_4}{\leftrightarrow} O - S + CO$ $H_2 + O - S \overset{K_5}{\leftrightarrow} S + H_2O$ $C - S + H_2O \overset{K_6}{\leftrightarrow} S + CO + H_2$	$-r_{CH_4} = \frac{k_2}{K_1} P_{CH_4} \left(1 + K_1 P_{CH_4}\right) \frac{K_4 P_{CO} P_{H_2}}{K_3 K_5 K_6 P_{CO} P_{H_2} P_{CO_2}}$	(18.14)	Pakhare et al. (2014)

*S and RDS represent the corresponding active site and the rate-determining step, whereas C_t is the total number of active sites phase was observed in in-situ Fourier transform infrared spectroscopy (FTIR) studies. In addition, adsorbed CH_4 species on catalyst surface were evidenced suggesting that dry reforming of methane over the abovementioned catalysts followed the ER mechanism. Thus, an ER kinetic model was proposed with the assumption of molecularly adsorbed CH_4 on active site and the subsequent CH_4 dissociation as the rate-determining step as shown in Eq. (18.12) of Table 18.4. The dissociated carbon species were further reacted with CO_2 in the gas phase to yield CO gas. This derived ER model demonstrated an excellent validation with experimental data.

Bao et al. (2017) examined NiCeMgAl catalyst for dry reforming of methane and developed an LHHW kinetic model (Eq. 18.13 in Table 18.4) based on the assumption of irreversibly adsorbed carbon gasification by surface –OH species as a rate-determining step and consideration of the significant contribution of the reverse water-gas shift reaction. Both CO_2 and CH_4 reactants were also assumed to adsorb dissociatively on the same active site of catalyst. The fitting of CH_4 consumption rate data to the derived kinetic model showed a high correlation coefficient of 0.97, indicative of satisfactory projection for dry reforming reaction rate.

In the kinetic and mechanistic studies of dry reforming of methane over $La_2RhZr_2O_7$ pyrochlores, Pakhare et al. (2014) conducted the deuterium kinetic isotope experiments with the employment of CD_4 during dry reforming and revealed that the rate-determining step was CH_4 activation or dissociation on active sites. Hence, an LHHW model involving dissociative CH_4 adsorption as a rate-determining step (Eq. 18.14 in Table 18.4) was proposed and in line with the obtained experimental data of kinetic and transient pulsing measurements.

18.6.2 Kinetics of Tri-reforming of Methane

Tri-reforming of methane is typically a synergistic combination of multiple reactions involving steam reforming of methane (SRM), dry reforming of methane, and partial oxidation of CH_4 (Song and Pan 2004). However, the kinetic and mechanistic studies of tri-reforming of methane are comparatively less than other reforming reactions owing to its complexity with the highest number of concomitant side reactions. Thus, mechanistic and kinetic investigation of tri-reforming of methane would turn out to be a crucial and alluring research area in the coming years.

The fundamental understanding of mechanism-derived kinetics is still vague and fewer studies have been carried out for tri-reforming of methane. Song and Pan (2004) conducted the first kinetic study of tri-reforming of methane over Ni-based catalysts. To identify the varied CO_2 conversion rate associated with reactant partial pressures, an empirical power rate law (Eq. 18.15) was simplified by retaining the CH_4 and O_2 partial pressures as constants. Therefore, the contribution of CH_4 and O_2 partial pressures would be reflected by other model parameters:

$$-r_i = A_{exp}\left(\frac{-E_{app}}{RT}\right)P_{CO_2}^{\alpha}P_{H_2O}^{\beta} \tag{18.15}$$

where $-r_i$, E_{app}, and A are the corresponding CH_4 or CO_2 reaction rates, apparent activation energy, and pre-exponential factor. P_{H2O} and P_{CO2} are the partial pressure of H_2O and CO_2 reactants, whereas α and β are reaction orders belonging to CO_2 and H_2O, respectively. The computed values of kinetic parameters obtained from Eq. (18.15) for tri-reforming of methane over different Ni-based catalysts are summarized in Table 18.5. From the experimental data and computational results, Song and Pan (2004) observed the negative reaction order values for H_2O reactant during the fitting of CO_2 consumption rates to the abovementioned power law model (Eq. 18.15) for all catalysts. This observation indicated the competitive adsorption between H_2O and CO_2 on the catalyst surface for reacting with CH_4 to yield H_2 and CO. As a result, an increase in P_{H2O} reasonably reduced the rate of CO_2 consumption. As seen in Table 18.5, reaction orders also varied considerably depending on catalysts employed.

To intrinsically scrutinize the interaction between catalyst surface and CO_2 molecule with respect to mechanistic steps, they also employed a simplified LHHW model, which was derived from dry reforming side reaction for minimizing the intricacy of tri-reforming of methane. These mechanistic steps including the adsorption of CH_4 and CO_2 reactants as well as surface reaction are provided in Table 18.6. As seen in the simplified kinetic model, the interaction between catalyst surface and CO_2 molecules was the major factor accounting for the equilibrium CO_2 adsorption constant, i.e., K_2. The enhancing interaction of CO_2 with catalyst surface would provoke an increase in the number of adsorbed CO_2 species at high P_{CO2}. Hence, these adsorbed CO_2 species occupied most of active sites available for CH_4 adsorption, thereby inducing a drop in CH_4 consumption rate. This could be indicative of negative CO_2 reaction order.

In the assessment of Ni-Mg/β-SiC at varied temperature and reactant feed ratios for tri-reforming of methane, García-Vargas et al. (2015b) observed that methane partial oxidation with the rapid achievement of full conversion was the predominant process among other concomitant main reactions since oxygen reactant was virtually absent in gaseous effluent from tri-reforming of methane. Based on this observation, a kinetic model merely in association with steam reforming, dry reforming, and water-gas shift reaction was proposed for capturing the individual rate of these reactions (see Table 18.7). The apparent activation energy of steam reforming, dry

Table 18.5 Kinetic parameters obtained from simplified power law model for tri-reforming of methane by Song and Pan (2004)

Kinetic parameters	Ni/Al_2O_3	Ni/MgO	Ni/MgO/CeZrO
Effect of P_{CO2}			
α, CH_4	0.79	−0.87	0
α, CO_2	1.90	0.53	0.98
Effect of P_{H2O}			
β, CH_4	−0.06	−0.64	0.03
β, CO_2	−1.57	−2.59	−1.08
Apparent activation energy, E_{app} (kJ/mol)			
CO_2	247.0	160.1	165.7
CH_4	69.1	219.6	67.4

Table 18.6 Tri-reforming of methane mechanistic steps for LHHW model proposed by Song and Pan (2004)

Elementary steps	Kinetic model
Adsorption of reactants $CH_4 + S \overset{K_1}{\leftrightarrow} CH_4 - S$ $CO_2 + S \overset{K_2}{\leftrightarrow} CO_2 - S$	$-r_{CH_4} = \frac{kK_1K_2P_{CH_4}P_{CO_2}}{\left(1 + K_1P_{CH_4} + K_2P_{CO_2}\right)^2}$
Reaction on catalysts surface $CH_4 - S + CO_2 - S \overset{k}{\rightarrow} 2H_2 + 2CO + 2S(RDS)$	

Note: k and S represent the rate constant and accessible active sites, respectively

Table 18.7 Kinetic expressions and estimated parameters for steam reforming, dry reforming, and water-gas shift reaction side reactions of tri-reforming of methane (García-Vargas et al. 2015a)

Reactions	Kinetic parameters	Kinetic models
Steam reforming of methane (SRM)	$E_{a_1} = 74.7$ kJ/mol $k_1^0 = 85.8$ mol s^{-1} kPa^{-1}	$r_{SRM} = k_1P_{CH_4}\left(1 - \frac{P_{H_2}^3P_{CO}}{K_{SRM}P_{CH_4}P_{H_2O}}\right)$ $K_{SRM} = 1.198 \times 10^{17} e^{\left(\frac{-26830}{T}\right)}$
Dry reforming of methane (DRM)	$E_{a_2} = 77.8$ kJ/mol $k_2^0 = 71.0$ mol s^{-1} kPa^{-1}	$r_{DRM} = k_2P_{CH_4}\left(1 - \frac{P_{H_2}^2P_{CO}^2}{K_{DRM}P_{CH_4}P_{H_2O}}\right)$ $K_{DRM} = 6.780 \times 10^{18} e^{\left(\frac{-31230}{T}\right)}$
Water-gas shift reaction (WGS)	$E_{a_3} = 54.3$ kJ/mol $k_3^0 = 149.9$ mol s^{-1} kPa^{-1}	$r_{WGS} = k_3\left(\frac{P_{H_2O}P_{CO}}{P_{H_2}} - \frac{P_{CO_2}}{K_{WGS}}\right)$ $K_{WGS} = 10^{\left(\frac{2078}{T} - 2.029\right)}$

reforming, and water-gas shift reaction is 74.7, 77.8, and 54.3 kJ/mol, respectively, and the kinetic models show a relatively good fitness to experimental data with mean deviations from 10.8% to 19.4%.

Unlike the kinetic expressions suggested by García-Vargas et al. (2015a), Chein et al. (2017) proposed that different kinetic rate models involved steam reforming, reverse CO_2 methanation (RCM), and water-gas shift reaction (Table 18.8) since they found that the presence of H_2O induced a great influence on tri-reforming reaction rate in their parametric investigation. They also reported that in the absence of water,

Table 18.8 Kinetic rate models for steam reforming, reverse CO_2 methanation, and water-gas shift reactions in tri-reforming of methane (Chein et al. 2017)

Reactions	Kinetic rate models
Steam reforming of methane (SRM)	$r_{SRM} = \dfrac{\dfrac{k_1}{P_{H_2}^{2.5}}\left(P_{CH_4}P_{H_2O} - \dfrac{P_{H_2}^3 P_{CO}}{K_{SRM}}\right)}{DEN^2}$ $K_{SRM} = 1.198\times10^{23} e^{\left(\frac{-26830}{T}\right)}$ $k_1 = 3.711\times10^{17} e^{\left(\frac{-240100}{RT}\right)}$ $DEN = 1 + K_{CH_4}P_{CH_4} + K_{CO}P_{CO} + K_{H_2}P_{H_2} + \dfrac{K_{H_2O}P_{H_2O}}{P_{H_2}}$
Water-gas shift reaction (WGS)	$r_{WGS} = \dfrac{\dfrac{k_2}{P_{H_2}}\left(P_{H_2O}P_{CO} - \dfrac{P_{H_2}P_{CO_2}}{K_{WGS}}\right)}{DEN^2}$ $K_{WGS} = 1.767\times10^{-2} e^{\left(\frac{4400}{T}\right)}$ $k_2 = 5.431 e^{\left(\frac{-67130}{RT}\right)}$
Reverse CO_2 methanation (RCM)	$r_{RCM} = \dfrac{\dfrac{k_3}{P_{H_2}^{3.5}}\left(P_{CH_4}P_{H_2O}^2 - \dfrac{P_{H_2}^4 P_{CO_2}}{K_{RCM}}\right)}{DEN^2}$ $K_{RCM} = K_{SRM}K_{WGS}$ $k_3 = 8.96\times10^{16} e^{\left(\frac{-24390}{RT}\right)}$

Note: Denominator *DEN*

dry reforming of methane became the predominant reaction inducing a low H_2/CO ratio of 1, while steam reforming of CH_4 was the main reaction with the absence of CO_2 reactant.

As partial conclusion, this section presents a comprehensive review on various kinetic models derived from the respective rate-determining step in association with different proposed mechanistic steps. Many kinetic models were widely reported into three forms, namely power law, ER, and LHHW models. The power law models could provide empirical reaction orders to reveal the dependence of reaction activity on reactant partial pressure. However, it could not reflect the essential influence factor of rate-determining step toward catalytic activity with respect to mechanistic pathway. Thus, numerous rate-determining steps presented in LHHW and ER models have been suggested for predicting the consumption rate of reactants. Generally, these rate-determining steps involve C–H bonds cleavage in CH_4 molecule on metal, dissociation of oxidizing agent (e.g., O_2, CO_2, or H_2O) on metal or support, formation of intermediate species and decomposition, oxidation of carbonaceous species, and several other parallel reactions. To implement the dry reforming and tri-reforming technologies for industrial application, the best-fit kinetic modeling associated with the intrinsic dry reforming and tri-reforming mechanism is indispensable for efficiently designing the catalyst system and optimizing the reactor design.

18.7 Industrial Applications

Despite the interest in scaling up this technology, problems related to catalyst activity and stability still need to be solved to advance the readiness level of this process. To the best of our knowledge, only very few reports are available on the literature regarding the scale-up of dry reforming of methane. Rego de Vasconcelos and Lavoie (2018) recently reported the scale-up of a dry reforming technology combined with renewable electricity. In this case, an electrified fixed-bed reactor is used in combination with steel wool as a catalyst. The tests performed showed CH_4 and CO_2 conversions equal to 100% over 200 h of time on stream. The advantages of this approach are the very low cost and the good catalytic performance of the catalyst as well as the use of renewable electricity to provide the energy to the reaction. The technology evolved from bench scale with capacity of treating less than 1 L/min of a mixture containing CH_4 and CO_2 to a pilot capable of treating 40 L/min of the same gaseous mixture. The technology developed is compatible for biogas, tail gas, and natural gas.

The Linde Group has also claimed the development of an innovative dry reforming process that convert large volumes of CO_2 into syngas using a nickel-based catalyst, which has been tested for more than 1000 h (Linde 2019). Besides CH_4 and CO_2, water is also used in this process to prevent catalyst deactivation by coke formation (Tullo 2016). The Linde Pilot Reformer research facility is located at Pullach (Germany) and is currently under scale up for commercial application.

Steam reforming of natural gas has been deployed at large industrial scale. This technology can be transposed to the reforming of landfill gas, as the case of the VABHYOGAZ3 project, financed by ADEME in France (ADEME 2016). This project aims to design a complete production chain of liquid hydrogen production from landfill gas containing up to 60% of CH_4. Landfill gas is firstly reformed into syngas at around 900 °C, with a large excess of steam (molar ratio of steam/methane of around 4:1). In fact, the landfill gas used in this project containing CH_4, CO_2, and O_2 is fed together with steam to the reforming catalytic reactor. The syngas is then converted into a mixture rich in H_2 and CO_2 by water-gas shift reaction. Pure hydrogen (99.99%) is obtained by the pressure swing adsorption process, which is then compressed to 350–700 bar into liquid hydrogen for injection in hydrogen-fueled vehicles. CO_2 obtained from the pressure swing adsorption process is valorized by producing sodium bicarbonate for flue gas treatment. The current production unit developed by the industrial partners of this project has a capacity of 10 kg of liquid hydrogen per day (ADEME 2016; Pham Minh et al. 2018a, b). The consortium of this project can deploy production units of 80–800 kg of liquid hydrogen per day from landfill gas with competitive cost of liquid hydrogen versus other production ways (e.g., electrolysis of water).

18.8 Conclusions

Biogas production from biological degradation of biomass and biowaste continuously increases worldwide. Heat and electricity production from biogas have been largely deployed. Current research and development on biogas valorization mostly consists in the production of value-added chemicals via biogas reforming. Both dry- and tri-reforming processes need high temperature (e.g., 800–1000 °C) and low pressure (e.g., atmospheric pressure) to reach high methane conversion. The utilization of a performing catalyst is indispensable to obtain exploitable reaction rate and good syngas selectivity. The main criteria of such a catalyst include high thermal stability, high coke resistance, high activity, and stability. Nickel is the most appropriate active metal for biogas reforming, while various supports have been found suitable. Significant efforts have also been devoted to kinetic and mechanism studies of dry- and tri-reforming processes. This is mandatory for reactor design for further deployment of these processes at large-scale.

References

Abdoulmoumine N, Adhikari S, Kulkarni A, Chattanathan, S (2015) A review on biomass gasification syngas cleanup. Appl. Ener. 155:294-307.

Abdullah B, Ghani NAA, Vo DVN (2017) Recent advances in dry reforming of methane over Ni-based catalysts. J Clean Prod 162:170–185

Abdulrasheed A, Jail A, Gambo Y, Ibrahim M, Hambali H, Hamid M (2019) A review on catalyst development for dry reforming of methane to syngas: recent advances. Renew Sust Energ Rev 108:175–193

Achinas S, Achinas V, Euverink GJW (2017) A technological overview of biogas production from biowaste. Engineering 3:299–307

ADEME (2016) Vabhyogaz 3—Valorisation du biogaz en hydrogène. https://occitanie.ademe.fr/sites/default/files/valorisation-biogaz-hydrogene-vabhyogaz.pdf. Accessed 17 May 2019

Amin MH, Patel J, Sage V, Lee WJ, Periasamy S, Dumbre D, Mozammel T, Prasad V, Samanta C, Bhargava SK (2015) Tri-reforming of methane for the production of syngas: review on the process, catalyst and kinetic mechanism. APCChE 2015 congress incorporating Chemeca, no. Sept: 1–10

Anchieta CG, Assaf EM, Assaf JM (2019) Effect of ionic liquid in Ni/ZrO_2 catalysts applied to syngas production by methane tri-reforming. Int J Hydrogen Energ 44:9316–9327

Angelidaki I, Treu L, Tsapekos P, Luo G, Campanaro S, Wenzel H, Kougias PG (2018) Biogas upgrading and utilization: current status and perspectives. Biotechnol Adv 36:452–466

Aramouni NAK, Touma JG, Tarboush BA, Zeaiter J, Ahmad MN (2018) Catalyst design for dry reforming of methane: analysis review. Renew Sust Energ Rev 82:2570–2585

Asimakopoulos K, Gavala, HN, Skiadas IV (2018) Reactor systems for syngas fermentation processes: A review. Chem. Eng. J. 348:732–744.

Aw MS, Črnivec IGO, Pintar A (2014a) Tunable ceria–zirconia support for nickel–cobalt catalyst in the enhancement of methane dry reforming with carbon dioxide. Catal Commun 52:10–15

Aw MS, Črnivec IGO, Djinovic P, Pintar A (2014b) Strategies to enhance dry reforming of methane: synthesis of ceria-zirconia/nickelecobalt catalysts by freeze-drying and NO calcination. Int J Hydrogen Energ 39:12636–12647

Aw MS, Zorko M, Djinovic P, Pintar A (2015) Insights into durable NiCo catalysts on β-SiC/ $CeZrO_2$ an γ-Al_2O_3/$CeZrO_2$ advanced supports prepared from facile methods for CH_4–CO_2 dry reforming. Appl Catal B 164:100–112

Ayodele BV, Khan MR, Lam SS, Cheng CK (2016) Production of CO-rich hydrogen from methane dry reforming over lanthania-supported cobalt catalyst: kinetic and mechanistic studies. Int J Hydrogen Energ 41:4603–4615

Aziz NIHA, Hanafiah MM, Gheewala SH (2019) A review on life cycle assessment of biogas production: challenges and future perspectives in Malaysia. Biomass Bioenergy 122:361–374

Bao Z, Lu Y, Yu F (2017) Kinetic study of methane reforming with carbon dioxide over NiCeMgAl bimodal pore catalyst. AICHE J 63:2019–2029

Bian Z, Das S, Wai MH, Hongmanorom P, Kawi S (2017) A review on bimetallic nickel-based catalysts for CO_2 reforming of methane. ChemPhysChem 18:3117–3134

Biogas Renewable Energy (2019). http://www.biogas-renewable-energy.info/biogas_composition.html. Accessed 8 May 2019

Bobadilla LF, Garcilaso V, Centeno MA, Odriozola JA (2017) Monitoring the reaction mechanism in model biogas reforming by in situ transient and steady-state DRIFTS measurements. ChemSusChem 10:1193–1201

Boukha Z, Kacimi M, Pereira MFR, Faria JL, Figueiredo JL, Ziyad M (2007) Methane dry reforming on Ni loaded hydroxyapatite and fluoroapatite. Appl Catal A Gen 317:299–309

Boukha Z, Yeste MP, Cauqui MA, González-Velasco JR (2019) Influence of Ca/P ratio on the catalytic performance of Ni/hydroxyapatite samples in dry reforming of methane. Appl Catal A Gen 580:34–45

Brown RC (2011) Thermochemical processing of biomass: conversion into fuels, chemicals and power. Wiley, Chichester, UK

Cao P, Adegbite S, Zhao H, Lester E, Wu T (2018) Tuning dry reforming of methane for F-T syntheses: a thermodynamic approach. Appl Energy 227:190–197

Chein RY, Hsu WH (2018) Thermodynamic analysis of syngas production via tri-reforming of methane and carbon gasification using flue gas from coal-fired power plants. J Clean Prod 200:242–258

Chein RY, Wang CY, Yu CT (2017) Parametric study on catalytic tri-reforming of methane for syngas production. Energy 118:1–17

Clemens H, Bailis R, Nyambane A, Ndung'u V (2018) Africa biogas partnership program: a review of clean cooking implementation through market development in East Africa. Energ Sust Develop 46:23–31

Cui Y, Zhang H, Xu H, Li W (2007) Kinetic study of the catalytic reforming of CH_4 with CO_2 to syngas over Ni/α-Al_2O_3 catalyst: the effect of temperature on the reforming mechanism. Appl Catal A Gen 318:79–88

Das S, Sengupta M, Bag A, Shah M, Bordoloi A (2018) Facile synthesis of highly disperse Ni–co nanoparticles over mesoporous silica for enhanced methane dry reforming. Nanoscale 10:6409–6425

Daza CE, Gallego J, Mondragón F, Moreno S, Molina R (2010) High stability of Ce-promoted Ni/mg-Al catalysts derived from hydrotalcites in dry reforming of methane. Fuel 89:592–603

Dębek R, Zubek K, Motak M, Galvez ME, Da Costa P, Grzybek T (2015) Ni–Al hydrotalcite-like material as the catalyst precursors for the dry reforming of methane at low temperature. C R Chim 18:1205–1210

Di Maio R, Fais S, Ligas P, Piegari E, Raga R, Cossu R (2018) 3D geophysical imaging for site-specific characterization plan of an old landfill. Waste Manag 76:629–642

Diez-Ramirez J, Dorado F, Martinez-Valiente A, Garcia-Vargas J, Sanchez P (2016) Kinetic, energetic and exergetic approach to the methane tri-reforming process. Int J Hydrogen Energ 41:19339–19348

Djinovic P, Pintar A (2017) Stable and selective syngas production from dry CH_4-CO_2 streamsover supported bimetallic transition metal catalysts. Appl Catal B 206:675–682

Djinovic P, Crnivec IGO, Erjavec B, Pintar A (2012) Influence of active metal loading and oxygen mobility on coke-free dry reforming of Ni–co bimetallic catalysts. Appl Catal B 125:259–270

Djinovic P, Crnivec IGO, Pintar A (2015) Biogas to syngas conversion without carbonaceous deposits via thedry reforming reaction using transition metal catalysts. Catal Today 253:155–162

Eklund B, Anderson EP, Walker BL, Burrows DB (1998) Characterization of landfill gas composition at the fresh kills municipal solid-waste landfill. Environ Sci Technol 32:2233–2237

Elfattah S, Eldrainy Y, Attia A (2016) Upgrade Egyptian biogas to meet the natural gas network quality standard. Alexandria Eng J 55:2279–2283

Estephane J, Aouad S, Hany S, El Khoury B, Gennequin C, El Zakhem H, El Nakat J, Aboukaïs A, Aad EA (2015) CO_2 reforming of methane over Ni–CO/ZSM5 catalysts. Aging and carbon deposition study. Int J Hydrogen Energ 40:9201–9208

Faungnawakij K, Kikuchi R, Eguchi K (2007) Thermodynamic analysis of carbon formation boundary and reforming performance for steam reforming of dimethyl ether. J Power Sources 164:73–79

García-Vargas JM, Valverde JL, Dorado F, Sanchez P (2014a) Influence of the support on the catalytic behaviour of Ni catalysts for the dry reforming reaction and the tri-reforming process. J Mol Catal A Chem 395:108–116

García-Vargas JM, Valverde JS, Díez J, Sánchez P, Dorado F (2014b) Influence of alkaline and alkaline-earth cocations on the performance of Ni/β-SiC catalysts in the methane tri-reforming reaction. Appl Catal B 148–149:322–329

García-Vargas JM, Valverde JL, Díez J, Dorado F, Sánchez P (2015a) Catalytic and kinetic analysis of the methane tri-reforming over a Ni–mg/β-SiC catalyst. Int J Hydrogen Energ 40:8677–8687

García-Vargas JM, Valverde JS, Díez J, Sánchez P, Dorado F (2015b) Preparation of Ni–mg/β-SiC catalysts for the methane tri-reforming: effect of the order of metal impregnation. Appl Catal B 164:316–323

Goula MA, Charisiou ND, Papageridis KN, Delimitis A, Pachatouridou E, Iliopoulou EF (2015) Nickel on alumina catalysts for the production of hydrogen rich mixtures via the biogas dry reforming reaction: influence of the synthesis method. Int J Hydrogen Energ 40:9183–9200

GRDF, Gaz Réseau Distribution France (2019). https://www.grdf.fr/dossiers/biomethane-biogaz/unites-injection-gaz-vert-biomethane-reseau. Accessed 11 May 2019

Grouset D, Ridart C (2018) Lowering energy spending together with compression, storage, and transportation costs for hydrogen distribution in the early market. In: Azzaro-Pantel C (ed) Hydrogen supply chains-design, deployment and operation. Academic Press, Waltham, MA, pp 207–270

Guharoy U, Le Saché E, Cai Q, Reina TR, Gu S (2018) Understanding the role of Ni-Sn interaction to design highly effective CO_2 conversion catalysts for dry reforming of methane. J CO2 Util 27:1–10

Hoo PY, Hashim H, Ho WS (2018) Opportunities and challenges: landfill gas to biomethane injection into natural gas distribution grid through pipeline. J Clean Prod 175:409–419

Horn R, Schlögl R (2015) Methane activation by heterogeneous catalysis. Catal Lett 145:23–39

Hou Z, Chen P, Fang H, Zheng X, Yashima T (2006) Production of synthesis gas via methane reforming with CO_2 on noble metals and small amount of noble-(Rh-) promoted Ni catalysts. Int J Hydrogen Energ 31:555–561

IRENA (2017) International Renewable Energy Agency: Renew Capacity Statistics 2016

Jafarbegloo M, Tarlani A, Mesbah AW, Sahebdelfar S (2015) Thermodynamic analysis of carbon dioxide reforming of methane and its practical relevance. Int J Hydrogen Energ 40:2445–2451

Jaffrin A, Bentounes N, Joan A, Makhlouf S (2003) Landfill biogas for heating greenhouses and providing carbon dioxide supplement for plant growth. Biosyst Eng 86:113–123

Jang WJ, Shim JO, Kim HM, Yoo SY, Roh HS (2019) A review on dry reforming of methane in aspect of catalytic properties. Catal Today 324:15–26

Jiang H, Li H, Xu H, Zhang Y (2007) Preparation of $Ni/Mg_xTi_{1-x}O$ catalysts and investigation on their stability in tri-reforming of methane. Fuel Process Technol 88:988–995

Jiang S, Lu Y, Wang S, Zhao Y, Ma X (2017) Insight into the reaction mechanism of CO_2 activation for CH_4 reforming over NiO-MgO: a combination of DRIFTS and DFT study. Appl Surf Sci 416:59–68

Kaparaju P, Rintala J (2013) Generation of heat and power from biogas for stationary applications: boilers, gas engines and turbines, combined heat and power (CHP) plants and fuel cells. In: Wellinger A, Murphy J, Baxter D (eds) The biogas handbook, science, production and applications. Woodhead Publishing, Oxford, Cambridge, Philadelphia, New Delhi, pp 404–427

Karam L, Hassan NE (2018) Advantages of mesoporous silica based catalysts in methane reforming by CO_2 from kinetic perspective. J Env Chem Eng 6:4289–4297

Kawi S, Kathiraser Y, Ni J, Oemar U, Li Z, Saw ET (2015) Progress in synthesis of highly active and stable nickel-based catalysts for carbon dioxide reforming of methane. ChemSusChem 8:3556–3575

Kaza S, Yao L, Bhada-Tata P, Van Woerden F (2018) What a waste 2.0: a global snapshot of solid waste management to 2050. The World Bank, Washington, DC

Khalil M, Berawi MA, Heryanto R, Rizalie A (2019) Waste to energy technology: the potential of sustainable biogas production from animal waste in Indonesia. Renew Sust Energ Rev 105:323–331

Khan IU, Othman MHD, Hashima H, Matsuura T, Ismail AF, Rezaei-DashtArzhandi M, Wan Azelee I (2017) Biogas as a renewable energy fuel—a review of biogas upgrading, utilization and storage. Energy Convers Manag 150:277–294

Kroll VCH, Swaan HM, Lacombe S, Mirodatos C (1997) Methane reforming reaction with carbon dioxide over Ni/SiO_2 catalyst. J Catal 398:387–398

Kumar R, Kumar K, Choudary NV, Pant KK (2019) Effect of support materials on the performance of Ni-based catalysts in tri-reforming of methane. Fuel Process Technol 186:40–52

Kvist T, Aryal N (2019) Methane loss from commercially operating biogas upgrading plants. Waste Manag 87:295–300

Linde (2019) "Innovative Dry Reforming Process." https://www.linde-engineering.com/en/innovations/innovate-dry-reforming/index.html. Accessed 15 May 2019

Lino AVP, Calderon YNC, Mastelaro VR, Assaf EM, Assaf JM (2019) Syngas for Fischer-Tropsch synthesis by methane tri-reforming using nickel supported on $MgAl_2O_4$ promoted with Zr, Ce and Ce-Zr. Appl Surf Sci 481:747–760

Liu K, Song C, Subramani V (2010) Hydrogen and syngas production and purification technologies. Wiley, Hoboken, USA

Majewski AJ, Wood J (2014) Tri-reforming of methane over $Ni@SiO_2$ catalyst. Int J Hydrogen Energ 39:12578–12585

Mittal S, Ahlgren EO, Shukla PR (2018) Barriers to biogas dissemination in India: a review. Energy Policy 112:361–370

Mo W, Ma F, Liu Y, Liu J, Zhong M, Nulahong A (2015) Preparation of porous Al_2O_3 by template method and its application in Ni-based catalyst for CH_4/CO_2 reforming to produce syngas. Int J Hydrogen Energ 40:16147–16158

Mohamedali M, Henni A, Ibrahim H (2018) Recent advances in supported metal catalysts for syngas production from methane. ChemEng 2:9

Mudhoo A (2012) Biogas production: pretreatment methods in anaerobic digestion. Wiley/Scrivener Publishing LLC, Hoboken, NJ/Salem, MA

Nanda S, Li K, Abatzoglou N, Dalai AK, Kozinski JA (2017) Advancements and confinements in hydrogen production technologies. In: Dalena F, Basile A, Rossi C (eds) Bioenergy systems for the future. Woodhead Publishing, Sawston, UK, pp 373–418

Néron A, Lantagne G, Marcos B (2012) Computation of complex and constrained equilibria by minimization of the Gibbs free energy. Chem Eng Sci 82:260–271

Nikoo MK, Amin NAS (2011) Thermodynamic analysis of carbon dioxide reforming of methane in view of solid carbon formation. Fuel Process Technol 92:678–691

Okonkwo C, Onokpite E, Onokwai A (2018) Comparative study of the optimal ratio of biogas production for, various organic wastes and weeds for digester/restarted digester. J King Saud Univ-Eng Sci 30:123–129

Osazuwa OU, Setiabudi HD, Abdullah S, Cheng CK (2017) Syngas production from methane dry reforming over $SmCoO_3$ perovskite catalyst: kinetics and mechanistic studies. Int J Hydrogen Energ 42:9707–9721

Pakhare D, Spivey J (2014) A review of dry (CO_2) reforming of methane over noble metal catalysts. Chem Soc Rev 43:7813–7837

Pakhare D, Schwartz V, Abdelsayed V, Haynes D, Shekhawat D, Poston J, Spivey J (2014) Kinetic and mechanistic study of dry (CO_2) reforming of methane over Rh-substituted $La_2Zr_2O_7$ pyrochlores. J Catal 316:78–92

Papadopoulou C, Matralis H, Verykios X (2012) Utilization of biogas as a renewable carbon source: dry reforming of methane. In: Guczi L, Erdôhelyi A (eds) Catalysis for alternative energy generation. Springer Nature, New York, NY, pp 57–127

Persson M, Jönsson O, Wellinger A (2009) IEA Bioenergy, Task 37—Energy from biogas and landfill gas

Pertiwiningrum A, Agus DKC, Wuri MA (2018) Renewable energy of biogas through integrated organic cycle system in tropical system. In: Gokten S, Kucukkocaoglu G (eds) Energy management for sustainable development. IntechOpen, London, UK, pp 99–117

Pham Minh D, Phan TS, Grouset D, Nzihou A (2018a) Thermodynamic equilibrium study of methane reforming with carbon dioxide, water and oxygen. J Clean Energ Technol 6:309–313

Pham Minh D, Siang TJ, Vo DVN, Phan TS, Ridart C, Nzihou A, Grouset D (2018b) Hydrogen production from biogas reforming: an overview of steam reforming, dry reforming, dual reforming, and tri-reforming of methane. In: Hydrogen supply chains. Elsevier, Amsterdam, pp 111–166

Phan TS, Sane AR, Rego de Vasconcelos B, Nzihou A, Sharrock P, Grouset D, Minh DP (2018) Hydroxyapatite supported bimetallic cobalt and nickel catalysts for syngas production from dry reforming of methane. Appl Catal B 224:310–321

Pino L, Vita A, Cipitì F, Laganà M, Recupero V (2011) Hydrogen production by methane tri-reforming process over Ni–ceria catalysts: effect of La-doping. Appl Catal B 104:64–73

Pradhan BB, Limmeechokchai L (2017) Electric and biogas stoves as options for cooking in Nepal and Thailand. Energy Procedia 138:470–475

Puigjaner L (2011) Syngas from waste emerging technologies. Springer, New York, NY

Pullen T (2015) Anaerobic digestion- making biogas-making energy. Routledge, New York, NY

Rahnama H, Farniaei M, Abbasi M, Rahimpour MR (2014) Modeling of synthesis gas and hydrogen production in a thermally coupling of steam and tri-reforming of methane with membranes. J Ind Eng Chem 20:1779–1792

Rego de Vasconcelos BR (2016) Phosphates-based catalyst for synthetic gas (syngas) production using CO_2 and CH_4. Ph.D. thesis, Ecole Nationale Supérieure des Mines d'Albi-Carmaux, Albi, France

Rego de Vasconcelos B, Lavoie JM (2018) Is dry reforming the solution to reduce natural gas carbon footprint? Int J Ener Prod Manag 3:44–56

Rego de Vasconcelos B, Pham Minh D, Lyczko N, Phan TS, Sharrock P, Nzihou A (2018a) Upgrading greenhouse gases (methane and carbon dioxide) into syngas using nickel-based catalysts. Fuel 226:195–203

Rego de Vasconcelos BR, Pham Minh D, Sharrock P, Nzihou A (2018c) Regeneration study of Ni/hydroxyapatite spent catalyst from dry reforming. Catal Today 310:107–115

Rostrup-Nielsen JR, Hansen JHB (1993) CO_2 reforming of methane over transition metals. J Catal 144:38–49

Sahota S, Shaha G, Ghosha P, Kapoor R, Sengupta S, Singh P, Vijay V, Sahay A, Vijay VK, Thakur IS (2018) Review of trends in biogas upgradation technologies and future perspectives. Biores Technol Rep 1:79–88
San José-Alonso D, Illán-Gómez MJ, Román-Martínez MC (2011) K and Sr promoted co alumina supported catalysts for the CO_2 reforming of methane. Catal Today 176:187–190
Sapountzi FM, Zhao C, Boréave A, Retailleau-Mevel L, Niakolas D, Neofytidis C, Vernoux P (2018) Sulphur tolerance of au-modified Ni/GDC during catalytic methane steam reforming. Cat Sci Technol 8:1578–1588
Scarlat N, Dallemand JF, Fahl F (2018) Biogas: developments and perspectives in Europe. Renew Energy 129:457–472
Schildhauer TJ, Biollaz MA (2016) Synthetic natural gas from coal, dry biomass, and power-to-gas applications. Wiley, Hoboken, NJ
Shah YT (2017) Chemical energy from natural gas and synthetic gas. CRC Press, Boca Raton, FL
Si LJ, Wang CZ, Sun NN, Wen X, Zhao N, Xiao FK, Wei W, Sun YH (2012) Influence of preparation conditions on the performance of Ni-CaO-ZrO_2 catalysts in the tri-reforming of methane. J Fuel Chem Technol 40:210–215
Singh SA, Madras G (2016) Sonochemical synthesis of Pt, Ru doped TiO_2 for methane reforming. Appl Catal A Gen 518:102–114
Singha RK, Shukla A, Yadav A, Adak S, Iqbal Z, Siddiqui N, Bal R (2016) Energy efficient methane tri-reforming for synthesis gas production over highly coke resistant nanocrystalline Ni–ZrO_2 catalyst. Appl Energy 178(2016):110–125
Song C, Pan W (2004) Tri-reforming of methane: a novel concept for catalytic production of industrially useful synthesis gas with desired H_2/CO ratios. Catal Today 98:463–484
Tang P, Zhu Q, Wu Z, Ma D (2014) Methane activation: the past and future. Energy Environ Sci 7:2580–2591
Themelis N, Ulloa P (2007) Methane generation in landfills. Renew Energy 32:1243–1257
Tullo AH (2016) Dry reforming puts CO_2 to work. Chem Eng News 94:30. https://cen.acs.org/content/cen/articles/94/i17/Dry-reforming-puts-CO2-work.html. Accessed 17 May 2019
Wang S, Lu GQM (1996) Carbon dioxide reforming of methane to produce synthesis gas over metal-supported catalysts: state of the art. Energ Fuels 10:896–904
Wei J, Iglesia E (2004) Isotopic and kinetic assessment of the mechanism of reactions of CH_4 with CO_2 or H_2O to form synthesis gas and carbon on nickel catalysts. J Catal 224:370–383
Weiland P (2010) Biogas production: current state and perspectives. Appl Microbiol Biotechnol 85:849–860
Wolfbeisser A, Sophiphun O, Bernardi J, Wittayakun J, Föttinger K, Rupprechter G (2016) Methane dry reforming over ceria-zirconia supported Ni catalysts. Catal Today 277:234–245
Xu J, Zhou W, Li Z, Wang J, Ma J (2009) Biogas reforming for hydrogen production over nickel and cobalt bimetallic catalysts. Int J Hydrogen Energ 34:6646–6654
Yasmin N, Grundmann P (2019) Adoption and diffusion of renewable energy—the case of biogas as alternative fuel for cooking in Pakistan. Renew Sust Energ Rev 101:255–264
Youcai Z, Ziyang L (2017) General structure of sanitary landfill. In: Pollution control and resource recovery: municipal solid wastes at landfill. Elsevier, Amsterdam, pp 1–10
Zhou H, Zhang T, Sui Z, Zhu YA, Han C, Zhu K, Zhou X (2018) A single source method to generate Ru-Ni-MgO catalysts for methane dry reforming and the kinetic effect of Ru on carbon deposition and gasification. Appl Catal B 233:143–159

Opportunities for Biodiesel Compatibility as a Modern Combustion Engine Fuel

19

Swarup Kumar Nayak, Purna Chandra Mishra, Sonil Nanda, Biswajeet Nayak, and Muhamad Mat Noor

Abstract

This chapter summarizes the feasibility of effective utilization of biodiesel in modern vehicle engines. The parameters discussed in this chapter include diesel engine characterization and diagnostics including performance, emissions, and combustion behavior. The lifecycle and economic analyses with future scope of biodiesel are also described. From the review, it is conferred that a huge proportion of biodiesel is produced from edible vegetable oils, which is a threat to the food supply. Biodiesel sources are focused upon non-edible oils and other feedstocks that do not compete with the food crops. Therefore, the selection of appropriate feedstock is essential to confirm the low-cost production of biodiesel. Concerning the engine characteristics and combustion diagnostics, it can be ensured that biodiesel improves engine performance and emission characteristics with little engine modifications such as injection timing, pressure, exhaust gas recirculation, etc. The review holds well on the possibility of using biodiesel in diesel engines, but still it is not economically viable and needs more research and technology advancements to make it competitive with other conventional fuels in the market.

S. K. Nayak · P. C. Mishra (✉)
School of Mechanical Engineering, Kalinga Institute of Industrial Technology (KIIT University), Bhubaneswar, Odisha, India
e-mail: pcmishrafme@kiit.ac.in

S. Nanda
Department of Chemical and Biological Engineering, University of Saskatchewan, Saskatoon, Saskatchewan, Canada

B. Nayak
Department of Mechanical Engineering, Einstein Academy of Technology and Management, Bhubaneswar, Odisha, India

M. M. Noor
Faculty of Mechanical Engineering, Universiti Malaysia Pahang, Pekan, Pahang, Malaysia

S. Nanda et al. (eds.), *Biorefinery of Alternative Resources: Targeting Green Fuels and Platform Chemicals*, https://doi.org/10.1007/978-981-15-1804-1_19

Keywords
Biodiesel · Transesterification · Greenhouse gas emissions · Engine performance · Non-edible oil

19.1 Introduction

Energy has turned into an essential component of humankind to keep the monetary development and maintain an exclusive requirement of livelihood and living wage. It is envisaged that the world could require more than half of the energy that is being utilized now by the year 2030, of which a significant portion may be utilized by rapidly developing countries like India and China (Shahid and Jamal 2011; Atabani et al. 2012; Nanda et al. 2015). The worldwide transportation and energy utilization is foreseen to increase by a standard of 1.8% every year by 2035 (USEIA 2010). Despite the fact that fossil fuel tends to abide by the best antecedent of vitality, the oil contribution in the world market is foreseen to decrease only at a small fraction from 35% in 2007 to 30% in 2035 (Shahid and Jamal 2011; USEIA 2010). It is accepted that the environmental degradation and ecosystem imbalance are currently the most important problems faced by the world related to fossil fuel extraction and utilization (Rana et al. 2018, 2019). It is anticipated that nearly 8 billion metric tons of CO_2 could be discharged into the environment until 2035 due to rapid industrial development and globalization (Atabani et al. 2011).

Globally, the transportation sector represents roughly 22% of aggregate CO_2 emissions in 2008 (Atabani et al. 2011). On-road transportation contributes to around 10% of the worldwide greenhouse gas (GHG) emissions (Nayak et al. 2017a, b). GHGs mainly consist of CO_2, CH_4, CO, CO, NO_x, CO, and aerosols. In 2008, roughly 30% of the U.S. GHG emissions originated from the transportation part. Thus, transportation sector is the second largest source of GHG emissions in the United States (Sharma and Singh 2009). According to the current appraisals, oil reserves in India can continue for the next 20–25 years. Recently, the per capita emission of CO_2 in India has increased from 0.5 to 1.7 metric tons per year.

Biofuels seem to be promising in terms of reducing the dependency on fossil fuels and mitigating the GHG emissions (Nanda et al. 2017). Lignocellulosic biomasses such as agricultural crop residues and woody biomass can be effectively converted to biofuels such as biodiesel, bio-oil, bioethanol, biobutanol, biohydrogen, biomethane, synthesis gas, etc. (Nanda et al. 2014, 2016). Biodiesels are produced as monoalkyl esters of long-chain fatty acids derived from vegetable oils, animal fats, and alcohols with or without using a catalyst (Nayak et al. 2017a; Reddy et al. 2018). Compared to fossil diesel, biodiesel produces negligible emissions of SO_x, CO, CO_2, particulate matter, and hydrocarbons. More free oxygen leads to complete combustion and reduced emission (Nayak et al. 2017b). The increasing cost of edible vegetable oils is one of the major concerns for many Asian countries. There is a need for dedicated feedstock research and identification of superior biomass species, which could produce appreciable quality and quantity of biodiesel.

The main motivation behind this chapter is to discuss the evolution of next-generation biodiesel and its compatibility in modern diesel engines with complete analysis of performance, emission, combustion, engine durability, and wear and tear analysis. After a comprehensive literature survey, it was viewed that minimal information is available on the usage of next-generation biodiesel in the modern or hybrid diesel engines. Hence, the objective of the present chapter is to display an extensive report on:

1. Biodiesel evolution and its compatibility in present diesel engines without any engine modification.
2. Synthesis of the available literature and summarize the opportunities and challenges.
3. Experimental data quantifying the performance, emissions, and combustion analysis of biodiesel.

19.2 Biodiesel Properties and Measurement Methods

Biodiesel standards are being prescribed globally to maintain its quality in the competitive market. Since biodiesel is mostly derived from oil seeds, some standard specifications are laid down to meet the fuel quality to ensure better fuel performance without any significant engine modification. Biodiesel produced from various feedstocks must adhere to the specifications laid down by the International Biodiesel Standard. In 2008, American Standards for testing materials (ASTM) published new biodiesel blend specification standards, which include ASTM 6751-3. Similarly, several other countries also published their own standards such as for European Union (EN 14214), Germany (Deutsches Institut für Normung), Czech Republic (Czech State Norm), India (Bureau of Indian Standards), and Italy (Italian National Standards Institute).

Some of the general parameters to determine the quality of biodiesel are density (g/cm^3), viscosity (mm^2/s), flash point (°C), fire point (°C), cetane number, carbon residue (%), iodine number, methanol/ethanol (mass %), acid value (mg KOH/g), ash content (w/w %), cloud point (°C), pour point (°C), etc., which also depend upon the type of feedstock selected and free fatty acid (FFA) composition (Demirbas 2007; Karmakar et al. 2010; Atadashi et al. 2011; Lin et al. 2011). Some general parameters published by different nations for testing the quality of biodiesel are depicted in Table 19.1. A synopsis of physicochemical properties of biodiesel derived from various feedstocks is shown in Table 19.2.

Viscosity is an imperative property of biodiesel because it leads to poor atomization of air-fuel mixture in the combustion chamber. Kinematic viscosity of liquid fuel is measured at 40 °C as per the specifications given in ASTM D445. Generally, kinematic viscosity of biodiesel is higher than that of diesel due to its significant molecular mass and chemical structure. High viscosity may lead to cold starting problem, which increases mechanical stress on the fuel and injection pump system and affect the fuel delivery qualities. High viscosity has greater impact on fuel

Table 19.1 Typical specification of biodiesel in the United States and European countries

Properties	Units	Diesel (ASTM D975)		Biodiesel			
				EN 14214–2008		ASTM D6751	
		Test method	Limits	Test method	Limits	Test method	Limits
Flash point	°C	D975	60–80	EN ISO 3679	Min. 100	D93	Min. 125
Cloud point	°C	D975	−15 to −5	–	–	D2500	−3 to −13
Pour point	°C	D975	−35 to −15	–	–	D97	−15 to −17
Cetane number	–	D4737	46	EN ISO 5165	Min. 50	D613	Min. 47
Density (at 15 °C)	kg/m^3	D1298	810–850	EN ISO 3675	850–890	D1298	880
Viscosity (at 40 °C)	mm^2/s	D445	2–5	EN ISO 3104	3–5	D445	1.9–6.0
Acid number	mg KOH/g	–	–	EN ISO 14104	Max. 0.5	D664	Max. 0.5
Cold filter plugging point	°C	D590	−9	EN ISO 14214	–	D6371	Max. 5
Oxidations stability	mg/L	D2274	26	EN ISO 14112	Min. 3.5 h	–	–
Carbon residue	% m/m	D4530	0.2	EN ISO 10370	Max. 0.4	D4530	Max. 0.05
Sulfated ash content	% mass	–	–	EN ISO 3987	Max. 0.02	D874	Max. 0.002
Ash content Water and sediment	% mass	D482; D2709	99/0.04	EN ISO 12937	500 mg/kg	D–/2709	0.005 vol%
Free glycerine	% mass	–	–	EN ISO 14105	Max. 0.03	D6584	Max. 0.02
Total glycerine	% mass	–	–	EN ISO 14105	0.24	D6584	0.25

(continued)

Table 19.1 (continued)

Properties	Units	Diesel (ASTM D975) Test method	Limits	Biodiesel EN 14214–2008 Test method	Limits	ASTM D6751 Test method	Limits
Phosphorus	% mass	–	–	EN ISO 14107	0.001	D4951	Max. 0.01
Carbon	wt%	D975	14	–	–	PS1 21	76
Hydrogen	wt%	D975	–	–	–	PS1 21	11
Oxygen	wt%	–	–	–	–	PS1 21	10
Total contamination	mg/kg	D22624	–	EN ISO 12662	23	D5452	25
Boiling point	°C	–	–	–	–	D7398	100–600
Saponification value	mg KOH/g	–	–	–	–	D5558	Max. 350

References: Çelikten et al. (2010), Shameer et al. (2016), Saba et al. (2016)

atomization, i.e., on fuel droplet size and injector spray geometry during injection. The maximum allowable viscosity limit for biodiesel as per ASTM D445 is 1.9–6 mm^2/s. Density of liquid is measured as per specification given in ASTM D1298 and EN ISO 3675/12185. The density of biodiesel varies with feedstocks and it is generally greater than fossil diesel.

Fire point is an extension of flash point in a way that reflects the condition at which vapor burn continuously for more than 4 s. Fire point is generally higher than flash point by 10–15 °C. The flash point and the fire point of the test fuel blends are measured as per the ASTM D93-66 standard and EN ISO 3697 (Xue et al. 2011; Nayak and Mishra 2018). Flash point of biodiesel is generally more than 150 °C while it is only 55–65 °C for diesel (Nayak and Mishra 2019). Poor cold flow properties lead to blockage in pipelines and pumps eventually prompting to fuel starvation. There are no specific standards derived to measure cold flow properties. Hence, every country can focus on its own particular limits depending upon the local climatic conditions. From the literature, it is revealed that biodiesel experiences certain negative effects of cold flow properties, which are much higher than that of diesel.

Pour point is the lowest temperature at which the fuel becomes semi-solid and loses its flow characteristics. Therefore, it is the measurement of minimum temperature at which the fuel can flow. The pour point is always lower than the cloud point. Both cloud point and pour point are measured by utilizing the ASTM D2500, EN ISO 23015, and D97 standards. Biodiesel has higher pour point and cloud point than fossil diesel (Xue et al. 2011).

Cold filter plugging point (CFPP) is the minimal temperature during which a particular volume of diesel can still go through a standardized filtration device in a particular period at specific conditions. This parameter is vital because in cold weather conditions, if CFPP is high then it will plug the fuel flow in pipelines,

Table 19.2 Typical physicochemical properties of biodiesel derived from various feedstocks

Fuel properties	Jatropha FAME	*Calophyllum inophyllum* FAME	*Madhuca* FAME	*Mesua* FAME	Rubber FAME	*Camelina sativa* FAME	*Canola sativa* FAME
Density at 15 °C (kg/m³)	879.5	888.6	874	898			
Viscosity at 40 °C (cSt)	4.8 g	7.724	3.98	6.2	5.81	4.15	4.42
Cetane number	51.6	51.9	65	54		52.8	
Iodine number	104	85					
Calorific value (MJ/kg)	39.23		36.8	42.23	36.5		
Acid neutralization value (mg KOH/g)	0.4	0.76	0.41	0.01		0.31	0.01
Pour point (°C)	2		6	3	−8	−4	−9
Flash point (°C)	135	151	208	112	130	>160	>160
Cloud point (°C)	2.7	38	–	–	4	3	−3.3
Cold filter plugging point (°C)	0		–	–	–	−3	−7
Water and sediment content (vol%)	<0.005	–	–	0.035	–	<0.005	<0.005
Ash content (w/w %)	0.012	0.026	0.01	0.01	–		
Sulfated ash (m/m %)	0.009		–	–		<0.005	<0.005
Free glycerine (m/m %)	0.006		–	–		0.002	0.006
Total glycerine (m/m %)	0.1	0.232	–	–		0.08	0.114
Conradson carbon residue, 100% (mass %)	0.025	0.434	0.02	0.25		0.075	0.03

References: Çelikten et al. (2010), Shameer et al. (2016), Saba et al. (2016)

pumps, or engines affecting their performances. CFPP is measured by using the ASTM D6371 procedures (Demirbas 2007).

Oxidation stability is a chemical reaction, which occurs by combination of the oxygen with lubricating oil. It is one of the main considerations that aides survey the nature of biodiesel. Oxidation occurs because of the vicinity of unsaturated fat chains, which quickly respond with the oxygen when it is exposed to air (Zuleta and Rios 2012). Chemical composition of biodiesel makes it more vulnerable to oxidative deterioration in comparison to that of diesel. Oxidation leads to an increase in the oil's viscosity and deposition of varnish and sludge. ASTM D6751 and EN 14214 standards use the Rancimat method (EN ISO 14112) to measure oxidation stability (Demirbas 2007; Zuleta and Rios 2012).

The cetane number is the evidence of the ignition capacity of fuel to auto-ignite rapidly after being injected. Better ignition quality of fuel is associated with a higher cetane number, which is one of the most important parameters for biodiesel quality. Higher cetane number shows a shorter time between the ignition and the start of fuel infusion into the combustion chamber (Nayak and Mishra 2019). The cetane number of diesels is 47 min for ASTM D613 and 51 min for EN ISO 5165 (Nayak and Mishra 2018). Biodiesel has a cetane number more than diesel, which brings higher combustion efficiency (Demirbas 2007).

Fatty acids are usually derived from triglycerides, and when they are in free form, i.e., not attached to any other molecules, they are called free fatty acid (FFA). The FFA are important source of fuel because when processed they yield large amount of adenosine triphosphate (ATP). More the amount of fatty acid, higher is the acid value. Generally, acid value is measured in terms of mg KOH/g of fatty acid methyl ester (FAME), which is measured by utilizing ASTM D664 and EN 14104 standards. Higher acid value results in the corrosion of the fuel supply system. Therefore, the acid value is approved to a maximum of 0.5 mg KOH/g (Zuleta and Rios 2012).

A specific amount of heat is released from the combustion of a specific quantity of fuel. Moisture content is one of the important parameters for determining the heating value of oils (Karmakar et al. 2010). Minimum heating value of biodiesel is about 35 MJ/kg as per the EN 14213 standards but there are still no specifications laid down by ASTM D6751 and EN 14214. The iodine value indicates the degree of unsaturation of the vegetable oils. A high degree of unsaturation leads to a higher chemical reactivity. The deposition to polymerized vegetable oils with low iodine value has lower value of carbon residue and higher oxidation stability. The test method used for determining the iodine value of biodiesel is ASTM D5768 (Lin et al. 2011; Karmakar et al. 2010).

19.3 Biodiesel Engine Performance and Characteristics

Most of the studies available from literature review are on the performance and emission characteristics of biodiesel-fueled diesel engine by determining the brake thermal efficiency (BTE), brake specific fuel consumption (BSFC), exhaust gas temperature (EGT), and emission parameters on basis of CO_2, CO, NO_x,

Table 19.3 Engine specifications for testing of biodiesel performance

Engine model	Number of cycles	Brake power (kW)	Speed (rpm)	Compression ratio	Displacement (m^3)	Cooling system	References
DI engine	1	–	–	–	–	Air	Devendra et al. (2015)
TDI engine	6	135	2500	–	5900	Water	Satputaley et al. (2017)
Kirloskar TV1	1	5.2	1500	–	661	Water	Pradhan et al. (2016)
ADCR CRDi engine	4	85	3700	17.5:1	2636	Water	Işık et al. (2017)
Isuzu 4HF1	4	88	1800	19:1	4334	Water	Mikulski et al. (2016)
NWK 22	4	18	1500	17:1	2400	Water	Jayaprabakar and Karthikeyan (2016)
DJC type DI engine	4	12	1800	19:1	–	Air	Lei et al. (2016)
Kirloskar AV1	1	3.7	1500	16.5:1	553	Water	Atmanli (2016)
4 stroke WC, DI engine	1	7.4	1500	19.5:1	–	Water	Lahane and Subramanian (2015)
4 stroke AC, DI engine	1	4.4	1500	19.5:1	–	Water	Gopal et al. (2014)
4 stroke AC, DI engine	6	81	2600	17.5:1	–	Air	Sanli et al. (2015)
4 stroke WC, DI engine	4	89	2000	16.4:1	–	Water	Tüccar et al. (2014)

4 stroke AC, DI engine	1	7.08	2400	18:1	–	Air	Tan et al. (2017)
4 stroke WC, DI engine	1	3.5	1500	17.5:1	–	Water	Kakati and Gogoi (2016)
4 stroke AC, DI engine	1	3.09	3800	18:1	–	Air	Al-lwayzy and Yusaf (2017)
4 stroke WC, DI engine	1	5.9	1500	17.5:1	–	Water	Ramalingam et al. (2016)

Note: Direct injection *DI* engine, Brake power *BP*, Turbocharged direct injection engine *TDI*, Common rail direct injection *CRDi* hydrocarbons, and smoke opacity (Çelikten et al. 2010). Engine performance is an indication of the degree of efficiency for the fuel engine to accomplish the conversion of the chemical energy contained in the fuel into the mechanical and kinetic energy. The various diesel engines used by different researchers are listed in Table 19.3. The engine brake power (BP) consists of two elements such as the brake mean effective pressure (BMEP), i.e., the force available to work and the speed at which it is working. Therefore, the output of the engine can be increased either by increasing the speed or BMEP. Table 19.4 summarizes the comprehensive review of various biodiesel varieties and their engine performance.

Satputaley et al. (2017) during their experimentation on direct injection (DI) engine fueled with microalgal methyl ester reported an increment of about 4.4% BSFC in contrast to conventional diesel fuel at the load of 5.15 kW. This might be because microalgal methyl ester conveys lower heating value. Thus, more amount of fuel is being consumed to provide it the power output. Pradhan et al. (2016) investigated a novel method for preparing biodiesel from waste oil, thereby fabricating a new infrared radiator reactor. Their experiments depicted a higher value of BSFC of 6.3% and 7.9% for both B10 and B20 blends, respectively, in contrast to diesel at optimal loading condition of 8 N. However, at the peak load, B100 depicted lower fuel consumption than other test fuels.

Waste cooking oil is a potential feedstock for biodiesel production (Reddy et al. 2016; Nanda et al. 2019). Işık et al. (2017) during their investigation upon direct injection engine using waste cooking oil and *n*-butanol found that B20 blend of *n*-butanol had lower BSFC in contrast to B20 blend of waste cooking oil. Such a trend of lower fuel consumption for B20 was observed due to the presence of higher concentration of oxygen. This also results in lowering the cylinder wall temperature, thereby hampering the combustion behavior of *n*-butanol fuel. Mikulski et al. (2016) conducted their experimentation upon a common rail direct injection (CRDi) engine using swine lard biodiesel. From the analysis, it was seen that BSFC for B25, B50, and B75 blends were higher by 3.2%, 8.5%, and 13.8%, respectively, in contrast to diesel fuel. It was also seen that the fuel conversion efficiency (FCE) for all B25, B50, and B75 blends were found to be lesser, i.e., 1.6%, 4.8%, and 7.8%, respectively, than that of diesel fuel. The reason for such a trend was due to the lower heating value and reduced ignition delay. Jayaprabakar and Karthikeyan (2016) investigated the diesel engine using rice bran biodiesel blends by varying the injection timing from 23°bTDC until 26°bTDC. From the experimentation, it was clear that rice bran blends depicted higher BTE than other test fuels because of its higher heating value with advanced injection timing.

19.4 Emission Characteristics from Biodiesel-Fueled Engines

Hydrocarbon emissions take place when the fuel molecules inside the combustion chamber do not burn completely or burn partially. Hydrocarbons along with oxides of nitrogen (NO_x), CO, and CO_2 deplete the ozone layer and are the chief cause of global warming and smog formation in urban areas. CO is a product of incomplete

Table 19.4 Review of comparative engine performance in edible plant oil, biodiesel, and fossil diesel

Test condition	Biodiesel	BTE (%)	BSFC (kg/kWh)	BSEC (kg/kWh)	EGT (°C)	BP (kW)	References
Constant speed; injection timing (20°, 23° and 26°)	Algae (B20)	5%↓ for 20°bTDC at 4.3 kW	0.35 kg/kWh↓ for 20°bTDC at 4.3 kW	–	–	–	Devendra et al. (2015)
Emission test cycle	Microalgae (B100)	1.67%↑	9.79 g/kWh↑	–	–	0.12 kW↓	Satputaley et al. (2017)
Constant speed; load (0–18 kg) step of 2 kg	Microalgae (B100)	2%↓ at 2 kg load	4.4%↑	5 MJ/kWh at 2 kg↑	–	–	Pradhan et al. (2016)
Constant speed; variable load (28, 84, 140, 196 and 224 nm)	Waste cooking oil (B100)	2%↓	Avg. 50 g/kWh↑	–	–	–	Işık et al. (2017)
Constant speed; variable load (2, 3.3 and 4.6 kW)	Waste cooking oil (B20)	2% at 4.6 kW↓	75 g/kWh at 4.6 kW↑	–	–	–	Jayaprabakar and Karthikeyan (2016)
Constant speed; variable load (1, 3.6 and 9 kW)	Waste fry oil (B100)	1.89%↑	8.64%↑	–	–	–	Lei et al. (2016)
Constant speed; variable load (2, 4, 6 and 8 kW)	Waste mustard oil (B100)	2% at 2 N↓; 6% at 8 N↓	0.0015 kg/kWh at 2 N↑; 0.001 kg/kWh at 8 N↑	–	–	6% at 8 N↑	Atmanli (2016)
Constant speed (1500 rpm)	Biodiesel-diesel blends	3.2%↓	–	Higher for biodiesel blend	–	–	Lahane and Subramanian (2015)
Constant speed; variable load	Waste cooking oil methyl ester	3.6% at high load ↓	0.5 g/kWh at optimum load ↑	–	–	–	Gopal et al. (2014)

(continued)

Table 19.4 (continued)

Test condition	Biodiesel	BTE (%)	BSFC (kg/kWh)	BSEC (kg/kWh)	EGT (°C)	BP (kW)	References
Variable speed (1100–1700 rpm)	Waste fry oil	Higher for pure esters	0.2 g/kWh↑	–	–	–	Sanli et al. (2015)
Variable speed (1200–2800 rpm)	Diesel-algae; biodiesel-butanol blends	–	0.46 g/kWh↑	–	–	2.26%↓	Tüccar et al. (2014)
Variable speed (1600–2400 rpm)	Diesel-biodiesel-bioethanol	8.9%↓ for all blends	3.2% at full load condition ↑	–	–	–	Tan et al. (2017)
Variable load	Kutkura fruit seed oil	Avg. 4% max.↑	Slightly lower for B20 blend ↓	–	–	–	Kakati and Gogoi (2016)
Variable speed (1770–3800 rpm)	Microalgae (*Chlorella protothecoides*) biodiesel	Avg. 5.7% max.↑	10.2%↑ for all blends at full load	–	6.1%↓ at full load	–	Al-lwayzy and Yusaf (2017)
Variable load	*Annona* biodiesel with 1,4-dioxane fuel additive	Avg. 3% at optimal load ↑	8.8%↑	–	Slightly higher for all additive blends	–	Ramalingam et al. (2016)

Note: Brake thermal efficiency *BTE*, Brake power *BP*, Brake specific energy consumption *BSEC*, Brake specific fuel consumption *BSFC*, Exhaust gas temperature *EGT*, Decreasing trend ↓, Increasing trend ↑

combustion or partial oxidation of carbon present in the fuel. These vehicular emissions are currently one of the most extensive and extending metropolitan environmental problems (Ramalingam et al. 2016).

Satputaley et al. (2017) during their experimentation on a direct injection engine fuelled with microalgal methyl ester depicted a reduction in hydrocarbon emission for about 2% in contrast to diesel and other test fuels for all load conditions. The reason for such a trend was due to the enhanced cetane number of microalgal methyl ester blends. Satputaley et al. (2017) also showed a reduction of about 38 ppm for microalgal oil while 22 ppm for microalgal methyl ester upon loading condition of 5.15 kW. However, according to most of the reports there is an increase in NO_x emission as depicted in Table 19.5. Moreover, considering the emission analysis, a reduction of CO emission (27%↓) was found upon 5.15 kW of load. This might be due to high oxygen content in the prepared test fuel, thereby converting CO into CO_2.

Jayaprabakar and Karthikeyan (2016) investigated a diesel engine using rice bran biodiesel blends and microalgae oil by varying injection timing from 23°bTDC to 26°bTDC. From their experimentation, it was revealed that hydrocarbon emission lowered by 5% for both test fuel blends. This was because of high oxygen concentration in biodiesel, which improved the process of combustion for advanced injection timing. They conducted their experimentation upon a CRDi engine using swine lard biodiesel. From their analysis, it was seen that B25 blend depicted similar trends in contrast to diesel fuel considering the fuel conversion efficiency of the mean value 1.7%. Moreover, B50 (4.8%↓) and B75 (7.3%↓) provided lower hydrocarbon emissions in comparison to diesel fuel. The reason for this reduction in fuel conversion efficiency was reported because of the lower calorific value of swine lard biodiesel blends.

Atmanli (2016) carried out the experimentation on comparative analysis of diesel-waste oil biodiesel with propanol, *n*-butanol, and 1-pentanol blends in a direct injection engine. The results depicted higher hydrocarbon emission for pure biodiesel for about 78.9%↑ more than conventional diesel fuel for all load conditions. Moreover, the blending of *n*-butanol and 1-pentanol with biodiesel by 20% depicts lower hydrocarbon emission of 17.4% and 17.6%, respectively. However, the addition of propanol of 20% to biodiesel blend enhanced hydrocarbon emission by 35.4% in contrast to other test fuel blends. The results also depicted an increment of CO emission of about 33.8% for waste oil biodiesel in contrast to conventional fuel. Moreover, the addition of propanol, *n*-butanol, and 1-pentanol to waste oil blends enhanced CO emissions more around 39.9%, 38.3%, and 12.7%, respectively, in comparison to that of diesel fuel. According to the literature, there is a decrease in CO emission as shown in Table 19.5.

Pradhan et al. (2016) investigated a novel method for preparing biodiesel from waste oil by fabricating a new infrared radiator reactor. Their experiment demonstrated low hydrocarbon emissions of 16.3%↓ for B10 biodiesel blends in contrast to other test fuels. This was due to enhanced combustion for waste mustard oil blends. The experiment depicted lower CO emissions of 75%↓ at low loads and 70.3%↓ at peak load conditions. Işık et al. (2017) during their experimentation stated an addition of both waste cooking oil and *n*-butanol of 10% with conventional

Table 19.5 Comparative studies on engine exhaust emissions from biodiesel and conventional fuel

Test condition	Biodiesel	NO_x	CO_2	CO	Hydrocarbons	Smoke	Particulate matter	Reference
ETC test cycle	Microalgae (B100)	2.57 g/kWh↑	–	0.215 g/kWh↓	0.03 g/kWh↓	–	–	Panwar et al. (2010)
Constant speed; variable load (0–18 kg) in steps of 2 kg	Microalgae (B100)	38 and 22 ppm↓	–	27%↓	Approx. 2% for all load ↓	Approx. 5%↓	–	Zhu et al. (2011)
UDC/NEDC (15, 30, 50 and 100 km/h)	Animal fat (B50)	14 ppm↑	–	–	–		–	Agarwal and Dhar (2013)
Constant speed; variable load (1–5 kW) in steps of 1 kW	Animal fat residue (B100)	140 ppm and minimum load ↑	–	17.7% at low load ↓	Avg. 8.9% at 1500 rpm↑		200 mg/s at low load ↓	Vedharaj et al. (2014)
Constant speed; variable load (150, 300, 450 and 600 nm)	Waste chicken fat (B100)	16% at full load	2.5%↑	12%↓	Avg. 9.2%↑		–	Hwang et al. (2014)
2 speeds (1500; 3000 rpm); variable load (50, 100 and 150 nm)	Swine lard (B25; B50 and B100)	12.6% at 1500 rpm↑	Avg. 0.3%, 1.8% and 13.5% at 1500 rpm↑	Avg. 27%, 38% and 37% at 1500 rpm↓	7.3% at 1500 rpm↓		–	Nayak and Pattanaik (2014)
Variable speed (900–2700 rpm); constant load	Trout oil (B10; B20, B40 and B50)	40 ppm for B20↑	Avg. 1%, 3% and 6% at 3000 rpm↓	Avg. 19%, 25% and 30% at 3000 rpm↓	45% at 3000 rpm↓	40% for B10↓	–	Dhar and Agarwal (2014)
Variable speed; constant load	Waste cooking oil (B100)	32.5 ppm for 1200 rpm↑	2% for 3600 at full load ↓	Avg. 40% for 3600 rpm at full load ↑	Low for all speed at 50% and 100% loads	–	–	Sharma and Murugan (2015)

Constant speed; variable load	Waste cooking oil (B10)	8.7%↑	2%↑	11.8% for low and medium load ↓	29% for full load ↓	7% for peak load ↓	–	Karthikayan et al. (2015)
Constant speed; variable load	Waste cooking oil (B100)	Avg. 18.3%↑	–	Avg. 31%↓	Avg. 57%↓	Approx. 10% for low and medium loads ↑	–	Dhar and Agarwal (2015)
Constant speed; variable load	Waste cooking oil (B75)	Avg. 6.5%↑	Avg. 13.3%↑	Avg. 46.1%↑	Avg. 23.5% for B50↑	–	–	Senthil et al. (2016)
Variable speed (100, 1400 and 1700 rpm)	Waste fry oil (B100)	Avg. 11.3%↑	Avg. 2.08%↑	Avg. 22.3%↓	Avg. 29.36%↓	–	–	Gangil et al. (2016)
Constant speed; variable load (2, 3.3 and 4.6 kW)	Waste cooking oil (B20)	9.01%↑	9.09%↓	10.5%↓	10% at 4.6 kW↑	–	–	Sanjid et al. (2016)
Constant speed; variable load (1, 3.6 and 9 kW)	Waste fry oil (B100)	Avg. 1.68%↓	–	Avg. 12.7%↑	Avg. 78.9%↑	–	–	Dharma et al. (2016)
Constant speed; variable load (2, 4, 6 and 8 kW)	Waste mustard oil (B100)	29% at 2 N↑	30% at 2 N↓	75% at 2 N↓	16.3% at 2 N↓	–	–	Lalvani et al. (2016)
Constant speed; variable load	Fish oil (B100)	0.5 g/kWh↑	Approx. 8 g/kWh at low and high load ↓	Approx. 8 g/kWh at low and high load ↓	Approx. 0.5 g/kWh at low load ↓	–	64 mg/m^3 at peak load ↓	Kim et al. (2016)

(continued)

Table 19.5 (continued)

Test condition	Biodiesel	NO_x	CO_2	CO	Hydrocarbons	Smoke	Particulate matter	Reference
Constant speed; variable load	Fish oil (B100)	Avg. 5.2%↓	Avg. 33.7%↓	Avg. 33.7%↓	Avg. 26.2%↓	Avg. 3%↑	–	Tamilselvan and Nallusamy (2015)
Variable speed (1000–2500 rpm); constant load	Mahua and Jatropha biodiesel blends	Approx. 10 ppm↑	Avg. 16.24%↓	Avg. 16.24%↓	Avg. 19.8%↓	Avg. 10%↓	–	Acharya et al. (2017a)
Constant speed (1500 rpm)	Biodiesel-diesel blends	42.8%↑	–	Avg. 0.005%↓	Avg. 57%↓	17.4%↓	–	Acharya et al. (2017b)
Constant speed; variable load	Waste cooking oil methyl ester	20% for full load ↑	–	Avg. 45.5%↓	Avg. 24.2%↓	Avg. 9.3%↓	–	Lee et al. (2017)
Variable speed (1100–1700 rpm)	Waste fry oil	28% at 1500 rpm↑	Avg. 13.3%↑	Avg. 10.5%↓	24.2%↓	–	–	Srithar et al. (2017)
Variable speed (1200–2800 rpm)	Diesel-algae blend; biodiesel-Butanol blends	18.3% at 2000 rpm↓	–	14%↓	Avg. 8.1%↓	Avg. 44.9%↓	–	Sivaramakrishnan (2018)
Constant speed; variable load	Micro algae methyl ester	–	–	Avg. 43.3%↓	Avg. 17.7%↓	Avg. 40%↓	–	Mahalingam et al. (2018)
Variable speed	*Ceiba pentandra*	20% at 1500 rpm↑	11.11%↓	–	12%↓	–	–	Nayak et al. (2014)

Note: European transient cycle *ETC*, Particulate matter *PM*, Hydrocarbons *HC,* Urban driving cycle *UDC*, New European driving cycle *NEDC* fuel resulting in higher hydrocarbon emissions (10%↑) more than conventional diesel fuels for all load conditions. Moreover, the addition of 20% waste cooking oil and *n*-butanol resulted in lower hydrocarbon emissions. This was because of low heating value, short ignition delay, and higher oxygen content of biodiesel in contrast to conventional diesel fuel.

19.5 Conclusions

This critical review of different biodiesel varieties including edible oils, non-edible oils, and animal fats summarized their physicochemical properties, engine performance, and emissions. The major problem associated with plant oil is its higher viscosity than diesel, which affects the spray characteristics leading to improper combustion. The biodiesel engine brake power may also fluctuate depending upon the type of feedstock and oil extraction process. The emissions of CO, hydrocarbons, and smoke reduce considerably but NO_x generally increase because of higher oxygen concentration in biodiesel. Many techniques have been undertaken to improve the performance and emission characteristics of biodiesel like preheating, blending, emulsification, transesterification, pyrolysis, hydrocracking, thermal cracking, etc. Some reports also revealed modification in engine design and configuration like injector pressure, injection timing, additional combustion chamber, and exhaust gas recirculation to increase biodiesel efficiency. More research and development will determine the large-scale commercial utilization of biodiesel in the newer vehicles at a global scale.

References

Acharya N, Nanda P, Panda S, Acharya S (2017a) A comparative study of stability characteristics of Mahua and Jatropha biodiesel and their blends. J King Saud Univ Eng Sci 31:184–190

Acharya N, Nanda P, Panda S, Acharya S (2017b) Analysis of properties and estimation of optimum blending ratio of blended mahua biodiesel. Eng Sci Technol Int J 20:511–517

Agarwal AK, Dhar A (2013) Experimental investigation of performance emission and combustion characteristics of karanja oil blends fuelled DICI engine. Renew Energy 52:283–291

Al-lwayzy SH, Yusaf T (2017) Diesel engine performance and exhaust gas emissions using microalgae *Chlorella protothecoides* biodiesel. Renew Energy 101:690–701

Atabani AE, Badruddin IA, Mekhilef S, Silitonga AS (2011) A review on global fuel economy standard, labels and technology in the transportation sector. Renew Sust Energ Rev 15:4586–4610

Atabani AE, Silitonga AS, Badruddin IA, Mahlia TMI, Masjuki HH, Mekhilef S (2012) A comprehensive review on biodiesel as an alternative energy resource and its characteristics. Renew Sust Energ Rev 16:2070–2093

Atadashi IM, Aroua MK, Abdul Aziz A (2011) Biodiesel separation and purification: a review. Renew Energy 36:437–443

Atmanli A (2016) Comparative analyses of diesel–waste oil biodiesel and propanol, *n*-butanol or 1-pentanol blends in a diesel engine. Fuel 176:209–215

Çelikten İ, Koca A, Arslan MA (2010) Comparison of performance and emissions of diesel fuel: rapeseed and soybean oil methyl esters injected at different pressures. Renew Energy 35:814–820

Demirbas A (2007) Importance of biodiesel as transportation fuel. Energ Policy 35:4661–4670
Devendra S, Singal SK, Garg MO, Maiti P, Mishra S, Ghosh PK (2015) Transient performance and emission characteristics of a heavy-duty diesel engine fuelled with microalga *Chlorella variabilis* and *Jatropha curcas* biodiesels. Energy Convers Manag 106:892–900
Dhar A, Agarwal AK (2014) Performance, emission and combustion characteristic of karanja biodiesel in transportation engine fuel. Fuel 119:70–80
Dhar A, Agarwal AK (2015) Effect of karanja biodiesel blends on particulate emission from a transportation engine. Fuel 141:154–163
Dharma S, Masjuki HH, Ong HC, Sebayang AH, Silitonga AS, Kusumo F, Mahlia TMI (2016) Optimization of biodiesel production process for mixed *Jatropha curcas–Ceiba pentandra* biodiesel using response surface methodology. Energ Convers Manage 115:178–190
Gangil S, Singh R, Bhavate P, Bhagat D, Modhera B (2016) Evaluation of engine performance and emission with methyl ester of Karanja oil. Perspect Sci 8:241–243
Gopal KN, Pal A, Sharma S, Samanchi C, Sathyanarayanan K, Elango T (2014) Investigation of emissions and combustion characteristics of a CI engine fueled with waste cooking oil methyl ester and diesel blends. Alexandria Eng J 53:281–287
Hwang J, Qi D, Jung Y, Bae C (2014) Effect of injection parameter on the combustion and emission characteristics in a common rail direct injection diesel engine fuelled with waste cooking oil biodiesel. Renew Energy 63:9–17
Işık MZ, Hasan B, Bahattin I, Huseyin A (2017) The effect of *n*-butanol additive on low load combustion, performance and emissions of biodiesel diesel blend in a heavy duty diesel power generator. J Energy Inst 90:174–184
Jayaprabakar J, Karthikeyan A (2016) Performance and emission characteristics of rice bran and alga biodiesel blends in a CI engine. Mater Today Proc 3:2468–2474
Kakati J, Gogoi TK (2016) Biodiesel production from Kutkura (*Meyna spinosa Roxb. Ex.*) fruit seed oil: its characterization and engine performance evaluation with 10% and 20% blends. Energ Convers Manage 121:152–161
Karmakar A, Karmakar S, Mukherjee S (2010) Properties of various plants and animal feedstock for biodiesel production. Bioresour Technol 101:7201–7210
Karthikayan S, Sankaranarayanan G, Karthikeyan R (2015) Green technology effect of injection pressure, timing and compression ratio in constant pressure heat addition cycle by an eco-friendly material. Ecotoxicol Environ Saf 121:63–66
Kim HJ, Park SH, Lee CS (2016) Impact of fuel spray angles and injection timing on the combustion and emission characteristics of a high-speed diesel engine. Energy 107:572–579
Lahane S, Subramanian KA (2015) Effect of different percentages of biodiesel–diesel blends on injection, spray, combustion, performance, and emission characteristics of a diesel engine. Fuel 139:537–545
Lalvani JIJ, Parthasarathy M, Dhinesh B, Annamalai K (2016) Pooled effect of injection pressure and turbulence inducer piston on performance, combustion, and emission characteristics of a DI diesel engine powered with biodiesel blend. Ecotoxicol Environ Saf 134:336–343
Lee S, Lee CS, Park S, Gupta JG, Maurya RK, Agarwal AK (2017) Spray characteristics, engine performance and emission analysis for Karanja biodiesel and its blends. Energy 119:138–151
Lei Z, Yao X, Cheung CS, Chun G, Zhen H (2016) Combustion, gaseous and particulate emission of a diesel engine fueled with *n*-pentanol (C5 alcohol) blended with waste cooking oil biodiesel. Appl Ther Eng 102:73–79
Lin L, Cunshan Z, Xiangqian S, Mingdong D (2011) Opportunities and challenge for biodiesel fuel. Appl Energy 88:1020–1031
Mahalingam A, Devarajan Y, Radhakrishnan S, Vellaiyan S, Nagappan B (2018) Emission analysis on mahua oil biodiesel and higher alcohol blends in diesel engine. Alexandria Eng J 57:2627–2631
Mikulski M, Duda K, Slawomir W (2016) Performance and emissions of a CRDI diesel engine fuelled with swine lard methyl esters–diesel mixture. Fuel 164:206–219
Nanda S, Mohammad J, Reddy SN, Kozinski JA, Dalai AK (2014) Pathways of lignocellulosic biomass conversion to renewable fuels. Biomass Conv Bioref 4:157–191

Nanda S, Azargohar R, Dalai AK, Kozinski JA (2015) An assessment on the sustainability of lignocellulosic biomass for biorefining. Renew Sust Energ Rev 50:925–941

Nanda S, Kozinski JA, Dalai AK (2016) Lignocellulosic biomass: a review of conversion technologies and fuel products. Curr Biochem Eng 3:24–36

Nanda S, Rana R, Zheng Y, Kozinski JA, Dalai AK (2017) Insights on pathways for hydrogen generation from ethanol. Sustain Energ Fuel 1:1232–1245

Nanda S, Rana R, Hunter HN, Fang Z, Dalai AK, Kozinski JA (2019) Hydrothermal catalytic processing of waste cooking oil for hydrogen-rich syngas production. Chem Eng Sci 195:935–945

Nayak SK, Mishra PC (2018) Pre-and post-mixed hybrid biodiesel blends as alternative energy fuels-an experimental case study on turbo-charged direct injection diesel engine. Energy 160:910–923

Nayak SK, Mishra PC (2019) Achieving high performance and low emission in a dual fuel operated engine with varied injection parameters and combustion chamber shapes. Energ Convers Manage 180:1–24

Nayak SK, Pattanaik BP (2014) Experimental investigation on performance and emission characteristics of a diesel engine fuelled with mahua biodiesel using additive. Energy Procedia 54:569–579

Nayak SK, Nayak SK, Mishra PC, Tripathy S (2014) Influence of compression ratio on combustion characteristics of a VCR engine using *Calophyllum inophyllum* biodiesel and diesel blends. J Mech Sci Technol 29:4047–4052

Nayak SK, Behera GR, Mishra PC, Kumar A (2017a) Functional characteristics of jatropha biodiesel as a promising feedstock for engine application. Energ Sour Part A 39:299–305

Nayak SK, Behera GR, Mishra PC, Sahu SK (2017b) Biodiesel vs diesel: a race for the future. Energ Sourc Part A 39:1453–1460

Panwar NL, Shrirame HY, Rathore NS, Jindal S, Kurchani AK (2010) Performance evaluation of a diesel engine fueled with methyl ester of castor seed oil. Appl Ther Eng 30:245–249

Pradhan P, Chakraborty S, Chakraborty R (2016) Optimization of infrared radiated fast and energy-efficient biodiesel production from waste mustard oil catalyzed by Amberlyst 15: engine performance and emission quality assessments. Fuel 173:60–68

Ramalingam S, Rajendran S, Ganesan P (2016) Improving the performance is better and emission reductions from Annona biodiesel operated diesel engine using 1,4- dioxane fuel additive. Fuel 185:804–809

Rana R, Nanda S, Kozinski JA, Dalai AK (2018) Investigating the applicability of Athabasca bitumen as a feedstock for hydrogen production through catalytic supercritical water gasification. J Environ Chem Eng 6:182–189

Rana R, Nanda S, Maclennan A, Hu Y, Kozinski JA, Dalai AK (2019) Comparative evaluation for catalytic gasification of petroleum coke and asphaltene in subcritical and supercritical water. J Energ Chem 31:107–118

Reddy SN, Nanda S, Kozinski JA (2016) Supercritical water gasification of glycerol and methanol mixtures as model waste residues from biodiesel refinery. Chem Eng Res Des 113:17–27

Reddy SN, Nanda S, Sarangi PK (2018) Applications of supercritical fluids for biodiesel production. In: Sarangi PK, Nanda S, Mohanty P (eds) Recent advancements in biofuels and bioenergy utilization. Springer Nature, Singapore, pp 261–284

Saba T, Estephane J, El Khoury B, El Khoury M, Khazma M, El Zakhem H, Aouad S (2016) Biodiesel production from refined sunflower vegetable oil over KOH/ZSM5 catalysts. Renew Energy 90:301–306

Sanjid A, Kalam MA, Masjuki HH, Varman M, Zulkifli MNW, Abedin MJ (2016) Performance and emission of multi-cylinder diesel engine using biodiesel blends obtained from mixed inedible feedstocks. J Clean Prod 112:4114–4122

Sanli H, Canakci M, Alptekin E, Turkcan A, Ozsezen AN (2015) Effects of waste frying oil based methyl and ethyl ester biodiesel fuels on the performance, combustion and emission characteristics of a DI diesel engine. Fuel 159:179–187

Satputaley SS, Zodpe DB, Deshpande NV (2017) Performance, combustion and emission study on CI engine using microalgae oil and microalgae oil methyl esters. J Energy Inst 90:513–521

Senthil M, Visagavel K, Saravanan CG, Rajendran K (2016) Investigation of red mud as catalyst in Mahua oil biodiesel production and its engine performance. Fuel Process Technol 149:7–14
Shahid EM, Jamal J (2011) Production of biodiesel: a technical review. Renew Sust Energ Rev 15:4732–4745
Shameer PM, Ramesh K, Sakthivel R, Purnachandran R (2016) Studies on correlation between NOx and in-cylinder temperature in a D.I diesel engine using fluke thermal imager for different alternate fuel blends. Asian J Res Soc Sci Human 6:373–389
Sharma A, Murugan S (2015) Combustion, performance and emission characteristics of a di diesel engine fuelled with non-petroleum fuel: a study on the role of fuel injection timing. J Energy Inst 88:364–375
Sharma YC, Singh B (2009) Development of biodiesel: current scenario. Renew Sust Energ Rev 13:1646–1610
Sivaramakrishnan K (2018) Investigation on performance and emission characteristics of a variable compression multi fuel engine fuelled with Karanja biodiesel diesel blend. Egypt J Pet 27:177–186
Srithar K, Balasubramanian AK, Pavendan V, Kumar BA (2017) Experimental investigation on mixing of two biodiesel as alternative fuel diesel engine. J King Saud Univ—Eng Sci 29:50–56
Tamilselvan P, Nallusamy N (2015) Performance, combustion and emission characteristics of a compression ignition engine operating on pine oil. Biofuels 6:273–281
Tan YH, Abdullah MO, Nolasco-Hipolito C, Zauzi NSA, Abdullah GW (2017) Engine performance and emissions characteristics of a diesel engine fueled with diesel biodiesel- bioethanol emulsions. Energy Convers Manag 132:54–64
Tüccar G, Özgür T, Aydın K (2014) Effect of diesel–microalgae biodiesel–butanol blends on performance and emissions of diesel engine. Fuel 132:47–52
USEIA, U.S. Energy Information Administration (2010) International Energy Outlook. http://www.eia.doe.gov/oiaf/ieo/pdf/0484%282010%29.pdf. Accessed 22 Apr 2011
Vedharaj S, Vallinayagnam R, Yang WM, Chou SK, Lee PS (2014) Effect of adding 1, 4-dioxine with kapok biodiesel on the characteristics of a diesel engine. Appl Energy 136:1166–1173
Xue J, Grift TE, Hunsan AC (2011) Effect of biodiesel on engine performance and emission. Renew Sust Energ Rev 15:1098–1116
Zhu L, Cheung CS, Zhang WG, Huang A (2011) Combustion, performance and emission characteristics of a DI diesel engine fueled with ethanol–biodiesel blends. Fuel 90:1743–1750
Zuleta EC, Rios AL (2012) Oxidation stability and cold flow behaviour of plam, sachainchi, Jatropha and castor oil biodiesel blends. Fuel Process Technol 102:3417–3423

Current Advancements in Microbial Fuel Cell Technologies

20

Latika Bhatia, Prakash K. Sarangi, and Sonil Nanda

Abstract

With the increase in global energy utilization and expected deficits in crude oil supply, there is an extensive and prompt enthusiasm in developing alternative green sustainable energy sources. In the recent years, the utilization of biomass-based energy in terms of fuel or power has become an imperative requirement on a global scale to enhance the environmental sustainability. Organic wastes are considered as the main feedstock for bioenergy and electricity production. Microbial fuel cells are enthralling bioelectrochemical devices that utilize microbial biomass as the main catalytic source to convert organic waste matter into electrical energy. Recent advancements in microbial fuel cells in terms of structural modification, substrates utilization, modes of operation, supplementation of different microbial communities, overcoming limitations, and exploring new applications toward clean environment are critically described in this chapter. The role of eukaryotic microorganisms in microbial fuel cells is discussed. The perspectives of different microbial species and their biocatalytic activities to generate electrical energy from organic wastes along with future scope and possibilities are also critically reviewed. This chapter provides glimpses of the technological advancements and utilities of the microbial fuel cells.

L. Bhatia
Department of Microbiology and Bioinformatics, Atal Bihari Vajpayee University, Bilaspur, Chhattisgarh, India

P. K. Sarangi (✉)
Directorate of Research, Central Agricultural University, Imphal, Manipur, India
e-mail: sarangi77@yahoo.co.in

S. Nanda
Department of Chemical and Biological Engineering, University of Saskatchewan, Saskatoon, Saskatchewan, Canada

S. Nanda et al. (eds.), *Biorefinery of Alternative Resources: Targeting Green Fuels and Platform Chemicals*, https://doi.org/10.1007/978-981-15-1804-1_20

Keywords
Microbial fuel cells · Biofuels · Bioenergy · Microorganisms

20.1 Introduction

In the recent years, an exponential increase in the energy utilization around the world is observed due to rapid industrialization and urbanization both in developing and developed countries (Nanda et al. 2014b, 2015a). Fossil fuels not only result in atmospheric pollution due to massive greenhouse gas emissions but also cause global warming, climate change, and threat to many natural ecosystems (Nanda et al. 2016). Many notable efforts have been made to explore convincing strategies in resolving the energy crisis by focusing on a variety of renewable sources such as solar, wind, tidal, geothermal, and waste biomass (e.g., lignocellulosic biomass, cattle manure, industrial effluents, food waste, waste cooking oil, etc.) (Nanda et al. 2013, 2015a, b, 2016a, b, 2019). The fuel cells are much recent approaches generating energy involving metal catalysts of high value (Rahimnejad et al. 2012a). There are many benefits associated with fuel cells, a major among them being zero emission of greenhouse gases (viz. SO_x, NO_x, CO_2, and CO). However, fuel cells also pose some disadvantages such as being expensive and generating high microbial biomass (Rahimnejad et al. 2015).

Biological fuel cell works on the principles of electrochemistry and physics along with the involvement of living organism that performs catalytic redox activity. There are two types of biological fuel cells, namely enzymatic fuel cell (EFC) and microbial fuel cell (MFC). Enzymatic fuel cells produce electrical energy with the aid of selective enzymes that work as a catalyst toward redox reactions, whereas microbial fuel cells generate electricity from organic compound with the aid of electroactive microorganisms (Santoro et al. 2017). In other words, MFCs are a form of renewable devices that hold the potential to transform chemical energy into electricity involving a variety of electrochemically active bacteria. In MFCs, a broad spectrum of carbon sources including organic substrate or even pollutants in wastewater are oxidized by electrochemically active bacteria, thereby transferring the generated electron to anodes.

Microbial fuel cells involve anaerobic metabolism of electrochemically active bacteria for electricity generation. These electrochemically active bacteria consume low-grade waste or organic pollutants that are too wet to be burned, thereby making the MFC technology highly advantageous. The breakdown of biomass is an important process for the production of target products by utilizing microorganisms and their enzymes. There are some contrasting features of MFCs over conventional low-temperature fuel cells such as biotic electro-catalyst at anodic side, near-ambient operating temperature (15–45 °C), utilization of complex biomass as anodic fuel, neutral pH condition, and low environmental impacts (He et al. 2005; Larrosa-Guerrero et al. 2010; Borole et al. 2011; Tremouli et al. 2016; Tee et al. 2017). The biotechnological implementation and computational tools are continuously making remarkable impact to overcome the critical issues of process engineering in MFCs. The chapter makes an overview of the recent progress and application of MFCs.

20.2 Brief History of Microbial Fuel Cells

The history of electricity generation by microorganisms is believed to be initially performed by Potter in 1910 (Potter 1911). However, such findings were not well established because of having only the only concept of electron transport by microorganisms from cellular metabolism process, which is also regarded as extracellular electron transfer. Over the time, many researchers attempted biocatalytic electricity generation. When the idea of current and power generation was established through the addition of electron acceptor, real advancements came out on this context. After significant development of MFC reactors during the late 1990s by utilizing wastewater as the substrate, the real progress on this technology started. Later years also witnessed many researches enhancing the overall energy output and efficiency of MFCs. Diversification was achieved by modifying the architecture of basic integrals like the electrodes or by researching with a variety of solutions or electrolyte being used.

Several strategies were framed to augment the comprehensive output and the electricity production efficiency. Many researchers globally have worked on the configuration and modification of the basic components of MFCs such as electrodes and utilized solution or fuel. In 2005, another alteration for augmenting electron transmission tool directly by extracellular conductive connections was discovered. Additionally, research sphere of MFCs broadened in 1999 when it was noticed that mediator was not an essential part within MFCs (Rahimnejad et al. 2015).

During the last decade, MFCs have established great attention among researchers and academicians globally being a novel technology toward a sustainable and renewable production of energy (Du et al., 2008). The electrochemical actions of microorganisms play vital roles for conversion of waste organic residues into electrical energy through oxidation and electron transfer. The first generation of MFC was based on the principle of electron mediators. In this type of MFC, direct transfer of electrons by microorganisms to anode was not possible, which necessitated the requirement of external electron acceptors. The focus of such type of MFC is the implementation of electron mediators by acceptance of electrons of bacterial origin and then subsequently releasing these electrons to the anode. Such mediator-assisted electron transfer process has some limitations of toxicity and instability (Du et al. 2007). Thus, with further research, second-generation MFCs were developed to increase the mediator-less electron transfer (Kim et al. 2002; Chaudhuri and Lovley 2003). *Shewanella putrefaciens*, *Geobacteraceae sulfurreducens*, *Geobacteraceae metallireducens*, and *Rhodoferax ferrireducens* are few examples of active bacteria used in MFCs (Kim et al. 2002; Bond and Lovley 2003; Holmes et al. 2004; Crittenden et al. 2006; Liu et al. 2007).

20.3 Process Mechanism of Microbial Fuel Cells

The microorganisms divert the electrons generated from the substrate for their metabolic requirement along with migration toward the anode. This process demands efficiency for sustainable generation of power. Depending on the mode of electrons transportation to the anode, MFCs can be classified into two groups, especially mediator-using MFCs and mediator-less MFCs (Pant et al. 2010). Some synthetic exogenous mediators are neutral red, methylene blue, and 2-hydroxy-1,4-naphthoquinone. The incorporation of exogenous mediators could help in this scenario if they are stable and non-toxic. The employment of native electron shuttles generated by the microorganisms can also circumvent the problem of electron flow. The secondary metabolites can be employed for MFC applications, as they are potential redox mediators. These metabolites are potential reversible electron acceptors as they carry electrons from bacterial cell to an anode. The way the chambers of anode and cathode are assembled, it classifies MFC prototype into either single-chambered (Fig. 20.1) or double-chambered (Fig. 20.2).

MFCs are described by biological and electrochemical parameters. The substrate-loading rates in continuous systems are biological parameters, whereas power density and cell voltage are chief electrochemical parameters. There are several factors influencing the performance of MFCs, a few of which include the following (Rahimnejad et al. 2015):

1. Oxygen supply and its consumption in cathode chamber.
2. Substrate oxidation in anode chamber.
3. Electron shuttle from anode compartment to anode surface.
4. Proton exchange membrane permeability.

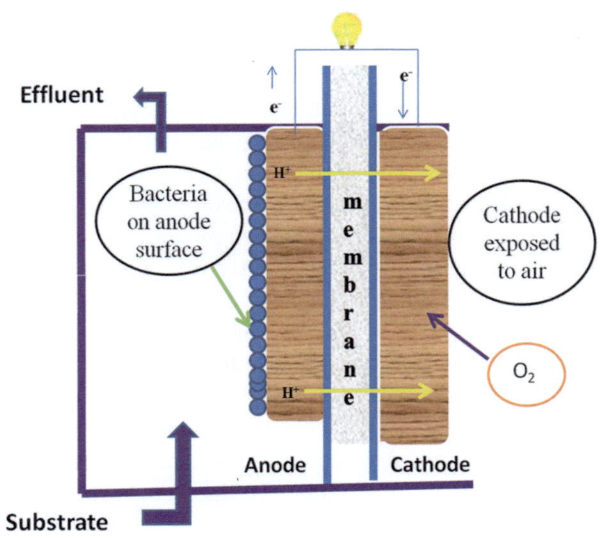

Fig. 20.1 A single-chambered microbial fuel cell

There are many electrochemically active bacteria acting as MFCs. These bacteria possess extracellular electron transport mechanism that helps transport of electron. Thus, these microorganisms can produce electricity from various energy sources. When microorganisms start metabolizing organic matter inside an anaerobic anode chamber, they generate the free electrons and protons. The electrons travel through the electron transport chain releasing energy, which supports microbial survival and metabolism. These electrons reach external terminal electron acceptors (resistor) toward the cathode through several extracellular electron transport mechanisms, thereby generating voltage. There is aerobic cathode chamber where protons are diffused simultaneously via the selective proton exchange membrane.

Water molecules are produced in this chamber due to the reduction of oxygen by electrons and protons, thereby completing the charge balance. This technology has proven itself as one of the optimal methods to undergo simultaneous bioremediation and power generation (Wang et al. 2014). Elevated endogenous secretion of pyocyanin mediators could be achieved by chemical alterations of the insulating interface junction across the cellular membrane of *Escherichia coli* (Hou et al. 2013) and genetically engineered *Pseudomonas aeruginosa* (Wang et al. 2013), thereby enhancing the power output in these MFCs. Human urine has also been explored as a source of energy for small-scale stacked MFCs (Ieropoulos et al. 2013). The performance of MFCs can be further enhanced by employing extracellular electron transport processes for better understanding of microbial species interactions.

Earlier studies have revealed that *Pseudomonas* sp. are prominent members of microbial population used in MFCs. These bacteria produce phenazine-based

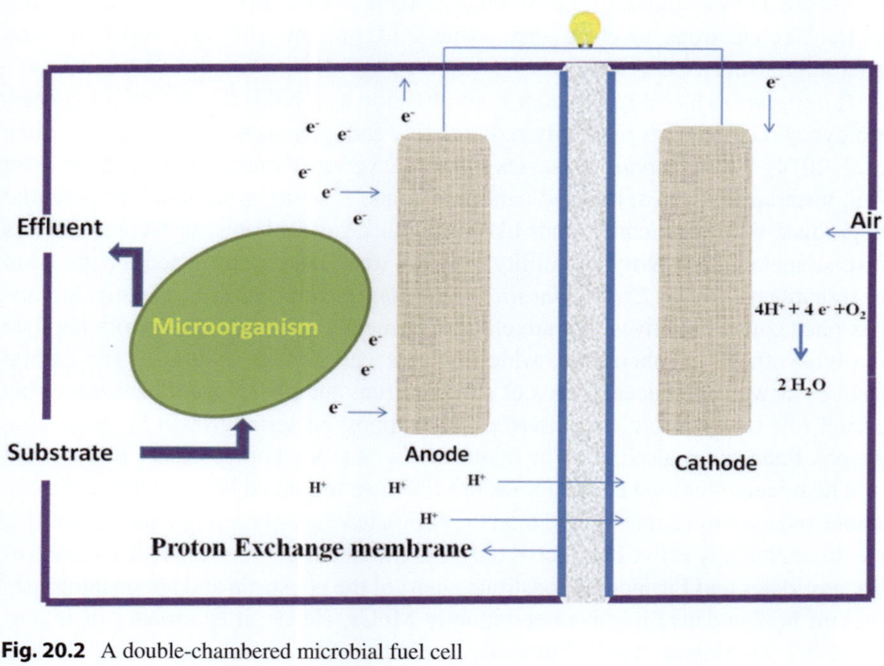

Fig. 20.2 A double-chambered microbial fuel cell

mediators that relocate electrons to the electrode. *Brevibacillus* sp. PTH1 is found along with *Pseudomonas* sp. in an acetate-fed MFC. *Pseudomonas* sp. CMR12a produces a significant amount of phenazine-1-carboxamide (PCN) and biosurfactants in the anode of MFCs. *Pseudomonas* sp. CMR12a_Reg is a regulatory mutant that is unable to synthesize PCN. The supernatants of this mutant did not show the above improvement effects. When rhamnolipids were used as biosurfactants at the concentration of 1 mg/L along with purified PCN, it enhanced the generation of electricity by *Brevibacillus* sp. PTH1. *Brevibacillus* manifests the ferric iron reduction that is supplied as goethite (FeOOH) with concomitant oxidization of acetate, thereby achieving transfer of electron extracellularly. This effect was missing when PCN was added alone. The possibilities indicate that this synergy of bacteria is a prominent mechanism in the anodic electron transfer of an MFC, thus aiding *Brevibacillus* sp. PTH1 to acquire its supremacy (Pham et al. 2008).

20.4 Microorganisms Used in Microbial Fuel Cells

MFCs are potential sources for the generation of clean energy as well as for pollutants remediation. The microbial system in anode has a potential to oxidize a broad spectrum of organic substrates including organic pollutants found in wastewater. In this process, the generation of electrons and its transfer to inertial solid electrodes occurs, thereby making it feasible to generate electricity from chemical energy (Osman et al. 2010). MFCs can also be used as biosensors and in secondary fuel production (Das and Mangwani 2010). *Geobacter*, *Shewanella*, *Pseudomonas*, and *E. coli* are a few examples of electrochemically active bacteria holding the potential to transfer electrons to electrodes. Table 20.1 presents different substrates and microorganisms used in MFCs.

P. aeruginosa has attracted enormous attention as it holds the potential to generate pyocyanin that acts as highly redox-active endogenous electron shuttles (Shen et al. 2014). *Pseudomonas* possesses metabolic versatility and can be isolated from soil, water, petroleum spills, and natural habitats. *P. aeruginosa* holds the potential to produce significant compounds like antibiotics, siderophores, surfactants, etc. Its diverse metabolic activity and utility in waste treatment has attracted the attention of researchers toward *P. aeruginosa*. In addition, *P. aeruginosa* is a source of various phenazine derivatives that are electrochemically active. Many factors regulate the biosynthesis of phenazine, which include cell density, nature of the carbon sources as well as concentrations of oxygen, iron, and phosphate. Phenazines play a vital role in anaerobic conditions as they support bacterial growth by generating energy. Phenazines also aid in the maintenance of redox homeostasis.

The mixed microbial communities in MFCs are beneficial because they are more stable and easy to be maintained than the MFCs having only a single pure culture of electrochemically active bacteria. Except the exoelectrogens, fermentative strains of Bacteroidetes and Firmicutes constitute a part of the consortia and are unanimously present in abundance in mixed community MFCs. Hence, it is a matter of importance to investigate the community structure for mixed-community MFCs.

Table 20.1 Different substrates and microorganisms used in microbial fuel cells

Substrate	Type of inoculum	Type of MFCs	Maximum power or current produced	References
1,2-dichloroethane	Microbial consortia from acetate-enriched MFC	Two-chambered MFC	0.008 mA/cm^2	Pham et al. (2009)
Acetate	Pre-acclimated bacteria from MFC	Cube-shaped chamber MFC	0.8 mA/cm^2	Logan et al. (2007)
Arabitol	Pre-acclimated bacteria from MFC	Single-chambered air cathode MFC	0.68 mA/cm^2	Catal et al. (2008)
Azo dye with glucose	Mixture of aerobic and anaerobic sludge	Single-chambered air cathode MFC	0.09 mA/cm^2	Sun et al. (2009)
Carboxymethylcellulose	Co-culture of *Clostridium cellulolyticum* and *Geobacter sulfurreducens*	Double-chambered MFC	0.05 mA/cm^2	Ren et al. (2008)
Cellulose particles	Pure culture of *Enterobacter cloacae*	MFC with carbon cloth anode	0.02 mA/cm^2	Rezaei et al. (2009)
Ethanol	Anaerobic sludge from wastewater plant	Double-chambered aqueous cathode MFC	0.025 mA/cm^2	Kim et al. (2007)
Furfural	Pre-acclimated bacteria from anode of a ferricyanide-cathode MFC	Single-chambered air cathode MFC	0.17 mA/cm^2	Luo et al. (2010)
Galactitol	Pre-acclimated bacteria from MFC	Single-chambered air cathode MFC	0.78 mA/cm^2	Catal et al. (2008)
Glucose	Mediator-less *Saccharomyces cerevisiae*	Double-chambered MFC	28 mW/m^2	Rahimnejad et al. (2012b)
Glucose	Mediator-less *Saccharomyces cerevisiae*	Air cathode MFC	3 mW/m^2	Sayed et al. (2012)

(continued)

Table 20.1 (continued)

Substrate	Type of inoculum	Type of MFCs	Maximum power or current produced	References
Glucose	Mixed bacterial culture of *Rhodococcus* and *Paracoccus*	Single-chambered air cathode MFC	0.70 mA/cm^2	Catal et al. (2008)
Glucose	Methylene blue-mediated *Saccharomyces cerevisiae*	Double-chambered MFC	150 mW/m^2	Ganguli and Dunn (2009)
Glucose	Methylene blue-mediated *Saccharomyces cerevisiae*	Double-chambered MFC	1500 mW/m^2	Ganguli and Dunn (2009)
Landfill leachate	Leachate and sludge	Double-chambered MFC	0.0004 mA/cm^2	Greenman et al. (2009)
Mannitol	Pre-acclimated bacteria from MFC	Single-chambered air cathode MFC	0.58 mA/cm^2	Catal et al. (2008)
Microalage	*Chlorella vulgaris*	Single-chambered air cathode MFC	0.2 mA/cm^2	Velasquez-Orta et al. (2009)
Phenol	Mixed aerobic activated sludge and anaerobic sludge	Double-chambered MFC	0.1 mA/cm^2	Luo et al. (2009)
Ribitol	Pre-acclimated bacteria from MFC	Single-chambered air cathode MFC	0.73 mA/cm^2	Catal et al. (2008)
Sodium formate	Anaerobic digested fluid from a sewage treatment plant	Double-chambered MFC	0.22 mA/cm^2	Ha et al. (2008)
Sodium fumarate	Pure culture of *Geobacter sulfurreducens*	Stainless steel cathode	2.05 mA/cm^2	Dumas et al. (2008)
Sorbitol	Pre-acclimated bacteria from MFC	Single-chambered air cathode MFC	0.62 mA/cm^2	Catal et al. (2008)
Starch	*Clostridium butyricum*	Double-chambered MFC	1.3 mA/cm^2	Niessen et al. (2004)

(continued)

Table 20.1 (continued)

Substrate	Type of inoculum	Type of MFCs	Maximum power or current produced	References
Sucrose	Anaerobic sludge from septic tank	Double-chambered mediator less MFC	0.19 mA/cm^2	Behera and Ghangrekar (2009)
Synthetic wastewater	Mediator-less *Saccharomyces cerevisiae*	Air cathode with graphite plate anode	25.5 mW/m^2	Raghavulu et al. (2001)
Synthetic wastewater with molasses and urea	Anaerobic mixture from wastewater plant	Double-chambered MFC	0.005 mA/cm^2	Kargi and Eker (2007)
Xylitol	Pre-acclimated bacteria from MFC	Single-chambered air cathode MFC	0.71 mA/cm^2	Catal et al. (2008)
Xylose	Mixed bacterial culture	Single-chambered air cathode MFC	0.74 mA/cm^2	Catal et al. (2008)
Xylose and humic acid	Domestic wastewater	Double-chambered MFC	0.06 mA/cm^2	Huang and Angelidaki (2008)

Direct-ethanol fuel cell (DEFC) is a type of fuel cell, which requires ethanol as the feedstock. Short-chain alcohols are vital category of next-generation biofuels, which can be produced through fermentative pathways (Nanda et al. 2014a, 2017a).

Pseudomonas putida has proven itself as a diverse microorganism for biofuels production as it is catabolically diverse and has accomplished metabolism and elaborated resilient power to assorted toxic materials (Udaondo 2012). *P. putida* possesses features like adaptable metabolism, immense innate resilience to noxious materials, and flexibility of metabolic engineering. Therefore, it has appeared to be a distinguished microorganism for high-output yield of next-generation biofuels. Moreover, *P. putida* gains attention as a promising and potential organism for ethanol formation at an industrial scale as it possesses important characteristics such as: (a) native elaborated protection to various stressors, which includes solvents, (b) susceptibility to genetic modification, (c) competence to multiply rapidly, and (d) generally recognized as safe (GRAS) (Martins dos Santos et al. 2004).

When the biosynthetic pathway from *C. acetobutylicum* was imposed in *P. putida*, the latter gained the potency to produce *n*-butanol (Nielsen 2009). As *P. putida* also possesses the capacity to degrade *n*-butanol, it becomes vital to knock out the analogous genes to enhance *n*-butanol production (Cuenca 2016). Moreover, the engineered strain of *P. putida* has the ability to grow under anaerobic conditions and employed in two-phase liquid extraction systems (Schmitz 2015; Basler et al. 2018).

P. putida strains are robust to utilize efficiently a wide range of carbon sources ranging from aliphatic to aromatic hydrocarbons. *P. putida* can be employed for biofuels production from lignocellulosic materials as it has been genetically modified for aromatic compounds formation from lignin (Johnson and Beckham 2015; Vardon et al., 2015). Lignin is a cross-linked phenylpropane polymer that binds cellulose and hemicellulose together in lignocellulosic biomass (Fougere et al. 2016; Rana et al. 2018). Bacterial ligninolytic systems as compared to fungal system are more specific in action.

P. putida has proven to be an efficient biocatalyst after being engineered to support the formation of a broad spectrum of compounds such as polyketides, non-ribosomal peptides, rhamnolipids, aromatics, and non-aromatics (Loeschcke and Thies 2015). Many strains of *Pseudomonas* possess the potential to efficiently utilize numerous industrial products and solvents as the sole source of carbon. Moreover, it can also tolerate high amount of toxic aromatic compounds (Basler et al. 2018).

Prema et al. (2015) researched on the production of biofuel using waste papers from *P. aeruginosa*. Cellulose structure (the main constituent of paper) was lysed into simple fermentable sugar. As a result, about 40% fermentable sugar was obtained after hydrolysis. *P. aeruginosa* converted starch of waste paper substrate into simple fermentable sugar. The saccharification was followed by the simultaneous fermentation of sugars to ethanol. *Pseudomonas, Enterobacter,* and *Bacillus* are reported to efficiently express transesterification activity. *Pseudomonas fluorescens, Pseudomonas cepacia*, and *Rhizomucor miehei* possess enzymes to support transesterification activity (Escobar-Niño et al. 2014). Roman et al. (2017) produced ethyl esters from coconut oil and ethanol by employing chitosan-entrapped lipase produced by *P. fluorescens*.

Growing energy crops is associated with intensive utilization of fertilizers and pesticides, which lead to many negative impacts on the environment. Microalga is one of the competent organisms in biofuel production as it provides immense biomass productivity, tremendous aggregation of lipids, huge sequestration of CO_2, ability to thrive in wastewater and no competition to agricultural lands (Yadav et al. 2019). A consortia of plant growth-promoting rhizobacteria (PGPR) such as *Pseudomonas*, *Azospirillum*, *Bacillus*, *Rhodococcus*, and *Acinetobacter* found in activated sludge are found to effectively support and flourish the growth of microalgae (Nanda and Abraham 2011, 2013). This can be mediated by two means, i.e., combining microalgae and bacteria cultivation in a single process or pretreating wastewater with bacteria before going for microalgal farming in isolated processes. Pretreating wastewater with bacteria provides more supportive situations for the growth of microalgae.

20.5 Major Applications for Microbial Fuel Cells

The metabolic system of bacteria is employed in MFCs for the generation of electricity by utilizing a broad spectrum of organic substrates. Marine sediments are needed to be explored to provide current for low power devices. A thorough understanding of microbiology, synthetic biology, fluid dynamics, and thermodynamics associated with the system could help in improving MFC technology. Some researchers are unfolding efficient MFC technology, which could lead to an efficient and complete degradation of wastes and toxic chemicals while producing clean electricity.

20.5.1 Wastewater Treatment and Electricity Generation

Wastewater treatment is in priority in applying MFC. Microorganisms possess exclusive metabolic assets because of which they hold immense potential to be employed in MFCs either as a pure culture or as consortia. Sewage sludge, food processing wastewater, and swine wastewater are some of the substrates rich in organic matter capable of feeding a wide spectrum of microorganisms employed in MFCs. Enormous amounts of growth promoters are found in MFC substrates that have the potential to accelerate the growth of active microorganisms employed in wastewater treatment. The energy demand on treatment plants could diminish because of simultaneous operation of bioenergy generation and waste bioremediation.

20.5.2 Biosensors

Organic content can be calculated in wastewater by its biological oxygen demand (BOD) using conventional methods. Many methods are not found suitable to control biological wastewater treatment processes and its on-line monitoring. MFC is a potential of BOD sensor. It is feasible to accurately measure BOD value at a broad spectrum of the concentration range of organic contents present in wastewater. With the aid of MFC-based biosensor, it is possible to monitor organic matter, toxicity, and microbial activity in various environments (Davila et al. 2011; Stein et al. 2011).

20.5.3 Secondary Fuel Production

The generation of secondary fuels like hydrogen has also been conducted through MFC. In this operation, proton and electron generated in the anodic chamber move to cathode, which then react with oxygen to form water. Thermodynamic principles do not favor the generation of hydrogen or it can be said that it is a tough process for a cell to convert proton and electron into hydrogen. The generation of hydrogen demands an enhancement in external potential applied at the cathode that can circumvent the thermodynamic barrier because of which proton and electron generated in the anodic reaction chamber join at the cathode to form hydrogen. MFC is an ecofriendly producer of hydrogen than that of the classical method of glucose fermentation (Nanda et al. 2017b). Wagner et al. (2009) reported hydrogen and methane production by using modified microbial electrolytic cells with increased external potential at the cathode.

Microbial electrolysis cell (MEC) is another modified version of MFCs, which supports the energy-rich production of chemicals (Call and Logan 2008, Rozendal et al. 2009; Wagner et al. 2009). Sediment bioremediation is supported when environments like river, lake, and marine systems apply MFCs technology (He et al. 2007; Donovan et al. 2008; Mathis et al. 2008). With the aid of microbial desalination cell (MDC), a modified version of MFC, it is now possible to curtail the salinity of brackish water or seawater along with the generation of electrical power from organic matters (Cao et al. 2009; Mehanna et al. 2010; Chen et al. 2012).

20.6 Current Advancements in Fuel Cell Technologies

The U.S. space program boosted MFC development in 1960s as a feasible technology for the disposal of organic wastes for space flights that could simultaneously generate power. There are many aspects on which MFC technology has been explored such as feasible implementation, anode and cathode performances, variety of substrates used in MFCs, etc. MFCs have been explored as a novel way of electricity generation along with bioremediation of wastes. Electricity production via MFC technology has gained momentum after the origin of phototrophic MFC and solar-powered MFC. Microbial electrolysis cell that has an anoxic cathode with elevated external potential is the manifestation of recent alterations in MFCs technology (Bullen et al. 2006).

It is important to note that electrochemically active bacteria play a prominent role in generating power in MFCs as they possess the ability to generate electrons during metabolism and transfer these electrons via its cell membrane to the anode. Hence, genetic engineering of the electrochemically active bacteria is becoming a key area of current research along with electrode material modification, operation parameters optimization, and scale-up of the reactor.

Recent decades have witnessed a significant improvement in MFC technology. However, many hindrances block the scale-up and practical application of this technology. Along with these challenges, MFCs have faced two impediment problems in power generation. Firstly, the power produced in MFCs and concentrations of substrate are directly related, although in a momentous role in each system. The generation of power is obstructed if the substrate concentration exceeds a specific value. Secondly, high internal resistance is another issue that needs to be tackled as it utilizes a significant amount of power produced in MFCs, thereby restricting the MFC output.

The concept of design of sediment-type microbial fuel cell (SMFC) is because detritus of plant and animal and anthropogenic organic materials generate soil and sediments where the organic carbon content varies from 0.4 to 2.2 wt%. Exoelectrogens can consume these materials, thereby liberating electrons that are then directly transported outside the cell. SMFC has an anode ingrained in the anaerobic sediment and linked through an electrical circuit and a cathode electrode dangled in overlying water (Xu et al. 2015).

Power production and organic load removal are the thrust areas that have been focused on optimizing MFC performance for several decades. These efforts have been fruitful in terms of rapid improvement with power generation in multiple magnitudes as it was a few decades back. However, the usage of cheap and sustainable materials for the MFCs fabrication needs attention, which will not only help in cost-effective scaling-up but also subside the problem of accumulation of toxic wastes generated from old electronic components, plastics, and batteries.

The reliance on artificial mediators is also one of the obstacles that needs to be overcome. Few chemicals aid in moving electrons from within the bacterial cell to the surface of anodes such as neutral red, methylene blue, and thionine. However, many a times these chemical mediators are not required as the microbial species could direct conductance. The current focus has shifted toward exploring cost-effective and easily available materials that could lead to an advancement in MFC technology.

Platinum and other catalysts are replacing the cathode electrode. There are a few materials that are comparatively cheaper than platinum and therefore provide a competitive advantage (Cheng and Wu 2013). Another component that needs to grab focus is the ion exchange membrane (IEM). Liquid feedstock is used in MFCs in the anode, which has a potential to carry charged ions like protons. Hence, the ion exchange membrane is not prerequisite if there is physical or electrochemical separation between the anode and cathode. One of the simple methods is the complete removal of the membrane from the MFC (Logan et al. 2007).

Membrane-less MFCs are cost-effective approach as it curtails the reactor cost. The demerit associated with this approach is that oxygen diffuses toward the anode electrode and there is a need to place the electrodes a certain distance apart. This would also permit electrodes to be placed closely while stopping a high flux of oxygen to the anode. Microporous filtration membranes, canvas, nylon-infused membrane, and paper are a few examples of reported porous materials. These cost-effective materials have been intensively studied for their capacity to enhance power generation along with their feasibility for field application. Ceramic is one of the most dependable and trusted porous materials that can be employed in MFCs (Zhuang et al. 2009).

The ambient environment is a prerequisite for MFC to operate as it works on the metabolism of mesophilic microorganisms that survive in lower and ambient temperatures. In this aspect, MFCs differ from the solid-oxide fuel cell (SOFCs). The use of ceramics as a material also helps MFCs to operate at extreme conditions proving as a suitable material for MFCs. The parameters like the clay type, porosity, wall thickness, and density can be customized for a target application. Moreover, adaptation can be brought in clay material to enhance microbial colonization (Winfield et al. 2013).

20.7 Conclusions

Microbial fuel cells are bioelectrochemical devices where chemical energy stored in organic waste substrates is transformed into electrical energy by utilizing the biocatalytic microorganisms. MFC is an ideal, renewable, and sustainable approach not only to treat organic wastes but also for electricity generation. MFCs can be with the mediator and mediator-less MFCs based on the movement of electrons produced by electrochemically active bacteria from the media to the anode. It also suffices the production of secondary fuel along with the bioremediation of lethal compounds. There is a huge spectrum of anaerobic bacteria involved in the reduction of contaminants, thereby transferring electrons to a solid electrode and generating electricity in MFCs.

The recent years have witnessed the expansion in the scope of MFCs, thereby broadening its potentials from electricity production toward other specialized applications. MFC technologies are optimal ways to approach the production of renewable energy while remediating the pollutants. Significant efforts are made to enhance the performance and efficiency of MFCs while reducing the operating costs so that large-scale application of MFCs is feasible. Further advancements in microbial metabolism in MFC systems as well as MFC applications at a much larger scale are important.

References

Basler G, Mitchell Thompson M, Ercek DT, Keasling J (2018) A *Pseudomonas putida* efflux pump acts on short-chain alcohols. Biotechnol Biofuels 11:136

Behera M, Ghangrekar MM (2009) Performance of microbial fuel cell in response to change in sludge loading rate at different anodic feed pH. Bioresour Technol 100:5114–5121

Bond DR, Lovley DR (2003) Electricity production by *Geobacter sulfurreducens* attached to electrodes. Appl Environ Microbiol 69:1548–1555

Borole AP, Reguera G, Ringeisen B, Wang Z-W, Feng Y, Kim BH (2011) Electroactive biofilms: current status and future research needs. Energy Environ Sci 4:4813–4834

Bullen RA, Arnot TC, Lakemanc JB, Walsh FC (2006) Biofuel cells and their development. Biosens Bioelectron 21:2015–2045

Call D, Logan BE (2008) Hydrogen production in a single chamber microbial electrolysis cell lacking a membrane. Environ Sci Technol 42:3401–3406

Cao X, Huang X, Liang P, Xiao K, Zhou Y, Zhang X, Logan BE (2009) A new method for water desalination using microbial desalination cells. Environ Sci Technol 43:7148–7152

Catal T, Xu S, Li K, Bermek H, Liu H (2008) Electricity production from polyalcohols in single-chamber microbial fuel cells. Biosens Bioelectron 24:855–860

Chaudhuri SK, Lovley DR (2003) Electricity generation by direct oxidation of glucose in mediatorless microbial fuel cells. Nat Biotechnol 21:1229–1232

Chen S, Liu G, Zhang R, Qin B, Luo Y (2012) Development of the microbial electrolysis desalination and chemical-production cell for desalination as well as acid and alkali productions. Environ Sci Technol 46:2467–2472

Cheng S, Wu J (2013) Air-cathode preparation with activated carbon as catalyst, PTFE as binder and nickel foam as current collector for microbial fuel cells. Bioelectrochemistry 92:22–26

Crittenden SR, Sund CJ, Sumner JJ (2006) Mediating electron transfer from bacteria to a gold electrode via a self-assembled monolayer. Langmuir 22:9473–9476

Cuenca MDS (2016) A *Pseudomonas putida* double mutant deficient in butanol assimilation: a promising step for engineering a biological bio fuel production platform. FEMS Microbiol Lett 363:fnw018

Das S, Mangwani N (2010) Recent developments in microbial fuel cells: a review. J Sci Ind Res 69:727–731

Davila D, Esquivel JP, Sabate N, Mas J (2011) Silicon-based microfabricated microbial fuel cell toxicity sensor. Biosens Bioelectron 26:2426–2430

Donovan C, Dewan A, Heo D, Beyenal H (2008) Batteryless, wireless sensor powered by a sediment microbial fuel cell. Environ Sci Technol 42:8591–8596

Du Z, Li H, Gu T (2007) A state of the art review on microbial fuel cells: a promising technology for wastewater treatment and bioenergy. Biotechnol Adv 25:464–482

Du Z, Li Q, Tong M, Li S, Li H (2008) Electricity generation using membrane-less microbial fuel cell during wastewater treatment. Chin J Chem Eng 16:772–777

Dumas C, Basseguy R, Bergel A (2008) Microbial electrocatalysis with *Geobacter sulfurreducens* biofilm on stainless steel cathodes. Electrochem Acta 53:2494–2500

Escobar-Niño A, Luna C, Luna D, Marcos AT, Cánovas D, Mellado E (2014) Selection and characterization of biofuel-producing environmental bacteria isolated from vegetable oil-rich wastes. PLoS One 9:e104063

Fougere D, Nanda S, Clarke K, Kozinski JA, Li K (2016) Effect of acidic pretreatment on the chemistry and distribution of lignin in aspen wood and wheat straw substrates. Biomass Bioenergy 91:56–68

Ganguli R, Dunn BS (2009) Kinetics of anode reactions for a yeast-catalysed microbial fuel cell. Fuel Cells 9:44–52

Greenman J, Gálvez A, Giusti L, Ieropoulos I (2009) Electricity from landfill leachate using microbial fuel cells: comparison with a biological aerated filter. Enzym Microb Technol 44:112–119

Ha PT, Tae B, Chang IS (2008) Performance and bacterial consortium of microbial fuel cell fed with formate. Energy Fuel 22:164–168

He Z, Minteer SD, Angenent LT (2005) Electricity generation from artificial wastewater using an upflow microbial fuel cell. Environ Sci Technol 39:5262–5267

He Z, Shao H, Angenent LT (2007) Increased power production from a sediment microbial fuel cell with a rotating cathode. Biosens Bioelectron 22:3252–3255

Holmes DE, Bond DR, O'Neil RA, Reimers CE, Tender LR, Lovley DR (2004) Microbial communities associated with electrodes harvesting electricity from a variety of aquatic sediments. Microb Ecol 48:178–190

Hou H, Chen X, Thomas AW (2013) Conjugated oligo electrolytes increase power generation in *E.coli* microbial fuel cells. Adv Mater 25:1593–1597

Huang L, Angelidaki I (2008) Effect of humic acids on electricity generation integrated with xylose degradation in microbial fuel cells. Biotechnol Bioeng 100:413–422

Ieropoulos IA, Ledezma P, Stinchcombe A, Papaharalabos G, Melhuish C, Greenman J (2013) Waste to real energy: the first MFC powered mobile phone. Phys Chem Chem Phys 15:15312–15316

Johnson CW, Beckham GT (2015) Aromatic catabolic pathway selection for optimal production of pyruvate and lactate from lignin. Metab Eng 28:240–247

Kargi F, Eker S (2007) Electricity generation with simultaneous wastewater treatment by a microbial fuel cell (MFC) with cu and cu–au electrodes. J Chem Technol Biotechnol 82:658–662

Kim HJ, Park HS, Hyun MS, Chang IS, Kim M, Kim BH (2002) A mediator-less microbial fuel cell using a metal reducing bacterium, *Shewanella putrefaciense*. Enzyme Microl Tech 30:145–152

Kim JR, Jung SH, Regan JM, Logan BE (2007) Electricity generation and microbial community analysis of alcohol powered microbial fuel cells. Bioresour Technol 98:2568–2577

Larrosa-Guerrero A, Scott K, Head IM, Mateo F, Ginesta A, Godinez C (2010) Effect of temperature on the performance of microbial fuel cells. Fuel 89:3985–3994

Liu ZD, Du ZW, Lian J, Zhu XY, Li SH, Li HR (2007) Improving energy accumulation of microbial fuel cells by metabolism regulation using *Rhodoferax ferrireducens* as biocatalyst. Lett Appl Microbiol 44:393–398

Loeschcke A, Thies S (2015) *Pseudomonas putida* a versatile host for the production of natural products. Appl Microbiol Biotechnol 99:6197–6214

Logan BE, Cheng S, Watson V, Estadt G (2007) Graphite fiber brush anodes for increased power production in air-cathode microbial fuel cells. Environ Sci Technol 41:3341–3346

Luo H, Liu G, Zhang R, Jin S (2009) Phenol degradation in microbial fuel cells. Chem Eng J 147:259–264

Luo Y, Liu G, Zhang R, Zhang C (2010) Power generation from furfural using the microbial fuel cell. J Power Sources 195:190–194

Martins dos Santos VAP, Heim S, Moore ER, Strätz M, Timmis KN (2004) Insights into the genomic basis of niche specificity of *Pseudomonas putida* KT2440. Environ Microbiol 6:1264–1286

Mathis BJ, Marshall CW, Milliken CE, Makkar RS, Creager SE, May HD (2008) Electricity generation by thermophilic microorganisms from marine sediment. Appl Microbiol Biotechnol 78:147–155

Mehanna M, Kiely PD, Call DF, Logan BE (2010) Microbial electrodialysis cell for simultaneous water desalination and hydrogen gas production. Environ Sci Technol 44:9578–9583

Nanda S, Abraham J (2011) Impact of heavy metals on microbial activity in the rhizosphere of *Jatropha multifida* and their effective remediation. Afr J Biotechnol 10:11948–11955

Nanda S, Abraham J (2013) Remediation of heavy metal contaminated soil. Afr J Biotechnol 12:3099–3109

Nanda S, Mohanty P, Pant KK, Naik S, Kozinski JA, Dalai AK (2013) Characterization of north American lignocellulosic biomass and biochars in terms of their candidacy for alternate renewable fuels. Bioenergy Res 6:663–677

Nanda S, Dalai AK, Kozinski JA (2014a) Butanol and ethanol production from lignocellulosic feedstock: biomass pretreatment and bioconversion. Energ Sci Eng 2:138–148

Nanda S, Mohammad J, Reddy SN, Kozinski JA, Dalai AK (2014b) Pathways of lignocellulosic biomass conversion to renewable fuels. Biomass Conv Bioref 4:157–191

Nanda S, Azargohar R, Dalai AK, Kozinski JA (2015a) An assessment on the sustainability of lignocellulosic biomass for biorefining. Renew Sust Energ Rev 50:925–941

Nanda S, Reddy SN, Hunter HN, Butler IS, Kozinski JA (2015b) Supercritical water gasification of lactose as a model compound for valorization of dairy industry effluents. Ind Eng Chem Res 54:9296–9306

Nanda S, Reddy SN, Mitra SK, Kozinski JA (2016) The progressive routes for carbon capture and sequestration. Energ Sci Eng 4:99–122

Nanda S, Dalai AK, Gökalp I, Kozinski JA (2016a) Valorization of horse manure through catalytic supercritical water gasification. Waste Manag 52:147–158

Nanda S, Isen J, Dalai AK, Kozinski JA (2016b) Gasification of fruit wastes and agro-food residues in supercritical water. Energ Convers Manage 110:296–306

Nanda S, Golemi-Kotra D, McDermott JC, Dalai AK, Gökalp I, Kozinski JA (2017a) Fermentative production of butanol: perspectives on synthetic biology. New Biotechnol 37:210–221

Nanda S, Rana R, Zheng Y, Kozinski JA, Dalai AK (2017b) Insights on pathways for hydrogen generation from ethanol. Sustain Energ Fuel 1:1232–1245

Nanda S, Rana R, Hunter HN, Fang Z, Dalai AK, Kozinski JA (2019) Hydrothermal catalytic processing of waste cooking oil for hydrogen-rich syngas production. Chem Eng Sci 195:935–945

Nielsen DR (2009) Engineering alternative butanol production platforms in heterologous bacteria. Metab Eng 11:262–273

Niessen J, Schröder U, Scholz F (2004) Exploiting complex carbohydrates for microbial electricity generation—a bacterial fuel cell operating on starch. Electrochem Commun 6:955–958

Osman MH, Shah AA, Walsh FC (2010) Recent progress and continuing challenges in bio-fuel cells. Part II: microbial. Biosens Bioelectron 26:953–963

Pant D, Bogaert GV, Diels L, Vanbroekhoven K (2010) A review of the substrates used in microbial fuel cells (MFCs) for sustainable energy production. Bioresour Technol 101:1533–1543

Pham TH, Boon N, Aelterman P, Clauwaert P, De Schamphelaire L, Vanhaecke L, De Maeyer K, Höfte M, Verstraete W, Rabaey K (2008) Metabolites produced by *Pseudomonas* sp. enable a gram-positive bacterium to achieve extracellular electron transfer. Appl Microbiol Biotechnol 77:1119–1129

Pham H, Boon N, Marzorati M, Verstraete W (2009) Enhanced removal of 1,2-dichloroethane by anodophilic microbial consortia. Water Res 43:2936–2946

Potter MC (1911) Electrical effects accompanying the decomposition of organic compounds. Proc Royal Soc B—Biol Sci 84:260–276

Prema D, Prabha ML, Gnanavel G (2015) Production of biofuel using waste papers from *Pseudomonas aeruginosa*. Int J ChemTech Res 8:1803–1809

Raghavulu V, Goud RK, Sarma PN, Mohan SV (2001) *S. cerevisiae* as anodic biocatalyst for power generation in biofuel cell: influence of redox condition and substrate load. Bioresour Technol 102:2751–2757

Rahimnejad M, Ghoreyshi A, Najafpour G, Younesi H, Shakeri M (2012a) A novel microbial fuel cell stack for continuous production of clean energy. Int J Hydrogen Energ 37:5992–6000

Rahimnejad M, Najafpour G, Ghoreyshi A, Talebnia F, Premier G, Bakeri G, Kim J, Oh SE (2012b) *Thionine* increases electricity generation from microbial fuel cell using *S. cerevisiae* and exoelectrogenic mixed culture. J Microbiol 50:575–580

Rahimnejad M, Adhami A, Darvari S, Zirepour A, Oh SE (2015) Microbial fuel cell as new technology for bioelectricity generation: a review. Alexandria Eng J 54:745–756

Rana R, Nanda S, Meda V, Dalai AK, Kozinski JA (2018) A review of lignin chemistry and its biorefining conversion technologies. J Biochem Eng Bioproc Technol 1:2

Ren Z, Steinburg LM, Regan JM (2008) Electricity production and microbial biofilm characterization in cellulose-fed microbial fuel cells. Water Sci Technol 58:617–622

Rezaei F, Xing D, Wagner R, Regan JM, Richard TL, Logan BE (2009) Simultaneous cellulose degradation and electricity production by *Enterobacter cloacae* in a microbial fuel cell. Appl Environ Microbiol 75:3673–3678

Roman DCC, Pérez H, Castro F, Orrego CE, Giraldo OH, Silveira EG (2017) Ethyl esters (biodiesel) production by *Pseudomonas fluorescens* lipase immobilized on chitosan with magnetic properties in a bioreactor assisted by electromagnetic field. Fuel 196:481–487

Rozendal RA, Leone E, Keller J, Rabaey K (2009) Efficient hydrogen peroxide generation from organic matter in a bioelectrochemical system. Electrochem Commun 11:1752–1755

Santoro C, Arbizzani C, Erable B, Ieropoulos I (2017) Microbial fuel cells: from fundamentals to applications. A review. J Power Sources 356:225–244

Sayed E, Tsujiguchi T, Nakagawa N (2012) Catalytic activity of baker's yeast in a mediator-less microbial fuel cell. Bioelectrochemistry 86:97–101

Schmitz S (2015) Engineering mediator based electroactivity in the obligate aerobic bacterium *Pseudomonas putida* KT2440. Front Microbiol 10:284

Shen HB, Yong XY, Chen YL, Liao ZH, Si RW, Zhou J, Wang SY, Yong YC, OuYang PK, Zheng T (2014) Enhanced bioelectricity generation by improving pyocyanin production and membrane permeability through sophorolipid addition in *Pseudomonas aeruginosa* inoculated microbial fuel cells. Bioresour Technol 167:490–494

Stein NE, Keesman KJ, Hamelers HVM, van Straten G (2011) Kinetic models for detection of toxicity in a microbial fuel cell based biosensor. Biosens Bioelectron 26:3115–3120

Sun J, Hu Y-Y, Bi Z, Cao Y-Q (2009) Simultaneous decolorization of azo dye and bioelectricity generation using a microfiltration membrane air-cathode single chamber microbial fuel cell. Bioresour Technol 100:3185–3192

Tee PF, Abdullah MO, Tan IAW, Amin MAM, Nolasco-Hipolito C, Bujang K (2017) Effects of temperature on wastewater treatment in an affordable microbial fuel cell-adsorption hybrid system. J Environ Chem Eng 5:178–188

Tremouli A, Martinos M, Lyberatos G (2016) The effects of salinity, pH and temperature on the performance of a microbial fuel cell. Waste Biomass Valor 8:2037–2043

Udaondo Z (2012) Analysis of solvent tolerance in *Pseudomonas putida* DOTT1E based on its genome sequence and a collection of mutants. FEBS Lett 586:2932–2938

Vardon DR, Franden MA, Johnson CW, Karp EM, Guarnieri MT, Linger JG, Salm MJ, Strathmann TJ, Beckham GT (2015) Adipic acid production from lignin. Energy Environ Sci 8:617–628
Velasquez-Orta S, Curtis TP, Logan BE (2009) Energy from algae using microbial fuel cells. Biotechnol Bioeng 103:1068–1076
Wagner RC, Regan JM, Oh SE, Zuo Y, Logan BE (2009) Hydrogen and methane production from swine wastewater using microbial electrolysis cells. Water Res 43:1480–1488
Wang VB, Chua S, Cao B (2013) Engineering PQS biosynthesis pathway for enhancement of bioelectricity production in *Pseudomonas aeruginosa* microbial fuel cells. PLoS One 8(5)
Wang VB, Chua SL, Cai Z (2014) A stable synergistic microbial consortium for simultaneous azo dye removal and bioelectricity generation. Bioresour Technol 155:71–76
Winfield J, Chambers LD, Rossiter J, Ieropoulos I (2013) Comparing the short and long term stability of biodegradable, ceramic and cation exchange membranes in microbial fuel cells. Bioresour Technol 148:480–486
Xu B, Ge Z, He Z (2015) Sediment microbial fuel cells for wastewater treatment: challenges and opportunities. Environ Sci: Water Res Technol 1:279–284
Yadav P, Reddy SN, Nanda S (2019) Cultivation and conversion of algae for wastewater treatment and biofuel production. In: Nanda S, Sarangi PK, Vo DVN (eds) Fuel processing and energy utilization. CRC Press, Florida, USA, pp 159–175
Zhuang L, Zhou S, Wang Y, Liu C, Geng S (2009) Membrane-less cloth cathode assembly (CCA) for scalable microbial fuel cells. Biosens Bioelectron 24:3652–3656

MIX
Papier aus verantwortungsvollen Quellen
Paper from responsible sources
FSC® C105338

If you have any concerns about our products,
you can contact us on
ProductSafety@springernature.com

In case Publisher is established outside the EU,
the EU authorized representative is:
Springer Nature Customer Service Center GmbH
Europaplatz 3, 69115 Heidelberg, Germany

Printed by Libri Plureos GmbH
in Hamburg, Germany